EDEXCEL A LEVEL

FURTHER MATHEMATICS
Statistics

JOHN DU FEU
SERIES EDITORS
ROGER PORKESS AND CATHERINE BERRY
CONSULTANT EDITORS
KEITH PLEDGER AND JAN DANGERFIELD

In order to ensure that this resource offers high-quality support for the associated Pearson qualification, it has been through a review process by the awarding body. This process confirms that this resource fully covers the teaching and learning content of the specification or part of a specification at which it is aimed. It also confirms that it demonstrates an appropriate balance between the development of subject skills, knowledge and understanding, in addition to preparation for assessment.

Endorsement does not cover any guidance on assessment activities or processes (e.g. practice questions or advice on how to answer assessment questions), included in the resource nor does it prescribe any particular approach to the teaching or delivery of a related course.

While the publishers have made every attempt to ensure that advice on the qualification and its assessment is accurate, the official specification and associated assessment guidance materials are the only authoritative source of information and should always be referred to for definitive guidance.

Pearson examiners have not contributed to any sections in this resource relevant to examination papers for which they have responsibility.

Examiners will not use endorsed resources as a source of material for any assessment set by Pearson.

Endorsement of a resource does not mean that the resource is required to achieve this Pearson qualification, nor does it mean that it is the only suitable material available to support the qualification, and any resource lists produced by the awarding body shall include this and other appropriate resources.

Hachette UK's policy is to use papers that are natural, renewable and recyclable products and made from wood grown in sustainable forests. The logging and manufacturing processes are expected to conform to the environmental regulations of the country of origin.

Orders: please contact Bookpoint Ltd, 130 Park Drive, Milton Park, Abingdon, Oxon OX14 4SE.
Telephone: (44) 01235 827720. Fax: (44) 01235 400401. Email education@bookpoint.co.uk
Lines are open from 9 a.m. to 5 p.m., Monday to Saturday, with a 24-hour message answering service.
You can also order through our website: www.hoddereducation.co.uk

ISBN: 978 1 5104 1455 6

© John du Feu, Roger Porkess, Catherine Berry, Jan Dangerfield and MEI 2018

First published in 2018 by

Hodder Education,
An Hachette UK Company
Carmelite House
50 Victoria Embankment
London EC4Y 0DZ

www.hoddereducation.co.uk

Impression number 10 9 8 7 6 5 4 3 2 1

Year 2022 2021 2020 2019 2018

All rights reserved. Apart from any use permitted under UK copyright law, no part of this publication may be reproduced or transmitted in any form or by any means, electronic or mechanical, including photocopying and recording, or held within any information storage and retrieval system, without permission in writing from the publisher or under licence from the Copyright Licensing Agency Limited. Further details of such licences (for reprographic reproduction) may be obtained from the Copyright Licensing Agency Limited, www.cla.co.uk

Cover photo © Robert Matton AB/Alamy Stock Photo

Typeset in Bembo Std, 11/13 pts. by Aptara®, Inc.

Printed in Italy

A catalogue record for this title is available from the British Library.

Contents

Getting the most from this book   v
Prior knowledge   vii

S1 AS   **1 Statistical problem solving**   1
    1.1 The problem solving cycle   2

S1 AS   **2 Discrete random variables**   20
    2.1 Conditions for discrete random variables   22
    2.2 Expectation and variance   26
    2.3 Linear functions of random variables   33

S1 AS   **3 Discrete probability distributions**   44
    3.1 The binomial distribution   45
    3.2 The Poisson distribution   49
    3.3 Applications of the Poisson distribution   57
    3.4 The discrete uniform distribution   64

S1 AS   **4 Chi-squared tests**   70
    4.1 The chi-squared test for a contingency table   71
    4.2 Goodness of fit tests   85

Practice questions: set 1   104

S2 AS   **5 Measuring correlation**   107
    5.1 Product moment correlation   108
    5.2 Rank correlation   134

S2 AS   **6 Linear regression**   146
    6.1 The least squares regression line   147
    6.2 Residual sum of squares   156

S2 AS   **7 Continuous random variables**   163
    7.1 Probability density function   165
    7.2 Measures of average and spread for a continuous distribution   174
    7.3 The expectation and variance of a function of X 183
    7.4 The cumulative distribution function, F   188
    7.5 The continuous uniform (rectangular) distribution   200

8 Conditional probability   209   **S2 AS**
    8.1 Screening tests   210

Practice questions: set 2   219

9 Further discrete distributions   223   **S1 AL**
    9.1 Repeating trials until success is achieved   224
    9.2 The negative binomial distribution   230
    9.3 Testing the parameter in a discrete distribution   234

10 The Central Limit Theorem   240   **S1 AL**
    10.1 The sampling distribution of the mean   242

11 Quality of tests   248   **S1 AL**
    11.1 The power of a test   248

12 Probability generating functions   260   **S1 AL**
    12.1 Probabilities defined by a probability generating function   261
    12.2 Expectation and variance   266
    12.3 The PGF of the sum of independent random variables   271

Practice questions: set 3   282

13 Expectation algebra and the Normal distribution   284   **S2 AL**
    13.1 The sums and differences of Normal variables   287
    13.2 Modelling discrete situations   289
    13.3 More than two independent random variables   293
    13.4 The distribution of the sample mean   300

14 Confidence intervals for the mean — 308
S2 AL

14.1 The theory of confidence intervals — 309
14.2 Interpreting sample data using the t-distribution — 321
14.3 Confidence intervals for the difference between two means — 335

15 Hypothesis tests on the mean — 341
S2 AL

15.1 Testing the mean and variance of a single sample — 342
15.2 Testing the difference of the means of two populations — 353

16 Estimation — 363
S2 AL

16.1 Bias in estimators — 364
16.2 Estimators for the population mean and variance — 370

17 Variance: confidence intervals and hypothesis tests — 379
S2 AL

17.1 The distribution of of S^2 — 380
17.2 Hypothesis test for the difference in population variances — 386

Practice questions: set 4 — 393

Answers — 396
Index — 436

Getting the most from this book

Mathematics is not only a beautiful and exciting subject in its own right but also one that underpins many other branches of learning. It is consequently fundamental to our national wellbeing.

This book covers the Statistics elements in the Edexcel AS and A Level Further Mathematics specifications. Students start these courses at a variety of stages. Some embark on AS Further Mathematics in Year 12, straight after GCSE, taking it alongside AS Mathematics, and so may have no prior experience of Statistics. In contrast, others only begin Further Mathematics when they have completed the full A Level Mathematics and so have already met the Statistics covered in Edexcel A Level Mathematics (Year 2). Between these two extremes are the many who have covered the Statistics in AS Mathematics but no more. This book has been written with all these users in mind. Those who already know some Statistics will find some revision material in the early chapters.

In the Edexcel Further Mathematics specification, two Statistics papers are available at AS level each of which is also available at A level, with additional material.

Chapter 1 is review material from AS Mathematics. Chapters 2 to 4 cover AS Further Statistics 1 and Chapters 5 to 7 cover AS Further Statistics 2.

Chapter 8 is review material from A level Mathematics and some extension material on Bayes' theorem. Chapters 9 to 13 cover the additional material for A level Further Statistics 1 and Chapters 14 to 17 complete the additional material for A level Further Statistics 2.

Between 2014 and 2016 A Level Mathematics and Further Mathematics were very substantially revised, for first teaching in 2017. Changes that particularly affect Statistics include increased emphasis on

- Problem solving
- Mathematical rigour
- Use of ICT
- Modelling.

This book embraces these ideas. A large number of exercise questions involve elements of problem solving and require rigorous logical argument.

Throughout the book the emphasis is on understanding and interpretation rather than mere routine calculations, but the various exercises do nonetheless provide plenty of scope for practising basic techniques. The exercise questions are split into three bands. Band 1 questions (indicated by a green bar) are designed to reinforce basic understanding; Band 2 questions (yellow bar) are broadly typical of what might be expected in an examination; Band 3 questions (red bar) explore around the topic and some of them are rather more demanding.

In addition to the exercise questions, there are four sets of Practice questions. The first of these covers the background material in Chapters 1 to 4, the second is based on the AS content in Chapters 5 to 8, the third set covers Chapters 9 to 12, and the fourth set, coming at the end of the book, is drawn from the complete Further Mathematics A Level.

There are places where the work depends on knowledge from earlier in the book or elsewhere and this is flagged up in the Prior knowledge boxes. This should be seen as an invitation to those who have problems with the particular topic to revisit it. At the end of each chapter there is a list of key points covered as well as a summary of the new knowledge (learning outcomes) that readers should have gained.

Several features are used to make this book easier to follow and more informative.

- **Call-out boxes** are used to provide additional explanation particularly in the worked examples.
- **Notes** are also used for additional explanation, particularly where it involves seeing broader or deeper aspects of the topic.
- **Caution boxes** are used to highlight points where it is easy to go wrong.
- **Discussion points** invite readers to talk about particular points with their fellow students and their teacher and so enhance their understanding. Short answers to discussion points are given, where appropriate, either in the text or at the back of the book.
- **Activities** are designed to help readers get into the thought processes of the new work that they are about to meet; having done an Activity, what follows will seem much easier.
- **Historical notes** provide readers with interesting information about the people who first worked on the topics, and insights into the world's cultural and intellectual development.

As a consequence of the changes to AS and A Level requirements in Further Mathematics, large parts of this book are new material, including sections written specifically to cover the Edexcel specifications. Where existing material has been used, it has typically been drawn from the well tried and tested earlier MEI books.

The Edexcel formulae booklet can be downloaded from Edexcel's website: https://qualifications.pearson.com/content/dam/pdf/A%20Level/Mathematics/2017/specification-and-sample-assesment/Pearson_Edexcel_A_Level_GCE_in_Mathematics_Formulae_Book.pdf.

Answers to all exercise questions and practice questions are provided at the back of the book, and also online at www.hoddereducation.co.uk/EdexcelFurtherMathsStatistics

Catherine Berry
Roger Porkess

Prior knowledge

This book is written on the assumption that readers are familiar with the statistics in GCSE Mathematics. Thus they should know a variety of elementary display techniques such as pictograms, tallies, pie charts, bar charts and scatter diagrams (including the ideas of correlation and a line of best fit). Summary measures which they are expected to know include mean, median, mode and range. Readers are also expected to be familiar with basic probability.

Chapter 1
This chapter sets up the framework in which much of statistics is carried out in everyday life. It is about statistical problem solving and so involves using display techniques and summary measures to shed light on real problems. Consequently, it draws on and extends the prior knowledge for the book. In particular, it introduces frequency charts and histograms, and variance and standard deviation. This chapter includes summary information about many of the terms that you will use throughout the book, including types of data, distributions and sampling.

Chapter 2
In this chapter on discrete random variables, you will use your prior knowledge of probability from GCSE, display techniques from GCSE and Chapter 1, and your knowledge of variance and standard deviation from Chapter 1.

Chapter 3
This chapter on discrete probability distributions covers the binomial, Poisson and uniform distributions, all of which are particular examples of discrete random variables. Consequently, it builds on and exemplifies Chapter 2. It also draws on background information from Chapter 1, for example, about the idea of a distribution.

Chapter 4
In this chapter you meet the use of χ^2 tests in several different circumstances. These include goodness of fit tests for the uniform, binomial and Poisson distributions which you met in Chapter 3.

Chapter 5
This chapter is about correlation and association in bivariate data and so readers will draw on their GCSE experience of scatter diagrams. The chapter includes Spearman's rank correlation coefficient and many readers will have met this in other subjects, for example geography.

Chapter 6
This chapter is about regression lines in bivariate data. It extends the basic idea that you met in GCSE and formalises it beyond just drawing a line by eye.

Chapter 7
This chapter is on continuous random variables and so it builds on and extends the ideas in Chapters 2 and 3 on discrete random variables to continuous variables. The cumulative distribution function is introduced in this chapter which generalises the idea of cumulative frequency from GCSE. You will need to be able to differentiate and integrate polynomials and other functions.

Chapter 8
This chapter is about conditional probability. It builds on and extends your knowledge of probability from GCSE.

Chapter 9
This chapter builds on Chapters 2 and 3 with the introduction of two more discrete distributions, the geometric and the negative binomial. The latter part of the chapter extends the ideas of hypothesis testing which you met in Chapters 4 and 5.

Chapter 10

This chapter requires recall of the Normal distribution and continuity corrections from A level Mathematics. It is about the Central Limit Theorem and, while this is new, some of the distributions to which it is applied were covered in Chapter 3.

Chapter 11

In this chapter the ideas of hypothesis testing are extended to possible errors. So it is based on an understanding of the basic methodology of hypothesis testing, such as that developed in Chapters 4, 5 and 9.

Chapter 12

Probability generating functions use algebra and calculus to find the mean and variance of the standard probability distributions from Chapters 3 and 9. You need to be able to differentiate exponential functions.

Chapter 13

This chapter starts with a review of the Normal distribution which is in A level Mathematics. It then builds on ideas of expectation and variance that were introduced in Chapter 2.

Chapter 14

This chapter is on confidence intervals for the mean. The work depends on the Normal distribution but goes on to situations where the *t*-distribution is used.

Chapter 15

The work in this chapter is closely related to that in Chapter 14, but it also requires a good understanding of hypothesis testing from earlier in the book.

Chapter 16

This chapter is about estimation and depends on expectation and variance, ideas which were introduced in Chapter 2 and developed in Chapter 13.

Chapter 17

The final chapter of the book is about confidence intervals for variance and associated hypothesis tests; it develops general ideas from earlier in the book, particularly Chapters 14 and 15.

1 Statistical problem solving

A judicious man looks at statistics, not to get knowledge but to save himself from having ignorance foisted on him.
Thomas Carlyle (1795–1881)

Discussion point
Do you agree with the 'not to get knowledge' part of Carlyle's statement?

Think of one example where statistics has promoted knowledge or is currently doing so.

How would statistics have been different in Carlyle's time from now?

1 The problem solving cycle

Statistics provides a powerful set of tools for solving problems. While many of the techniques are specific to statistics they are nonetheless typically carried out within the standard cycle.

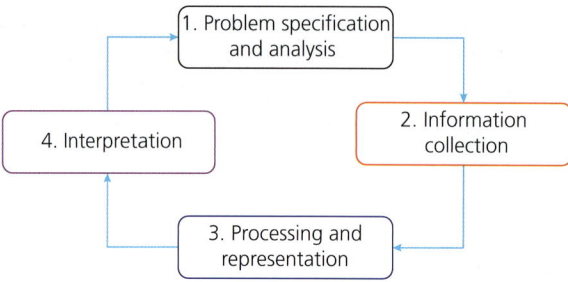

Figure 1.1

This chapter reviews the techniques that are used at the various stages, with particular emphasis on the information collection and the processing and representation elements.

Problem specification and analysis, and interpretation

The problem-solving cycle begins with a problem. That may seem like stating the obvious but it is not quite. Much of the work you do in statistics involves applying statistical techniques to statistical problems. By contrast, the problems tackled in this cycle are drawn from real life. They often require the use of statistics, but as a means to the end of providing an answer to the original problem, situation or context. Here are some examples:

- Is a particular animal in danger of extinction?
- How many coaches should a train operating company put on a particular train?
- Will a new corner shop be viable in a particular location?
- What provision of Intensive Care and High Dependency places should a hospital's neonatal unit make?

To answer questions like these you need data, but before collecting them it is essential to plan the work. Too often poor planning results in inappropriate data being collected. So, at the outset, you need to know:

- what data you are going to collect
- how you are going to collect the data
- how you are going to analyse the data
- how much data you will need
- how you are going to present the results
- what the results will mean in terms of the original problem.

Thus planning is essential and this is the work that is carried out in the first stage, problem specification and analysis, and at the end of the process interpretation is required; this includes the possible conclusion that the problem has not been addressed satisfactorily and the whole cycle must be repeated.

Both planning and interpretation depend on knowledge of how data are collected, processed and represented and so the second and third stages of the cycle are the focus of this chapter.

Information collection

The information needed in statistics is usually in the form of data so the information collection stage in the cycle is usually called **data collection**. This is an important part of statistics and this section outlines the principles involved, together with the relevant terminology and notation.

Terminology and notation

Data collection often requires you to take a sample, a set of items which are drawn from the relevant population and should be representative of it. The complete population may be too large for it to be practical, or economical, to consider every item.

A sample provides a set of data values of a random variable, drawn from all such possible values, the **parent population**. The parent population can be finite, such as all professional netball players, or infinite, such as the points where a dart can land on a dart board.

A representation of the items available to be sampled is called the **sampling frame**. This could, for example, be a list of the sheep in a flock, a map marked with a grid or an electoral register. In many situations no sampling frame exists nor is it possible to devise one, for example for the cod in the North Atlantic. The proportion of the available items that are actually sampled is called the **sampling fraction**. A 100% sample is called a **census**.

The term **random** is often used in connection with data collection. For a process to be described as random, each item in the population has a probability of being included in the sample. In many situations these probabilities are equal, but this is not essential. Most calculators give random numbers and these allow you to select an item from a list at random.

A parent population, often just called the population, is described in terms of its parameters, such as its mean, μ, and variance, σ^2. By convention, Greek letters are used to denote these population parameters.

A value derived from a sample is written in Roman letters, such as $\bar{x}$ and s. Such a number is the value of a sample statistic (or just statistic). When sample statistics are used to estimate the parent population parameters they are called estimates.

Thus if you take a random sample for which the mean is $\bar{x}$, you can use $\bar{x}$ to estimate the population mean, μ. Thus if in a particular sample $\bar{x} = 25.9$, you can use 25.9 as an estimate of the population mean. You would, however, expect the true value of μ to be somewhat different from 25.9.

An estimate of a parameter derived from sample data will, in general, differ from its true value. The difference is called the **sampling error**. To reduce the sampling error, you want your sample to be as representative of the parent population as you can make it. This, however, may be easier said than done.

> **Note**
>
> Sampling fraction
> $= \dfrac{\text{Sample size}}{\text{Population size}}$

> **Note**
>
> Imagine you want to select a day of the year at **random**. You can number them 1 to 365. Then set your calculator to generate a three-digit number. If it is 365 or less, that gives you your day. If it is over 365, reject it and choose another number.

The problem solving cycle

Sampling

There are several reasons why you might want to take a sample. These include:

- to help you understand a situation better
- as part of a pilot study to inform the design of a larger investigation
- to estimate the values of the parameters of the parent population
- to avoid the work involved in cleaning and formatting all the data in a large set
- to conduct a hypothesis test.

> This is often the situation when you are collecting data as part of the problem.

At the outset you need to consider how your sample data will be collected and the steps you can take to ensure their quality. You also need to plan how you will interpret your data. Here is a checklist of questions to ask yourself when you are taking a sample.

> **Discussion point**
> Give examples of cases where the answers to the first six of these questions are 'no'.

- Are the data relevant to the problem?
- Are the data unbiased?
- Is there any danger that the act of collection will distort the data?
- Is the person collecting the data suitable?
- Is the sample of a suitable size?
- Is a suitable sampling procedure being followed?
- Is the act of collecting the data destructive?

> Sample size is important. The larger the sample, the more accurate will be the information it gives you.

> For example, bringing rare deep sea creatures to the surface for examination may result in their deaths.

There are many sampling techniques. The list that follows includes the most commonly used. In considering them, remember that a key aim when taking a sample is that it should be **representative** of the parent population being investigated.

Simple random sampling

In a **simple random sampling procedure**, every possible sample of a given size is equally likely to be selected. It follows that in such a procedure every member of the parent population is equally likely to be selected. However, the converse is not true. It is possible to devise a sampling procedure in which every member is equally likely to be selected but some samples are not possible; an example occurs with **systematic sampling** which is described later.

Simple random sampling is fine when you can do it, but you must have a sampling frame. To carry out simple random sampling, the population must first be numbered, usually from 1 to n. Random numbers between 1 and n are then generated, and the corresponding members of the population are selected. If any repeats occur, more random numbers have to be generated to replace them. Note that, in order to carry out a hypothesis test or to construct a confidence interval (see Chapter 13), the sample taken should be a simple random sample and so some of the sampling methods below are not suitable for these purposes.

> **Example from real life**

Jury selection

The first stage in selecting a jury is to take a simple random sample from the electoral role.

Stratified sampling

Sometimes it is possible to divide the population into different groups, or strata. In **stratified sampling**, you would ensure that all strata were sampled. In **proportional stratified sampling**, the numbers selected from each of the strata are proportional to their size. The selection of the items to be sampled within each stratum is done at random, often using simple random sampling. Stratified sampling usually leads to accurate results about the entire population, and also gives useful information about the individual strata.

Example from real life

Opinion polls

Opinion polls, such as those for the outcome of an election, are often carried out online. The polling organisation collects sufficient other information to allow respondents to be placed in strata. They then use responses from the various strata in proportion to their sizes in the population.

Cluster sampling

Cluster sampling also starts with sub groups of the population, but in this case the items are chosen from one or several of the subgroups. The subgroups are now called clusters. It is important that each cluster should be reasonably representative of the entire population. If, for example, you were asked to investigate the incidence of a particular parasite in the puffin population of northern Europe, it would be impossible to use simple random sampling. Rather, you would select a number of sites and then catch some puffins at each place. This is cluster sampling. Instead of selecting from the whole population you are choosing from a limited number of clusters.

Example from real life

Estimating the badger population size

An estimate of badger numbers in England, carried out between 2011 and 2013, was based on cluster sampling using 1411 1 km² squares from around the country. The number of badger setts in each square was counted. The clusters covered about 1% of the area of the country. There had been earlier surveys in 1985–88 and 1994–97 but with such long time intervals between them the results cannot be used to estimate the short-term variability of population over a period of years rather than decades. There are two places where local populations have been monitored over many years. So the only possible estimate of short-term variability would depend on just two clusters.

Systematic sampling

Systematic sampling is a method of choosing individuals from a sampling frame. If the items in the sampling frame are numbered 1 to n, you would choose a random starting point such as 38 and then every subsequent kth value, for example sample numbers 38, 138, 238 and so on. When using systematic sampling you have to beware of any cyclic patterns within the frame. For example, suppose that a school list is made up class by class, each of exactly 25 children, in order of merit, so that numbers 1, 26, 51, 76, 101, ... in the frame are those at the top of their class. If you sample every 50th child starting with number 26, you will conclude that the children in the school are very bright.

> **Example from real life**
>
> ### Rubbish on beaches
>
> Information was collected, using systematic sampling, about the amount and type of rubbish on the high water line along a long beach.
>
> The beach was divided up into 1 m sections and, starting from a point near one end, data were recorded for every 50th interval.

Quota sampling

Quota sampling is the method often used by companies employing people to carry out opinion surveys. An interviewer's quota is always specified in stratified terms, for example how many males and how many females. The choice of who is sampled is then left up to the interviewer and so is definitely non-random.

> **Example from real life**
>
> If you regularly take part in telephone interviews, you may notice that, after learning your details, the interviewer seems to lose interest. That is probably because quota sampling is being used and the interviewer already has enough responses from people in your category.

Opportunity sampling

Opportunity sampling (also known as 'convenience sampling') is a very cheap method of choosing a sample where the sample is selected by simply choosing people who are readily available. For example, an interviewer might stand in a shopping centre and interview anybody who is willing to participate.

> **Example from real life**
>
> ### Credit card fraud
>
> A barrister asked a mathematician to check that his argument was statistically sound in a case about credit card fraud. The mathematician wanted to find out more about the extent of suspected fraud. By chance, he was about to attend a teachers' conference and so he took the opportunity to ask delegates to fill in a short questionnaire about their relevant personal experience, if any. This gave him a rough idea of its extent and so achieved its aim.

Self-selected sampling

Self-selected sampling is a method of choosing a sample where people volunteer to be a part of the sample. The researcher advertises for volunteers, and accepts any that are suitable.

> **Example from real life**
>
> ### A medical study
>
> Volunteers were invited to take part in a long-term medical study into the effects of particular diet supplements on heart function and other conditions. They would take a daily pill which might have an active ingredient or might be a placebo, but they would not know which they were taking. Potential participants were then screened for their suitability. At six-monthly intervals those involved were asked to fill in a questionnaire about their general health and lifestyle. The study was based on a self-selected sample of some 10 000 people.

Other sampling techniques

This is by no means a complete list of sampling techniques. Survey design and experimental design cover the formulation of the most appropriate sampling procedures in particular situations. They are major topics within statistics but beyond the scope of this book.

Processing and representation

At the start of this stage you have a set of raw data; by the end, you have worked them into forms that will allow people to see the information that this set contains, with particular emphasis on the problem in hand. Four processes are particularly important.

- Cleaning the data, which involves checking outliers, errors and missing items.
- Formatting the data so that they can be used on a spreadsheet or statistics package.
- Presenting the data using suitable diagrams, which is described below.
- Calculating summary measures, which is also described below.

Describing data

> **Note**
> In this example, the random variable happens to be discrete.
> Random variables can be discrete or continuous.

The data items you collect are often values of **variables** or of **random variables**. The number of goals scored by a football team in a match is a variable because it varies from one match to another; because it does so in an unpredictable manner, it is a random variable. Rather than repeatedly using the phrase 'The number of goals scored by a football team in a match' it is usual to use an upper case letter like X to represent it. Particular values of a random variable are denoted by a lower case letter; often (but not always) the same letter is used. So if the random variable X is 'The number of goals scored by a football team in a match', for a match when the team scores 5 goals, you could say $x = 5$.

The number of times that a particular value of a random variable occurs is called its **frequency**.

When there are many possible values of the variable, it is convenient to allocate the data to groups. An example of the use of **grouped data** is the way people are allocated to age groups.

The pattern in which the values of a variable occur is called its **distribution**. This is often displayed in a diagram with the variable on the horizontal scale and a measure of frequency or probability on the vertical scale. If the diagram has one peak, the distribution is **unimodal**; if the peak is to the left of the middle, the distribution has **positive skew** and if it is to the right, it has **negative skew**. If the distribution has two distinct peaks, it is **bimodal**.

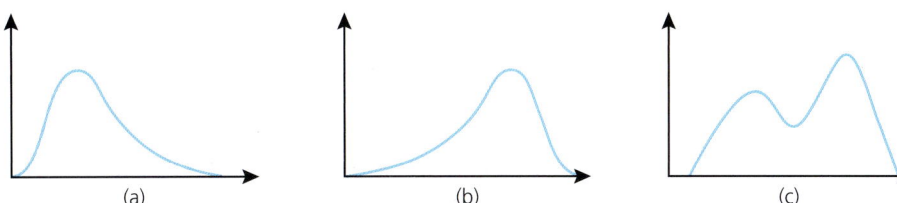

Figure 1.2 (a) Positive skew (b) negative skew (c) a bimodal distribution

> **Note**
>
> **Identifying outliers**
>
> There are two common tests:
>
> - Is the item more than 2 standard deviations from the mean?
> - Is the item more than 1.5 × the interquartile range beyond the nearer quartile?

A data item which is far away from the rest is called an **outlier**. An outlier may be a mistake, for example a faulty reading from an experiment, or it may be telling you something really important about the situation you are investigating. When you are cleaning your data, it is essential to look at any outliers and decide which of these is the case, and so whether to reject or accept them.

The data you collect can be of a number of different types. You always need to know what type of data you are working with as this will affect the ways you can display them and what summary measures you can use.

Categorical (or qualitative) data come in classes or categories, like types of fish or brands of toothpaste. Categorical data are also called **qualitative**, particularly if they can be described without using numbers.

Common displays for categorical data are pictograms, dot plots, tallies, pie charts and bar charts. A summary measure for the most typical item of categorical data is the modal class.

> **Notes**
>
> A pie chart is used for showing proportions of a total.
>
> There should be gaps between the bars in a bar chart.

Ranked data are the positions of items within their group when they are ordered according to size, rather than their actual measurements or scores. For example the competitors in a competition could be given their positions as 1st, 2nd, 3rd, etc. Ranked data are extensively used in the branch of statistics called **Exploratory Data Analysis**; this is beyond the scope of this book, but some of the measures and displays for ranked data are more widely used and are relevant here.

The median divides the data into two groups, those with high ranks and those with low ranks. The **lower quartile** and the **upper quartile** do the same for these two groups so, between them, the two quartiles and the median divide the data into four equal-sized groups according to their ranks. These three measures are sometimes denoted by Q_1, Q_2 and Q_3. These values, with the highest and lowest value can be used to create a box plot (or box and whisker diagram).

> **Note**
>
> You have to be aware when working out the median as to whether n is odd or even. If it is odd, for example if $n = 9$, $\frac{n+1}{2}$ works out to be a whole number but that is not so if n is even. For example if $n = 10$, $\frac{n+1}{2} = 5\frac{1}{2}$. In that case, the data set does not have a single middle value; those ranked 5 and 6 are equally spaced either side of the middle and so the median is half way between their values.

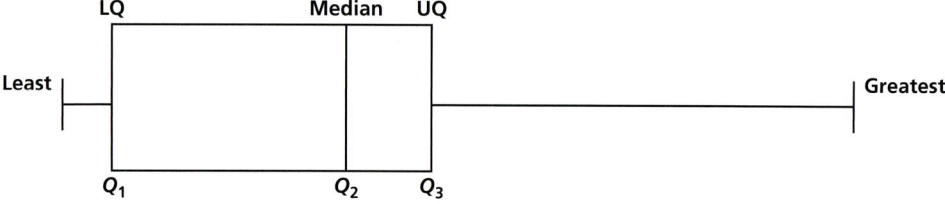

Figure 1.3 Box plot (or box and whisker diagram)

The median is a typical middle value and so is sometimes called an **average**. More formally, it is a **measure of central tendency**. It often provides a good representative value. The median is easy to work out if the data are stored on a spreadsheet since that will do the ranking for you. Notice that extreme values have little, if any, effect on the median. It is described as resistant to outliers. It is often useful when some data values are missing but can be estimated.

Interquartile range and semi interquartile range are measures of spread for ranked data, as is the range.

Drawing a stem-and-leaf diagram can be helpful when ranking data.

Numerical (or quantitative) data occur when each item has a numerical value (and not just a rank), like the number of people travelling in a car or the values of houses.

Numerical data are described as discrete if items can take certain particular numerical values but not those in between. The number of eggs a song bird lays (0, 1, 2, 3, 4, ...), the number of goals a hockey team scores in a match (0, 1, 2, 3, ...) and the sizes of women's clothes in the UK (... 8, 10, 12, 14, 16, ...) are all examples of discrete variables. If there are many possible values, it is common to group discrete data.

By contrast, **continuous** numerical data can take any appropriate value if measured accurately enough.

Distance, mass, temperature and speed are all continuous variables. You cannot list all the possible values.

If you are working with continuous data you will always need to **group** them. This includes two special cases:

- The variable is actually discrete but the intervals between values are very small. For example, cost in euros is a discrete variable with steps of €0.01 (i.e. 1 cent) but this is so small that the variable may be regarded as continuous.
- The underlying variable is continuous but the measurements of it are rounded (for example, to the nearest mm), making your data discrete. All measurements of continuous variables are rounded and, providing the rounding is not too coarse, the data should normally be treated as continuous. A particular case of rounding occurs with people's ages; this is a continuous variable but is usually rounded down to the nearest completed year.

> **Note**
> In frequency charts and histograms, the values of the variables go at the ends of the bars.

> **Note**
> If you are using a frequency chart, the class intervals should all be equal. For a histogram, they don't have to be equal. So, if you have continuous data grouped into classes of unequal width, you should expect to use a histogram.

Displaying numerical data

Commonly used displays for discrete data include a vertical line chart and a stem-and-leaf diagram. A frequency table can be useful in recording, sorting and displaying discrete numerical data.

A frequency chart and a histogram are the commonest ways of displaying continuous data. Both have a continuous horizontal scale covering the range of values of the variable. Both have vertical bars.

- In a frequency chart, frequency is represented by the height of a bar. The vertical scale is Frequency.
- In a histogram, frequency is represented by the area of a bar. The vertical scale is Frequency density.

Look at this frequency chart and histogram. They show the time, t minutes, that a particular train was late at its final destination in 150 journeys.

On both graphs, the interval 5–10 means $5 < t \leqslant 10$, and, similarly, for other intervals. A negative value of t means the train was early.

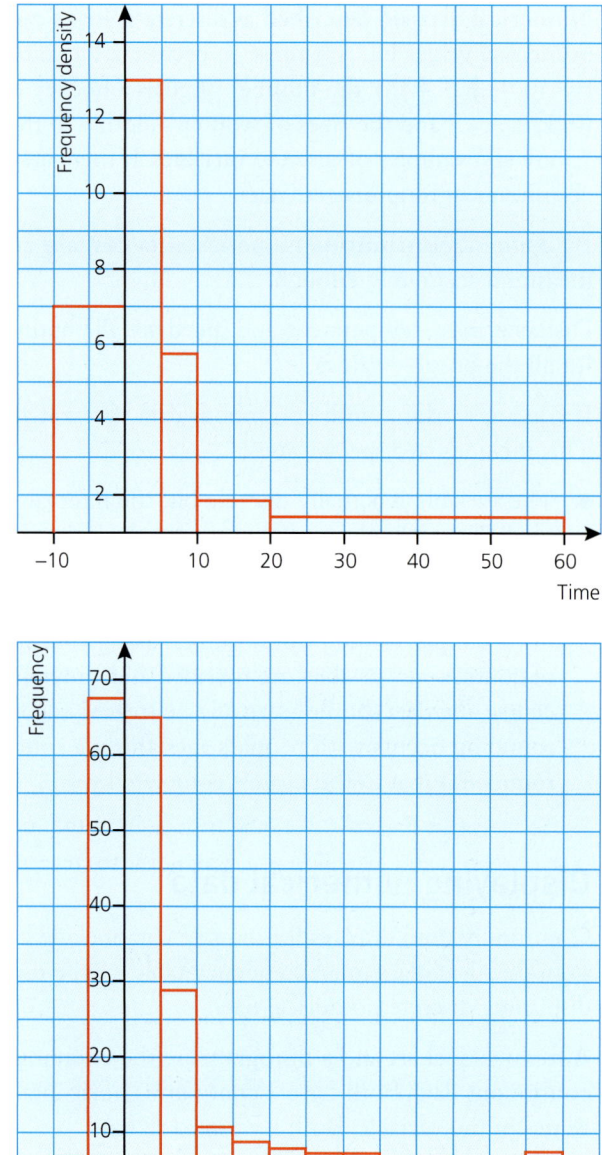

Figure 1.4

Discussion point
What is the same about the two displays and what is different?

Numerical data can also be displayed on a **cumulative frequency curve**. To draw the cumulative frequency curve, you plot the cumulative frequency (vertical axis) against the upper boundary of each class interval (horizontal axis). Then you join the points with a smooth curve. This lends itself to using the median, quartiles and other percentiles as summary measures.

Summary measures for numerical data

Summary measures for both discrete and continuous numerical data include the following.

Table 1.1

Central tendency	Spread	Position in the data
Mean	Range	Lower quartile
Weighted mean	Interquartile range	Median
Mode	Standard deviation	Upper quartile
Mid-range	Variance	Percentile
Median		
Modal class (grouped data)		

> **Note**
> In practice, many people would just enter the data into their calculators and read off the answer. However, it is important to understand the ideas that underpin the calculation.

Standard deviation

Standard deviation is probably the most important measure of spread in statistics. The calculation of standard deviation, and of variance, introduce important notation which you will often come across. This is explained in the example that follows.

Example 1.1

Alice enters dance competitions in which the judges give each dance a score between 0 and 10. Here is a sample of her recent scores.

Table 1.2

| 7 | 4 | 9 | 8 | 7 | 8 | 8 | 10 | 9 | 10 |

Calculate the mean, variance and standard deviation of Alice's scores.

Solution

Alice received 10 scores, so the number of data items, $n = 10$.

In the following table her scores are denoted by $x_1, x_2, ..., x_{10}$, with the general term x_i.

The mean score is $\bar{x}$.

Note

An **alternative** but equivalent form of S_{xx} is given by
$$S_{xx} = \Sigma x_i^2 - n\bar{x}^2$$
In this case, $S_{xx} = 668 - 10 \times 8^2 = 668 - 640 = 28$.

This is, as expected, the same value as that found above.

Table 1.3

	x_i	$x_i - \bar{x}$	$(x_i - \bar{x})^2$	x_i^2
x_1	7	-1	1	49
x_2	4	-4	16	16
x_3	9	1	1	81
x_4	8	0	0	64
x_5	7	-1	1	49
x_6	8	0	0	64
x_7	8	0	0	64
x_8	10	2	4	100
x_9	9	1	1	81
x_{10}	10	2	4	100
Σ	80	0	28	668

The quantity $(x_i - \bar{x})$ is the **deviation** from the mean. Notice that the total of the deviations $\Sigma(x_i - \bar{x})$ is zero. It has to be so because $\bar{x}$ is the mean, but finding it gives a useful check that you haven't made a careless mistake so far.

The value of 668 for Σx_i^2 was found in the right hand column of the table.

The mean is given by $\bar{x} = \Sigma \dfrac{x_i}{n} = \dfrac{80}{10} = 8.0$.

The variance is given by $s^2 = \dfrac{S_{xx}}{n}$, where $S_{xx} = \Sigma(x_i - \bar{x})^2$

In this case, $S_{xx} = 28$.

So the variance is $s^2 = \dfrac{28}{10} = 2.8$

The standard deviation is $s = \sqrt{\text{variance}} = \sqrt{2.8} = 1.673$

Notes

Some authors use $s^2 = \dfrac{S_{xx}}{n-1}$. This is STDEV.S on a spreadsheet.

For large values of n this makes very little difference.

ACTIVITY 1.1

Using a spreadsheet, enter the values of x in cells B2 to B11. Then, using only spreadsheet commands, and without entering any more numbers, obtain the values of $(x - \bar{x})$ in cells C2 to C11, of $(x - \bar{x})^2$ in cells D2 to D11 and of x^2 in E2 to E11. Still using only the spreadsheet commands, find the standard deviation using both of the given formulae.

Now, as a third method, use the built in functions in your spreadsheet, for example =AVERAGE and =STDEV.PA to calculate the mean and standard deviation directly.

Notation

The notation in the example is often used with other variables.

So, for example, $S_{yy} = \Sigma(y_i - \bar{y})^2 = \Sigma y_i^2 - n\bar{y}^2$.

In Chapters 5 and 6, you will meet an equivalent form for bivariate data,

$$S_{xy} = \Sigma(x_i - \bar{x})(y_i - \bar{y}) = \Sigma x_i y_i - n\bar{x}\bar{y}.$$

You can also extend the notation to cases where the data are given in frequency tables.

For example, Alice's dance scores could have been written as the frequency table below.

Note

Bivariate data cover two variables, such as the birth rate and life expectancy of different countries.

When you are working with bivariate data you are likely to be interested in the relationship between the two variables, how this can be seen on a scatter diagram and how it can be quantified.

Table 1.4

x_i	4	7	8	9	10
f_i	1	2	3	2	2

The number of items is then given by $n = \Sigma f_i$

The mean is $\bar{x} = \dfrac{\Sigma f_i x_i}{n}$

The sum of squared deviations is $S_{xx} = \Sigma f_i (x_i - \bar{x})^2 = \Sigma f_i x_i^2 - n\bar{x}^2$

As before, the variance is $s^2 = \dfrac{S_{xx}}{n}$

and the standard deviation is $s = \sqrt{\text{variance}} = \sqrt{\dfrac{S_{xx}}{n}}$.

Exercise 1.1

① A club secretary wishes to survey a sample of members of his club. He uses all members present at a meeting as a sample

(i) Explain why this sample is likely to be biased.

Later the secretary decides to choose a random sample of members. The club has 253 members and the secretary numbers the members from 1 to 253. He then generates random 3-digit numbers on his calculator. The first six random numbers are 156, 965, 248, 156, 073 and 181. The secretary uses each number, where possible, as the number of a member in the sample.

(ii) Find possible numbers for the first four members in the sample. [OCR]

② This stem-and-leaf diagram shows the mean GDP per person in European countries, in thousands of US$. The figures are rounded to the nearest US$ 1000.

Table 1.5

	Europe
0	4 7 8 8 8
1	1 1 2 4 6 8 9
2	0 1 2 3 3 3 4 4 5 6 8 8
3	0 0 1 6 6 7 7 8 8
4	0 1 1 1 1 3 3 5 6
5	4 5 7
6	1 6
7	
8	0 9

> **Note**
> Key 3|7 = US$ 37 000
> $n = 49$

(i) The mean per capita income for the UK is US$ 37 300. What is the rank of the UK among European countries (where rank 1 = largest GDP)?

(ii) Find the median and quartiles of the data.

(iii) Use the relevant test to identify any possible outliers.

(iv) Describe the distribution.

(v) Comment on whether these data can be used as a representative sample for the GDP of all the countries in the world.

③ Debbie is a sociology student. She is interested in how many children women have during their lifetimes. She herself has one sister and no brothers. She asks the other 19 students in her class 'How many children has your mother had?' Their answers follow; the figure for herself is included.

Table 1.6

1	1	2	3	1	4	2	1	2	2
2	2	2	1	0	1	2	3	8	3

Debbie says

'Thank you for your help. I conclude that the average woman has exactly 2.15 children.'

(i) Name the sampling method that Debbie used.

(ii) Explain how she obtained the figure 2.15.

(iii) State four things that are wrong with her method and her stated conclusion.

④ A supermarket chain is considering opening an out-of-town shop on a greenfield site. Before going any further they want to test public opinion and carry out a small pilot investigation. They employ three local students to ask people 'Would you be in favour of this development?' Each of the students is told to ask 30 adult men, 30 adult women and 40 young people who should be under 19 but may be male or female.

Their results are summarised in this table.

Table 1.7

Interviewer	Men			Women			Young people		
	Yes	No	Don't know	Yes	No	Don't know	Yes	No	Don't know
A	5	20	5	18	12	0	12	10	18
B	12	14	4	20	8	2	11	11	18
C	9	18	3	17	9	4	10	15	15

(i) Name the sampling method that has been used.

The local development manager has to give a very brief report to the company's directors and this will include his summary of the findings of the pilot survey.

(ii) List the points that he should make.

The directors decide to take the proposal to the next stage and this requires a more accurate assessment of local opinion.

(iii) What sampling method should they use?

⑤ A certain animal is regarded as a pest. There have been two surveys, eight years apart, to find out the size of the population in the UK. After the second survey a newspaper carried an article which included these words.

> **This animal is out of control. Its numbers have doubled in just 8 years.**

The actual population, which no one knows, is shown on the following graph for 1995–2015. The unit on the vertical scale is 100 000 animals.

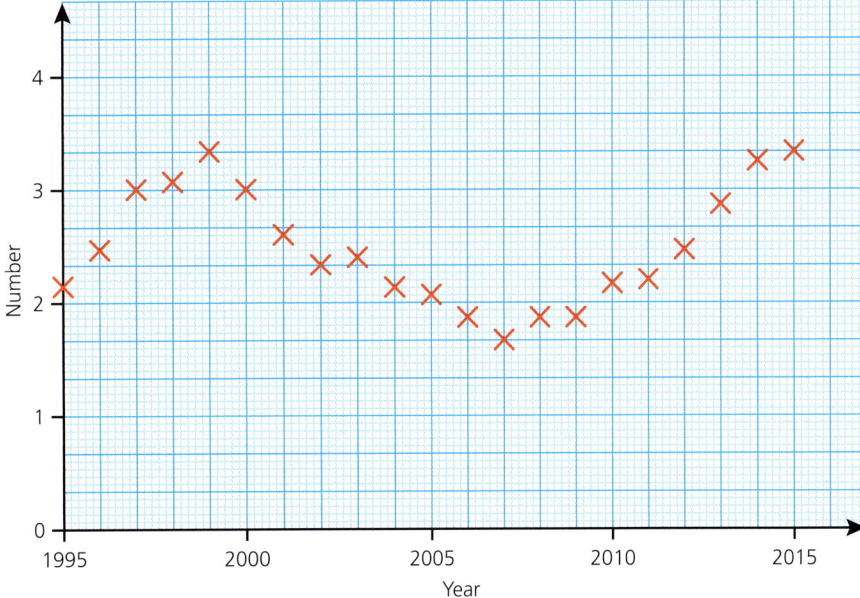

Figure 1.5 Graph of population

(i) Describe the apparent pattern of the size of the population.

(ii) In which years does it seem that the surveys were carried out?

(iii) Suggest conclusions which might have been reached if the two surveys had been in (a) 1999 and 2007 (b) 1999 and 2015.

Historic data of the sale of furs of Arctic mammals, such as lynx and hares, by the Hudson Bay Company indicate a 10 to 11 year cycle in their population numbers over many years.

(iv) Suppose the data are available on a spreadsheet. Describe how a systematic sample might be taken from the data on the spreadsheet. Comment on the problems that might result.

⑥ A study is conducted on the breeding success of a type of sea bird. Four islands are selected and volunteers monitor nests on them, counting the number of birds that fledge (grow up to fly away from the nest).

The results are summarised in the table below.

Table 1.8

Island	Number of fledglings				
	0	1	2	3	>3
A	52	105	31	2	0
B	10	81	55	6	0
C	67	33	2	0	0
D	29	65	185	11	0

(i) Describe the sample that has been used.

(ii) Explain why you are unable to give an accurate value for the sampling fraction.

(iii) Estimate the mean number of fledglings per nest and explain why this figure may not be very close to that for the whole population.

(iv) Ornithologists estimate that there are about 120 000 breeding pairs of these birds. Suggest appropriate limits within which the number of fledglings might lie, showing the calculations on which your answers are based.

(7) The Highways Authority proposes to impose new parking restrictions on a small town. The Town Council fears that it will be bad for trade, and so for the town's prosperity. They plan to object, but first they need data so that they can estimate the cost to the town.

The council call a special meeting. Their room can only take 20 more people (in addition to the councillors) and they invite

   8 of the 41 shops

   7 of the 33 restaurants and cafes

   5 of the 27 hotels and bed and breakfasts.

(i) Describe the sort of sample they have selected. Explain how they decided on the numbers from the various groups.

(ii) They use a random selection procedure to decide who actually gets invited. Describe two possible ways they might do this.

(iii) Explain why those selected are not a simple random sample.

At the meeting, those present are asked to estimate the annual cost, to the nearest £1000, to their businesses if the parking restrictions go ahead. Their replies, in thousands of pounds, are given in the table below.

Table 1.9

Shops	2	1	5	0	12	10	8	3
Restaurants and cafes	1	2	2	1	3	2	1	-
Hotels and B&B	15	0	0	10	1	-	-	-

(iv) Use these figures to estimate the total cost of the parking restrictions to the town.

(v) Comment on the likely accuracy of the estimate and suggest measures that might be taken to improve it.

(8) A health authority takes part in a national study into the health of women during pregnancy. One feature of this is that pregnant women are invited to volunteer for a fitness programme in which they exercise every day. Their general health is monitored and the days on which their babies arrive are recorded and shown on the histogram below.

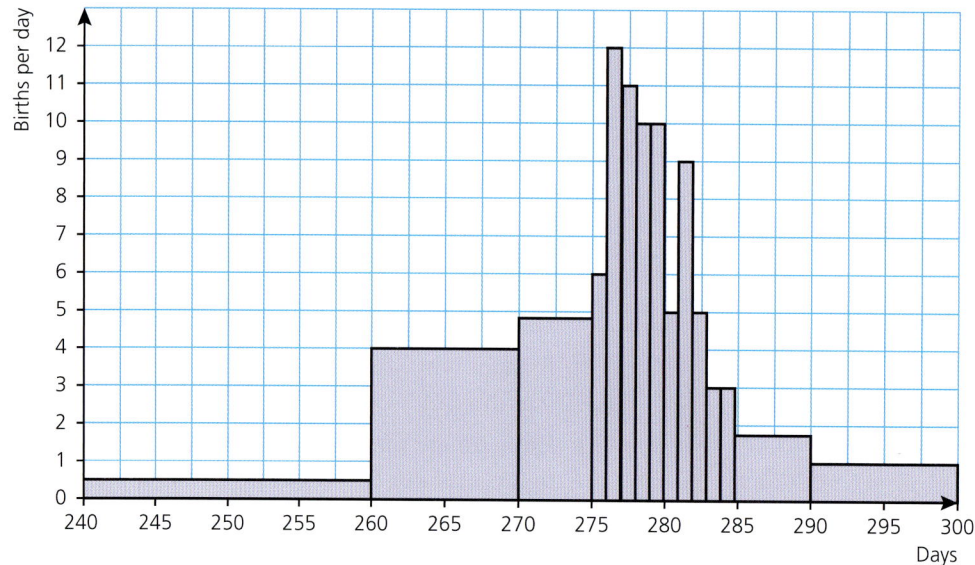

Figure 1.6

(i) The women who take part in the programme constitute a sample. What sort of sample is it?

(ii) Describe the parent population from which the sample is drawn and give one reason why it may not be completely representative. Comment on the difficulties in selecting a representative sample for this study.

(iii) Use the histogram to find how many women participated in the study.

The 'due date' of a mother to be is set at 280 days. Babies born before 260 days are described as 'pre-term'. Those born after 287 days are 'post-term'.

(iv) Give your answers to parts (a), (b) and (c) to 1 decimal place.

  (a) Find the percentage of the babies that arrived on their due dates.

  (b) Find the percentage that were pre-term.

  (c) Estimate the percentage of babies that were post-term.

  (d) Explain why your answers to parts (a), (b) and (c) do not add up to 100.

⑨ (i) Births, deaths and marriages are listed for England and Wales (and separately for other parts of the UK). As part of a pilot study, three research students, A, B and C, selected samples of the ages of women who died in 2016 from the list, using random numbers.

Summary data for their samples are:

Table 1.10

A	$n = 25$	$\Sigma x = 2038$	$\Sigma x^2 = 168545$
B	$n = 35$	$\Sigma x = 2720$	$\Sigma x^2 = 214893$
C	$n = 40$	$\Sigma x = 3230$	$\Sigma x^2 = 264936$

  (a) Describe the samples that the research students took.

  (b) Find the mean and standard deviation for each of the samples. Explain why they are not all the same.

  (c) Find the mean and standard deviation when the three samples are put together. Comment on the likely accuracy of your answer.

(ii) State the two formulae for S_{xx} and show algebraically that they are equivalent.

⑩ A local police force records the number of people arrested per day during January, February and March one year. The results are as follows.

Table 1.11

No. of arrests	Frequency
0	55
1	24
2	6
3	2
4	0
5	2
6, 7	0
8	1
>8	0

(i) Find the mean and standard deviation of the number of arrests per day.

(ii) The figure 8 is an outlier. It was the result of a fight on a train that stopped in the area. It is suggested that the data should not include that day. What percentage changes would that make to the mean and standard deviation?

(iii) Find the percentage error if the standard deviation (with the outlier excluded) is worked out using the formula
$$s = \sqrt{\frac{S_{xx}}{n-1}} \text{ instead of } s = \sqrt{\frac{S_{xx}}{n}}.$$

(iv) The standard deviation of a sample is worked out using a divisor $(n-1)$ instead of n. Find the smallest value of n for which the error in doing so is less than 1%.

KEY POINTS

1. The problem solving cycle has four stages:
   - problem specification and analysis
   - information collection
   - processing and representation
   - interpretation.
2. Information collection often involves taking a sample.
3. There are several reasons why you might wish to take a sample:
   - to help you understand a situation better
   - as part of a pilot study to inform the design of a larger investigation
   - to estimate the values of the parameters of the parent population
   - to avoid the work involved in cleaning and formatting all the data in a large set
   - to conduct a hypothesis test.
4. Sampling procedures include:
   - simple random sampling
   - stratified sampling
   - cluster sampling
   - systematic sampling
   - quota sampling
   - opportunity sampling
   - self-selected sampling.
5. For processing and representation, it is important to know the type of data you are working with i.e.:
   - categorical or qualitative data
   - ranked data
   - discrete numerical data
   - continuous numerical data
   - bivariate data.
6. Display techniques and summary measures must be appropriate for the type of data.

7 Notation for mean and standard deviation.

Mean $\bar{x} = \Sigma \dfrac{x_i}{n}$

Sum of square deviations $S_{xx} = \Sigma(x_i - \bar{x})^2 = \Sigma x_i^2 - n\bar{x}^2$

Variance $s^2 = \dfrac{S_{xx}}{n}$

Standard deviation $s = \sqrt{\text{variance}} = \sqrt{\dfrac{S_{xx}}{n}}.$

LEARNING OUTCOMES

When you have completed this chapter you should be able to:

- use statistics within a problem solving cycle
- explain why sampling may be necessary in order to obtain information about a population, and give desirable features of a sample, including the size of the sample
- know a variety of sampling methods, the situations in which they might be used and any problems associated with them
- explain the advantage of using a random sample when inferring properties of a population
- display sample data appropriately
- calculate and interpret summary measures for sample data.

2 Discrete random variables

Probability theory is nothing but common sense reduced to calculation.
Pierre Simon Laplace

An archery competition is held each month. In the first round of the competition, each competitor has five tries at hitting a small target. Those who hit the target at least three times get through to the next round. In April, there are 250 competitors in the first round. The frequencies of the different numbers of possible successes are as follows.

Table 2.1

Number of successes	0	1	2	3	4	5
Frequency	65	89	48	21	11	16

The numbers of successes are necessarily discrete. A discrete frequency distribution is best illustrated by a vertical line chart, as in Figure 2.1. This shows you that the distribution has positive skew, with the bulk of the data at the lower end of the distribution.

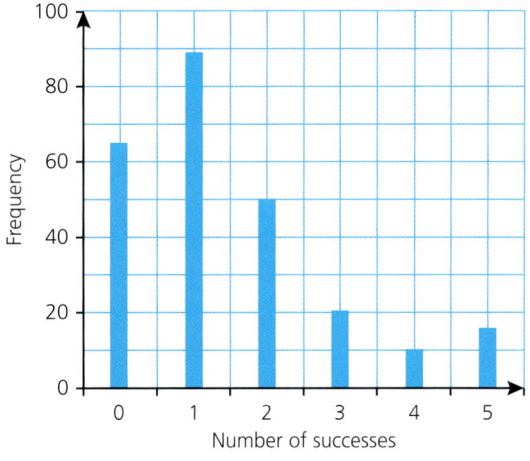

Figure 2.1

> **Note**
> You can draw a diagram to show this **probability distribution**. It is identical in shape to Figure 2.1 but with probability rather than frequency on the vertical axis.

The survey involved 250 competitors. This is a reasonably large sample and so it is reasonable to use the results to estimate the **probabilities** of the various possible outcomes: 0, 1, 2, 3, 4, 5 successes. You divide each frequency by 250 to obtain the **relative frequency**, or probability, of each outcome (number of successes).

Table 2.2

Outcome (Number of successes)	0	1	2	3	4	5
Probability	0.260	0.356	0.192	0.084	0.044	0.064

You now have a **mathematical model** to describe a particular situation. In statistics, you are often looking for models to describe and explain the data you find in the real world. In this chapter, you are introduced to some of the techniques for working with models for discrete data. Such models are called **discrete probability distributions**.

> For example: the number of rolls to get a six on a dice.

The number of successes is a **random variable** since the actual value of the outcome is variable and can only be predicted with a given probability, i.e. the outcomes occur at random. The random variable is **discrete** since the number of successes is an integer (between 0 and 5).

In the archery competition, the maximum number of successes is five so the variable is **finite**. For example, if each competitor had ten tries, then the maximum would be ten. In this case, there would be eleven possible outcomes (including zero). Two well-known examples of finite discrete random variables are the **binomial distribution** and the **negative binomial distribution**.

By contrast, if you considered the number of times you need to roll a pair of dice to get a double six, there is no theoretical maximum, and so the distribution is **infinite**. Two well-known examples of infinite discrete random variables are the **geometric distribution** and the **Poisson distribution**. You will study these discrete distributions in Chapter 3 and Chapter 9.

1 Conditions for discrete random variables

- A random variable is denoted by an upper case letter, such as X, Y, or Z.
- The particular values that the random variable takes are denoted by lower case letters, such as x, y, z and r.
- In the case of a discrete variable these are sometimes given suffixes such as $r_1, r_2, r_3, \ldots$
- Thus $P(X = r_1)$ means the probability that the random variable X takes a particular value r_1.
- If a finite discrete random variable can take n distinct values $r_1, r_2, \ldots, r_n$, with associated probabilities $p_1, p_2, \ldots, p_n$, then the sum of the probabilities must equal 1.
- In that case, $p_1 + p_2 + \ldots + p_n = 1$.
- This can be written more formally as

$$\sum_{1}^{n} p_k = \sum_{1}^{n} P(X = r_k) = 1.$$

- If there is no ambiguity, then

$$\sum_{1}^{n} P(X = r_k)$$

is often abbreviated to

$$\sum P(X = r).$$

> **Note**
> You will often see the expression $P(X = r)$ in a table heading.

> The various outcomes cover all possibilities; they are exhaustive.

Example 2.1

The probability distribution of a random variable X is given by

$P(X = r) = kr^2$ for $r = 3, 4, 5$

$P(X = r) = 0$ otherwise.

(i) Find the value of the constant k.

(ii) Illustrate the distribution and describe the shape of the distribution.

(iii) Two successful values of X are generated independently of each other. Find the probability that

   (a) both values of X are the same

   (b) the total of the two values of X is greater than 8.

Solution

(i) Tabulating the probability distribution for X gives:

Table 2.3

r	3	4	5
$P(X = r)$	$9k$	$16k$	$25k$

Since X is a random variable,

$$\Sigma(P(X=r)) = 1$$
$$9k + 16k + 25k = 1$$
$$50k = 1$$
$$k = 0.02$$

Hence $P(X = r) = 0.02r^2$ for $r = 3, 4, 5$ which gives the following probability distribution.

Table 2.4

r	3	4	5
$P(X = r)$	0.18	0.32	0.50

(ii) The vertical line chart in Figure 2.2 illustrates this distribution. It has negative skew.

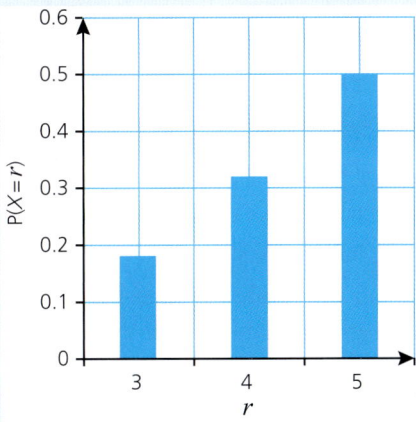

Figure 2.2

(iii) (a) P(both values of X are the same)
$$= (0.18)^2 + (0.32)^2 + (0.5)^2$$
$$= 0.0324 + 0.1024 + 0.25$$
$$= 0.3848$$

(b) P(total of the two values is greater than 8)

> The ways of getting a total greater than 8 are: 4 and 5, 5 and 4, 5 and 5.

$$= 0.32 \times 0.5 + 0.5 \times 0.32 + 0.5 \times 0.5$$
$$= 0.16 + 0.16 + 0.25$$
$$= 0.57$$

Exercise 2.1

① A fair five-sided spinner has faces labelled 1, 2, 3, 4, 5. The random variable X represents the score when the spinner is spun.

(i) Copy and complete the table below to show the probability distribution of X.

Table 2.5

r	1	2	3		
$P(X = r)$	0.2				

(ii) Illustrate the distribution.

(iii) Find the values of
  (a) $P(X > 2)$
  (b) $P(X$ is even$)$
  (c) $P(X > 5)$.

② The probability distribution of a discrete random variable X is given by

$P(X = r) = kr$ for $r = 1, 2, 3, 4$
$P(X = r) = 0$ otherwise.

(i) Copy and complete the table below to show the probability distribution of X in terms of k.

Table 2.6

r	1	2	3	
$P(X = r)$				4k

(ii) Use the fact that the sum of the probabilities is equal to 1 to find the value of k.

(iii) Find the values of
  (a) $P(X = 4)$
  (b) $P(X < 4)$.

③ A fair three-sided spinner has faces labelled 1, 2 and 3. The random variable X is given by the sum of the scores when the spinner is spun three times.

(i) Find the probability distribution of X.
(ii) Illustrate the distribution and describe the shape of the distribution.
(iii) Find the values of
  (a) $P(X > 6)$
  (b) $P(X$ is odd$)$
  (c) $P(|X - 4| < 2)$.

④ The random variable Y is given by the absolute difference when the spinner in Question 1 is spun twice.

(i) Find the probability distribution of Y.
(ii) Illustrate the distribution and describe the shape of the distribution.
(iii) Find the values of
  (a) $P(Y < 2)$
  (b) $P(Y$ is even$)$.

5. Two ordinary dice are thrown. The random variable X is the product of the numbers shown on the dice.
   (i) Find the probability distribution of X.
   (ii) What is the probability that any throw of the two dice results in a value of X which is an even number?

6. The probability distribution of a discrete random variable X is given by
   $$P(X = r) = \frac{kr}{4} \text{ for } r = 2, 3, 4, 5$$
   $$P(X = r) = 0 \text{ otherwise.}$$
   (i) Find the value of k and tabulate the probability distribution.
   (ii) If two successive values of X are generated independently find the probability that
      (a) the two values are equal
      (b) the first value is less than the second value.

7. A curiously shaped four-faced spinner produces scores, X, for which the probability distribution is given by
   $$P(X = r) = k(r^2 + 2r + 10) \text{ for } r = 0, 1, 2, 3, 4$$
   $$P(X = r) = 0 \text{ otherwise.}$$
   (i) Find the value of k and illustrate the distribution.
   (ii) Show that, when this spinner is spun twice, the probability of obtaining one non-zero score which is exactly twice the other is very nearly 0.18.

8. Four fair coins are tossed.
   (i) By considering the set of possible outcomes, HHHH, HHHT, etc., tabulate the probability distribution of X, the number of heads occurring.
   (ii) Illustrate the distribution and describe the shape of the distribution.
   (iii) Find the probability that there are more heads than tails.
   (iv) Without further calculation, state whether your answer to part (iii) would be the same if five fair coins were tossed. Give a reason for your answer.

9. A doctor is investigating the numbers of children, X, which women have in a country. She notes that the probability that a woman has more than five children is negligible. She suggests the following model for X
   $$P(X = 0) = 0.3$$
   $$P(X = r) = k(12 + 3r - r^2) \quad \text{for } r = 1, 2, 3, 4, 5$$
   $$P(X = r) = 0 \quad \text{otherwise.}$$
   (i) Find the value of k and write the probability distribution as a table.
   (ii) Find the probability that two women chosen at random both have more than three children.

10. A motoring magazine correspondent conducts a survey of the numbers of people per car travelling along a stretch of motorway. He denotes the number by the random variable X which he finds to have the following probability distribution.

    Table 2.7

r	1	2	3	4	5	6+
$P(X = r)$	0.57	0.28	a	0.04	0.01	negligible

    (i) Find the value of a.

He wants to find an algebraic model for the distribution and suggests the following model.

$P(X = r) = k2^{-r}$ for $r = 1, 2, 3, 4, 5$

$P(X = r) = 0$ otherwise.

   (ii) Find the value of k for this model.

   (iii) Compare the algebraic model with the probabilities he found, illustrating both distributions on one diagram. Do you think it is a good model?

11. In a game, each player throws four ordinary six-sided dice. The random variable X is the smallest number showing on the dice, so, for example, for scores of 2, 5, 3 and 4, $x = 2$.

   (i) Find the probability that $X = 6$, i.e. $P(X = 6)$.

   (ii) Find $P(X \geq 5)$ and deduce that $P(X = 5) = \frac{15}{1296}$.

   (iii) Find $P(X \geq r)$ and so deduce $P(X = r)$, for $r = 4, 3, 2, 1$.

   (iv) Illustrate and describe the probability distribution of X.

12. A box contains six black pens and four red pens. Three pens are taken at random from the box.

   (i) Illustrate the various outcomes on a probability tree diagram.

   (ii) The random variable X represents the number of black pens obtained. Find the probability distribution of X.

2 Expectation and variance

The next round of the archery competition is held in May. In this round there are 200 competitors altogether. The organisers of the competition would like to increase the number of people getting through to the next round, without changing the rules. They decide to give each of the competitors a relaxation session before their attempt. The number of successes for each competitor this time is as follows.

> **Discussion point**
>
> How would you compare the results in the competitions?

Table 2.8

Number of successes	0	1	2	3	4	5
Frequency	36	54	60	19	15	16

The competition involves 200 people. This is again a reasonably large sample and so, once again, it is reasonable to use the results to estimate the probabilities of the various possible outcomes: 0, 1, 2, 3, 4, 5 successes, as before.

Table 2.9

Outcome (Number of successes)	0	1	2	3	4	5
Probability (Relative frequency)	0.18	0.27	0.30	0.095	0.075	0.08

One way to compare the two probability distributions, in April and in May, is to calculate a measure of central tendency and a measure of spread.

Just as you can calculate the mean and variance of a frequency distribution, you can also do something very similar for a probability distribution.

- The most useful measure of central tendency for a probability distribution is the **mean** or **expectation** of the random variable. This is denoted by μ.
- The most useful measure of spread for a probability distribution is the **variance**, σ^2, or its square root the **standard deviation**, σ.

The expectation is given by $E(X) = \mu = \Sigma r P(X = r)$. Its calculation is shown below using the probability distribution for the archers in May (after the relaxation session) as an example.

Table 2.10

r	P(X = r)	rP(X = r)
0	0.18	0
1	0.27	0.27
2	0.30	0.60
3	0.095	0.285
4	0.075	0.30
5	0.08	0.4
Totals	1	1.855

> **Note**
> You will find it helpful to set your work out systematically in a table like this.

In this case:
$$E(X) = \mu = \Sigma r P(X = r)$$
$$= 0 \times 0.18 + 1 \times 0.27 + 2 \times 0.30 + 3 \times 0.095 + 4 \times 0.075 + 5 \times 0.08$$
$$= \mathbf{1.855}$$

There are two common ways of giving the variance.

Either $\text{Var}(X) = \sigma^2 = E(X^2) - [E(X)]^2$

or $E(X - \mu)^2 = \Sigma (r - \mu)^2 P(X = r)$

> This version can be remembered as 'The expectation of the squares minus the square of the expectation'. It can also be written as $E(X^2) - \mu^2$.

To see how variance is calculated the same probability distribution is used as an example.

> **Note**
> You will use these statistics later to compare the distribution of numbers of successes with and without the relaxation session.

The table below shows the working for the variance in May after the relaxation session, using both of the methods above.

Table 2.11

> The mean, μ, was found above. It is **1.855**.

r	P(X = r)	r²P(X = r)	r	P(X = r)	(r − μ)² P(X = r)
0	0.18	0	0	0.18	0.6194
1	0.27	0.27	1	0.27	0.1974
2	0.30	1.2	2	0.30	0.0063
3	0.095	0.855	3	0.095	0.1245
4	0.075	1.2	4	0.075	0.3451
5	0.08	2	5	0.08	0.7913
Totals	1	**5.525**	Totals	1	**2.0840**

S1 AS Expectation and variance

> **Discussion point**
> Using these two statistics, judge the success or otherwise of the relaxation session.

(a) $\text{Var}(X) = \Sigma r^2 P(X=r) - [\Sigma r P(X=r)]^2$
$= 5.525 - 1.855^2$
$= 2.084$

(b) $\text{Var}(X) = \Sigma(r-\mu)^2 P(X=r)$
$= 2.084$

The standard deviation of X is therefore $\sqrt{2.084} = 1.44$

In practice, the computation is usually easier in Method (a), especially when the expectation is not a whole number.

ACTIVITY 2.1
Show that the expectation and variance of the probability distribution in April (without the relaxation session) are 1.49 and 1.97, respectively.

Example 2.2

The discrete random variable X has the following probability distribution:

Table 2.12

r	0	1	2	3
$P(X=r)$	0.2	0.3	0.4	0.1

Find

(i) $E(X)$

(ii) $\text{Var}(X)$ using (a) $E(X^2) - \mu^2$ (b) $E([X-\mu]^2)$.

Solution

(i) **Table 2.13**

r	$P(X=r)$	$rP(X=r)$
0	0.2	0
1	0.3	0.3
2	0.4	0.8
3	0.1	0.3
Totals	1	1.4

> To find $E(X)$ you simply multiply each value of r by its probability and then add.

$E(X) = \mu = \Sigma r P(X=r)$
$= 0 \times 0.2 + 1 \times 0.3 + 2 \times 0.4 + 3 \times 0.1$
$= 1.4$

> **Discussion point**
> Look carefully at both methods for calculating the variance. Are there any situations where one method might be preferred to the other?

(ii)

Table 2.14

r	$P(X = r)$	$r^2 P(X = r)$
0	0.2	0
1	0.3	0.3
2	0.4	1.6
3	0.1	0.9
Totals	1	2.8

r	$P(X = r)$	$(r-\mu)^2 P(X = r)$
0	0.2	0.392
1	0.3	0.048
2	0.4	0.144
3	0.1	0.256
Totals	1	0.84

(a) $\text{Var}(X) = \sigma^2 = E(X^2) - \mu^2$
$= 2.8 - 1.4^2$
$= 0.84$

(b) $\text{Var}(X) = \sigma^2 = E([X - \mu]^2)$
$= \Sigma (r - \mu)^2 P(X = r)$
$= 0.84$

Notice that the two methods of calculating the variance in part (ii) give the same result, since one formula is just an algebraic rearrangement of the other. In practice, you would not need to use both methods and so only either the left-hand or the right-hand table would be needed to calculate the variance, according to which method you are using.

ACTIVITY 2.2

Use a spreadsheet to find the variance of X for the following probability distribution.

r	1	2	3	4
$P(X = r)$	0.25	0.22	0.08	0.45

You should use both methods of calculating the variance and check that they give the same result.

As well as being able to carry out calculations for the expectation and variance, you often need to solve problems in context. The following example illustrates this idea.

Example 2.3

Laura has one pint of milk on three days out of every four and none on the fourth day. A pint of milk costs 40p. Let X represent her weekly milk bill.

(i) Find the probability distribution of her weekly milk bill.

(ii) Find the mean (μ) and standard deviation (σ) of her weekly milk bill.

(iii) Find

(a) $P(X > \mu + \sigma)$

(b) $P(X < \mu - \sigma)$.

Expectation and variance

Solution

(i) The delivery pattern repeats every four weeks.

Table 2.15

M	Tu	W	T	F	Sa	Su	Number of pints	Milk bill
✓	✓	✓	✗	✓	✓	✓	6	£2.40
✗	✓	✓	✓	✗	✓	✓	5	£2.00
✓	✗	✓	✓	✓	✗	✓	5	£2.00
✓	✓	✗	✓	✓	✓	✗	5	£2.00

Tabulating the probability distribution for X gives the following.

Table 2.16

r (£)	2.00	2.40
$P(X = r)$	0.75	0.25

(ii)
$$E(X) = \mu = \Sigma r\, P(X = r)$$
$$= 2 \times 0.75 + 2.4 \times 0.25$$
$$= 2.1$$
$$\text{Var}(X) = \sigma^2 = E(X^2) - \mu^2$$
$$= 4 \times 0.75 + 5.76 \times 0.25 - 2.1^2$$
$$= 0.03$$
$$\Rightarrow \sigma = \sqrt{0.03} = 0.173 \text{ (correct to 3 s.f.)}$$

Hence her mean weekly milk bill is £2.10, with a standard deviation of about 17 p.

(iii) (a) $P(X > \mu + \sigma) = P(X > 2.27) = 0.25$

(b) $P(X < \mu - \sigma) = P(X < 1.93) = 0$

Exercise 2.2

① Find by calculation the expectation of the outcome with the following probability distribution.

Table 2.17

Outcome	1	2	3	4	5
Probability	0.1	0.2	0.4	0.2	0.1

How otherwise might you have arrived at your answer?

② The spreadsheet shows part of a discrete probability distribution, together with some calculations of $r \times P(X = r)$ and $r^2 \times P(X = r)$. Using only spreadsheet commands, and without entering any more numbers, obtain the remaining values in columns B, C and D. Hence find the mean and variance of this distribution.

	A	B	C	D
1	r	P(X=r)	r×P(X=r)	r²×P(X=r)
2	1	0.20		
3	2	0.30		
4	3	0.10	0.30	
5	4	0.05		
6	5	0.20		5.00
7	6			
8	SUM			

Figure 2.3

③ The probability distribution of the discrete random variable X is given by

$$P(X = r) = \frac{2r - 1}{16} \text{ for } r = 1, 2, 3, 4$$
$$P(X = r) = 0 \text{ otherwise.}$$

(i) Find $E(X)$.

(ii) Denote $E(X)$ by μ. Find $P(X < \mu)$.

④ (i) A discrete random variable X can take only the values 4 and 5, and has expectation 4.2.

By letting $P(X = 4) = p$ and $P(X = 5) = 1 - p$, solve an equation in p and so find the probability distribution of X.

(ii) A discrete random variable Y can take only the values 50 and 100. Given that $E(Y) = 80$, write out the probability distribution of Y.

⑤ The random variable Y is given by the absolute difference between the scores when two ordinary dice are thrown.

(i) Find $E(Y)$ and $Var(Y)$.

(ii) Find the values of the following.

   (a) $P(Y > \mu)$ (b) $P(Y > \mu + 2\sigma)$

⑥ Three fair coins are tossed. Let X represent the number of tails.

(i) Find $E(X)$.

(ii) Show that this is equivalent to $3 \times \frac{1}{2}$.

(iii) Find $Var(X)$.

(iv) Show that this is equivalent to $3 \times \frac{1}{4}$.

If instead ten fair coins are tossed, let Y represent the number of tails.

(v) Write down the values of $E(Y)$ and $Var(Y)$.

⑦ An unbiased tetrahedral dice has faces labelled 2, 4, 6 and 8. If the dice lands on the face marked 2, the player has to pay £5. If it lands on the face marked with a 4 or a 6, the player wins £2. If it lands on the face labelled 8, then no money changes hands.

Let X represent the amount 'won' each time the player throws the dice. (A loss is represented by a negative X value.)

(i) Copy and complete the following probability distribution for X.

Table 2.18

r	−5	0	2
P(X = r)			

(ii) Find $E(X)$ and $Var(X)$.

What does the sign of $E(X)$ indicate?

(iii) How much less would the player have to pay, when the dice lands on the face marked 2, so that he would break even in the long run?

⑧ 85% of first-class mail arrives the day after being posted, the rest takes one day longer.

For second class mail the corresponding figures are 10% and 40%, while a further 35% take three days and the remainder four days.

Two out of every five letters go by first class mail.

Let X represent the delivery time for letters.

(i) Explain why $P(X = 1) = 0.4$.

(ii) Copy and complete the following probability distribution for X.

Table 2.19

r	1	2	3	4
$P(X = r)$	0.4			

(iii) Calculate $E(X)$ and $Var(X)$.

⑨ (i) A discrete random variable X can take only the values 3, 4 and 5, has expectation 4 and variance 0.6.

By letting $P(X = 3) = p$, $P(X = 4) = q$ and $P(X = 5) = 1 - p - q$, solve a pair of simultaneous equations in p and q and so find the probability distribution of X.

(ii) A discrete random variable Y can take only the values 20, 50 and 100. Given that $E(Y) = 34$ and $Var(Y) = 624$, write out the probability distribution of Y.

⑩ A random number generator in a computer game produces values which can be modelled by the discrete random variable X with probability distribution given by

$P(X = r) = kr!$ for $r = 0, 1, 2, 3, 4$

$P(X = r) = 0$ otherwise

where k is a constant and $r! = r \times (r - 1) \times \cdots \times 2 \times 1$ with $0! = 1$.

(i) Show that $k = \frac{1}{34}$, and illustrate the probability distribution with a sketch.

(ii) Find the expectation and variance of X.

Two independent values of X are generated. Let these values be X_1 and X_2.

(iii) Show that $P(X_1 = X_2)$ is a little greater than 0.5.

(iv) Given that $X_1 = X_2$, find the probability that X_1 and X_2 are each equal to 4.

⑪ A traffic surveyor is investigating the lengths of queues at a particular set of traffic lights during the daytime, but outside rush hours. He counts the number of cars, X, stopped and waiting when the lights turn green on 100 occasions, with the following results.

Table 2.20

X	0	1	2	3	4	5	6	7	8	9+
f	3	10	13	16	18	17	12	9	2	0

(i) Use these figures to estimate the probability distribution of the number of cars waiting when the lights turn green.

(ii) Use your probability distribution to estimate the expectation and variance of X.

A colleague of the surveyor suggests that the probability distribution might be modelled by the expression

$P(X = r) = kr(8 - r)$ for $r = 0, 1, 2, 3, 4, 5, 6, 7, 8$
$P(X = r) = 0$ otherwise.

(iii) Find the value of k.

(iv) Find the expectation and variance of X given by this model.

(v) State, with reasons, whether it is a good model.

3 Linear functions of random variables

Sometimes you need to find the expectation and variance of a linear function of a random variable $E(aX + b)$ and $Var(aX + b)$.

Example 2.4

Figure 2.4

Kiara is an employee at an estate agency. The number of properties that she sells during a month is denoted by the random variable X. The probability distribution of X is as follows.

Table 2.21

Number of houses	0	1	2	3	4	5+
Probability	0.07	0.33	0.4	0.12	0.08	0

(i) Calculate $E(X)$.

Kiara earns £750 per month plus £400 for each house that she sells. The random variable Y is the amount (in £) that Kiara earns each month.

(ii) Show that $Y = 400X + 750$. Draw up a probability table for Y.

(iii) Calculate $E(Y)$.

(iv) Calculate $400E(X) + 750$ and comment on your result.

Solution

(i) $E(X) = 0 \times 0.07 + 1 \times 0.33 + 2 \times 0.4 + 3 \times 0.12 + 4 \times 0.08 = 1.81$

(ii) Table 2.22

Earnings, Y	750	1150	1550	1950	2350
Probability	0.07	0.33	0.4	0.12	0.08

(iii) $E(Y) = 750 \times 0.07 + 1150 \times 0.33 + 1550 \times 0.4$
$\qquad + 1950 \times 0.12 + 2350 \times 0.08$
$\qquad = 1474$

(iv) $400E(X) + 750 = 400 \times 1.81 + 750 = 1474$

Clearly $E(Y) = 400E(X) + 750$, both having the value 1474.

Linear functions of random variables

Expectation: general results

In Example 2.4 you found that $E(Y) = E(400X + 750) = 400E(X) + 750$.

The working was numerical, showing that both expressions came out to be 1474. In fact, the following rules for expectation apply to any random variable X where a, b, c and d are constants.

> **Note**
> These are very similar to the rules for the mean of a frequency distribution.

- $E(aX + b) = aE(X) + b$
- $E(cX) = cE(X)$
- $E(d) = d$

Example 2.5

The random variable X has the following probability distribution.

Table 2.23

r	1	2	3	4
P(X = r)	0.6	0.2	0.1	0.1

(i) Find Var(X) and the standard deviation of X.

(ii) Find Var$(3X)$ and the standard deviation of $3X$. Comment on the relationship between Var$(3X)$ and Var(X) and likewise for the standard deviations.

(iii) Find Var$(3X + 7)$ and compare this with your answer to part (ii).

Solution

(i) $E(X) = 1 \times 0.6 + 2 \times 0.2 + 3 \times 0.1 + 4 \times 0.1$
$= 1.7$

$E(X^2) = 1 \times 0.6 + 4 \times 0.2 + 9 \times 0.1 + 16 \times 0.1$
$= 3.9$

$\text{Var}(X) = E(X^2) - [E(X)]^2$
$= 3.9 - 1.7^2$
$= 1.01$

The standard deviation of $X = \sqrt{1.01} = 1.005$

(ii) $\text{Var}(3X) = E[(3X)^2] - \mu^2$
$= E(9X^2) - [E(3X)]^2$
$= 9E(X^2) - [3E(X)]^2$
$= 9 \times 3.9 - (3 \times 1.7)^2$
$= 35.1 - 26.01$
$= 9.09$

> You can use the formula for $E(aX + b)$ above to rewrite $E(9X^2)$ as $9E(X^2)$ and likewise for $E(3X)$.

This shows that $\text{Var}(3X) = 9 \times \text{Var}(X)$ ← Note that $9 = 3^2$.

The standard deviation of $3X = \sqrt{9.09} = 3.015 = 3 \times$ s.d. of X

> **Note**
> Notice that $E(9X^2 + 42X + 49)$ is written as $E(9X^2) + E(42X) + E(49)$. This illustrates the general rule that
> $E(X_1 + X_2 + \cdots)$
> $= E(X_1) + E(X_2) + \cdots$

(iii) $\text{Var}(3X+7) = E\left[(3X+7)^2\right] - \left[E(3X+7)\right]^2$
$= E(9X^2 + 42X + 49) - \left[3E(X) + 7\right]^2$
$= E(9X^2) + E(42X) + E(49) - \left[3 \times 1.7 + 7\right]^2$
$= 9E(X^2) + 42E(X) + 49 - 12.1^2$
$= 9 \times 3.9 + 42 \times 1.7 + 49 - 146.41$
$= 9.09$

This has the same value as $\text{Var}(3X)$ so is also equal to $9 \times \text{Var}(X)$

Variance: general results

In Example 2.5 you found that $\text{Var}(3X+7) = \text{Var}(3X) = 9 \times \text{Var}(X)$.

As with expectation, the working was numerical, showing that both expressions came out to be 9.09. In fact, the following rules for variance apply to any random variable X where a, b, and c are constants.

- $\text{Var}(aX) = a^2 \text{Var}(X)$
- $\text{Var}(aX + b) = a^2 \text{Var}(X)$
- $\text{Var}(c) = 0$ ← Notice that the variance of a constant is zero. It can only take one value so there is no spread.

It may seem surprising that $\text{Var}(aX) = a^2 \text{Var}(X)$ rather than simply $a \text{Var}(X)$, but the former relationship then gives the result that the standard deviation of $a(X)$ is equal to $a \times$ the standard deviation of X, as you would expect from common sense.

Notice also that adding a constant does not make any difference to the variance, which is again as you would expect.

Finally, the variance of a constant is zero. This result is obvious – a constant does not have any variation.

Sums and differences of random variables

Sometimes you may need to add or subtract a number of independent random variables. This process is illustrated in the next example.

Linear functions of random variables

Example 2.6

A cricket bat manufacturer makes the blades and the handles separately. The blades are made in five lengths (in cm), with a uniform discrete distribution:

38, 40, 42, 44, 46.

The lengths (in cm) of the handles of the bats also form a discrete uniform distribution:

22, 24, 26.

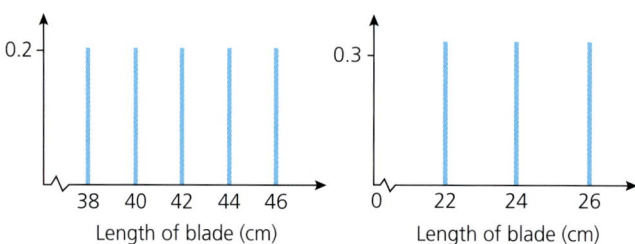

Figure 2.5

The blades and handles can be joined together to make bats of various lengths, and it may be assumed that the lengths of the two sections are independent. The different combinations occur with equal probabilities.

(i) How many different (total) bat lengths are possible?

(ii) Work out the mean and variance of random variable X_1, the length (in cm) of the blades.

(iii) Work out the mean and variance of random variable X_2, the length (in cm) of the handles.

(iv) Work out the mean and variance of random variable $X_1 + X_2$, the total length of the bats.

(v) Verify that
$$E(X_1 + X_2) = E(X_1) + E(X_2)$$
and $\mathrm{Var}(X_1 + X_2) = \mathrm{Var}(X_1) + \mathrm{Var}(X_2)$.

Solution

(i) The number of different bat lengths is 7. This can be seen from the sample space diagram below.

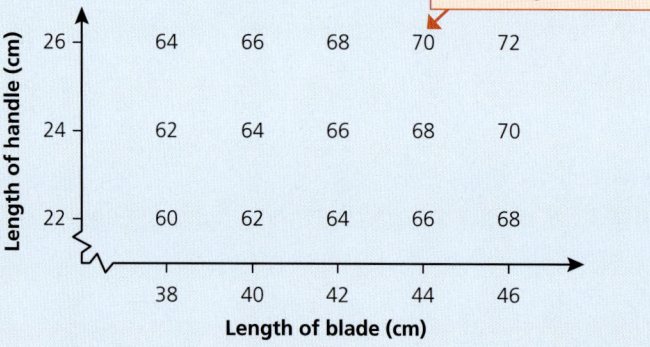

Figure 2.6

(ii) Table 2.24

Length of blade (cm)	38	40	42	44	46
Probability	0.2	0.2	0.2	0.2	0.2

$$E(X_1) = \mu_1 = \Sigma xp = (38 \times 0.2) + (40 \times 0.2) + (42 \times 0.2)$$
$$+ (44 \times 0.2) + (46 \times 0.2)$$
$$= 42 \text{ cm}$$
$$\text{Var}(X_1) = E(X_2^2) - \mu_2^2$$
$$E(X_1^2) = (38^2 \times 0.2) + (40^2 \times 0.2) + (42^2 \times 0.2)$$
$$+ (44^2 \times 0.2) + (46^2 \times 0.2)$$
$$= 1772$$
$$\text{Var}(X_1) = 1772 - 42^2 = 8$$

(iii) Table 2.25

Length of handle (cm)	2	24	26
Probability	$\frac{1}{3}$	$\frac{1}{3}$	$\frac{1}{3}$

$$E(X_2) = \mu_2 = 22 \times \tfrac{1}{3} + 24 \times \tfrac{1}{3} + 26 \times \tfrac{1}{3} = 24 \text{ cm}$$
$$\text{Var}(X_2) = E(X_2^2) - \mu_2^2$$
$$E(X_2^2) = \left(22^2 \times \tfrac{1}{3}\right) + \left(24^2 \times \tfrac{1}{3}\right) + \left(26^2 \times \tfrac{1}{3}\right) = 578.667 \text{ to 3 d.p.}$$
$$\text{Var}(X_2) = 578.667 - 24^2 = 2.667 \text{ to 3 d.p.}$$

(iv) The probability distribution of $X_1 + X_2$ can be obtained from Figure 2.5.

Table 2.26

Total length of cricket bat (cm)	60	62	64	66	68	70	72
Probability	$\frac{1}{15}$	$\frac{2}{15}$	$\frac{3}{15}$	$\frac{3}{15}$	$\frac{3}{15}$	$\frac{2}{15}$	$\frac{1}{15}$

$$E(X_1 + X_2) = \left(60 \times \tfrac{1}{15}\right) + \left(62 \times \tfrac{2}{15}\right) + \left(64 \times \tfrac{3}{15}\right) + \left(66 \times \tfrac{3}{15}\right)$$
$$+ \left(68 \times \tfrac{3}{15}\right) + \left(70 \times \tfrac{2}{15}\right) + \left(72 \times \tfrac{1}{15}\right)$$
$$= 66 \text{ cm}$$
$$\text{Var}(X_1 + X_2) = E\left[(X_1 + X_2)^2\right] - 66^2$$

$$E\left[(X_1+X_2)^2\right] = \left(60^2 \times \tfrac{1}{15}\right) + \left(62^2 \times \tfrac{2}{15}\right) + \left(64^2 \times \tfrac{3}{15}\right) + \left(66^2 \times \tfrac{3}{15}\right)$$
$$+ \left(68^2 \times \tfrac{3}{15}\right) + \left(70^2 \times \tfrac{2}{15}\right) + \left(72^2 \times \tfrac{1}{15}\right)$$
$$= \tfrac{65500}{15} = 4366.667 \text{ to 3 d.p.}$$
$$\mathrm{Var}(X_1 + X_2) = 4366.667 - 66^2 = 10.667 \text{ to 3 d.p.}$$

(v) $E(X_1 + X_2) = 66 = 42 + 24 = E(X_1) + E(X_2)$, as required.

$\mathrm{Var}(X_1 + X_2) = 10.667 = 8 + 2.667 = \mathrm{Var}(X_1) + \mathrm{Var}(X_2)$, as required.

> **Note**
> Notice that the standard deviations of X_1 and X_2 do not add up to the standard deviation of $X_1 + X_2$.
> i.e. $\sqrt{8} + \sqrt{2.667} \neq \sqrt{10.667}$
> or $2.828 + 1.633 \neq 3.266$

General results

Example 2.6 has illustrated the following general results for the sums and differences of random variables.

For any two random variables X_1 and X_2

- $E(X_1 + X_2) = E(X_1) + E(X_2)$
- $E(X_1 - X_2) = E(X_1) - E(X_2)$

If the variables X_1 and X_2 are independent

- $\mathrm{Var}(X_1 + X_2) = \mathrm{Var}(X_1) + \mathrm{Var}(X_2)$
- $\mathrm{Var}(X_1 - X_2) = \mathrm{Var}(X_1) + \mathrm{Var}(X_2)$

Note that if X_1 and X_2 are not independent, then the relationship between $\mathrm{Var}(X_1 + X_2) = \mathrm{Var}(X_1) + \mathrm{Var}(X_2)$ does not hold. In this case, the relationship between $\mathrm{Var}(X_1 + X_2)$ and the individual variances of X_1 and X_2 is beyond the scope of this book.

> Replacing X_2 by $-X_2$ in this result gives
> $E(X_1 + (-X_2))$
> $= E(X_1) + E(-X_2)$

> Replacing X_2 by $-X_2$ in this result gives
> $\mathrm{Var}(X_1 + (-X_2))$
> $= \mathrm{Var}(X_1) + \mathrm{Var}(-X_2)$
> $= \mathrm{Var}(X_1) + (-1)^2 \mathrm{Var}(X_2)$
> $= \mathrm{Var}(X_1) + \mathrm{Var}(X_2)$

Linear combinations of two or more independent random variables

> A 'linear combination' of random variables is an expression of the form
> $a_1 X_1 + a_2 X_2 + \cdots + a_n X_n$.

These results can also be generalised to include linear combinations of random variables.

For any random variables X and Y

$$E(aX + bY) = aE(X) + bE(Y)$$

where a and b are constants.

If X and Y are independent

$$\mathrm{Var}(aX + bY) = a^2 \mathrm{Var}(X) + b^2 \mathrm{Var}(Y)$$

These results may be extended to any number of random variables.

Exercise 2.3

① The probability distribution of random variable X is as follows.

Table 2.27

x	1	2	3	4	5
$P(X = x)$	0.1	0.2	0.3	0.3	0.1

(i) Find (a) $E(X)$ (b) $Var(X)$.

(ii) Verify that $Var(2X) = 4 Var(X)$.

② The probability distribution of a random variable X is as follows.

Table 2.28

x	0	1	2
$P(X = x)$	0.5	0.3	0.2

(i) Find (a) $E(X)$ (b) $Var(X)$.

(ii) Verify that $Var(5X + 2) = 25 Var(X)$.

③ The expectations and variances of independent random variables A, B and C are 35 and 9, 30 and 8 and 25 and 6, respectively. Write down the expectations and variances of

(i) $A + B + C$

(ii) $5A + 4B$

(iii) $A + 2B + 3C$

(iv) $4A - B - 5C$.

④ Prove that $Var(aX - b) = a^2 Var(X)$, where a and b are constants.

⑤ A coin is biased so that the probability of obtaining a tail is 0.75. The coin is tossed four times and the random variable X is the number of tails obtained.

Find

(i) $E(2X)$

(ii) $Var(3X)$.

⑥ A discrete random variable W has the following distribution.

Table 2.29

w	1	2	3	4	5	6
$P(W = w)$	0.1	0.2	0.1	0.2	0.1	0.3

Find the mean and variance of

(i) $W + 7$

(ii) $6W - 5$.

⑦ The random variable X is the number of heads obtained when four unbiased coins are tossed. Construct the probability distribution for X and find

(i) $E(X)$

(ii) $Var(X)$

(iii) $Var(3X + 4)$.

S1 AS — Linear functions of random variables

⑧ A mining company is prospecting for new deposits of a particular metal ore. There are five sites for which it has acquired a licence to prospect. At each site there is a probability of 0.3 of finding an amount of ore which makes it worth opening a new mine. It explores the sites, one at a time, until it finds one which has a suitable amount of ore, after which it stops exploring. The random variable X represents the number of sites explored. The probability distribution for X is as follows.

Table 2.30

r	1	2	3	4	5
$P(X = r)$	0.3	0.21	0.147	0.1029	0.2401

(i) Explain why $P(X = 2) = 0.21$.

(ii) Find $E(X)$ and $Var(X)$.

It costs £600 000 to begin prospecting, however many sites are investigated. It then costs £275 000 per site to carry out the prospecting.

(iii) Find the expectation and variance of the total cost of the exploration.

⑨ A game at a charity fair costs £1 to play. The player rolls two fair dice and receives in return 10p plus 25 times the lower of the scores on the dice.

(i) Find the probability distribution of X, the lower of the scores.

(ii) Find the expectation and the variance of the players winnings in one go at the game.

(iii) In the long run, will the game make or lose money for the charity?

⑩ The probability distributions of two independent random variables X and Y are as follows.

Table 2.31

x	1	2	3	4
$P(X = x)$	0.2	0.4	0.3	0.1

Table 2.32

y	0	1	2
$P(Y = y)$	0.2	0.3	0.5

(i) Find   (a) $E(X)$ (b) $E(Y)$ (c) $Var(X)$ (d) $Var(Y)$.

(ii) Show the probability distribution of $Z = X + Y$ in a table.

(iii) Using your table, verify that   (a) $E(Z) = E(X) + E(Y)$
    (b) $Var(Z) = Var(X) + Var(Y)$.

(iv) Show the probability distribution of $W = X - Y$ in a table.

(v) Using your table, verify that   (a) $E(W) = E(X) - E(Y)$
    (b) $Var(W) = Var(X) + Var(Y)$.

⑪ An unbiased six-sided dice is thrown.

(i) Find the expectation and variance of the score on the dice.

(ii) The dice is thrown twice. Find the expectation and variance of the total score.

(iii) The dice is thrown twice. Find the expectation and variance of the differences in the two scores.

(iv) The dice is thrown ten times. Find the expectation and variance of the total score.

⑫ In a game at a charity fair, three coins are spun and a dice is thrown. The game costs one pound to play. The amount in pence that the player wins is ten times the score on the dice plus 20 times the number of heads that occur. Find the expectation and variance of the amount won by the player.

⑬ The random variable X represents the number of tails which occur when two fair coins are spun.
  (i) Find $E(X)$ and $Var(X)$.
  (ii) Find $E(10X - 5)$ and $Var(10X - 5)$.
  (iii) Find the expectation and variance of the total of 50 observations of X.
  (iv) Explain the similarities and differences between your answers to part (iii) and $E(50X)$ and $Var(50X)$.

⑭ At a garden centre, railway sleepers for edging a border come in three similar lengths 120 cm, 125 cm and 130 cm. If a sleeper is selected at random, the probabilities that its length is 120 cm, 125 cm and 130 cm are 0.5, 0.3 and 0.2, respectively. You can assume that there are sufficient sleepers that these probabilities do not change when a few are selected.
  (i) Find the expectation and variance of the length of a randomly selected sleeper.

In a hurry to edge a border in her garden, Jane selects four sleepers at random.

  (ii) Her border is 5.1 m long. Find the probability that she has enough length of sleepers to edge the border.
  (iii) Find the expectation and variance of the total length of the four sleepers.

⑮ A fair five-sided spinner has faces labelled 1, 2, 3, 4, 5. Martin plays a game in which he gains or loses money according to the number that the spinner comes to rest on when it is spun. He wins £5 if it lands on a 1, loses £2 if it lands on a 2, 3 or 4 and neither gains nor loses if it lands on a 5. Let X represent the amount 'won' each time the spinner is spun.
  (i) Copy and complete the following probability distribution for X.

  Table 2.33

r	−2	0	
P(X = r)			

  (ii) Find $E(X)$ and $Var(X)$.
  (iii) How much less would the player have to pay, when the spinner lands on the face marked 2, so that he would break even in the long run?

⑯ A family eats a lot of bread. Most days, at least one loaf of bread is bought by the family. The random variable X represents the number of loaves bought by the family on a day. The probability distribution of X is shown in the table.

Table 2.34

r	0	1	2
P(X = r)	0.3	0.35	0.35

(i) Find $E(X)$ and Var(X).

The number of loaves bought on any day depends on the number of loaves bought on the previous day. The table below shows the joint distribution of the numbers of loaves X_1 and X_2 bought on two successive days.

Table 2.35

		Second day X_2			
		0	1	2	Total
First day X_1	0	0	0.05	0.25	0.3
	1	0.05	0.2	0.1	0.35
	2	0.25	0.1	0	0.35
	Total	0.3	0.35	0.35	1

(ii) Copy and complete the following probability distribution for $Y = X_1 + X_2$, the total number bought on the two days.

Table 2.36

R	0	1	2	3	4
$P(Y = r)$	0	0.1	0.7		

(iii) Find $E(Y)$, and show that Var$(Y) = 0.29$.

(iv) Explain why Var(Y) is not equal to Var(X_1) + Var(X_2).

KEY POINTS

1 For a **discrete random variable**, X, which can take only the values $r_1, r_2, \ldots, r_n$, with probabilities $p_1, p_2, \ldots, p_n$, respectively:

- $p_1 + p_2 + \cdots + p_n = \sum_{n=1}^{n=k} p_k = \sum_{1}^{n} P(X = r_k) = 1 \quad p_k \geq 0$

2 A discrete probability distribution is best illustrated by a **vertical line chart**.

- The **expectation** $= E(X) = \mu = \Sigma r P(x = r)$.
- The **variance** $= \text{Var}(X) = \sigma^2$

$$= E(X - \mu)^2 = \Sigma(r - \mu)^2 P(X = r)$$

or $E(X^2) - [E(X)]^2 = \Sigma r^2 P(X = r) - [\Sigma r P(X = r)]^2$.

3 For a discrete random variable, X, which can assume only the values $x_1, x_2, \ldots, x_n$ with probabilities $p_1, p_2, \ldots, p_n$, respectively:

- $\Sigma p_i = 1 \quad p_i \geq 0$
- $E(X) = \Sigma x_i p_i = \Sigma x_i \times P(X = x_i)$
- Var$(X) = E(X^2) - [E(X)]^2$.

4 For any random variables X and Y and constants a, b and c:
- $\mathrm{E}(c) = c$
- $\mathrm{E}(aX) = a\mathrm{E}(X)$
- $\mathrm{E}(aX + c) = a\mathrm{E}(X) + c$
- $\mathrm{E}(X \pm Y) = \mathrm{E}(X) \pm \mathrm{E}(Y)$
- $\mathrm{E}(aX + bY) = a\mathrm{E}(X) + b\mathrm{E}(Y)$.

5 For two random variables X and Y, and constants a, b and c:
- $\mathrm{Var}(c) = 0$
- $\mathrm{Var}(aX) = a^2 \mathrm{Var}(X)$
- $\mathrm{Var}(aX + c) = a^2 \mathrm{Var}(X)$.

and, if X and Y are independent,
- $\mathrm{Var}(X \pm Y) = \mathrm{Var}(X) + \mathrm{Var}(Y)$
- $\mathrm{Var}(aX + bY) = a^2 \mathrm{Var}(X) + b^2 \mathrm{Var}(Y)$.

LEARNING OUTCOMES

When you have completed this chapter you should be able to:
- use probability functions, given algebraically or in tables
- calculate the numerical probabilities for a simple distribution
- draw and interpret graphs representing probability distributions
- calculate the expectation (mean), $\mathrm{E}(X)$, and understand its meaning
- calculate the variance, $\mathrm{Var}(X)$, and understand its meaning
- use the result $\mathrm{E}(aX + b) = a\mathrm{E}(X) + b$ and understand its meaning
- use the result $\mathrm{Var}(aX + b) = a^2 \mathrm{Var}(X)$ and understand its meaning
- find the mean of any linear combination of random variables and the variance of any linear combination of independent random variables.

3 Discrete probability distributions

Doublethink means the power of holding two contradictory beliefs in one's mind simultaneously, and accepting both of them.
— George Orwell

A learner driver has done no revision for the driving theory test. He decides to guess each answer. There are 50 questions in the test and four answers to each question. Exactly one of these answers is correct. He answers all of them.

> **Discussion point**
> How many questions on average would you expect the learner driver to get correct?

1 The binomial distribution

- The probability of getting a question correct is clearly 0.25 as there are four responses of which one is correct.
- The probability of getting a question wrong is therefore 0.75.
- If the student gets r questions correct out of 50 then he must get the remaining $50 - r$ wrong.
- You might think that the probability of getting r questions correct out of 50 is $0.25^r \times 0.75^{50-r}$.
- However, there are many possible combinations, in fact $^{50}C_r$ of them, so you need to multiply the above term by $^{50}C_r$.

> For example the student could get the first r questions correct and the remaining ones wrong, or the first $r-1$ correct as well as the last one and the rest wrong. In all the number of possibilities is $^{50}C_r$.

So the probability of getting r questions correct out of 50 is equal to $^{50}C_r \times 0.25^r \times 0.75^{50-r}$.

This situation is an example of the binomial distribution. For a binomial distribution to be appropriate, the following conditions must apply.

- You are conducting trials on random samples of a certain size, denoted by n.
- On each trial, the outcomes can be classified as either **success** or **failure**.

> In this example, the possible outcomes are **correct** (success) and **wrong** (failure).

In addition, the following assumptions are needed if the binomial distribution is to be a good model and give reliable answers.

- The outcome of each trial is independent of the outcome of any other trial.
- The probability of success is the same on each trial.

> This is particularly true when, in contrast to answering a question involving coins, dice, cards etc., you are modelling a situation drawn from real life.

The probability that the number of successes, X, has the value r, is given by

$$P(X = r) = {^nC_r} p^r q^{n-r} \text{ for } r = 0, 1, \ldots, n$$

$$P(X = r) = 0 \text{ otherwise.}$$

You can either use this formula to find binomial probabilities or you can find them directly from your calculator without using the formula.

The notation $B(n,p)$ is often used to mean that the distribution is binomial, with n trials each with probability of success p.

> The probability of success is usually denoted by p and that of failure by q, so $p + q = 1$.

Example 3.1

On average, 5% of patients do not turn up for their appointment at a dental clinic. There are 30 appointments per day.

(i) Find the probability that on a random chosen day
 (a) everybody turns up,
 (b) exactly two people do not turn up,
 (c) at least three people do not turn up.

(ii) What modelling assumption do you have to make to answer the questions in part (i)? Do you think that this assumption is reasonable?

> **Note**
> The distribution which you need to use is B(30, 0.05)

> **USING ICT**
> Notice that you can find the answers to all three parts of this example directly from your calculator without using the formula.

Solution

(i) The probabilities, to 4 s.f., are

(a) $0.95^{30} = 0.2146$

(b) $^{30}C_2 \times 0.05^2 \times 0.95^{28} = 0.2586$

(c) $1 - {}^{30}C_2 \times 0.05^2 \times 0.95^{28} - {}^{30}C_1 \times 0.05^1 \times 0.95^{29} - 0.95^{30} = 0.1878$

(ii) You have to assume that the people who miss appointments do so independently of each other. It may not be true, as perhaps two members of the same family may both have an appointment and so would probably either both attend or neither. Also fewer people may attend when the weather is bad.

Expectation and variance of the binomial distribution

> **Note**
> The results are:
> $E(X_1 + X_2) = E(X_1) + E(X_2)$
> $Var(X_1 + X_2)$
> $= Var(X_1) + Var(X_2)$
> Note that these results can be extended to any number of random variables.

In order to find the expectation and variance of X, you can sum a series, involving binomial coefficients but this is fairly awkward. However, you can instead think of X as the sum of n independent variables $X_1, X_2, \ldots, X_n$. Each of these variables is a binomial random variable which takes the value 1 with probability p and the value 0 with probability $1 - p$. To find the expectation and variance of X, you first find the expectation and variance of one of these binomial random variables. You can then use the results from the previous chapter (see note on left) to find the expectation and variance of X.

$$E(X_i) = 0 \times (1-p) + 1 \times p = p$$

$$E(X_i^2) = 0^2 \times (1-p) + 1^2 \times p = p$$

$$Var(X_i) = E(X_i^2) - [E(X_i)]^2 = p - p^2 = p(1-p) = pq \text{ where } q = 1 - p$$

You now use the results that

$$E(X_1 + X_2 + \cdots + X_n) = E(X_1) + E(X_2) + \cdots + E(X_n)$$
$$= p + p + \cdots + p = np \text{ and}$$

$$Var(X_1 + X_2 + \cdots + X_n) = Var(X_1) + Var(X_2) + \cdots + Var(X_n)$$
$$= p(1-p) + p(1-p) + \cdots + p(1-p)$$
$$= np(1-p) = npq$$

> Thus if $X \sim B(n, p)$
> $E(X) = \mu = np$ and
> $Var(X) = \sigma^2 = np(1-p) = npq$

Example 3.2

Using ICT

You can use a spreadsheet to check these formulae for a particular binomial distribution, for example B(6, 0.25). Note that using the formulae, the mean $= np = 6 \times 0.25 = 1.5$ and the variance $= npq = 6 \times 0.25 \times 0.75 = 1.125$.

In order to check these results, take the following steps:

1 Enter the values of X into cells B1 to H1.

2 Enter the formula provided by your spreadsheet to find the P(X = 0), for example =BINOM.DIST(B1,6,0.25,FALSE) into cell B2.

3 Copy this formula into cells C2 to H2 to find P(X = 1) to P(X = 6).

> The number in cell G2 should come out to be 0.0044. Check that you have got this right before continuing.

	A	B	C	D	E	F	G	H	I
1	r	0	1	2	3	4	5	6	SUM
2	P(X=r)	0.1780	0.3560						

Figure 3.1

4 Enter the formula = B1*B2 into cell B3 to calculate $0 \times P(X = 0)$.

5 Copy this formula into cells C3 to H3 to calculate $r \times P(X = r)$ for the remaining cells.

	A	B	C	D	E	F	G	H	I
1	r	0	1	2	3	4	5	6	SUM
2	P(X=r)	0.1780	0.3560						
3	r×P(X=r)	0.0000	0.3560						

Figure 3.2

6 Enter the formula = B1^2*B2 into cell B4 to calculate $0^2 \times P(X = 0)$.

7 Copy this formula into cells C3 to H3 to calculate $r^2 \times P(X = r)$ for the remaining cells.

	A	B	C	D	E	F	G	H	I
1	r	0	1	2	3	4	5	6	SUM
2	P(X=r)	0.1780	0.3560						
3	r×P(X=r)	0.0000	0.3560						
4	r^2×P(X=r)	0.0000	0.3560						

Figure 3.3

8 Use the SUM function to sum the values in each of rows 2, 3 and 4. Note that the sum of row 2 should be 1 (as you would expect).

9 Enter the formula = I3 into cell B6 to find the mean (which is simply the sum of the cells B3 to H3). If you have done it correctly the answer should be 1.5.

10 Enter the formula = I4 − I3^2 into cell B7 to calculate the variance. This answer should be 1.125.

ACTIVITY 3.1

Use a spreadsheet to confirm that, for B(8, 0.625), the procedures used in the example above give the same answers as the standard binomial formulae for the mean (*np*) and variance (*npq*).

S1 AS The binomial distribution

Exercise 3.1

1. A fair coin is spun ten times.
   - (i) State the distribution of the number of heads that occur.
   - (ii) Find the probability of exactly four heads occurring.
   - (iii) Find the probability of at least five heads occurring.

2. The random variable $X \sim B(15, 0.4)$
   - (i) Find $E(X)$.
   - (ii) Find $P(X = 3)$.
   - (iii) Find $P(X > 3)$.

3. Ten fair dice are rolled. The number of times that a six occurs is denoted by X.
   - (i) State the distribution of X.
   - (ii) What is the expected number of sixes?
   - (iii) Find $P(X = 1)$.
   - (iv) Find $P(X > 2)$.

4. In an airport departure lounge a stall offers the chance to win a sports car. The game is to draw a card at random from each of four normal packs. If all four cards are aces, then the car is won.
   - (i) What is the probability of winning the car?
   - (ii) How many goes would be expected before the car was won?
   - (iii) The stall charges £5 per go and the car costs £35 000. What is the expected profit per go?

5. The random variable $X \sim B(25, 0.1)$.
   - (i) State
     - (a) the number of trials
     - (b) the probability of success in any trial
     - (c) the probability of failure in any trial.
   - (ii) Find the probability of at most 4 successes.
   - (iii) Find the value of the integer a such that $P(X < a) < 0.99$ and $P(X \leq a) > 0.99$.

6. A pottery company manufactures bowls in batches of 100. The probability of a bowl being faulty is known to be 0.03.
   - (i) What is the probability that in a batch of bowls there are less than five faulty bowls?
   - (ii) What is the probability that in two batches of bowls there are at most eight faulty bowls?
   - (iii) What is the probability that in each of two batches of bowls there are at most four faulty bowls?
   - (iv) Explain why your answer to part (ii) is different from your answer to part (iii).

7. Mark travels to work by bus each morning. The probability of his bus being late is 0.25.
   - (i) What is the probability that his bus is late twice in a five-day working week?
   - (ii) What is the probability that it is late at most once in a five-day working week?

48

(iii) Fred buys a quarterly ticket which allows him to travel for 13 weeks. How many times would he expect the bus to be late in this thirteen-week period?

⑧ A manufacturer supplies DVD players to retailers in batches of 20. It has 5% of the players returned because they are faulty.

(i) Write down a suitable model for the distribution of the number of faulty DVD players in a batch.

Find the probability that a batch contains

(ii) no faulty DVD players,

(iii) more than 4 faulty DVD players.

(iv) Find the mean and variance of the number of faulty DVD players in a batch. [Edexcel 6684 Q1 Jan 2010]

⑨ In an election, 20% of people support the Progressive Party. A random sample of eight voters is taken.

(i) Find the probability that at least two of them support the Progressive Party.

(ii) Find the mean and variance of the number of Progressive Party supporters in the sample.

⑩ The random variable $X \sim B(4, 0.3)$. Prove that the mean and variance of X are 1.2 and 0.84, respectively,

(i) using the method above just before the start of this exercise

(ii) using the definitions of expectation and variance.

⑪ The random variable $X \sim B(n, p)$. The values of the mean and variance are 24 and 14.4, respectively.

(i) Find $P(X = 30)$.

(ii) Find $P(X > 30)$.

2 The Poisson distribution

Electrics Express – next day delivery

Since the new website went live, with next day delivery for all items, the number of orders has increased dramatically. We have taken on more staff to cope with the demand for our products. Orders come in from all over the place. It seems impossible to predict the pattern of demand, but one thing we do know is that currently we receive an average of 150 orders per hour.

S1 AS The Poisson distribution

The appearance of this update on the Electrics Express website prompted a statistician to contact Electrics Express. She offered to analyse the data and see what suggestions she could come up with.

For her detailed investigation, she considered the distribution of the number of orders per minute. For a random sample of 1000 single-minute intervals during the last month, she collected the following data.

> There were five occasions on which there were more than seven orders. These were grouped into a single category and treated as if all five of them were eight.

Table 3.1

Number of orders per minute	0	1	2	3	4	5	6	7	> 7
Frequency	70	215	265	205	125	75	30	10	5

Summary statistics for this frequency distribution are as follows.

$$n = 1000, \quad \Sigma xf = 2525 \quad \text{and} \quad \Sigma x^2 f = 8885$$

$$\Rightarrow \bar{x} = 2.525 \quad \text{and} \quad s = 1.58 \text{ (to 3 s.f.)}$$

She also noted that:

- orders made on the website appear at random and independently of each other
- the average number of orders per minute is about 2.5 which is equivalent to 150 per hour.

She suggested that the appropriate probability distribution to model the number of orders was the Poisson distribution.

The particular Poisson distribution, with an average number of 2.5 orders per minute, is defined as an **infinite** discrete random variable given by

$$P(X = r) = e^{-2.5} \times \frac{2.5^r}{r!} \quad \text{for } r = 0, 1, 2, 3, 4, \ldots$$

where

- X represents the random variable 'number of orders per minute'
- e is the mathematical constant 2.718 …
- $e^{-2.5}$ can be found from your calculator as 0.082 (to 2 s.f.)
- $r!$ means r **factorial**, for example $5! = 5 \times 4 \times 3 \times 2 \times 1 = 120$.

> **USING ICT**
>
> You can find Poisson probabilities directly from your calculator, without using this formula.

You can use the formula to calculate the values of the corresponding probability distribution, together with the **expected** frequencies it would generate. For example,

$$P(X = 4) = e^{-2.5} \times \frac{2.5^4}{4!}$$
$$= 0.13360\ldots$$
$$= 0.134 \text{ (to 3 s.f.)}$$

The table shows the observed frequencies for the orders on the website, together with the expected frequencies for a Poisson distribution with a mean of 2.5.

Table 3.2

Number of orders per minute (r)	0	1	2	3	4	5	6	7	> 7
Observed frequency	70	215	265	205	125	75	30	10	5
$P(X = r)$	0.082	0.205	0.257	0.214	0.134	0.067	0.028	0.010	0.004
Expected frequency	82	205	257	214	134	67	28	10	4

> **Note**
> Note that the final probability is found by subtracting the other probabilities from 1.
> Note also that the total of the expected frequencies is 1001 due to rounding.

The closeness of the observed and expected frequencies (see Figure 3.4) implies that the Poisson distribution is indeed a suitable model in this instance.

Note also that the sample mean, $\bar{x} = 2.525$ is very close to the sample variance, $s^2 = 2.509$ (to 4 s.f.). You will see later that, for a Poisson distribution, the expectation and variance are the same. So the closeness of these two summary statistics provides further evidence that the Poisson distribution is a suitable model.

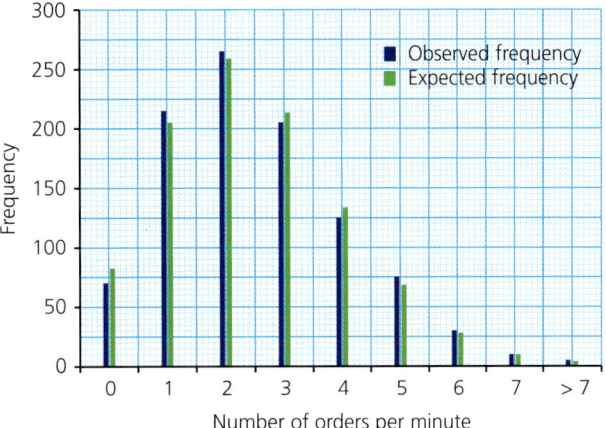

Figure 3.4

> **Note**
> As with the discrete random variables you met in Chapter 2, the Poisson distribution may be illustrated by a vertical line chart.

This is an example of the Poisson distribution. When you learnt about the binomial distribution, you needed to consider whether certain conditions and certain modelling assumptions were needed. The same is true for the Poisson distribution.

The following **conditions** are needed for the Poisson distribution to apply.

- The variable is the frequency of events that occur in a fixed interval of time or space.
- The events occur randomly.

There is rarely any doubt as to whether these conditions are satisfied. However, for the Poisson distribution to be a good model the following are also needed.

- Events occur independently of one another.
- The mean number of events occurring in each interval of the same size is the same.

There will often be doubt as to whether either or both of these are satisfied, and often the best you can say is that you must assume them to be true. So they will usually be modelling assumptions.

> So the variable takes values 0, 1, 2, 3, …

> That is, events do not occur at regular or predictable intervals.

> Whether or not one event occurs does not affect the probability of whether another event occurs.

> So the probability of an event occurring in an interval of a given size is the same. This condition can also be written as 'events occur at a constant average rate'.

> The mean number of events per interval is often denoted by λ (pronounced 'lambda'). λ is the parameter of the Poisson distribution.

So if X represents the number of events in a given interval, then

$$P(X = r) = e^{-\lambda} \times \frac{\lambda^r}{r!} \quad \text{for } r = 0, 1, 2, 3, 4, \ldots$$

The Poisson distribution has an infinite number of outcomes, so only part of the distribution can be illustrated. The shape of the Poisson distribution depends on the value of the parameter λ. If λ is small, the distribution has positive skew, but as λ increases the distribution becomes progressively more symmetrical. Three typical Poisson distributions are illustrated in Figure 3.5.

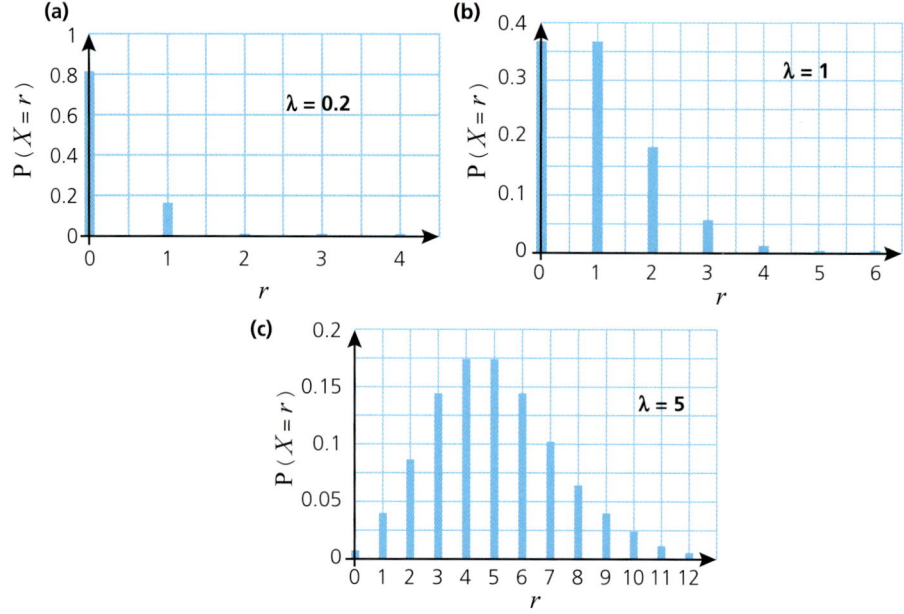

Figure 3.5 The shape of the Poisson distribution for (a) $\lambda = 0.2$ (b) $\lambda = 1$ (c) $\lambda = 5$

There are many situations in which events happen singly and the average number of occurrences per given interval of time or space is uniform and is known or can be easily found. Such events might include:

- the number of goals scored by a team in a football match
- the number of telephone calls received per minute at a call centre
- the number of accidents in a factory per week
- the number of particles emitted in a minute by a radioactive substance
- the number of typing errors per page in a document
- the number of flaws per metre in a roll of cloth
- the number of micro-organisms in 1 ml of pond water.

Modelling with the Poisson distribution

Often you will not be able to say that the conditions and assumptions for a Poisson distribution are met exactly, but you will nevertheless be able to get useful probabilities from the Poisson distribution. Sometimes it will be clear that some or all of the conditions and assumptions at not satisfied, and so you can confidently say that the Poisson distribution is not a good model.

Example 3.3

Discuss whether the Poisson distribution provides a good probability model for the variable X in each of the following scenarios.

(i) X is the number of cars that pass a point in the grandstand in a Formula 1 race in an interval of 3 minutes.

(ii) X is the number of coins found in $1\,m^3$ of earth during the investigation of an archaeological site.

(iii) X is the number of cars that pass a given point on a main road in a ten-second period between 6 am and noon on a weekday.

(iv) X is the number of separate incidents reported to a Fire Brigade control room in a 1-hour period.

Solution

> Cars pass at a roughly constant rate, but this is not the same as 'constant **average** rate'. 'Constant rate' implies no variation.

(i) Cars in a Formula 1 race will pass the grandstand at fairly regular and predictable intervals, so the 'random' condition does not hold and a Poisson distribution is almost certainly not a good model.

(ii) It is quite likely that coins will be found in groups, or even in a hoard, and in this case the independence condition would not hold. If on the other hand you are in part of the site where single coins might have been lost on an occasional basis then independence could be assumed and the Poisson distribution might be a good model.

(iii) It is likely that the mean number of cars passing during rush hours would not be the same as the mean number passing at other times, so the 'constant average rate' assumption is unlikely to hold. This may mean that the Poisson distribution is not a good model.

(iv) In general incidents such as these are likely to occur independently and at a uniform rate, at least within a relatively short time interval. However, circumstances might exist which negate this, for instance in the case of a series of deliberate attacks.

Discussion point

The managers of the new Avonford maternity hospital need to know how many beds are needed. At a meeting, one of the managers suggests that the number of births per day in the region covered by the hospital could be modelled by a Poisson distribution.

(i) What assumptions are needed for the Poisson distribution to be a good model?
(ii) Are these assumptions likely to hold?
(iii) What else would the managers need to consider when planning the number of beds?

> **Note**
> For a Poisson distribution with parameter λ,
> mean $= E(X) = \lambda$,
> variance $= \text{Var}(X) = \lambda$

The mean and variance of the Poisson distribution are both equal to the parameter λ.

You can see these results in the example about Electrics Express. The Poisson parameter was $\lambda = 2.5$, the mean of the number of orders placed per minute on the website was 2.525 and the variance was 2.512.

When modelling data with a Poisson distribution, the closeness of the mean and variance is one indication that the model fits the data well.

When you have collected the data, go through the following steps in order to check whether the data may be modelled by a Poisson distribution.

- Work out the mean and variance and check that they are roughly equal.
- Use the sample mean to work out the Poisson probability distribution and a suitable set of expected frequencies.
- Compare these expected frequencies with your observations.

The Poisson distribution

Example 3.4   The number of defects in a wire cable can be modelled by the Poisson distribution with a uniform rate of 1.5 defects per kilometre.

Find the probability that

(i) a single kilometre of wire will have exactly three defects.

(ii) a single kilometre of wire will have at least five defects.

Solution

Let X represent the number of defects per kilometre.

> You are told that defects occur with a uniform rate of 1.5 defects per kilometre. From this, you can infer that the value of the mean, λ, is 1.5.

$$P(X = r) = e^{-1.5} \times \frac{1.5^r}{r!} \quad \text{for } r = 0, 1, 2, 3, 4, \ldots$$

(i) $P(X = 3) = e^{-1.5} \times \frac{1.5^3}{3!}$

$= 0.125510\ldots$

$= 0.126$ (to 3 s.f.)

(ii) $P(X \geq 5) = 1 - P(X \leq 4)$

$= 1 - 0.98142\ldots$

$= 0.0186$

You can use the term you have obtained to work out the next one. Although calculators can work out every term, sometimes you might still find it useful to understand this process. For the Poisson distribution with parameter λ

$P(X = 0) = e^{-\lambda}$

$P(X = 1) = \lambda e^{-\lambda} = \lambda P(X = 0)$ Multiply the previous term by λ

$P(X = 2) = e^{-\lambda} \times \frac{\lambda^2}{2!} = \frac{\lambda}{2} P(X = 1)$ Multiply the previous term by $\frac{\lambda}{2}$

$P(X = 3) = e^{-\lambda} \times \frac{\lambda^3}{3!} = \frac{\lambda}{3} P(X = 2)$ Multiply the previous term by $\frac{\lambda}{3}$

In general, you can find $P(X = r)$ by multiplying your previous probability, $P(X = r - 1)$, by $\frac{\lambda}{r}$.

> **Note**
> The process of finding the next term from the previous one can be described as a **recurrence relation**.

Example 3.5   Jasmit is considering buying a telephone answering machine. He has one for five days' free trial and finds that 22 messages are left on it. Assuming that this is typical of the use it will get if he buys it, find:

(i) the mean number of messages per day

(ii) the probability that on one particular day there will be exactly six messages

(iii) the probability that on one particular day there will be more than six messages.

Solution

(i) Converting the total for five days to the mean for a single day gives

$$\text{daily mean} = \frac{22}{5} = 4.4 \text{ messages per day}$$

(ii) Calling X the number of messages per day

$$P(X = 6) = e^{-4.4} \times \frac{4.4^6}{6!}$$

$$= 0.124 \text{ (to 3 s.f.)}$$

(iii) $P(X \leq 6) = 0.8436$

and so

$P(X > 6) = 1 - 0.8436$

$= 0.156$ (to 3 s.f.)

Exercise 3.2

① If $X \sim \text{Poisson}(1.75)$, calculate
  (i) $P(X = 2)$
  (ii) $P(X > 0)$.

② If $X \sim \text{Poisson}(3.1)$, calculate
  (i) $P(X = 3)$
  (ii) $P(X < 2)$
  (iii) $P(X \leq 2)$.

③ The number of cars passing a house in a residential road between 10 a.m. and 11 a.m. on a weekday is a random variable, X. Give a condition under which X may be modelled by a Poisson distribution.

Suppose that $X \sim \text{Poisson}(3.4)$. Calculate $P(X \geq 4)$.

④ The number of wombats that are killed on a particular stretch of road in Australia in any one day can be modelled by a Poisson (0.42) random variable.
  (i) Calculate the probability that exactly two wombats are killed on a given day on this stretch of road.
  (ii) Find the probability that exactly four wombats are killed over a five-day period on this stretch of road.

⑤ A typesetter makes 1500 mistakes in a book of 500 pages. On how many pages would you expect to find (i) 0, (ii) 1, (iii) 2, (iv) 3 or more mistakes? State any assumptions in your workings.

⑥ 350 raisins are put into a mixture which is well stirred and made into 100 small buns. Estimate how many of these buns will
  (i) be without raisins
  (ii) contain five or more raisins.

  In a second batch of 100 buns, exactly one has no raisins in it.
  (iii) Estimate the total number of raisins in the second mixture.

⑦ In which of the following scenarios is it likely that X can be well modelled by a Poisson distribution? For those scenarios where X is probably not a good model, give a reason.
  (i) X is the number of aeroplanes landing at Heathrow Airport in a randomly chosen period of 1 hour.
  (ii) X is the number of foxes that live in a randomly chosen urban region of area 1 km².
  (iii) X is the number of tables booked at a restaurant on a randomly chosen evening.

(iv) X is the number of particles emitted by a radioactive substance in a period of 1 minute.

8 A ferry takes cars on a short journey from an island to the mainland. On a representative sample of weekday mornings, the numbers of vehicles, X, on the 8 a.m. sailing were as follows.

20 24 24 22 23 21 20 22 23 22
21 21 22 21 23 22 20 22 20 24

(i) Show that X does not have a Poisson distribution.

In fact 20 of the vehicles belong to commuters who use that sailing of the ferry every weekday morning. The random variable Y is the number of vehicles other than those 20 who are using the ferry.

(ii) Investigate whether Y may reasonably be modelled by a Poisson distribution.

The ferry can take 25 vehicles on any journey.

(iii) On what proportion of days would you expect at least one vehicle to be unable to travel on this particular sailing of the ferry because there was no room left and so have to wait for the next one?

9 A garage uses a particular spare part at an average rate of five per week.
Assuming that usage of this spare part follows a Poisson distribution, find the probability that

(i) exactly five are used in a particular week
(ii) at least five are used in a particular week
(iii) exactly ten are used in a two-week period
(iv) at least ten are used in a two-week period
(v) exactly five are used in each of two successive weeks.
(vi) If stocks are replenished weekly, determine the number of spare parts which should be in stock at the beginning of each week to ensure that, on average, the stock will be insufficient on no more than one week in a 52-week year.

10 Accidents on a particular stretch of motorway occur at an average rate of 1.5 per week.

(i) Write down a suitable model to represent the number of accidents per week on this stretch of motorway.

Find the probability that

(ii) there will be 2 accidents in the same week
(iii) there is at least one accident per week for 3 consecutive weeks
(iv) there are more than 4 accidents in a 2-week period.
[Edexcel 6684 Q2 January 2006]

11 Weak spots occur at random in the manufacture of a certain cable at an average rate of 1 per 120 metres. If X represents the number of weak spots in 120 m of cable, write down the distribution of X.

Lengths of this cable are wound on to drums. Each drum carries 60 m of cable. Find the probability that a drum will have three or more weak spots.

A contractor buys six such drums. Find the probability that two have just one weak spot each and the other four have none.

3 Applications of the Poisson distribution

New crossing near leisure centre?

A recent traffic survey has revealed that the number of vehicles using the main road near the leisure centre has reached levels where crossing the road has become hazardous.

The survey, carried out by a leisure centre staff member, suggested that the numbers of vehicles travelling in both directions along the main road has increased so much during the past year that pedestrians are almost taking their lives into their own hands when crossing the road.

Between 2 p.m. and 3 p.m., usually one of the quietest periods of the day, the average number of vehicles travelling into town is 3.5 per minute and the average number of vehicles travelling out of town is 5.7 per minute. A new crossing is a must.

If it can be shown that there is a greater than 1 in 4 chance of more than ten vehicles passing per minute, then there is a good chance of getting a pelican crossing.

Assuming that the flows of vehicles, into and out of town, can be modelled by independent Poisson distributions, you can model the flow of vehicles in both directions as follows.

Let X represent the number of vehicles travelling into town between 2 p.m. and 3 p.m. then $X \sim$ Poisson (3.5).

Let Y represent the number of vehicles travelling out of town between 2 p.m. and 3 p.m. then $Y \sim$ Poisson (5.7).

Let T represent the number of vehicles travelling in either direction between 2 p.m. and 3 p.m. then $T = X + Y$.

You can find the probability distribution for T as follows.

$$P(T=0) = P(X=0) \times P(Y=0)$$
$$= 0.0302 \times 0.0033$$
$$= 0.0001$$

$$P(T=1) = P(X=0) \times P(Y=1) + P(X=1) \times P(Y=0)$$
$$= 0.0302 \times 0.0191 + 0.1057 \times 0.0033$$
$$= 0.0009$$

> There are two ways of getting a total of 1. They are 0 and 1, 1 and 0.

S1 AS   Applications of the Poisson distribution

> There are three ways of getting a total of 2. They are 0 and 2, 1 and 1, 2 and 0.

$$P(T=2) = P(X=0) \times P(Y=2) + P(X=1) \times P(Y=1) + P(X=2) \times P(Y=0)$$
$$= 0.0302 \times 0.0544 + 0.1057 \times 0.0191 + 0.1850 \times 0.0033$$
$$= 0.0043$$

and so on.

You can see that this process is very time consuming. Fortunately, you can make life a lot easier by using the fact that if X and Y are two independent Poisson random variables, with means λ and μ, respectively, then $T = X + Y$ and T is a Poisson random variable with mean $\lambda + \mu$.

> **Note**
>
> $X \sim$ Poisson (λ) and $Y \sim$ Poisson (μ)
> $\Rightarrow X + Y \sim$ Poisson $(\lambda + \mu)$

Using $T \sim$ Poisson (9.2) gives the required probabilities straight away.

$$P(T=0) = 0.0001$$
$$P(T=1) = 0.0009$$
$$P(T=2) = 0.0043$$

You can now use the distribution for T to find the probability that the total traffic flow exceeds ten vehicles per minute.

$$P(T > 10) = 1 - P(T \leq 10)$$
$$= 1 - 0.6820$$
$$= 0.318$$

Since there is a greater than 25% chance of more than ten vehicles passing per minute, the case for the pelican crossing has been made, based on the Poisson probability models.

Example 3.6

A rare disease causes the death, on average, of 3.8 people per year in England, 0.8 in Scotland and 0.5 in Wales. As far as is known, the disease strikes at random and cases are independent of one another.

What is the probability of 7 or more deaths from the disease on the British mainland (i.e. England, Scotland and Wales together) in any year?

Solution

Notice first that:
- $P(7 \text{ or more deaths}) = 1 - P(6 \text{ or fewer deaths})$
- each of the three distributions fulfils the conditions for it to be modelled by the Poisson distribution.

You can therefore add the three distributions together and treat the result as a single Poisson distribution.

> **Note**
>
> You may only add Poisson distributions in this way if they are independent of each other.

The overall mean is given by   3.8   +   0.8   +   0.5   =   5.1
                               England Scotland Wales   Total

giving an overall distribution of Poisson (5.1).

The probability of 6 or fewer deaths is 0.7474.

So the probability of 7 or more deaths is given by $1 - 0.7474 = 0.2526$.

Example 3.7

On a lonely Highland road in Scotland, cars are observed passing at the rate of six per day and lorries at the rate of two per day. On the road, there is an old cattle grid which will soon need repair. The local works department decides that if the probability of more than 15 vehicles per day passing is less than 1%, then the repairs to the cattle grid can wait until next spring; otherwise it will have to be repaired before the winter.

When will the cattle grid have to be repaired?

Solution

Let C be the number of cars per day, L be the number of lorries per day and V be the number of vehicles per day.

Assuming that a car or a lorry passing along the road is a random event and that the two are independent

$$C \sim \text{Poisson}(6), L \sim \text{Poisson}(2)$$
and so $V \sim \text{Poisson}(6+2)$
$$\Rightarrow V \sim \text{Poisson}(8).$$
$$P(V \leqslant 15) = 0.9918.$$

The required probability is $P(V > 15) = 1 - P(V \leqslant 15)$
$$= 1 - 0.9918$$
$$= 0.0082.$$

This is just less than 1% and so the repairs are left until spring.

Link between binomial and Poisson distributions

In certain circumstances, you can use either the binomial distribution or the Poisson distribution as a model to calculate the probabilities you need. The example below illustrates this.

Example 3.8

It is known that, nationally, one person in a thousand is allergic to a particular chemical used in making a wood preservative. A firm that makes this wood preservative employs 500 people in one of its factories.

(i) Use the binomial distribution to estimate the probability that more than two people at the factory are allergic to the chemical.

(ii) What assumption are you making?

(iii) Using the fact that the mean of a binomial distribution is np, find the Poisson probability $P(Y > 2)$ where $Y \sim \text{Poisson}(np)$.

Solution

(i) Let X be the number of people in a random sample of 500 who are allergic to the chemical.

$$X \sim B(500, 0.001)$$
$$P(X > 2) = 1 - P(X \leq 2)$$
$$= 1 - 0.985669\ldots$$
$$= 0.0143$$

(ii) The assumption made is that people with the allergy are just as likely to work in the factory as those without the allergy. In practice, this seems rather unlikely: you would not stay in a job that made you unwell.

(iii) The mean of the binomial is $np = 500 \times 0.001 = 0.5$.

Using $Y \sim \text{Poisson}(0.5)$

$$P(Y > 2) = 1 - P(Y \leq 2)$$
$$= 1 - 0.985612\ldots$$
$$= 0.0144.$$

These two probabilities are very similar, so this suggests that sometimes the Poisson distribution and the binomial distribution give similar results. Four comparisons of binomial and Poisson probabilities are illustrated in Figure 3.6. In each case, the binomial probabilities are shown in blue and the Poisson probabilities in red.

If p is small, $q \approx 1$, so npq and the mean and variance are roughly equal, as required by a Poisson distribution.

> **Note**
>
> Note that if the binomial parameters are n and p then the corresponding Poisson distribution has parameter $\lambda = np$.

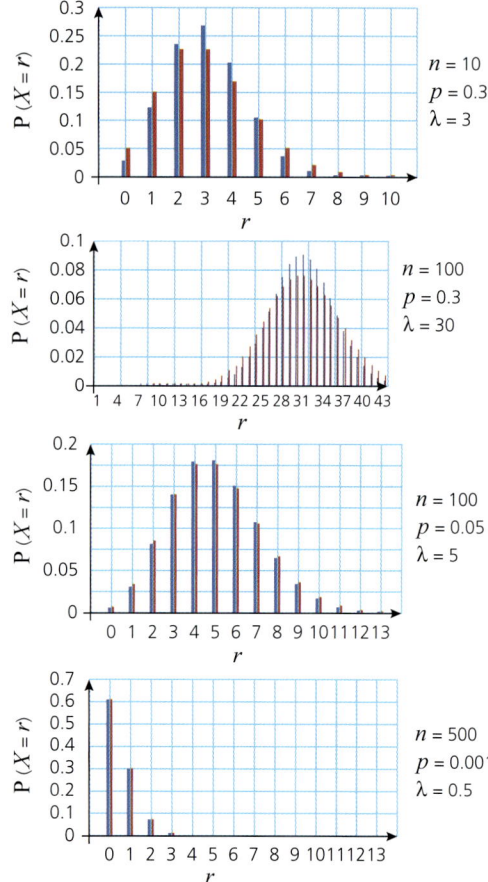

Figure 3.6 Comparison between the Poisson and binomial distributions for various values of n and p

Exercise 3.3

① You are given that $X \sim B(200, 0.04)$.

(i) Find $P(X = 5)$.

(ii) State the mean of a Poisson distribution which is likely to give a similar result to the probability found in part (i).

(iii) Use this mean to find the corresponding Poisson probability and compare it with your answer to part (i).

② The spreadsheet below shows two distributions, $B(10, 0.5)$ and Poisson(λ).

	A	B	C	D	E	F	G	H	I	J	K	L
1	r	0	1	2	3	4	5	6	7	8	9	10
2	Probability B(10, 0.5)	0.0010	0.0098	0.0439	0.1172	0.2051	0.2461	0.2051	0.1172	0.0439	0.0098	0.0010
3	Probability Poisson(λ)	0.0067	0.0337	0.0842	0.1404	0.1755	0.1755	0.1462	0.1044	0.0653	0.0363	0.0181

Figure 3.7

(i) The Poisson(λ) distribution is used to approximate the $B(10, 0.5)$ distribution. Write down the value of λ.

(ii) Plot a vertical line chart to compare the $B(10, 0.5)$ and the Poisson(λ) distributions.

(iii) Comment on whether the Poisson distribution (λ) is a good approximation to the $B(10, 0.5)$ distribution.

The spreadsheet shows two other distributions, $B(100, 0.5)$ and Poisson(μ).

	A	B	C	D	E	F	G	H	I	J	K	L
1	r	0	1	2	3	4	5	6	7	8	9	10
2	Probability B (100, 0.05)	0.0059	0.0312	0.0812	0.1396	0.1781	0.1800	0.1500	0.1060	0.0649	0.0349	0.0167
3	Probability Poisson (μ)	0.0067	0.0337	0.0842	0.1404	0.1755	0.1755	0.1462	0.1044	0.0653	0.0363	0.0181

Figure 3.8

(iv) The Poisson(μ) distribution is used to approximate the $B(100, 0.5)$ distribution. Write down the value of μ.

(v) Figure 3.9 shows a vertical line chart comparing the $B(100, 0.5)$ and the Poisson(μ) distributions. Compare this with the vertical line chart which you have drawn for $B(10, 0.05)$ and Poisson(λ) and comment on the differences.

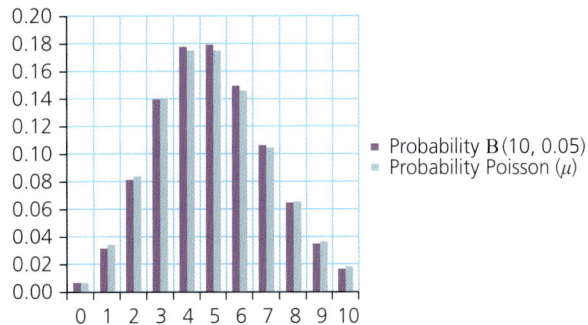

Figure 3.9

③ It is known that 0.3% of items produced by a certain process are defective. A random sample of 2000 items is selected.

(i) Use a binomial distribution to find the probability that there are at least five defective items in the sample.

(ii) Use a Poisson distribution to find the probability that there are at least five defective items in the sample.

(iii) Explain why your answers are similar although you are using two different distributions.

4. At a coffee shop both hot and cold drinks are sold. The number of hot drinks sold per minute may be assumed to be a Poisson variable with mean 0.7 and the number of cold drinks sold per minute may be assumed to be an independent Poisson variable with mean 0.4.
   (i) Calculate the probability that in a given one-minute period exactly one hot drink and one cold drink are sold.
   (ii) Calculate the probability that in a given three-minute period fewer than three drinks altogether are sold.
   (iii) In a given one-minute period exactly three drinks are sold. Calculate the probability that these are all hot drinks.

5. (i) Write down the conditions under which the Poisson distribution may be used as an approximation to the binomial distribution.

   A call centre routes incoming telephone calls to agents who have specialist knowledge to deal with the call. The probability of the caller being connected to the wrong agent is 0.01.
   (ii) Find the probability that 2 consecutive calls will be connected to the wrong agent.
   (iii) Find the probability that more than 1 call in 5 consecutive calls are connected to the wrong agent.

   The call centre receives 1000 calls each day.
   (iv) Find the mean and variance of the number of wrongly connected calls.
   (v) Use a Poisson approximation to find, to 3 decimal places, the probability that more than 6 calls each day are connected to the wrong agent.
   [Edexcel 6684 Q5 June 2007]

6. Telephone calls reach a departmental administrator independently and at random, internal ones at a mean rate of two in any five-minute period, and external ones at a mean rate of one in any five-minute period.
   (i) Find the probability that, in a five-minute period, the administrator receives
      (a) exactly three internal calls
      (b) at least two external calls
      (c) at most five calls in total.
   (ii) Given that the administrator receives a total of four calls in a five-minute period, find the probability that exactly two were internal calls.
   (iii) Find the probability that in any one-minute interval no calls are received.

7. During a weekday, cars pass a census point on a quiet side road independently and at random times. The mean rate for westward travelling cars is two in any five-minute period, and for eastward travelling cars is three in any five-minute period.

   Find the probability
   (i) that there will be no cars passing the census point in a given two-minute period
   (ii) that at least one car from each direction will pass the census point in a given two-minute period
   (iii) that there will be exactly ten cars passing the census point in a given ten-minute period.

⑧ A ferry company has two small ferries, A and B, that run across a river. The number of times that ferry A needs maintenance in a week has a Poisson distribution with mean 0.5, while, independently, the number of times that ferry B needs maintenance in a week has a Poisson distribution with mean 0.3.

Find, to three decimal places, the probability that in the next three weeks

(i) ferry A will not need maintenance at all

(ii) each ferry will need maintenance exactly once

(iii) there will be a total of two occasions when one or other of the two ferries will need maintenance.

⑨ Two random variables, X and Y, have independent Poisson distributions given by $X \sim \text{Poisson}(1.4)$ and $Y \sim \text{Poisson}(3.6)$, respectively.

(i) Using the distributions of X and Y only, calculate

(a) $P(X + Y = 0)$

(b) $P(X + Y = 1)$

(c) $P(X + Y = 2)$.

The random variable T is defined by $T = X + Y$.

(ii) Write down the distribution of T.

(iii) Use your distribution from part (ii) to check your results in part (i).

⑩ A boy is watching vehicles travelling along a motorway. All the vehicles he counts are either cars or lorries; the numbers of each may be modelled by two independent Poisson distributions. The mean number of cars per minute is 8.3 and the mean number of lorries per minute is 4.7.

(i) For a given period of one minute, find the probability that he sees

(a) exactly seven cars   (b) at least three lorries.

(ii) Calculate the probability that he sees a total of exactly ten vehicles in a given one-minute period.

(iii) Find the probability that he observes fewer than eight vehicles in a given period of 30 seconds.

⑪ The numbers of emissions per minute from two radioactive substances, A and B, are independent and have Poisson distributions with means 2.8 and 3.25, respectively.

Find the probabilities that in a period of one minute there will be

(i) at least three emissions from substance A

(ii) one emission from one of the two substances and two emissions from the other substance

(iii) a total of five emissions.

⑫ The number of cats rescued by an animal shelter each day may be modelled by a Poisson distribution with parameter 2.5, while the number of dogs rescued each day may be modelled by an independent Poisson distribution with parameter 3.2.

(i) Calculate the probability that on a randomly chosen day the shelter rescues

(a) exactly two cats   (b) exactly three dogs

(c) exactly five cats and dogs in total.

(ii) Given that one day exactly five cats and dogs were rescued, find the conditional probability that exactly two of these animals were cats.

13. A petrol station has service areas on both sides of a motorway, one to serve north-bound traffic and the other for south-bound traffic. The number of north-bound vehicles arriving at the station in one minute has a Poisson distribution with mean 1.4, and the number of south-bound vehicles arriving in one minute has a Poisson distribution with mean 2.3, the two distributions being independent.

(i) Find the probability that in a one-minute period
   (a) exactly three vehicles arrive
   (b) more than four vehicles arrive at this petrol station, giving your answers correct to three places of decimals.

Given that in a particular one-minute period five vehicles arrive, find

(ii) the probability that they are all from the same direction
(iii) the most likely combination of north-bound and south-bound vehicles.

14. A sociologist claims that only 3% of all suitably qualified students from inner city schools go on to university. The sociologist selects a random sample of 200 such students. Use a Poisson distribution to estimate the probability that

(i) exactly five go to university
(ii) more than five go to university.
(iii) If there is at most a 5% chance that more than n of the 200 students go to university, find the lowest possible value of n.

Another group of 100 students from inner city schools is also chosen. Find the probability that

(iv) exactly five of each group go to university
(v) exactly ten of all the chosen students go to university.

4 The discrete uniform distribution

You have now met the binomial distribution and the Poisson distribution. There are many other discrete distributions such as the hypergeometric distribution, the negative binomial, the uniform distribution and the geometric distribution. You will find out about the geometric distribution and the negative binomial distribution in Chapter 9.

The uniform distribution

When fair six-sided dice are thrown, there are only six outcomes for each of the dice, all equally likely. You can write the probability distribution for each of them formally as

$$P(X = r) = \frac{1}{6} \quad \text{for } r = 1, 2, 3, 4, 5, 6$$
$$P(X = r) = 0 \quad \text{otherwise,}$$

where X represents the score on one of the dice.

Tabulating the probability distribution for X gives

Table 3.3

r	1	2	3	4	5	6
$P(X=r)$	$\frac{1}{6}$	$\frac{1}{6}$	$\frac{1}{6}$	$\frac{1}{6}$	$\frac{1}{6}$	$\frac{1}{6}$

The vertical line chart in Figure 3.10 illustrates this distribution.

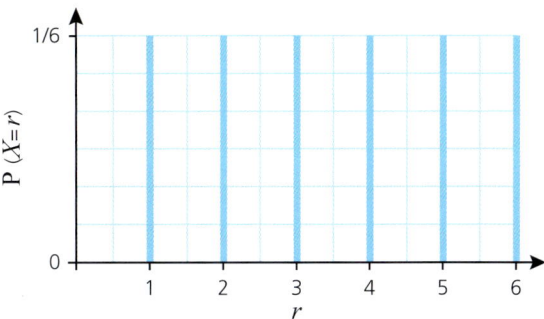

Figure 3.10

This is an example of the uniform probability distribution (sometimes also known as the rectangular distribution).

In general, the uniform probability distribution over the values $\{1, 2, ..., n\}$ is defined as follows:

$$P(X=r) = \frac{1}{n} \quad \text{for } r = 1, 2, ..., n$$
$$P(X=r) = 0 \quad \text{otherwise.}$$

However, the lowest value may be k rather than 1 in which case the distribution over the values $\{k, k+1, ..., n+k-1\}$ is as follows:

$$P(X=r) = \frac{1}{n} \quad \text{for } r = k, k+1, ..., n+k-1$$
$$P(X=r) = 0 \quad \text{otherwise.}$$

Expectation and variance of the uniform distribution

You can find the expectation and variance of the score X on one of the fair dice, using the formulae which you met in Chapter 2.

$$E(X) = \mu = \Sigma r P(X=r)$$
$$= 1 \times \frac{1}{6} + 2 \times \frac{1}{6} + \cdots + 6 \times \frac{1}{6}$$
$$= 21 \times \frac{1}{6} = \frac{21}{6} = \frac{7}{2}$$
$$\text{Var}(X) = \sigma^2 = E(X^2) - \mu^2$$
$$= 1^2 \times \frac{1}{6} + 2^2 \times \frac{1}{6} + \cdots + 6^2 \times \frac{1}{6} - \left(\frac{7}{2}\right)^2$$
$$= (1 + 4 + 9 + 16 + 25 + 36) \times \frac{1}{6} - \left(\frac{7}{2}\right)^2$$
$$= \frac{91}{6} - \frac{49}{4} = \frac{35}{12}$$

S1 AS — The discrete uniform distribution

These results can be generalised for a uniform random variable X taking the values $1, 2, \ldots, n$.

> Note that you are using the formula
> $$\sum_{1}^{n}(r) = \frac{n(n+1)}{2}.$$

$$E(X) = \sum_{1}^{n} r P(X = r)$$

$$= \sum_{1}^{n}\left(r \times \frac{1}{n}\right)$$

$$= \frac{1}{n}\sum_{1}^{n}(r)$$

$$= \frac{1}{n} \times \frac{n(n+1)}{2}$$

$$= \frac{n+1}{2}$$

To find the variance, you use the formula $\operatorname{Var}(X) = E(X^2) - [E(X)]^2$

$$E(X^2) = \sum_{1}^{n} r^2 P(X = r)$$

$$= \sum_{1}^{n} r^2 \frac{1}{n}$$

$$= \frac{1}{n}\sum_{1}^{n} r^2$$

> Note that you are using the formula
> $$\sum_{1}^{n}(r^2) = \frac{1}{6} n(n+1)(2n+1).$$

$$= \frac{1}{n} \times \frac{1}{6} n(n+1)(2n+1)$$

$$= \frac{1}{6}(n+1)(2n+1)$$

$$\operatorname{Var}(X) = E(X^2) - [E(X)]^2$$

$$= \frac{1}{6}(n+1)(2n+1) - \left(\frac{n+1}{2}\right)^2$$

$$= \frac{n+1}{12}[2(2n+1) - 3(n+1)]$$

$$= \frac{n+1}{12}(4n + 2 - 3n - 3)$$

$$= \frac{(n+1)(n-1)}{12}$$

$$= \frac{n^2 - 1}{12}$$

> **Note**
>
> For a discrete uniform distribution,
> $E(X) = \frac{n+1}{2}$, $\operatorname{Var}(X) = \frac{n^2-1}{12}$.

Example 3.9

A five-sided fair spinner has faces labelled 10, 11, 12, 13, 14. The score on the spinner when it is rolled once is denoted by Z.

(i) Prove that $E(Z) = 12$.

(ii) Prove that $\operatorname{Var}(Z) = 2$.

> **Note**
>
> Although it may seem obvious by symmetry that $E(Z) = 12$, the question asks for a **proof**. You therefore need to use the formula for expectation $E(X) = \sum r P(X = r)$.

Solution

(i) $E(Z) = \sum_{10}^{14} r P(Z = r)$.

$$= \sum_{10}^{14}\left(r \times \frac{1}{n}\right)$$

$$= 10 \times \frac{1}{5} + 11 \times \frac{1}{5} + 12 \times \frac{1}{5} + 13 \times \frac{1}{5} + 14 \times \frac{1}{5}$$

$$= 60 \times \frac{1}{5}$$

$$= 12$$

(ii) $\quad E(Z^2) = \sum_{r=1}^{n} r^2 P(Z=r)$

$= 100 \times \frac{1}{5} + 121 \times \frac{1}{5} + 144 \times \frac{1}{5} + 169 \times \frac{1}{5} + 196 \times \frac{1}{5}$

$= (100 + 121 + 144 + 169 + 196) \times \frac{1}{5}$

$= 730 \times \frac{1}{5}$

$= 146$

$\text{Var}(X) = E(X^2) - [E(X)]^2$

$= 146 - 144$

$= 2$

Exercise 3.4

1. Two fair six-sided dice are rolled.
   (i) Find the probability that the score on the first dice is at least 5.
   (ii) Find the probability that the score on at least one of the two dice is at least 5.

2. A fair three-sided spinner has sectors labelled 1, 2 and 3. The spinner is spun until a 3 is scored. The number of spins required to get a 3 is denoted by X.
   (i) Find $P(X=1)$.
   (ii) Find $P(X=5)$.
   (iii) Find $P(X>5)$.

3. A fair four-sided spinner has sectors labelled 2, 3, 4 and 5. The score on the spinner when it is spun once is denoted by X.
   (i) Find $P(X>3)$.
   (ii) Find $E(X)$.
   (iii) Find $\text{Var}(X)$.

4. Five fair six-sided dice are each rolled once.
   (i) Find the expectation and variance of the score on one of the dice.
   (ii) Find the expectation and variance of the total score on the five dice.

5. A fair seven-sided spinner has sectors labelled 7, 8, …,13. Find the expectation and variance of the score on the spinner.

6. The random variable X denotes the score on a fair three-sided spinner which has sectors labelled 1, 2 and 3. The random variable Y denotes the score on a fair three-sided spinner which has sectors labelled 10, 20 and 30.
   (i) Find $E(X)$ and $\text{Var}(X)$.
   (ii) Find $E(Y)$ and $\text{Var}(Y)$.
   (iii) The spinner with sectors labelled 1, 2, 3 is spun ten times. Find the expectation and variance of the total score.
   (iv) Compare your answers to parts (ii) and (iii).

The discrete uniform distribution

KEY POINTS

Binomial distribution

1. The binomial distribution may be used to model situations in which these conditions hold
   - You are conducting trials on random samples of a certain size, n.
   - On each trial the outcomes can be classified as either success or failure.

   For the binomial distribution to be a good model, these assumptions are required.
   - The outcome of each trial is independent of the outcome of any other trial.
   - The probability of success is the same on each trial.

2. For a binomial random variable X, where $X \sim B(n, p)$
   - $P(X = r) = {}^nC_r q^{n-r} p^r$ for $r = 0, 1, 2, ..., n$.

3. For $X \sim B(n, p)$
   - $E(X) = np$
   - $Var(X) = npq$.

Poisson distribution

1. The Poisson distribution may be used in situations in which:
   - the variable is the frequency of events occurring in fixed intervals of time or space.

2. For the Poisson distribution to be a good model:
   - events occur randomly
   - events occur independently
   - events occur at a uniform average rate.

3. For a Poisson random variable X, where $X \sim \text{Poisson}(\lambda)$
   - $P(X = r) = e^{-\lambda} \times \dfrac{\lambda^r}{r!}$ for $r = 0, 1, 2, ...$

4. For Poisson(λ)
   - $E(X) = \lambda$
   - $Var(X) = \lambda$.

5. The sum of two independent Poisson distributions
   - If $X \sim \text{Poisson}(\lambda)$ and $Y \sim \text{Poisson}(\mu)$, then $X + Y \sim \text{Poisson}(\lambda + \mu)$.

Uniform distribution

1. The uniform distribution may be used to model situations in which:
   - all outcomes are equally likely.

2. For a uniform random variable X
   - $P(X = r) = \dfrac{1}{n}$ for $r = 1, 2, ..., n$.

3. For a uniform random variable X
   - $E(X) = \dfrac{n+1}{2}$
   - $Var(X) = \dfrac{n^2 - 1}{12}$.

LEARNING OUTCOMES

When you have completed this chapter you should be able to:

- recognise situations under which the binomial distribution is likely to be an appropriate model
- calculate probabilities using a binomial distribution
- know and be able to use the mean and variance of a binomial distribution
- recognise situations under which the Poisson distribution is likely to be an appropriate model
- calculate probabilities using a Poisson distribution
- know and be able to use the mean and variance of a Poisson distribution
- know that the sum of two or more independent Poisson distributions is also a Poisson distribution
- recognise situations in which both the Poisson distribution and the binomial distribution might be appropriate models
- recognise situations under which the discrete uniform distribution is likely to be an appropriate model
- calculate probabilities using a discrete uniform distribution
- calculate the mean and variance of any given discrete uniform distribution.

4 Chi-squared tests

The fact that the criterion which we happen to use has a fine ancestry of statistical theorems does not justify its use. Such justification must come from empirical evidence that it works.

W. A. Shewhart

What kind of films do you enjoy?

To help it decide when to show trailers for future programmes, the management of a cinema asks a sample of its customers to fill in a brief questionnaire saying which type of film they enjoy. It wants to know whether there is any relationship between people's enjoyment of horror films and action movies.

Discussion point
How do you think the management should select the sample of customers?

1 The chi-squared test for a contingency table

The management of the cinema takes 150 randomly selected questionnaires and records whether those patrons enjoyed or did not enjoy horror films and action movies.

Table 4.1

Observed frequency, f_o	Enjoyed horror films	Did not enjoy horror films
Enjoyed action movies	51	41
Did not enjoy action movies	15	43

> **Note**
> You will meet larger contingency tables later in this chapter.

This method of presenting data is called a 2 × 2 **contingency table**. It is used where two variables (here 'attitude to horror films' and 'attitude to action movies') have been measured on a sample, and each variable can take two different values ('enjoy' or 'not enjoy').

The values of the variables fall into one or other of two categories. You want to determine the extent to which the variables are **related**.

It is conventional, and useful, to add the row and column totals in a contingency table: these are called the **marginal totals** of the table.

Table 4.2

Observed frequency, f_o	Enjoyed horror films	Did not enjoy horror films	Total
Enjoyed action movies	51	41	92
Did not enjoy action movies	15	43	58
Total	66	84	150

A formal version of the cinema management's question is, 'Is enjoyment of horror films independent of enjoyment of action movies?'. You can use the sample data to investigate this question.

You can estimate the probability that a randomly chosen cinema-goer will enjoy horror films as follows. The number of cinema-goers in the sample who enjoyed horror films is 51 + 15 = 66.

So the proportion of cinema-goers who enjoyed horror films is $\frac{66}{150}$.

In a similar way, you can estimate the probability that a randomly chosen cinema-goer will enjoy action movies. The number of cinema-goers in the sample who enjoyed action movies is 51 + 41 = 92.

So the proportion of cinema-goers who enjoyed action movies is $\frac{92}{150}$.

> Notice how you use the marginal totals 66 and 92 which were calculated previously.

If people enjoyed horror films and action movies independently with the probabilities you have just estimated, then you would expect to find, for instance:

The chi-squared test for a contingency table

Number of people enjoying both types

$= 150 \times$ P(a random person enjoying both types)

$= 150 \times$ P(enjoying horror) $\times$ P(enjoying action)

$= 150 \times \dfrac{66}{150} \times \dfrac{92}{150}$

$= \dfrac{6072}{150}$

$= 40.48.$

In the same way, you can calculate the number of people you would expect to correspond to each cell in the table.

Table 4.3

Expected frequency, f_e	Enjoyed horror films	Did not enjoy horror films	Total
Enjoyed action movies	$150 \times \dfrac{66}{150} \times \dfrac{92}{150} = 40.48$	$150 \times \dfrac{84}{150} \times \dfrac{92}{150} = 51.52$	92
Did not enjoy action movies	$150 \times \dfrac{66}{150} \times \dfrac{58}{150} = 25.52$	$150 \times \dfrac{84}{150} \times \dfrac{58}{150} = 32.48$	58
Total	66	84	150

Note that it is an inevitable consequence of this calculation that these expected figures have the same marginal totals as the sample data.

You are now in a position to test the original hypotheses, which you can state formally as:

H_0: enjoyment of the two types of film is independent.

H_1: enjoyment of the two types of film is not independent.

The expected frequencies were calculated assuming the null hypothesis is true. You know the actual sample frequencies and the aim is to decide whether those from the sample are so different from those calculated theoretically that the null hypothesis should be rejected.

A statistic which measures how far apart a set of observed frequencies is from the set expected under the null hypothesis is the χ^2 (chi-squared) statistic. It is given by the formula:

> The χ^2 test statistic is denoted by X^2.

$$X^2 = \sum \dfrac{(f_o - f_e)^2}{f_e} = \sum \dfrac{(\text{observed frequency} - \text{expected frequency})^2}{\text{expected frequency}}$$

You can use this here: the observed and expected frequencies are summarised below.

Table 4.4

Observed frequency, f_o	Enjoyed horror	Did not enjoy horror
Enjoyed action	51	41
Did not enjoy action	15	43

Expected frequency, f_e	Enjoyed horror	Did not enjoy horror
Enjoyed action	40.48	51.52
Did not enjoy action	25.52	32.48

The χ^2 statistic is:

$$X^2 = \Sigma \frac{(f_o - f_e)^2}{f_e} = \frac{(51-40.48)^2}{40.48} + \frac{(41-51.52)^2}{51.52} + \frac{(15-25.52)^2}{25.52} + \frac{(43-32.48)^2}{32.48}$$

$$= \frac{(10.52)^2}{40.48} + \frac{(10.52)^2}{51.52} + \frac{(10.52)^2}{25.52} + \frac{(10.52)^2}{32.48} = 12.626$$

> Note that the four numbers on the top lines (numerators) in this calculation are equal. This is not by chance; it will always happen with a 2 × 2 table. It provides you with a useful check and short cut when you are working out X^2.

Following the usual hypothesis-testing methodology, you want to know whether a value for this statistic at least as large as 12.626 is likely to occur by chance when the null hypothesis is true. The critical value at the 10% significance level for this test statistic is 2.706.

> You will see how to find critical values for a χ^2 test later in this chapter.

Since 12.626 > 2.706, you reject the null hypothesis, H_0, and conclude that people's enjoyment of the two types of film is not independent or that the enjoyment of the two is **associated**.

> **Note**
> Notice that you cannot conclude that enjoying one type of film **causes** people to enjoy the other. The test is of whether enjoyment of the two types is associated. It could be that a third factor, such as bloodthirstiness, causes both, but you do not know. The test tells you nothing about causality.

The diagram below shows you the relevant χ^2 distribution for this example, the critical region and the test statistic.

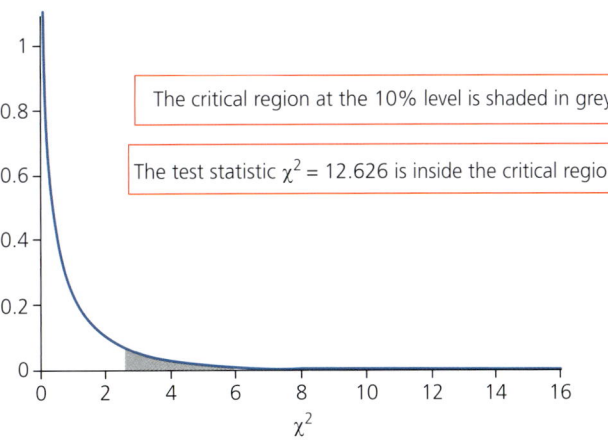

> The critical region at the 10% level is shaded in grey
>
> The test statistic $\chi^2 = 12.626$ is inside the critical region

Figure 4.1

> The information about the χ^2 distribution is for your interest – you do not need to use it to carry out the tests in this chapter.
>
> A standard Normal variable is drawn from a Normal population with mean 0 and variance 1.

The chi-squared distribution

The χ^2 distribution with n degrees of freedom is the distribution of the sum of the squares of n independent standard Normal random variables.

You can use it to test how well a set of data match a given distribution. A number of examples of such tests are covered in this chapter.

These tests include that used in the example of the cinema-goers: that is, whether the two classifications used in a contingency table are independent of one another. The hypotheses for such a test are:

H_0: The two variables whose values are being measured are independent in the population.

H_1: The two variables whose values are being measured are not independent in the population.

In order to carry out this test, you need to know more about the χ^2 distribution.

Figure 4.1 is an example of a χ^2 distribution. The shape of the χ^2 distribution curve depends on the number of free variables involved, the degrees of freedom, v. To find the value for v in this case, you start off with the number of cells which must be filled and then subtract one degree of freedom for each restriction, derived from the data, which is placed on the frequencies. In the cinema example above, you are imposing the requirements that the total of the frequencies must be 150, and that the overall proportions of people enjoying horror films and action movies are $\frac{66}{150}$ and $\frac{92}{150}$, respectively.

Hence $v = 4$ (number of cells)

$\qquad - 1$ (total of frequencies is fixed by the data)

$\qquad - 2$ (proportions of people enjoying each type are estimated from the data)

$\qquad = 1$.

So Figure 4.1 shows the shape of the χ^2 distribution for 1 degree of freedom.

In general, for an $m \times n$ contingency table, the degrees of freedom are:

$v = m \times n$ (number of cells)

$\qquad - (m + n - 1)$ (Row and column totals are fixed but row totals and column totals have the same sum.)

$\qquad = mn - m - n + 1$

$\qquad = (m - 1)(n - 1)$.

As you will see later in the chapter, the calculation of the degrees of freedom varies from one χ^2 test to another.

Figure 4.2 shows the shape of the chi-squared distribution for $v = 1, 2, 3, 6,$ and 10 degrees of freedom.

> **Note**
>
> As you can see, the shape of the chi-squared distribution depends very much on the number of degrees of freedom. So the critical region also depends on the number of degrees of freedom.

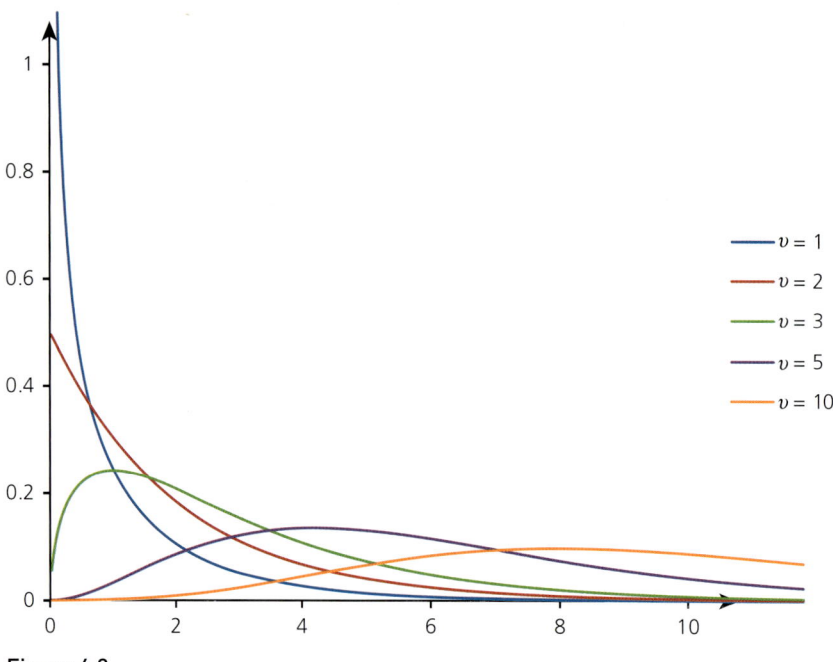

Figure 4.2

You can see in Figure 4.3 a typical χ^2 distribution curve together with the critical region for a significance level of $p\%$. An extract from a table of critical values of the χ^2 distribution for various degrees of freedom is also shown. The possible use of the left-hand tail probabilities (99%, 95% etc.) is discussed later in this chapter.

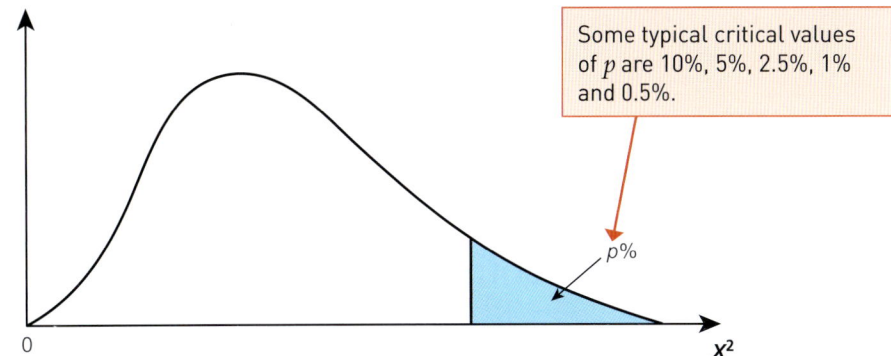

Some typical critical values of p are 10%, 5%, 2.5%, 1% and 0.5%.

p%	99	97.5	95	90	10	5.0	2.5	1.0	0.5
v = 1	.0001	.0010	.0039	.0158	2.706	3.841	5.024	6.635	7.879
2	.0201	.0506	0.103	0.211	4.605	5.991	7.378	9.210	10.60
3	0.115	0.216	0.352	0.584	6.521	7.815	9.348	11.34	12.84
4	0.297	0.484	0.711	1.064	7.779	9.488	11.14	13.28	14.86
5	0.554	0.831	1.145	1.610	9.236	11.07	12.83	15.09	16.75
6	0.872	1.237	1.635	2.204	10.64	12.59	14.45	16.81	18.55
7	1.239	1.690	2.167	2.833	12.02	14.07	16.01	18.48	20.28
8	1.646	2.180	2.733	3.490	13.36	15.51	17.53	20.09	21.95
9	2.088	2.700	3.325	4.168	14.68	16.92	19.02	21.67	23.59

Figure 4.3

Properties of the test statistic X^2

You have seen the test statistic is given by

$$X^2 = \sum_{\text{All classes}} \frac{(f_o - f_e)^2}{f_e}$$

> **Note**
>
> An alternative notation which is often used is to call the expected frequency in the ith class E_i and the observed frequency in the ith class O_i.
> In this notation
> $$X^2 = \sum_i \frac{(O_i - E_i)^2}{E_i}$$

Here are some points to notice.

- It is clear that as the difference between the expected values and the observed values increases then so will the value of this test statistic. Squaring the top gives due weight to any particularly large differences. It also means that all values are positive.
- Dividing $(f_e - f_o)^2$ by f_e has the effect of standardising that element, allowing for the fact that, the larger the expected frequency within a class, the larger will be the difference between the observed and the expected.
- The usual convention in statistics is to use a Greek letter for a parent population parameter and the corresponding Roman letter for the equivalent sample statistic. Unfortunately, when it comes to χ^2, there is no Roman

S1 AS — The chi-squared test for a contingency table

equivalent to the Greek letter χ since it translates into CH. Since X looks rather like χ a sample statistic from a χ^2 population is denoted by X^2. (In the same way Christmas is abbreviated to χmas but written Xmas.)

> For example:
> | Population parameters | Sample statistics |
> |---|---|
> | Greek letters | Roman letters |
> | μ | m |
> | σ | s |
> | ρ | r |

Continuing with tests on contingency tables

Example 4.1

The 4 × 3 contingency table below shows the type of car (saloon, sports, hatchback or SUV) owned by 360 randomly chosen people, and the age category (under 30, 30–60, over 60) into which the owners fall.

Table 4.5

Observed frequency, f_o	Age of driver			Total
	under 30	30–60	over 60	
Saloon	10	67	57	134
Sports car	19	14	3	36
Hatchback	32	47	34	113
SUV	7	56	14	77
Total	68	184	108	360

> **Note**
> The marginal totals are not essential in a contingency table, but it is conventional – and convenient – to add them. They are very helpful for subsequent calculations.

(i) Write down appropriate hypotheses for a test to investigate whether type of car and owner's age are independent.

(ii) Calculate expected frequencies assuming that the null hypothesis is true.

(iii) Calculate the value of the test statistic X^2.

(iv) Find the critical value at the 5% significance level.

(v) Complete the test.

(vi) Comment on how the ownership of different types of car depends on the age of the owner.

Solution

(i) H_0: Car type is independent of owner's age.

H_1: Car type is not independent of owner's age.

(ii) Table 4.6

Expected frequency, f_e	Age of driver			Total
	under 30	30–60	over 60	
Saloon	25.311	68.489	40.200	134
Sports car	6.800	18.400	10.800	36
Hatchback	21.344	57.756	33.900	113
SUV	14.544	39.356	23.100	77
Total	68	184	108	360

> You need to calculate the expected frequencies in the table assuming that the null hypothesis is true.
>
> Use the probability estimates given by the marginal totals.
>
> For instance the expected frequency for hatchback and owner's age is over 60 is given by
>
> $360 \times \dfrac{113}{360} \times \dfrac{108}{360}$
>
> $= \dfrac{113 \times 108}{360} = 33.900$

Note

This illustrates the general result for contingency tables:

Expected frequency for a cell

$= \dfrac{\text{product of marginal totals for that cell}}{\text{number of observations}}$

Note

You need to check that all the frequencies are large enough to make the χ^2 distribution a good approximation to the distribution of the X^2 statistic. The usual rule of thumb is to require all the expected frequencies to be greater than 5.

This requirement is (just) satisfied here. However, you might be cautious in your conclusions if the X^2 statistic is very near the relevant critical value. If some of the cells have small expected frequencies, you should either collect more data or amalgamate some of the categories if it makes sense to do so. For instance, two adjacent age ranges could reasonably be combined, but two car types probably could not.

(iii) The value of the X^2 statistic is $X^2 = \sum \dfrac{(f_o - f_e)^2}{f_e}$

The contributions of the various cells to this are shown in the table below.

Table 4.7

Contribution to test statistic	Age of driver		
	under 30	30–60	over 60
Saloon	9.262	0.032	7.021
Sports car	21.888	1.052	5.633
Hatchback	5.319	2.003	0.000
SUV	3.913	7.039	3.585

> An example of the calculation is
>
> $\dfrac{(10 - 25.311)^2}{25.311} = 9.262$
>
> for the top left cell.

Total $= 9.262 + 0.032 + 7.021 + 21.888 + \cdots + 3.585$

$X^2 = 66.749$

> The number of rows, m, is 4
> The number of columns, n, is 3

(iv) The degrees of freedom are given by $v = (m-1)(n-1)$.

$v = (4-1) \times (3-1) = 6$

From the χ^2 tables, the critical value at the 5% level with six degrees of freedom is 12.59.

(v) The observed X^2 statistic of 66.749 is greater than the critical value of 12.59. So the null hypothesis is rejected and the alternative hypothesis is accepted at the 5% significance level:

that car type is not independent of owner's age,

or that car type and owner's age are associated.

(vi) In this case, the under-30 age group own fewer saloon cars and SUVs, more hatchbacks and many more sports cars than expected. Other cells with relatively large contributions to the X^2 statistic correspond to SUVs being owned more often than expected by 30–60-year-olds, and less often than expected by older or younger drivers, and over-60s owning more saloon cars and fewer sports cars than expected.

Note
You reject the null hypothesis if the test statistic is **greater** than the critical value.

Note
You should always refer to the size of the contributions when commenting on the way that one variable is associated with the other (assuming, of course, that the conclusion to your test is that there is association).

USING ICT

Statistical software

You can use statistical software to carry out a χ^2 test for a contingency table. In order for the software to process the test, you need to input the information in the table of observed frequencies. This consists of category names and the observed frequencies, so, in this case, it is the information in this table.

Table 4.8

Observed frequency, f_o	Age of driver		
	under 30	30–60	over 60
Saloon	10	67	57
Sports car	19	14	3
Hatchback	32	47	34
SUV	7	56	14

The software then carries out all the calculations. Here is a typical output.

Chi-squared test

	under 30	30–60	over 60
Saloon	25.3111 9.2619 10	68.4889 0.0324 67	40.2 7.0209 57
Sports car	6.8 21.8882 19	18.4 1.0522 14	10.8 5.633 3
Hatchback	21.3444 5.3195 32	57.7556 2.003 47	33.9 0.0003 34
SUV	14.5444 3.9134 7	39.3556 7.0394 56	23.1 3.5848 14

Result
Chi-squared test
df 6
X^2 .... 66.7493
p 0.0000

Figure 4.4

> **Discussion points**
> The output includes the following information.
> → the expected frequencies
> → the contributions to the X^2 statistic
> → the degrees of freedom
> → the value of the X^2 statistic
> → the p-value for the test
>
> Identify where each piece of information is displayed.
>
> What other information is contained in the output box?

> Notice that the *p*-value is stated to be 0.0000. This requires some interpretation.
> - The other output figures are given either to 4 decimal places or as whole numbers.
> - So you can conclude that $p = 0.0000$ to 4 decimal places and therefore that $p < 0.00005$.
> - So the result is significant even at the 0.01% significance level.

Using a spreadsheet

You can also use a spreadsheet to do the final stages of this test. To set it up, you would need to take the following steps.

- Enter the same information as before: the variable categories and the observed frequencies.
- Use a suitable formula to calculate the expected frequencies.
- Combine classes as necessary if any expected frequencies are below 5.
- Use a suitable formula to calculate the contributions to the test statistic.
- Find the sum of the contributions.
- Find the *p*-value using the formula provided with the spreadsheet, for example =CHISQ.DIST.RT(H1,6).

Cell H1 contains the value of the test statistic, X^2.

There are 6 degrees of freedom.

In this case, a typical spreadsheet gives the value of *p* as 1.894E-12, ie 1.894×10^{-12}, so much less than the upper bound of 0.00005 inferred from the statistical software.

Exercise 4.1

① A group of 330 students, some aged 13 and the rest aged 16, is asked 'What is your usual method of transport from home to school?' The frequencies of each method of transport are shown in the table.

Table 4.9

	Age 13	Age 16
Walk	43	35
Cycle	24	42
Bus	64	49
Car	41	32

(i) Find the total of each row and each column.

A student is going to carry out a test to determine whether method of transport is independent of age.

(ii) Show that the expected frequency for age 16 students who walk is 37.35.

(iii) Show that the expected frequency for age 13 students who cycle is 34.40.

(iv) Would you expect the method of transport to be independent of age?

② A random sample of 80 students studying for a first aid exam was selected. The students were asked how many hours of revision they had done for the exam. The results are shown in the table, together with whether or not they passed the exam.

Table 4.10

	Pass	Fail
Less than 10 hours	13	18
At least 10 hours	42	7

(i) Find the expected frequency for each cell for a test to determine whether hours of revision is independent of passing or failing.

(ii) Find the corresponding contributions to the chi-squared test statistic.

③ A group of 281 voters is asked to rate how good a job they think the Prime Minister is doing. Each is also asked for the highest educational qualifications they have achieved. The frequencies with which responses occurred are shown in the table.

Table 4.11

Rating of PM	Highest qualifications achieved			
	None	GCSE or equivalent	A-level or equivalent	Degree or equivalent
Very poor	11	37	13	6
Poor	12	17	22	8
Moderate	7	11	25	10
Good	10	17	17	9
Very good	19	16	8	6

Use these figures to test whether there is an association between rating of the Prime Minister and highest educational qualification achieved.

④ A medical insurance company office is the largest employer in a small town. When 37 randomly chosen people living in the town were asked where they worked and whether they belonged to the town's health club, 21 were found to work for the insurance company, of whom 15 also belonged to the health club, while 7 of the 16 not working for the insurance company belonged to the health club.

Test the hypothesis that health club membership is independent of employment by the medical insurance company.

⑤ In a random sample of 163 adult males, 37 suffer from hay-fever and 51 from asthma, both figures including 14 men who suffer from both. Test whether the two conditions are associated.

⑥ In a survey of 184 London residents brought up outside the south-east of England, respondents were asked whether, job and family permitting, they would like to return to their area of origin. Their responses are shown in the table.

Table 4.12

Region of origin	Would like to return to	Would not like to return to
South-west	16	28
Midlands	22	35
North	15	31
Wales	8	6
Scotland	14	9

Test the hypothesis that desire to return is independent of region of origin.

7. A sample of 80 men and 150 women selected at random are tested for colour-blindness. Twelve of the men and five of the women are found to be colour-blind. Is there evidence at the 1% level that colour-blindness is sex-related?

8. Depressive illness is categorised as type I, II or III. In a group of depressive psychiatric patients, the length of time for which their symptoms are apparent is observed. The results are shown below.

Table 4.13

Length of depressive episode	Type of symptoms		
	I	II	III
Brief	15	22	12
Average	30	19	26
Extended	7	13	21
Semi-permanent	6	9	11

Is the length of the depressive episode independent of the type of symptoms?

9. The human resources manager of a large firm is investigating whether there is any association between the length of service of the employees and the type of training they receive from the firm. A random sample of 200 employee records is taken from the last few years and is classified according to these criteria. Length of service is classified as short (meaning less than 1 year), medium (1–3 years) and long (more than 3 years). Type of training is classified as being merely an initial 'induction course', proper initial on the job training but little, if any, more, and regular and continuous training. The data are as follows.

Table 4.14

Type of training	Length of service		
	Short	Medium	Long
Induction course	14	23	13
Initial on-the-job	12	7	13
Continuous	28	32	58

The output from a statistical package for these data is shown below.

Chi-squared test

	Short	Medium	Long
Induction course	13.5000 0.018519 14	15.500 3.6290 23	21.000 3.0476 13
Initial on-the-job	8.6400 1.3067 12	9.9200 0.85952 7	13.440 0.01440 13
Continuous	313.860 0.46766 28	36.580 0.57344 32	49.560 1.4373 58

Result

Chi-squared test

df 4

X^2 .... 11.354

p 0.022859

Figure 4.5

Use the output to examine at the 5% level of significance whether these data provide evidence of association between length of service and type of training, stating clearly your null and alternative hypotheses.

Discuss your conclusions.

⑩ In the initial stages of a market research exercise to investigate whether a proposed advertising campaign would be worthwhile, a survey of newspaper readership was undertaken. 100 people selected at random from the target population were interviewed. They were asked how many newspapers they read regularly. They were also classified as to whether they lived in urban or rural areas. The results were as follows.

Table 4.15

Number of newspapers read regularly	Urban	Rural
None	15	11
One	22	18
More than one	27	7

The output from a statistical package for these data is shown below.

Chi-squared test

	Urban	Rural
None	16.64 0.1616 15	9.36 0.2874 11
One	25.6 0.5063 22	14.4 0.9 18
More than one	21.76 1.2618 27	12.24 2.2433 7

Result

Chi-squared test

df 2

X^2 .... 5.3603

p 0.0686

Figure 4.6

(i) Use the output to examine at the 10% level of significance whether these data provide evidence of an association between the categories. State clearly the null and alternative hypotheses you are testing.

(ii) Justify the degrees of freedom for the test.

(iii) Discuss the conclusions reached from the test.

⑪ Public health officers are monitoring air quality over a large area. Air quality measurements using mobile instruments are made frequently by officers touring the area. The air quality is classified as poor, reasonable, good or excellent. The measurement sites are classified as being in residential areas, industrial areas, commercial areas or rural areas. The table shows a sample of frequencies over an extended period. The row and column totals and the grand total are also shown.

Table 4.16

Measurement site	Air quality				Row totals
	Poor	Reasonable	Good	Excellent	
Residential	107	177	94	22	400
Industrial	87	128	74	19	308
Commercial	133	228	148	51	560
Rural	21	71	24	16	132
Column totals	348	604	340	108	1400

Examine at the 5% level of significance whether or not there is any association between measurement site and air quality, stating carefully the null and alternative hypotheses you are testing. Report briefly on your conclusions.

⑫ The Director of Studies at a large college believed that students' grades in Mathematics were independent of their grades in English. She examined the results of a random group of candidates who had studied both subjects and she recorded the number of candidates in each of the 6 categories shown.

Table 4.17

	Maths grade A or B	Maths grade C or D	Maths grade E or U
English grade A or B	25	25	10
English grade C to U	15	30	15

(i) Stating your hypotheses clearly, test the Director's belief using a 10% level of significance. You must show each step of your working.

The Head of English suggested that the Director was losing accuracy by combining the English grades C to U in one row. He suggested that the Director should split the English grades into two rows, grades C or D and grades E or U as for Mathematics.

(ii) State why this might lead to problems in performing the test.

[Edexcel 6691 Q2 June 2007]

⑬ The bank manager at a large branch was investigating the incidence of bad debts. Many loans had been made during the past year; the manager inspected the records of a random sample of 100 loans, and broadly classified them as satisfactory or unsatisfactory loans and as having been made to private individuals, small businesses or large businesses. The data were as follows.

Table 4.18

	Satisfactory	Unsatisfactory
Private individual	22	5
Small business	34	11
Large business	21	3

(i) Discuss any problems which could occur in carrying out a χ^2 test to examine if there is any association between whether or not the loan was satisfactory and the type of customer to whom the loan was made.

(ii) State suitable null and alternative hypotheses for the test described in part (i).

(iii) Carry out a test at the 5% level of significance without combining any groups.

(iv) Explain which groups it might be best to combine and carry out the test again with these groups combined.

⑭ A survey of a random sample of 44 people is carried out. Their musical preferences are categorised as pop, classical or jazz. Their ages are categorised as under 20, 20 to 39, 40 to 59 and 60 or over. A test is to be carried out to examine whether there is any association between musical preference and age group. The results are as follows.

Table 4.19

		Musical preference		
		Pop	Classical	Jazz
Age group	Under 20	8	4	1
	20–39	3	3	0
	40–59	2	4	3
	60 or over	1	7	8

(i) Calculate the expected frequencies for 'Under 20' and '60 or over' for pop music.

(ii) Explain why the test would not be valid using these four age categories.

(iii) State which categories it would be best to combine in order to carry out the test.

(iv) Using this combination, carry out the test at the 5% significance level.

(v) Discuss briefly how musical preferences vary between the combined age groups, as shown by the contributions to the test statistic.

[MEI adapted]

2 Goodness of fit tests

The χ^2 test is commonly used to see if a proposed model fits observed data.

Goodness of fit test for the uniform distribution

Discussion point
Why is it the model that should fit the data and not the other way round?

Discussion point
Do you think that Roald Drysdale actually used biased dice to fool people?

THE AVONFORD STAR

The extraordinary world of Roald Drysdale and the dice that Roald rolled

Roald Drysdale claims to be able to influence the world around him by mind alone.

One demonstration of this is his influence over throwing dice. 'You must realise that objects we take to be inanimate do in fact have a spirit,' he explained. 'Take these dice. At the moment their spirits are willing them to produce a certain result. I can't change that, but by concentrating my thought field on the dice, I can enhance their spirits and help them to produce the outcome that they are seeking.'

The power of the mind – Roald Drysdale claims to have psychic influence

One evening the spirits of the dice were clearly willing them to show 1, as you can see from these remarkable results from 120 throws.

Table 4.20

Score	6	5	4	3	2	1
Frequency	12	16	15	23	24	30

Do these figures provide evidence that Roald had influenced the outcomes, or is this just the level of variation you would expect to occur naturally? Clearly, a formal statistical test is required.

If Roald's claim had been that he could make a particular number, say 6, turn up more often than the other numbers, you could use a binomial test.

However, all he said was that he could 'enhance the spirits of the dice' by making some number come up more than the others. So all six outcomes are involved and a different test is needed.

The expected distribution of the results, based on the null hypothesis that the outcomes are not biased, is easily obtained. The probability of each outcome is $\frac{1}{6}$ and so the expectation for each number $120 \times \frac{1}{6} = 20$.

Table 4.21

Outcome	1	2	3	4	5	6
Expected frequency, f_e	20	20	20	20	20	20

You would not, however, expect exactly this result from 120 throws. Indeed, you would be very suspicious if somebody claimed to have obtained it, and might well disbelieve it. You expect random variation to produce small differences in the frequencies. The question is whether the quite large differences in Roald's case can be explained in this way or not.

Goodness of fit tests

When this is written in the formal language of statistical tests, it becomes:

$H_0: p = \frac{1}{6}$ for each outcome.

$H_1: p \neq \frac{1}{6}$ for each outcome.

> Notice that these are the same hypotheses that you would use for a test on whether the dice are biased.

The expected frequencies are denoted by f_e and the observed frequencies by f_o. To measure how far the observed data are from the expected, you clearly need to consider the difference between the observed frequencies, f_o, and the expected, f_e. The measure which is used as a test statistic for this is denoted by X^2 and given by:

$$X^2 = \sum_{\text{All classes}} \frac{(f_o - f_e)^2}{f_e}$$

You have already met this statistic when carrying out tests involving contingency tables earlier in this chapter. In this case, the calculation of X^2 is as follows.

Table 4.22

Outcome	6	5	4	3	2	1
Observed frequency, f_o	12	16	15	23	24	30
Expected frequency, f_e	20	20	20	20	20	20
Difference, $f_o - f_e$	−8	−4	−5	3	4	10
$(f_o - f_e)^2$	64	16	25	9	16	100
$\dfrac{(f_o - f_e)^2}{f_e}$	3.2	0.8	1.25	0.45	0.8	5

$X^2 = 3.2 + 0.8 + 1.25 + 0.45 + 0.8 + 5 = 11.5$

The test statistic X^2 has the χ^2 (chi-squared) distribution. Critical values for this distribution are given in tables but, before you can use them, you have to think about two more points.

What is to be the significance level of the test?

This should really have been set before any data were collected. Because many people will be sceptical about Roald's claim, it would seem advisable to make the test rather strict, and so the 1% significance level is chosen.

How many degrees of freedom are involved?

As mentioned in the section on contingency tables, the shape of the χ^2 distribution curve depends on the number of free variables involved, the degrees of freedom, v. To find the value for v you start off with the number of cells which must be filled and then subtract one degree of freedom for each restriction, derived from the data, which is placed on the frequencies.

In this case, there are six classes (corresponding to scores of 1, 2, 3, 4, 5 and 6) but since the total number of throws is fixed (120) the frequency in the last class can be worked out if you know those of the first five classes.

$$v \quad = \quad 6 \quad - \quad 1$$

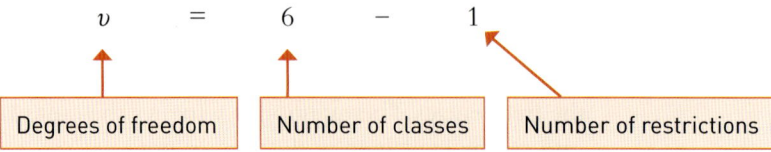

Degrees of freedom | Number of classes | Number of restrictions

Looking in the tables for the 1% significance level and $v = 5$ gives a critical value of 15.09; see Figure 4.7.

Since $11.5 < 15.09$, H_0 is accepted.

There is no reason at this significance level to believe that any number was any more likely to come up than any other. Roald's powers are not proved.

> **Note**
>
> As with the test for a contingency table, the expected frequency of any class must be at least five. If a class has an expected frequency of less than five, then it must be grouped together with one or more other classes until their combined expected frequency is at least five.
>
> When a particular distribution is fitted to the data, it may be necessary to estimate one or more parameters of the distribution. This, together with the restriction on the total, will reduce the number of degrees of freedom:
>
> v = number of classes – number of estimated parameters – 1

p%	99	97.5	95	90	10	5.0	2.5	1.0	0.5
$v = 1$	.0001	.0010	.0039	.0158	2.706	3.841	5.024	6.635	7.879
2	.0201	.0506	0.103	0.211	4.605	5.991	7.378	9.210	10.60
3	0.115	0.216	0.352	0.584	6.521	7.815	9.348	11.34	12.84
4	0.297	0.484	0.711	1.064	7.779	9.488	11.14	13.28	14.86
5	0.554	0.831	1.145	1.610	9.236	11.07	12.83	15.09	16.75
6	0.872	1.237	1.635	2.204	10.64	12.59	14.45	16.81	18.55
7	1.239	1.690	2.167	2.833	12.02	14.07	16.01	18.48	20.28
8	1.646	2.180	2.733	3.490	13.36	15.51	17.53	20.09	21.95
9	2.088	2.700	3.325	4.168	14.68	16.92	19.02	21.67	23.59

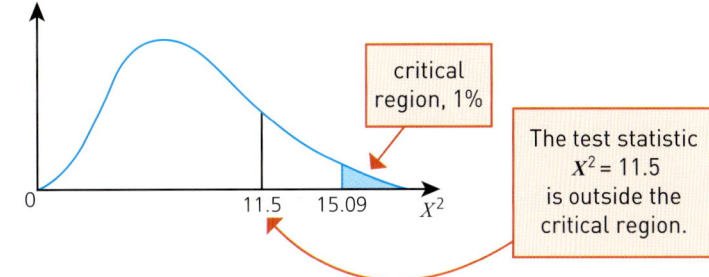

Figure 4.7

> **Note**
>
> This is a one-tailed test with only the right-hand tail under consideration. The interpretation of the left-hand tail (where the agreement seems to be too good) is discussed later in the chapter.

Goodness of fit test for the Poisson distribution

Example 4.2

The number of telephone calls made to a counselling service is thought to be modelled by the Poisson distribution. Data are collected on the number of calls received during one-hour periods, as shown in the table. Use these data to test at the 5% significance level whether a Poisson model is appropriate.

Table 4.23

No. of calls per hour	0	1	2	3	4	5	6	Total
Frequency	6	13	26	14	7	4	0	70

Solution

H_0: the number of calls can be modelled by the Poisson distribution.
H_1: the number of calls cannot be modelled by the Poisson distribution.

Nothing is known about the form of the Poisson distribution, so the data must be used to estimate the Poisson parameter.

From the data, the mean number of calls per hour is

$$\frac{0 \times 6 + 1 \times 13 + 2 \times 26 + 3 \times 14 + 4 \times 7 + 5 \times 4}{70} = \frac{155}{70} = 2.214$$

The Poisson distribution with parameter $\lambda = 2.214$ is as follows.

Table 4.24

x	$P(X = x)$	$70 \times P(X = x)$	Expected frequency
0	$e^{-2.214}$	70×0.1093	7.65
1	$P(X = x) \times \frac{2.214}{1}$	70×0.2419	16.93
2	$P(X = x) \times \frac{2.214}{2}$	70×0.2678	18.74
3	$P(X = x) \times \frac{2.214}{3}$	70×0.1976	13.84
4	$P(X = x) \times \frac{2.214}{4}$	70×0.1094	7.66
5	$P(X = x) \times \frac{2.214}{5}$	70×0.0484	3.39
≥ 6	$1 - P(X < 6)$	70×0.0256	1.79

> The expected frequencies are not rounded to the nearest whole number. To do so would invalidate the test. Expected frequencies do not need to be integers.

> The expected frequency for the last class is worked out as $1 - P(X < 6)$ and not as $P(X = 6)$, which would have cut off the right-hand tail of the distribution. The classes need to cover all **possible** outcomes, not just those that occurred in your survey.

The expected frequencies for the last two classes are both less than 5 but if they are put together to give an expected value of 5.2, the problem is overcome.

The table for calculating the test statistic is shown below.

Table 4.25

No. of calls, X	0	1	2	3	4	5+	Total
Observed frequency, f_o	6	13	26	14	7	4	70
Expected frequency, f_e	7.65	16.93	18.74	13.84	7.66	5.18	
$(f_o - f_e)$	−1.65	−3.93	7.26	0.16	−0.66	−1.18	
$\frac{(f_o - f_e)^2}{f_e}$	0.3544	0.9126	2.8080	0.0020	0.0567	0.2702	4.440

$X^2 = 0.3544 + 0.9126 + 2.8080 + 0.0020 + 0.0567 + 0.2702 = 4.4040$

The degrees of freedom are

v = number of classes − number of estimated parameters − 1

Table 4.26

No. of calls, X	0	1	2	3	4	5+	Total
Observed frequency, f_o	6	13	26	14	7	4	70

$v \quad = \quad 6 \quad - \quad 1 \quad - \quad 1 \quad = 4$

- The number of classes is 6 because 2 of the original 7 classes have been combined.
- λ was estimated as 2.214, one restriction.
- The total frequency (70) is one restriction.

From the tables, the critical value for a significance level of 5% and 4 degrees of freedom is 9.488.

The calculated test statistic, $X^2 = 4.404$. Since $4.404 < 9.488$, H_0 is accepted.

The data are consistent with a Poisson distribution for the number of calls.

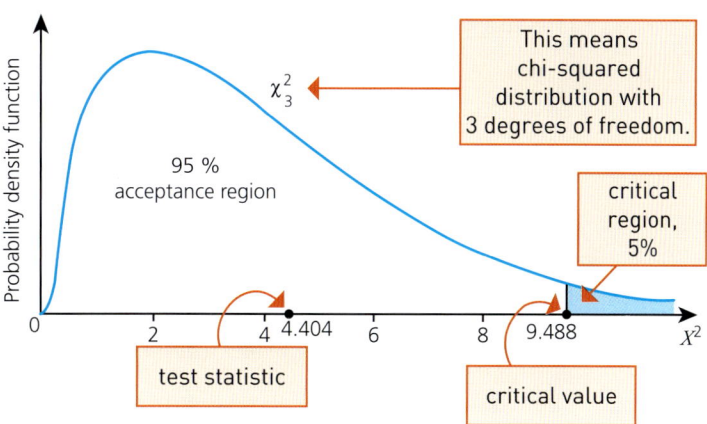

Figure 4.8

S1 AS Goodness of fit tests

> ### 💻 USING ICT
>
> #### A spreadsheet
>
> You can also use a spreadsheet to carry out the steps for a goodness of fit test for the Poisson model. To set it up, you would need to take the following steps.
>
> - Enter the variable categories and the observed frequencies. In this example they are:
>
> **Table 4.27**
>
> | No. of calls, X | 0 | 1 | 2 | 3 | 4 | 5+ |
> |---|---|---|---|---|---|---|
> | Observed frequency, f_o | 6 | 13 | 26 | 14 | 7 | 4 |
>
> - Find the mean value. *(In this case, the mean works out to be 2.21429.)*
> - Calculate the Poisson probabilities and use them to work out the expected frequencies. *(This includes the final class which is open-ended.)*
> - Combine classes as necessary if any expected frequencies are below 5.
> - Calculate the individual contributions to the test statistic.
> - Find the sum of the contributions.
> - Find the degrees of freedom.
> - Find the p-value using the formula provided with the spreadsheet, for example = CHISQ.DIST.RT.
>
> A typical display for this example is shown below.
>
> | | A | B | C | D | E | F | G | H |
> |---|---|---|---|---|---|---|---|---|
> | 1 | No. of calls. | 0 | 1 | 2 | 3 | 4 | 5 | >=6 |
> | 2 | Observed | 6 | 13 | 26 | 14 | 7 | 4 | 0 |
> | 3 | Expected | 7.646 | 16.931 | 18.745 | 13.836 | 7.659 | 3.392 | 1.792 |
> | 4 | | | | | | | | |
> | 5 | Mean | 2.21429 | | | | | | |
> | 6 | | | | | | | | |
> | 7 | No. of calls. | 0 | 1 | 2 | 3 | 4 | >=5 | Sum |
> | 8 | Observed | 6 | 13 | 26 | 14 | 7 | 4 | 70 |
> | 9 | Expected | 7.65 | 16.93 | 18.74 | 13.84 | 7.66 | 5.18 | 70.00 |
> | 10 | Contribution | 0.354423 | 0.912643 | 2.808036 | 0.001955 | 0.056694 | 0.270227 | 4.403979 |
> | 11 | d of freedom | 4 | | | | | | |
> | 12 | p-value | 0.3541 | | | | | | |
> | 13 | | | | | | | | |
>
> **Figure 4.9**
>
> You can adapt these steps to use a spreadsheet to test for goodness of fit of other distributions.

Goodness of fit test for the binomial distribution

Example 4.3

An egg packaging firm has introduced a new box for its eggs. Each box holds six eggs. Unfortunately, it finds that the new box tends to mark the eggs. Data on the number of eggs marked in 100 boxes are collected.

Table 4.28

No. of marked eggs	0	1	2	3	4	5	6	Total
No. of boxes, f_o	3	3	27	29	10	7	21	100

It is thought that the distribution may be modelled by the binomial distribution. Carry out a test on the data at the 0.5% significance level to determine whether the data can be modelled by the binomial distribution.

Solution

H_0: The number of marked eggs can be modelled by the binomial distribution.

H_1: The number of marked eggs cannot be modelled by the binomial distribution.

The binomial distribution has two parameters, n and p. The parameter n is clearly 6, but p is not known and so must be estimated from the data.

From the data, the mean number of marked eggs per box is

$$\frac{0\times 3+1\times 3+2\times 27+3\times 29+4\times 10+5\times 7+6\times 21}{100} = 3.45$$

Since the population mean is np you may estimate p by putting $6p = 3.45$

estimated $p = 0.575$ and estimated $q = 1 - p = 0.425$.

These parameters are now used to calculate the expected frequencies of $0, 1, 2, \ldots, 6$ marked eggs per box in 100 boxes.

Table 4.29

x	$P(X = x)$		Expected frequency, f_e $100 \times P(X = x)$
0	0.425^6	0.0059	0.59
1	$6 \times 0.575^1 \times 0.425^5$	0.0478	4.78
2	$15 \times 0.575^2 \times 0.425^4$	0.1618	16.18
3	$20 \times 0.575^3 \times 0.425^3$	0.2919	29.19
4	$15 \times 0.575^4 \times 0.425^2$	0.2962	29.62
5	$6 \times 0.575^5 \times 0.425^1$	0.1603	16.03
6	0.575^6	0.0361	3.61

In this case, there are three classes with an expected frequency of less than 5. The class for $x = 0$ is combined with the class for $x = 1$, bringing the expected frequency just over 5, and the class for $x = 6$ is combined with that for $x = 5$.

Table 4.30

No. of marked eggs, x	0, 1	2	3	4	5, 6	Total
Observed frequency, f_o	6	27	29	10	28	100
Expected frequency, f_e	5.37	16.18	29.19	29.62	19.64	100
$(f_o - f_e)$	0.68	10.82	−0.19	−19.62	8.36	
$\dfrac{(f_o - f_e)^2}{f_e}$	0.07	7.24	0.00	13.00	3.56	

The test statistic, $X^2 = 0.07 + 7.24 + 0.00 + 13.00 + 3.56 = 23.87$

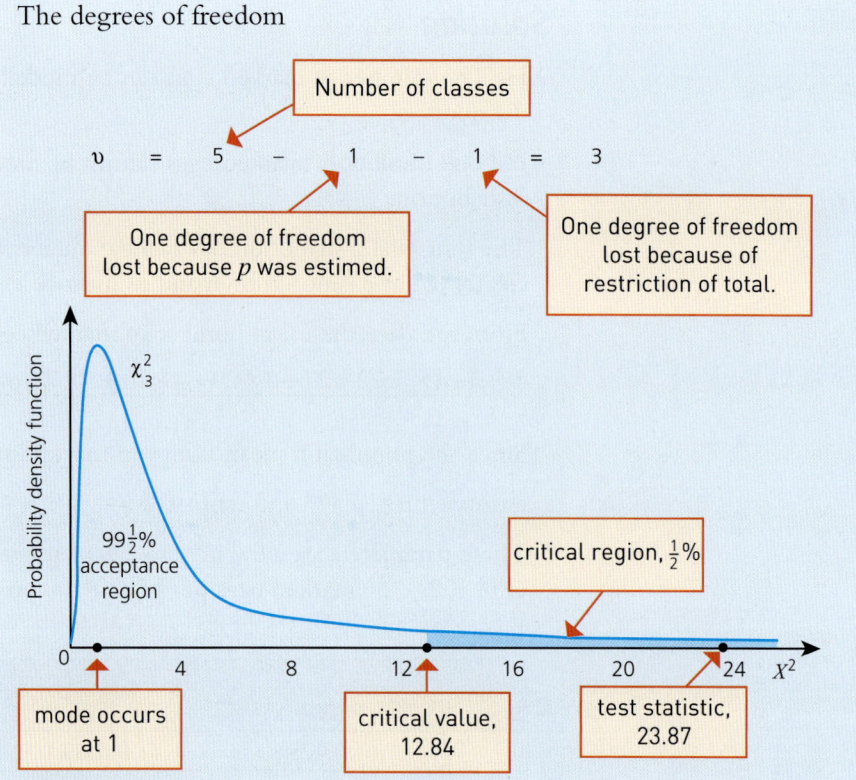

Figure 4.10

From the tables, for the 0.5% significance level and $v = 3$, the critical value of χ_3^2 is 12.84.

Since $23.87 > 12.84$, H_0 is rejected.

The data indicate that the binomial distribution is not an appropriate model for the number of marked eggs. If you look at the distribution of the data, as illustrated in Figure 4.11, you can easily see why, since it is bimodal.

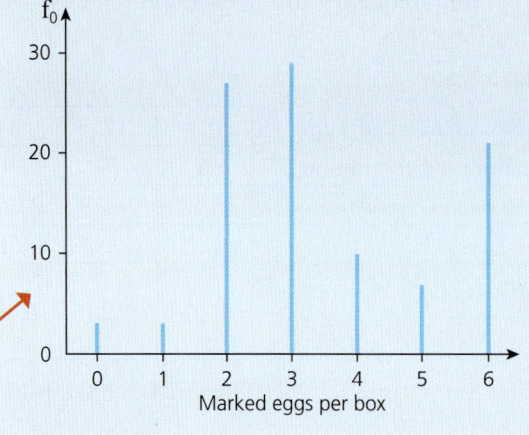

You can see that the distribution has two peaks. They are not both exactly the same height, but the second peak is sufficiently high to suggest bimodality.

Figure 4.11

Example 4.4

It is generally believed that a particular genetic defect is carried by 10% of people. A new and simple test becomes available to determine whether somebody is a carrier of this defect, using a blood specimen. As part of a research project, 100 hospitals are asked to carry out this test anonymously on the next 30 blood samples they take. The results are as follows.

Table 4.31

Number of positive tests	0	1	2	3	4	5	6	7+
Frequency, f_o	11	29	26	20	9	3	1	1

Do these figures support the model that 10% of people carry this defect, independently of any other condition, at the 5% significance level?

Solution

H_0: the model that 10% of people carry this defect is appropriate.

H_1: the model that 10% of people carry this defect is not appropriate.

The expected frequencies may be found using the binomial distribution $B(30, 0.1)$.

Table 4.32

No. of positive tests	0	1	2	3	4	5	6	7+
Expected frequency, f_e	4.24	14.13	22.77	23.61	17.71	10.23	4.74	2.58

Since the expected frequencies for 6 and 7+ positive tests are less than 5, you need to combine these two classes.

The calculation then proceeds as follows.

Table 4.33

No. of positive tests	0, 1	2	3	4	5	6+	Total
Observed frequency, f_o	40	26	20	9	3	2	100
Expected frequency, f_e	18.37	22.77	23.61	17.71	10.23	7.32	100.01
$(f_o - f_e)$	21.63	3.23	−3.61	−8.71	−7.23	−5.32	
$\dfrac{(f_o - f_e)^2}{f_e}$	25.47	0.46	0.55	4.28	5.11	3.87	

The test statistic $X^2 = 25.47 + 0.46 + 0.55 + 4.28 + 5.11 + 3.87$

$\qquad\qquad\qquad = 39.74$

The degrees of freedom,

$\qquad v \qquad = \qquad 6 \qquad - \qquad 0 \qquad - \qquad 1 \qquad = 5$

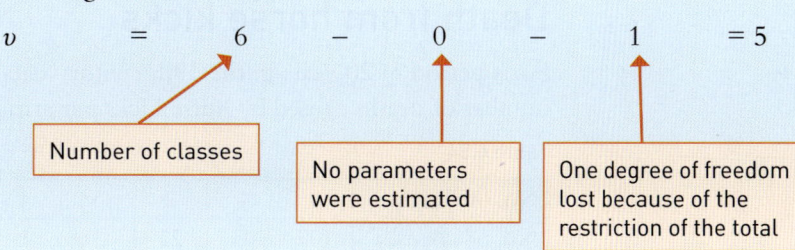

Number of classes

No parameters were estimated

One degree of freedom lost because of the restriction of the total

Goodness of fit tests

> **Note**
>
> 1. Although this example is like the previous one in that both used the binomial distribution as a model, the procedure is different. In this case, the given model included the information $p = 0.1$ and so you did not have to estimate the parameter p. Consequently, a degree of freedom was not lost from doing so.
>
> 2. The value of X^2 was very large in comparison with the critical value. What went wrong with the model?
>
> You will find that if you use the data to estimate p, it does not work out to be 0.1 but a little under 0.07. Fewer people are carriers of the defect than was believed to be the case. If you work through the example again with the model $p = 0.07$, you will find that the fit is good enough for you to start looking at the left-hand tail of the distribution.

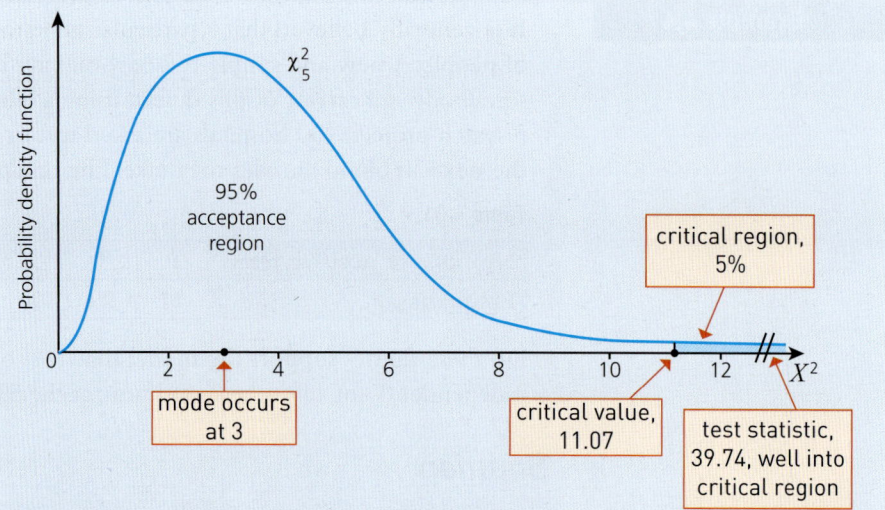

Figure 4.12

From the tables, for 5% significance level and $\nu = 5$, the critical value of χ_5^2 is 11.07. Since 39.74 > 11.07, H_0 is rejected.

The data indicate that the binomial distribution with $p = 0.1$ is not an appropriate model.

> **ACTIVITY 4.1**
>
> Carry out a goodness of fit test for the data in Example 4.4 using the p-value estimated from the data.

The left-hand tail

The χ^2 test is conducted as a one-tailed test, looking to see if the test statistic gives a value to the right of the critical value, as in the previous examples.

However, examination of the left-hand tail also gives information. In any modelling situation you would expect there to be some variability. Even when using the binomial to model a clear binomial situation, like the number of heads obtained in throwing a coin a large number of times, you would be very surprised if the observed and expected frequencies were identical. The left-hand tail may lead you to wonder whether the fit is too good to be credible. The following is a very famous example of the Poisson distribution.

Death from horse kicks

For a period of 20 years in the 19th century data were collected of the annual number of deaths caused by horse kicks per army corps in the Prussian army.

Table 4.34

No. of deaths	0	1	2	3	4
No. of corps, f_o	109	65	22	3	1

These data give a mean of 0.61. The variance of 0.6079 is almost the same, suggesting that the Poisson model may be appropriate.

The Poisson distribution (0.61) gives these figures.

Table 4.35

No. of deaths	0	1	2	3	4 or more
No. of corps, f_e	108.7	66.3	20.2	4.1	0.7

This looks so close that there seems little point in using a test to see if the data will fit the distribution. However, proceeding to the test, and remembering to combine classes to give expected values of at least 5, the null and alternative hypotheses are as follows.

Table 4.36

No. of deaths	0	1	2 or more
Observed frequency, f_o	109	65	26
Expected frequency, f_e	108.7	66.3	25
$(f_o - f_e)$	0.3	−1.3	1
$\dfrac{(f_o - f_e)^2}{f_e}$	0.001	0.025	0.040

H_0: Deaths from horse kick can be modelled by a Poisson distribution.

H_1: Deaths from horse kick cannot be modelled by a Poisson distribution.

The test statistic $X^2 = 0.001 + 0.025 + 0.040$

$= 0.066$

The degrees of freedom are given by

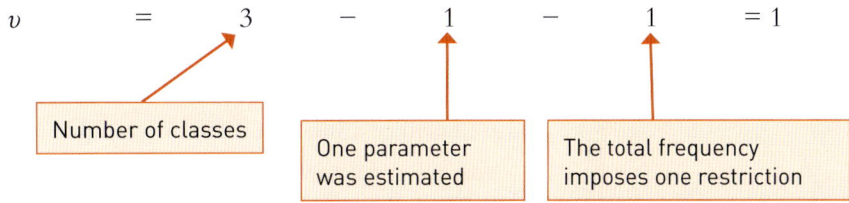

$\upsilon \;=\; 3 \;-\; 1 \;-\; 1 \;=\; 1$

- Number of classes
- One parameter was estimated
- The total frequency imposes one restriction

> **Discussion point**
> Looking at the expected values of the Poisson distribution in comparison with the observed values suggests that the fit is very good indeed. Is it perhaps suspiciously good? Might the data have been fixed?

The critical value of χ^2 for 1 degree of freedom at the 5% significance level is 3.841.

As $0.066 < 3.841$ it is clear that the null hypothesis, that the Poisson distribution is an appropriate model, should be accepted.

The tables relating to the left-hand tail of the χ^2 distribution give critical values for this situation. For example, the value for 95% significance level for $\upsilon = 1$ is 0.0039. This means that if the null hypothesis is true, you would expect a value for X^2 less than 0.0039 from no more than 5% of samples.

In this case, the test statistic is $X^2 = 0.066$. This is greater than the 95% critical value, and so you can conclude that a fit as good as this will occur with more than 1 sample in 20. That may well help allay your suspicions.

If your test statistic does lie within the left-hand critical region, you should check the data to ensure that the figures are genuine and that all the procedures have been carried out properly. There are three situations you should particularly watch out for.

Goodness of fit tests

- The model was constructed to fit a set of data. It is then being tested by seeing how well it fits the same data. Once the model is determined, new data should be used to test it.
- Some of the data have been omitted in order to produce a better fit.
- The data are not genuine.

Although looking at the left-hand tail of the χ^2 distribution may make you suspicious of the quality of the data, it does not provide a formal hypothesis test that the data are not genuine. Thus the term **critical region** is not really appropriate to this tail: **warning region** would be better.

The χ^2 test is a **distribution-free test**; this means that there are no modelling assumptions associated with the test itself. This has the advantage that the test can be widely used, but, on the other hand, it is not a very sensitive test. The test for rank correlation which you will meet in Chapter 5 is another example of a distribution-free test: one can test for rank correlation without knowing anything about the underlying bivariate distribution, but one frequently finds there is no significant evidence of correlation.

> ❗ If you obtain a *p*-value for a goodness of fit test, you need to take a little care interpreting it. The lower the *p*-value, the less likely it is that the sample could have been drawn from a distribution for which the null hypothesis is true; conversely, the higher the *p*-value, the more likely it is that the sample could have been drawn from such a distribution.
>
> Quite often you will want to show that a certain distribution is appropriate and so you will associate a high *p*-value with a successful outcome. By contrast, in many other hypothesis tests, you are hoping to show that the null hypothesis is false, and so something special has happened; in such cases, a low *p*-value can be regarded as a successful outcome.

Exercise 4.2

1. Find the expected frequencies of $0, 1, 2, 3, \geqslant 4$ successes if 80 observations are taken of a binomial random variable with $n = 20$ and $p = 0.09$.

2. You are given that $X \sim \text{Poisson}(1.6)$. Find the expected frequencies of $x = 0, 1, 2, 3, 4, \geqslant 5$ if 60 observations of X are taken.

3. Part of a large simulation study requires the provision of many simulated observations that can be taken as coming from the Poisson distribution with parameter 2. There is a suspicion that this part of the simulation is not working properly.

    Two hundred of the simulated observations are recorded in the form of a frequency table as follows.

    Table 4.37

Simulated observations	0	1	2	3	4	$\geqslant 5$
Frequency	20	40	52	49	27	12

    Carry out a χ^2 test, at the 5% level of significance, to examine whether the required Poisson distribution can be assumed for the simulated observations. [MEI adapted]

4. A typist makes mistakes from time to time in a 200-page book. The number of pages with mistakes are as follows.

Table 4.38

Mistakes	0	1	2	3	4	5
Pages	18	62	84	30	5	1

(i) Test at the 5% significance level whether the Poisson distribution is an appropriate model for these data.

(ii) What factors would make it other than a Poisson distribution?

⑤ A biologist crosses two pure varieties of plant, one with pink flowers, the other white. The pink is dominant so that the flowers of the second generation should be in the ratio pink : white = 3 : 1.

They plant the seeds in batches of 5 in 32 trays and count the numbers of plants with pink and with white flowers in each tray.

Table 4.39

White flowers	0	1	2	3	4	5
Frequency	9	12	7	2	1	1

(i) What distribution would you expect for the number of plants with white flowers?

(ii) Use these figures to test at the 2.5% significance level whether the distribution is that which you expected.

⑥ The number of goals scored by a football team is recorded for 100 games. The results are summarised in the table below.

Table 4.40

Number of goals	Frequency
0	40
1	33
2	14
3	8
4	5

(i) Calculate the mean number of goals scored per game.

The manager claimed that the number of goals scored per match follows a Poisson distribution. He used the answer in part (i) to calculate the expected frequencies given in the following table.

Table 4.41

Number of goals	Expected frequency
0	34.994
1	r
2	s
3	6.752
≥ 4	2.221

(ii) Find the value of r and the value of s giving your answers to 3 decimal places.

(iii) Stating your hypotheses clearly, use a 5% level of significance to test the manager's claim.
[Edexcel 6691 Q5 June 2009]

Goodness of fit tests

7 An examination board is testing a multiple-choice question. They get 100 students to try the question and their answers are as follows.

Table 4.42

Choice	A	B	C	D	E
Frequency	32	18	10	28	12

Are there grounds, at the 10% significance level, for the view that the question was so hard that the students guessed the answers at random?

8 A student on a geography field trip has collected data on the size of rocks found on a scree slope. The student counts the number of large rocks (that is, heavier than a stated weight) found in a 2 m square at the top, middle and bottom of the slope.

Table 4.43

	Top	Middle	Bottom
Number of large rocks	5	10	18

(i) Test at the 5% significance level whether these data are consistent with the hypothesis that the size of rocks is distributed evenly on the scree slope.

(ii) What does your test tell the student about the theory that large rocks will migrate to the bottom of the slope?

9 A researcher is investigating the feeding habits of bees. She sets up a feeding station some distance from a beehive and, over a long period of time, records the numbers of bees arriving each minute. For a random sample of 100 one-minute intervals she obtains the following results.

Table 4.44

Number of bees	0	1	2	3	4	5	6	7	≥ 8
Number of intervals	6	16	19	18	17	14	6	4	0

(i) Show that the sample mean is 3.1 and find the sample variance. Do these values support the possibility of a Poisson model for the number of bees arriving each minute? Explain your answer.

(ii) Show that, if a Poisson model with the mean in part (i) fits the data, the expected frequency for two bees arriving is 21.65.

(iii) Use the spreadsheet below to complete a test of the goodness of fit of a Poisson model to the data.

	A	B	C	D	E	F	G	H	I	J
1	Number of bees	0	1	2	3	4	5	6	7	>=8
2	Number of intervals	6	16	19	18	17	14	6	4	0
3	Probability	0.0450	0.1397	0.2165	0.2237	0.1733	0.1075	0.0555	0.0246	
4	Expected frequency	4.505	13.97	21.65	22.37	17.33	10.75	5.553	2.459	
5										
6	Attempts	<=1	2	3	4	5	>=6	Sum		
7	Observed frequency	22	19	18	17	14	10	100		
8	Probability	0.1847	0.2165	0.2237	0.1733	0.1075	0.0943	1.0000		
9	Expected frequency	18.47	21.65	22.37	17.33	10.75	9.433	100		
10	Contribution	0.6746	0.3235	0.8529	0.0065	0.9842	0.0340	2.8756		
11										
12	Mean		3.1							
13	X^2 statistic		2.8756							
14	d of f		4							
15	p-value		0.5789							

Figure 4.13

[MEI adapted]

⑩ A university student working in a small seaside hotel in the summer holidays looks at the records for the previous holiday season of 30 weeks. She records the number of days in each week on which the hotel had to turn away visitors because it was full. The data she collects are as follows.

Table 4.45

Number of days visitors turned away	0	1	2	3	4	5+	Total
Number of weeks	11	13	4	1	1	0	30

(i) Calculate the mean and variance of the data.

(ii) The student thinks that these data can be modelled by the binomial distribution. Carry out a test at the 5% significance level to see if the binomial distribution is a suitable model.

(iii) What other distribution might be used to model these data? Give your reasons.

⑪ Morag is writing a book. Every so often she uses the spell check facility in her word processing software, and, for interest, records the number of mistakes she has made on each page. In the first 20 pages, the results were as follows.

Table 4.46

No. of mistakes/page	0	1	2	3	4+	Total
Frequency	9	6	4	1	0	20

(i) Explain why it is not possible to use the χ^2 test on these data to decide whether the occurrence of spelling mistakes may be modelled by the Poisson distribution.

In the next 30 pages Morag's figures are as follows.

Table 4.47

No. of mistakes/page	0	1	2	3	4+	Total
Frequency	14	7	7	0	2	30

(ii) Use the combined figures, covering the first 50 pages, to test whether the occurrence of Morag's spelling mistakes may be modelled by the Poisson distribution. Use the 5% significance level.

(iii) If the distribution really is Poisson, what does this tell you about the incidence of spelling mistakes? Do you think this is realistic?

⑫ In a survey of five towns, the population of the town and the number of petrol-filling stations were recorded as follows.

Table 4.48

Town	Population (to nearest 10 000)	Number of filling stations
A	4	22
B	3	16
C	7	35
D	6	27
E	12	60
Totals	32	160

An assistant researcher, who wanted to find out whether the petrol stations were evenly distributed between the towns, performed a χ^2 test on the number of filling stations, with a null hypothesis that there was no difference in the number of filling stations in each town. She found that her X^2 value was 36.69. Without repeating her calculation, state with reasons what her conclusion was.

⑬ A local council has records of the number of children and the number of households in its area. It is therefore known that the average number of children per household is 1.40. It is suggested that the number of children per household can be modelled by the Poisson distribution with parameter 1.40. In order to test this, a random sample of 1000 households is taken, giving the following data.

Table 4.49

Number of children	0	1	2	3	4	5+
Number of households	273	361	263	78	21	4

(i) Find the corresponding expected frequencies obtained from the Poisson distribution with parameter 1.40.

(ii) Carry out a χ^2 test, at the 5% level of significance, to determine whether or not the proposed model should be accepted. State clearly the null and alternative hypotheses being tested and the conclusion which is reached. [MEI]

⑭ A total of 100 random samples of 6 items are selected from a production line in a factory and the number of defective items in each sample is recorded. The results are summarised in the table below.

Table 4.50

Number of defective items	0	1	2	3	4	5	6
Number of samples	6	16	20	23	17	10	8

(i) Show that the mean number of defective items per sample is 2.91.

A factory manager suggests that the data can be modelled by a binomial distribution with $n = 6$. He uses the mean from the sample above and calculates expected frequencies as shown in the table below.

Table 4.51

Number of defective items	0	1	2	3	4	5	6
Expected frequency	1.87	10.54	24.82	a	22.01	8.29	b

(ii) Calculate the value of a and the value of b, giving your answers to 2 decimal places.

(iii) Test, at the 5% level, whether or not the binomial distribution is a suitable model for the number of defective items in samples of 6 items. State your hypotheses clearly. [Edexcel 6691 Q6 June 2012]

⑮ A research station is doing some work on the germination of a new variety of genetically modified wheat.

They planted 120 rows containing 7 seeds in each row.

The number of seeds germinating in each row was recorded. The results are as follows.

Table 4.52

Number of seeds germinating in each row	0	1	2	3	4	5	6	7
Observed number of rows	2	6	11	19	25	32	16	9

(i) Write down two reasons why a binomial distribution may be a suitable model.

(ii) Show that the probability of a randomly selected seed from this sample germinating is 0.6.

The research station used a binomial distribution with probability 0.6 of a seed germinating. The expected frequencies were calculated to 2 decimal places. The results are as follows.

Table 4.53

Number of seeds germinating in each row	0	1	2	3	4	5	6	7
Expected number of rows	0.20	2.06	s	23.22	t	31.35	15.68	3.36

(iii) Find the value of s and the value of t.

(iv) Stating your hypotheses clearly, test, at the 1% level of significance, whether or not the data can be modelled by a binomial distribution. [Edexcel 6691 Q5 June 2014]

⑯ A man accuses a casino of having two loaded dice. He throws them 360 times with the following outcomes for their sum at each throw.

Table 4.54

Total	2	3	4	5	6	7	8	9	10	11	12
Freq.	5	15	30	35	45	61	53	45	31	24	16

(i) State null and alternative hypotheses for a test to investigate whether the dice are fair.

(ii) Show that the expected frequency for totals of 2 and 5 are 10 and 40, respectively.

(iii) Does the man have grounds for his accusation at the 5% significance level?

⑰ The manager of a large supermarket has recently moved from one store to another. At the previous store, it was known from surveys that 42% of the customers lived within 5 miles of the store, 35% lived between 5 and 10 miles from the store and the remaining 23% lived more than 10 miles from the store. The manager wishes to test whether the same proportions apply at the new store.

One Saturday morning, the first 100 customers to enter the store after 11.00 am were asked how far from the store they lived. The results, grouped into the same categories as for the previous store, were as follows.

Table 4.55

Distance (miles)	0–5	5–10	more than 10
Number of customers	34	48	18

(i) Assuming that this is a random sample of all the customers at the store, test at the 5% level of significance whether the proportions at the manager's new store may be taken to be the same as those at the previous store. Discuss your conclusions briefly.

(ii) Discuss whether this is likely to be a random sample. [MEI]

18 Ten cuttings were taken from each of 100 randomly selected garden plants. The numbers of cuttings that did not grow were recorded.

The results are as follows.

Table 4.56

No. of cuttings which did not grow	0	1	2	3	4	5	6	7	8, 9 or 10
Frequency	11	21	30	20	12	3	2	1	0

(i) Show that the probability of a randomly selected cutting, from this sample, not growing is 0.223.

A gardener believes that a binomial distribution might provide a good model for the number of cuttings, out of 10, that do not grow.

He uses a binomial distribution, with the probability 0.2 of a cutting not growing. The calculated expected frequencies are as follows.

Table 4.57

No. of cuttings which did not grow	0	1	2	3	4	5 or more
Expected frequency	r	26.84	s	20.13	8.81	t

(ii) Find the values of r, s and t.

(iii) State clearly the hypotheses required to test whether or not this binomial distribution is a suitable model for these data.

The test statistic for the test is 4.17 and the number of degrees of freedom used is 4.

(iv) Explain fully why there are 4 degrees of freedom.

(v) Stating clearly the critical value used, carry out the test using a 5% level of significance. [Edexcel 6691 Q6 June 2008]

KEY POINTS

1 Contingency tables

To test whether the variables in an $m \times n$ contingency table are independent the steps are as follows.

(i) The null hypothesis is that the variables are independent, the alternative is that they are not.

(ii) Calculate the marginal (row and column) totals for the table.

(iii) Calculate the **expected frequency** in each cell.

(iv) The X^2 statistic is $\sum \frac{(f_o - f_e)}{f_e}$ where f_o is the observed frequency and f_e is the expected frequency in each cell.

(v) The **degrees of freedom**, v, for the test is $(m-1)(n-1)$ for an $m \times n$ table.

(vi) Read the critical value from the χ^2 tables (alternatively, use suitable software) for the appropriate degrees of freedom and significance level. If X^2 is less than the significance level, the null hypothesis is accepted; otherwise it is rejected.

(vii) If two variables are not independent, you say that there is an **association** between them.

2 Goodness of fit tests

To test whether a distribution models a situation, the steps are as follows.

(i) Select your model, binomial, Poisson, etc.

(ii) Set up null and alternative hypotheses and choose the significance level.

(iii) Collect data. Record the observed frequency for each outcome.

(iv) Calculate the expected frequencies arising from the model.

(v) Check that the expected frequencies are all at least 5. If not, combine classes.

(vi) Calculate the test statistic.

The X^2 statistic is $\sum \frac{(f_o - f_e)^2}{f_e}$.

(vii) Find the degrees of freedom, v, for the test using the formula.
v = number of classes – number of estimated parameters – 1
where the number of classes is counted **after** any necessary combining has been done.

(viii) Read the critical value from the χ^2 tables for the appropriate degrees of freedom and significance level. If X^2 is less than the significance level, the null hypothesis is accepted; otherwise it is rejected.

(ix) Draw conclusions from the test – state what the test tells you about the model.

LEARNING OUTCOMES

When you have completed this chapter you should be able to:

➤ interpret bivariate categorical data in a contingency table

➤ apply the χ^2 test to a contingency table

➤ carry out a χ^2 test for goodness of fit of a uniform, binomial, geometric or Poisson model

➤ interpret the results of a χ^2 test using tables of critical values or the output from software.

Practice questions: set 1

1. The Highways Authority proposes to impose new parking restrictions on a small town. The Town Council fear that it will be bad for trade, and so for the town's prosperity. They plan to object, but first they need data so that they can estimate the cost to the town.

   The council call a special meeting. Their room can only take 20 more people (in addition to the councillors) and they invite

       8 of the 41 shops

       7 of the 33 restaurants and cafes

       5 of the 27 hotels and bed and breakfasts.

   (i) Describe the sort of sample they have selected. Explain how they decided on the numbers from the various groups. [3 marks]

   (ii) They use a random selection procedure to describe who actually gets invited. Describe two possible ways they might do this. [2 marks]

   Table 1

Shops	2	1	5	0	12	10	8	3
Restaurants and cafes	1	2	2	1	3	2	1	-
Hotels and B&B	15	0	0	10	1	-	-	-

   (iii) Use these figures to estimate the total cost of the parking restrictions to the town. [3 marks]

   (iv) Comment on the likely accuracy of the estimate and suggest measures that might be taken to improve it. [2 marks]

2. In a game at a fairground stall, the player rolls 3 fair dice and counts the number of sixes obtained. This number is denoted by X. The amount the player wins, in pounds, is denoted by W. The values of X and the corresponding values of W are shown in Table 1.

   Table 2

Score, X	0	1	2	3
Amount won, W (£)	0	1	10	100

   (i) Find the probability distribution of W. Find The expected value and standard deviation of W. [6 marks]

   Players pay £2 to play the game. This money is always kept by the stall holder.

   (ii) Find the expected profit the stall holder makes when 50 people play the game. [2 marks]

3. A boy is watching vehicles travelling along a motorway. All the vehicles he counts are either cars or lorries. The number of each may be modelled by two independent Poisson distributions. The mean number of cars per minute is 8.3 and the mean number of lorries per minute is 4.7.

   (i) For a given period of one minute, find the probability that he sees

       (a) exactly seven cars [1 mark]

       (b) at least three lorries. [1 mark]

   (ii) Calculate the probability that he sees a total of exactly ten vehicles in a given one-minute period. [4 marks]

(iii) Find the probability that he observes fewer than eight vehicles in a given period of 30 seconds. [3 marks]

(4) A research project investigated whether there was any association between the 'birth order of students and the subject they chose to study at university.

Birth order divides students into four groups as follows.

- Only child (someone having no siblings)
- First born (the oldest of a set of siblings)
- Middle born (someone having both older and younger siblings)
- Last born (the youngest of a set of siblings)

The subjects that students study are classified as Science, Humanities and Other.

A random sample of students at a large university was surveyed and the results were summarised, as shown in the following screenshot of a spreadsheet.

	A	B	C	D	E
1		Science	Humanities	Other	Total
2	Only child	8	5	6	19
3	First born	35	54	61	150
4	Middle born	8	21	8	37
5	Last born	31	26	33	90
6	Total	82	106	108	296

Figure 3

(i) Explain what these figures suggest about the numbers of children per family in this sample. [2 marks]

(ii) What does a comparison of the figures in cells E3 and E5 suggest? [1 mark]

(iii) State the null and alternative hypotheses for a chi-squared test on these data. [1 mark]

The spreadsheet is used to calculate the expected frequencies and the contributions to the test statistic, as shown in the following screenshot.

	A	B	C	D
7				
8	Exp freq	Science	Humanities	Other
9	Only child	5.26	6.80	6.93
10	First born	41.55	53.72	54.73
12	Middle born	10.25	13.25	13.50
11	Last born	24.93	32.23	32.84
13				
14	Contribs	Science	Humanities	Other
15	Only child	1.42	0.48	0.13
16	First born	1.03	0.00	0.72
17	Middle born	0.49	4.53	2.24
18	Last born	0.48	1.20	0.00

Figure 4

(iv) Give spreadsheet formulae for cells B9 and B15. [2 marks]

(v) Given that the sum of the contributions to the test statistic is 13.73, complete the test using a 5% significance level. [3 marks]

(vi) Discuss briefly what the two largest contributions to the test statistic indicate. [2 marks]

S1 AS — Practice questions: set 1

(5) Todd is a driving test examiner. He tests 6 candidates each day, and he wants to know whether or not the binomial distribution provides a good model for numbers who pass per day.

The table shows, for a random sample of 100 days, the numbers of Todd's candidates who pass.

Table 5

Number who pass	0	1	2	3	4	5	6
Number of days	9	12	33	18	20	7	1

(i) Find the mean pass rate for Todd's candidates over these 100 days [2 marks]

(ii) Carry out an appropriate chi-squared test, using a 10% significance level, to investigate whether or not the binomial distribution provides a good model [8 marks]

(6) LCD display panels consist of a very large number, typically many millions, of pixels. A very small proportion of pixels will be faulty. Let X denote the number of faulty pixels on a newly manufactured LCD panel of a particular size and type.

(i) State two assumptions required to justify modelling X using a Poisson distribution. [2 marks]

You are now given that the required assumptions hold and that the average number of faulty pixels per panel at the time of manufacture is 1.2.

(ii) Find the probability that a newly manufactured panel has

(a) no faulty pixels,

(b) fewer than 4 faulty pixels. [2 marks]

Newly manufactured panels are checked before sale, and those with 4 or more faulty pixels are not sold. Let Y be the number of faulty pixels on a panel that is sold.

(iii) Show that $P(Y = 0) = 0.3117$ correct to 4 decimal places. [1 mark]

(iv) Copy and complete the table for the distribution of Y.

Table 6

r	0	1	2	3
$P(Y = r)$	0.3117			

[3 marks]

(v) Calculate the mean and standard deviation of Y. [4 marks]

5 Measuring correlation

It is now proved beyond doubt that smoking is one of the leading causes of statistics.

John Peers

Discussion point

1. Across the world, countries with high life expectancy tend to have low birth rates and vice-versa. How would you describe this in mathematical language? Why do you think this happens?
2. Suggest some factors that affect the proportion of new-born babies, in one country, who have all four of their grandparents still alive.

1 Product moment correlation

The data in this table refer to the 22 American mainland countries. They cover their population in millions, birth rate per 1000 people, life expectancy in years and mean GDP per capita in thousands of US$.

Table 5.1

Country	Population	Life expectancy	Birth rate	GDP per capita
Argentina	43.0	77.51	16.88	18.6
Belize	0.3	68.49	25.14	8.8
Bolivia	10.6	68.55	23.28	5.5
Brazil	202.7	73.28	14.72	12.1
Canada	34.8	81.67	10.29	43.1
Chile	17.4	78.44	13.97	19.1
Colombia	46.2	75.25	16.73	11.1
Costa Rica	4.8	78.23	16.08	12.9
Ecuador	15.7	76.36	18.87	10.6
El Salvador	6.1	74.18	16.79	7.5
Guatemala	14.6	71.74	25.46	5.3
Guyana	0.7	67.81	15.90	8.5
Honduras	8.6	70.91	23.66	4.8
Mexico	120.3	75.43	19.02	15.6
Nicaragua	5.8	72.72	18.41	4.5
Panama	3.6	78.3	18.61	16.5
Paraguay	6.7	76.8	16.66	6.8
Peru	3.0	73.23	18.57	11.1
Suriname	0.6	71.69	16.73	12.9
United States	318.9	79.56	13.42	52.8
Uruguay	3.3	76.81	13.18	16.6
Venezuela	28.9	74.39	19.42	13.6

This is an example of a multivariate data set. For each country, the values of four variables are given.

Multivariate analysis is an important part of statistics but is beyond the scope of this book apart for the special case of bivariate data, where just two variables are considered.

Bivariate data are usually displayed on a scatter diagram like Figure 5.1. In this, life expectancy is plotted on the horizontal axis and birth rate on the vertical axis.

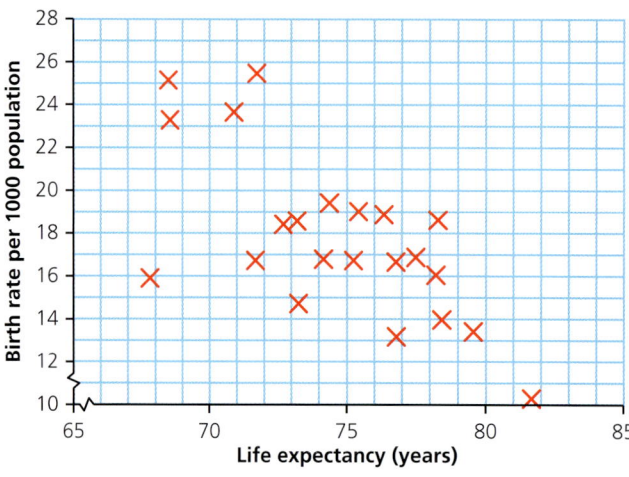

Figure 5.1

This diagram suggests that there is a relationship between birth rate and life expectancy. In general, such a relationship is called an **association.**

Sometimes, two special conditions apply.

- Both the variables are random.
- The relationship is linear.

> In such a case, the stronger the association, the closer the points on the scatter diagram will lie to a straight line.

Under these conditions the association is described as **correlation**. So correlation is a special case of association.

- If high values of both variables tend to occur together, and the same for low values, the correlation is **positive**.
- If, as in this example, high values of one variable are associated with low values of the other, the correlation is **negative**.

If the points on the scatter diagram lie exactly on a straight line, the correlation is described as **perfect**. However, it is much more common for the data to lie close to a straight line but not exactly on it, as in this case. The better the fit, the higher the level of correlation.

Dependent and independent variables

The scatter diagram in Figure 5.1 was drawn with the life expectancy on the horizontal axis and birth rate on the vertical axis. It could have been drawn the other way around. It is not obvious that either of the variables is dependent on the other.

By contrast, the weight of the passengers in an aeroplane is dependent on how many of them there are, but the reverse is not true; the number getting on is not determined by their weight. So, in this case, the number of passengers is described as the **independent variable** and their weight as the **dependent** variable. It is normal practice to plot the dependent variable on the vertical (y) axis and the independent variable on the horizontal (x) axis.

Here are some more examples of dependent and independent variables.

Independent variable	Dependent variable
The number of goals scored by a premier league football team in a season	The number of points the team has in the league table
The amount of rain falling on a field	The weight of a crop yielded while the crop is growing
The number of people visiting a bar	The volume of beer sold in an evening

Controlled variables

Sometimes, one or both of the variables is **controlled**, so that the variable only assumes a set of predetermined values; for example, the times at which temperature measurements are taken at a meteorological station. Controlled variables are **non-random**. Situations in which the independent variable is controlled and the dependent variable is random form the basis of **regression** analysis.

Interpreting scatter diagrams

You can often judge if correlation is present just by looking at a scatter diagram.

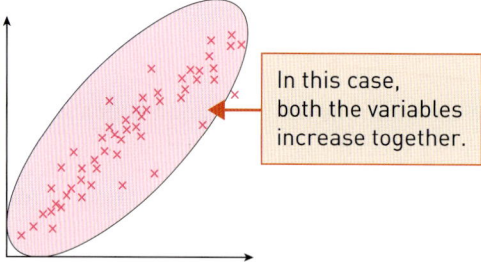

In this case, both the variables increase together.

Figure 5.2 Positive correlation

Notice that in Figure 5.2 almost all of the observation points can be contained within an ellipse. This shape often arises when both variables are random. You should look for it before going on to do a calculation of Pearson's product moment correlation coefficient (see page 111). The narrower the elliptical profile, the greater the correlation.

In this case, as one variable increases the other variable decreases.

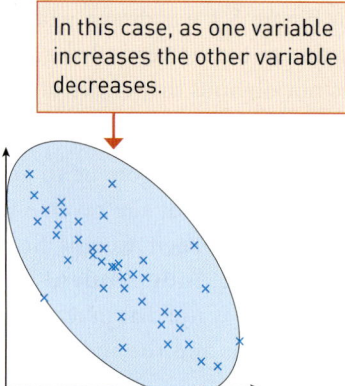

Figure 5.3 Negative correlation

In Figure 5.3 the points again fall into an elliptical profile and this time there is negative correlation. The fatter ellipse in this diagram indicates weaker correlation than in the case shown in Figure 5.2.

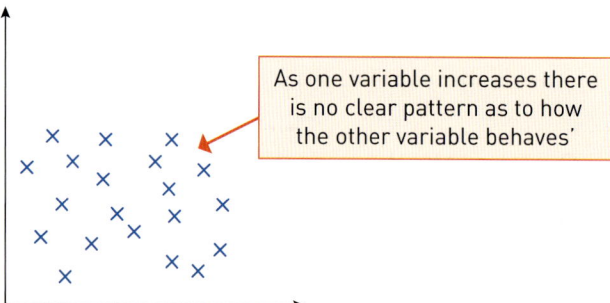

As one variable increases there is no clear pattern as to how the other variable behaves'

Figure 5.4 No correlation

In the case illustrated in Figure 5.4, the points fall randomly in the (x, y) plane and there appears to be no association between the variables.

> ⚠ You should be aware of some distributions which at first sight appear to indicate linear association, and so correlation, but in fact do not.

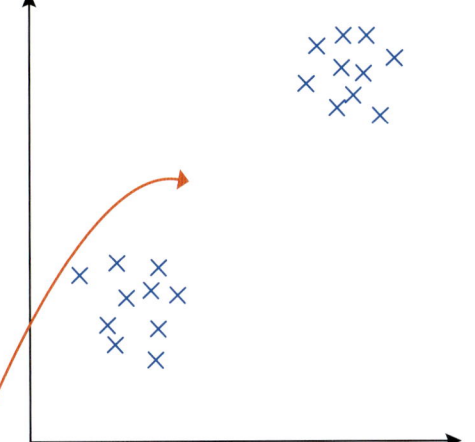

Figure 5.5 Two islands

> Notice that this distribution looks nothing like an ellipse.
> This scatter diagram is probably showing two quite different groups, neither of them having any correlation.

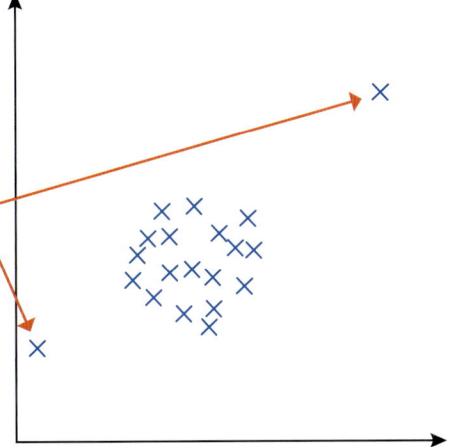

Figure 5.6 Outliers

> This is a small data set with no correlation.
> However, the two outliers give the impression that there is positive linear association, and so correlation.

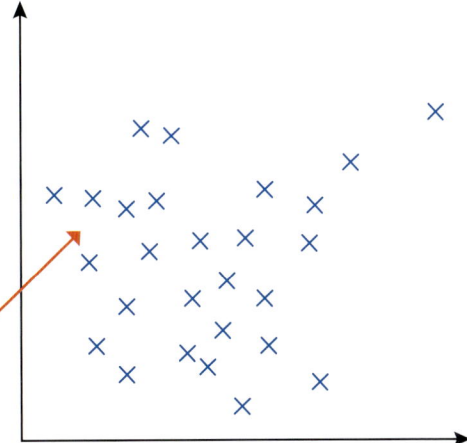

Figure 5.7 A funnel-shaped distribution

> At first sight you might think you could enclose these points with an ellipse but the more you look at it the fatter the ellipse would need to be.
> The point is that if you select just a few of the points on the scatter diagram you can obtain a false impression.

Product moment correlation

Discussion points

→ Here are the scatter diagrams for some of the other pairs of variables in the data for the American countries.

→ Comment on any insights the diagrams give you and the presence of outliers. Say whether they show any correlation or association.

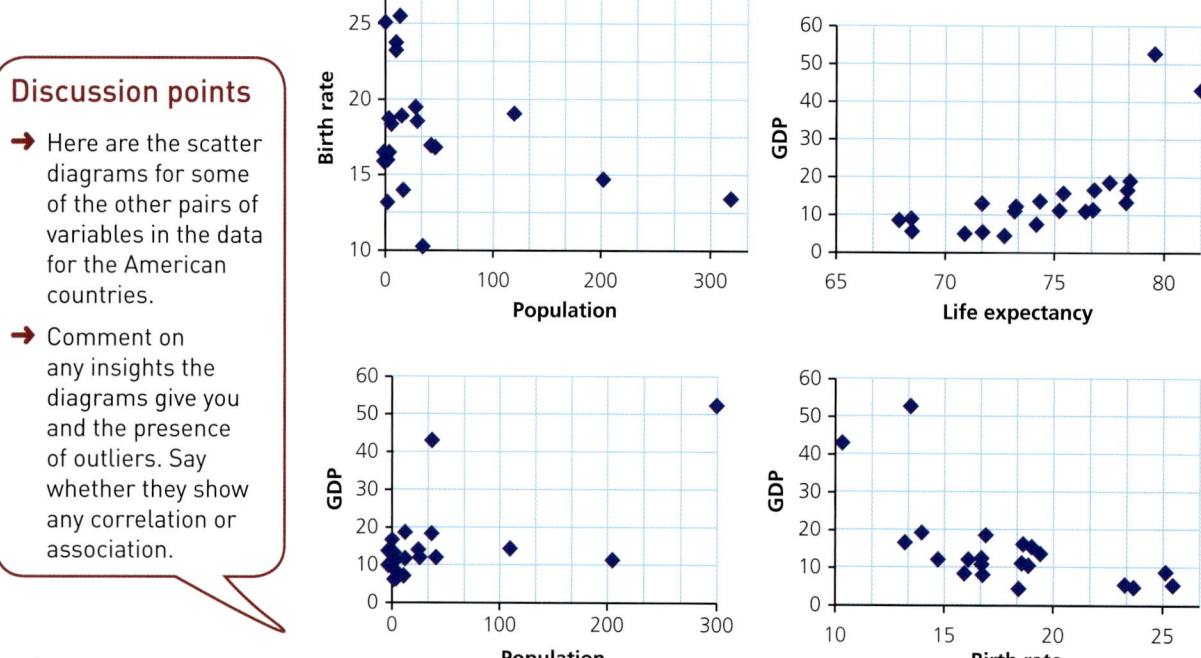

Figure 5.8

Correlation

The scatter diagram in Figure 5.1 revealed that there may be a mutual association between life expectancy and birth rate. This section shows how such a relationship can be quantified.

Figure 5.9 shows this scatter diagram again. Life expectancy is denoted by x and birth rate by y.

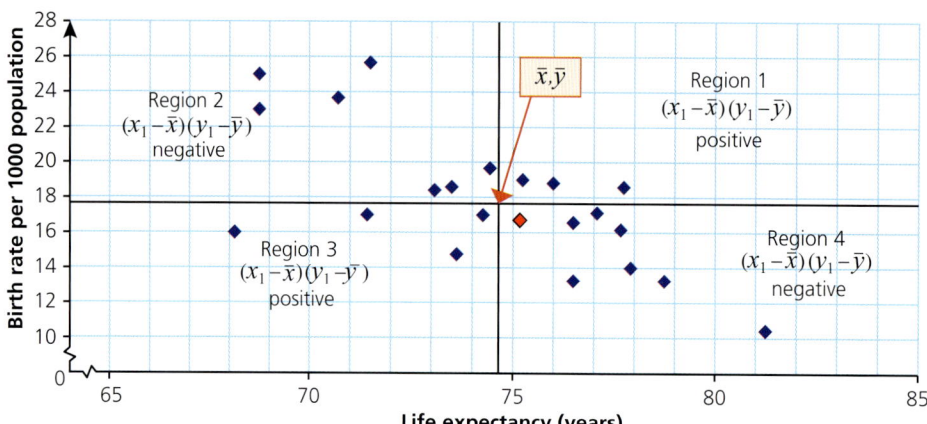

Figure 5.9

The mean life expectancy is

$$\bar{x} = \frac{\Sigma x_i}{n} = \frac{1641.35}{22} = 74.6\ldots$$

The mean birth rate is

$$\bar{y} = \frac{\Sigma y_i}{n} = \frac{391.79}{22} = 17.8\ldots$$

(x_i, y_i) are the various data points, for example (75.25, 16.73) for Colombia (marked in red); n is the number of such points, in this case 22.

You will see that the point $(\bar{x}, \bar{y})$ has also been plotted on the scatter diagram and lines drawn through this point parallel to the axes. These lines divide the scatter diagram into four regions.

You can think of the point $(\bar{x}, \bar{y})$ as the middle of the scatter diagram and so treat it as a new origin. Relative to $(\bar{x}, \bar{y})$, the co-ordinates of the various points are all of the form $(x_i - \bar{x}, y_i - \bar{y})$.

The table below gives the values of $(x_i - \bar{x})$, $(y_i - \bar{y})$ and $(x_i - \bar{x})(y_i - \bar{y})$.

Table 5.2

			$(x_i - \bar{x})$	$(y_i - \bar{y})$	$(x_i - \bar{x})(y_i - \bar{y})$
Argentina	77.51	16.88	2.903	−0.929	−2.696
Belize	68.49	25.14	−6.117	7.331	−44.845
Bolivia	68.55	23.28	−6.057	5.471	−33.139
Brazil	73.28	14.72	−1.327	−3.089	4.098
Canada	81.67	10.29	7.063	−7.519	−53.105
Chile	78.44	13.97	3.833	−3.839	−14.714
Colombia	75.25	16.73	0.643	−1.079	−0.694
Costa Rica	78.23	16.08	3.623	−1.729	−6.263
Ecuador	76.36	18.87	1.753	1.061	1.861
El Salvador	74.18	16.79	−0.427	−1.019	0.435
Guatemala	71.74	25.46	−2.867	7.651	−21.935
Guyana	67.81	15.9	−6.797	−1.909	12.973
Honduras	70.91	23.66	−3.697	5.851	−21.631
Mexico	75.43	19.02	0.823	1.211	0.997
Nicaragua	72.72	18.41	−1.887	0.601	−1.135
Panama	78.3	18.61	3.693	0.801	2.960
Paraguay	76.8	16.66	2.193	−1.149	−2.519
Peru	73.23	18.57	−1.377	0.761	−1.048
Suriname	71.69	16.73	−2.917	−1.079	3.146
United States	79.56	13.42	4.953	−4.389	−21.738
Uruguay	76.81	13.18	2.203	−4.629	−10.198
Venezuela	74.39	19.42	−0.217	1.611	−0.349
Total	1641.35	391.79	0	0	−209.54

> **Note**
> The point (x, y) has a moment of $(x - 74.6)$ about the vertical axis and of $(y - 17.8)$ about the horizontal axis. So $(x - 74.6)(y - 17.8)$ is called its product moment.

If you look at the data point for Colombia (shown in red in Figure 5.9), you can see that $(x_i - \bar{x})(y_i - \bar{y})$ is negative. This point is in Region 4. In fact, this will be true for all of the points in Region 4 and also Region 2. In both Region 1 and Region 3, $(x_i - \bar{x})(y_i - \bar{y})$ will be positive.

When there is positive correlation most or all of the data points will fall in Regions 1 and 3 and so you would expect the sum of these products to be positive and large.

When there is negative correlation as in this case, most or all of the points will be in Regions 2 and 4 and so you would expect the sum of these products to be negative and large.

When there is little or no correlation, the points will be scattered round all four regions. Those in Regions 1 and 3 will result in positive values of $(x_i - \bar{x})(y_i - \bar{y})$ but when you add these to the negative values from the points in Regions 2 and 4 you would expect most of them to cancel each other out. Consequently, the total value of all the terms should be small.

> **Note**
> In Table 5.2, S_{xy} is the total of the last column, so $S_{xy} = -209.54$.

In general, the sum of all these terms is denoted by S_{xy}.

$$S_{xy} = \Sigma(x_i - \bar{x})(y_i - \bar{y})$$

By itself the actual value of S_{xy} does not tell you very much because:

- no allowance has been made for the number of items of data
- no allowance has been made for the spread within the data
- no allowance has been made for the units of x and y.

Pearson's product moment correlation coefficient

To allow for both the number of items and the spread within the data, together with the units of x and y, the value of S_{xy} is divided by the square root of the product of S_{xx} and S_{yy}.

The sample product moment correlation coefficient is denoted by r and is given by

$$r = \frac{S_{xy}}{\sqrt{S_{xx}S_{yy}}} = \frac{\Sigma(x_i - \bar{x})(y_i - \bar{y})}{\sqrt{\Sigma(x_i - \bar{x})^2 \times \Sigma(y_i - \bar{y})^2}}$$

$$= \frac{\Sigma x_i y_i - n(\bar{x})(\bar{y})}{\sqrt{(\Sigma x_i^2 - n\bar{x}^2)(\Sigma y_i^2 - n\bar{y}^2)}}$$

> **Note**
> You may use either formulation, since they are algebraically equivalent. Example 5.1, overleaf, gives both methods, so you may judge for yourself.

where $S_{xx} = \Sigma(x_i - \bar{x})^2$, $S_{yy} = \Sigma(y_i - \bar{y})^2$.

The quantity, r, provides a standardised measure of correlation. Its value always lies within the range -1 to $+1$. (If you calculate a value outside this range, you have made a mistake.) A value of $+1$ means perfect positive correlation; in that case, all the points on a scatter diagram would lie exactly on a straight line with positive gradient. Similarly, a value of -1 means perfect negative correlation.

These two cases are illustrated in Figure 5.10.

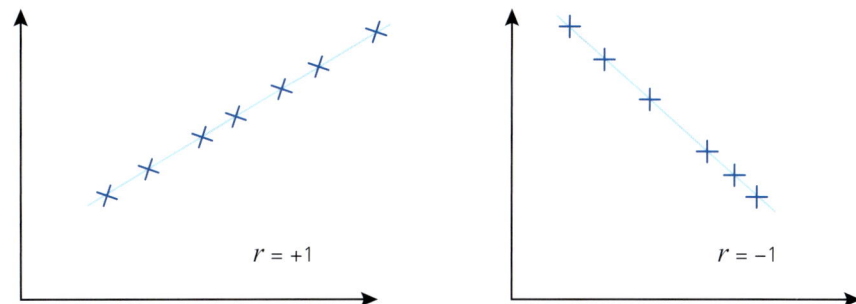

Figure 5.10 (i) Perfect positive correlation (ii) Perfect negative correlation

> S_{xx} and S_{yy} are the sums of the squares and are not divided by n. Similarly, S_{xy} is the sum of the terms $(x_i - \bar{x})(y_i - \bar{y})$ and is not divided by n.

In cases of little or no correlation, r takes values close to zero. The nearer the value of r is to $+1$ or -1, the stronger the correlation.

Example 5.1

A gardener wishes to know if plants which produce only a few potatoes produce larger ones. He selects five plants at random, counts how many eating sized potatoes they have produced, and weighs the largest one.

Table 5.3

Number of potatoes, x	5	5	7	8	10
Weight of largest, y (grams)	240	232	227	222	215

Calculate the sample product moment correlation coefficient, r, and comment on the result.

Solution
Method 1

Table 5.4

x	y	$x - \bar{x}$	$y - \bar{y}$	$(x - \bar{x})^2$	$(y - \bar{y})^2$	$(x - \bar{x})(y - \bar{y})$
5	240	−2	12.8	4	163.84	−25.6
5	232	−2	4.8	4	23.04	−9.6
7	227	0	−0.2	0	0.04	0
8	222	1	−5.2	1	27.04	−5.2
10	215	3	−12.2	9	148.84	−36.6
Totals 35	1136	0	0	18	362.80	−77.0

$\bar{x} = \dfrac{\Sigma x}{n} = \dfrac{35}{5} = 7$

$\bar{y} = \dfrac{\Sigma y}{n} = \dfrac{1136}{5} = 227.2$

$S_{xx} = \Sigma(x_i - \bar{x})^2 = 18$

$S_{yy} = \Sigma(y_i - \bar{y})^2 = 362.80$

$S_{xy} = \Sigma(x_1 - \bar{x})(y_1 - \bar{y}) = -77.0$

> **Weight and mass**
>
> In mechanics and physics the word 'weight' is used exclusively to mean the force of gravity on a body and the units of weight are those of force, such as newton. The word 'mass' is used to denote the amount of material in a body, with units such as kilogram, tonne and pound.
>
> In everyday use, the term weight is often used where mass would be used in mechanics.
>
> In statistics, the data used often come from everyday contexts or from other subjects. Consequently, it is very often the case that the word 'weight' is used to describe what would be called 'mass' in mechanics. Since statistics is a practical subject, using data to solve problems from a wide variety of contexts, this practice is accepted in this book. So the terms weight and mass are used interchangeably and should be interpreted according to the situation.

$$r = \frac{S_{xy}}{\sqrt{S_{xx}S_{yy}}} = \frac{\Sigma(x_i - \bar{x})(y_i - \bar{y})}{\sqrt{\Sigma(x_i - \bar{x})^2 \times \Sigma(y_i - \bar{y})^2}}$$

$$= \frac{-77}{\sqrt{18 \times 362.8}}$$

$$= -0.953 \text{ (to 3 s.f.)}$$

Method 2

$n = 5$

Table 5.5

x	y	x^2	y^2	xy
5	240	25	57 600	1200
5	232	25	53 824	1160
7	227	49	51 529	1589
8	222	64	49 284	1776
10	215	100	46 225	2150
Totals 35	1136	263	258 462	7875

$S_{xy} = \Sigma xy - n(\bar{x})(\bar{y})$
$= 7875 - 5 \times 7 \times 227.2$
$= -77$

$S_{yy} = \Sigma y^2 - n\bar{y}^2$
$= 258\,462 - 5 \times 227.2^2$
$= 362.8$

$\bar{x} = \frac{\Sigma x}{n} = \frac{35}{5} = 7$

$S_{xx} = \Sigma x^2 - n\bar{x}^2 = 263 - 5 \times 7^2 = 18$

$\bar{y} = \frac{\Sigma y}{n} = \frac{1136}{5} = 227.2$

$$r = \frac{S_{xy}}{\sqrt{S_{xx}S_{yy}}} = \frac{\Sigma x_i y_i - n(\bar{x})(\bar{y})}{\sqrt{(\Sigma x_i^2 - n\bar{x}^2)(\Sigma y_i^2 - n\bar{y}^2)}}$$

$$= \frac{-77}{\sqrt{18 \times 362.8}}$$

$$= -0.953 \text{ (to 3 s.f.)}$$

Conclusion: There is very strong negative linear correlation between the variables. Large potatoes seem to be associated with small crop sizes.

Carrying out the calculation

Two methods of carrying out the calculation are shown here.

- Either method will give the correct value of the product moment correlation coefficient, r.
- Method 2 is generally used when the values of the data give 'awkward' values for the sample means, making the calculations in Method 1 rather unwieldy.
- Sometimes you are given summary data: $\Sigma x, \Sigma y, \Sigma x^2, \Sigma y^2, \Sigma xy$. In these cases, Method 2 is the easier to use; it may be that your calculator has provided such summary data.

However, if as in this case you are given the raw data, then the usual method of finding r is to use the statistical facilities on a calculator.

Sometimes you are simply given summary statistics (values of Σx, Σx^2, Σxy, etc.) in which case you have to use the above formulae for S_{xx} etc.

USING ICT

A spreadsheet

You can use a spreadsheet to calculate the product moment correlation coefficient. The spreadsheet in Figure 5.11 shows the output for Example 5.1.

To set up the spreadsheet you need to:

1 Enter the data – here in columns A and B.
2 Use the formula provided by your spreadsheet to find the value of r, for example = PEARSON(A5:A9, B5:B9)
3 If you also want to see a scatter diagram of your data you can do this by highlighting the two columns of data and then choosing the relevant options, for example Insert followed by Scatter diagram.

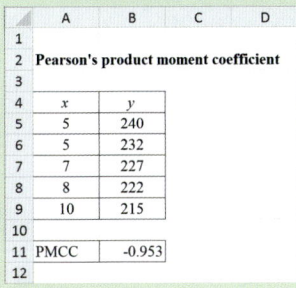

Figure 5.11

Interpreting the product moment correlation coefficient

The product moment correlation coefficient provides a measure of the correlation between the two variables. There are two different ways in which it is commonly used and interpreted: as a test statistic and as a measure of effect size.

Using r as a test statistic

When the data cover the whole of a population, the correlation coefficient tells you all that there is to be known about the level of correlation between the variables in the population. The population correlation coefficient is denoted by ρ.

> The symbol ρ is the Greek letter rho, pronounced 'row' as in 'row a boat'.

However, it is often the case that you do not know the level of correlation in a population and take a reasonably small sample to find out.

You use your sample data to calculate a value of the correlation coefficient, r. You know that if the value of r turns out to be close to $+1$ or -1, you can be reasonably confident that there is correlation, and that if r is close to 0 there is probably little or no correlation. What happens in a case such as $r = 0.6$?

To answer this question you have to understand what r is actually measuring. The data which you use when calculating r are actually a **sample** taken from a parent bivariate distribution, rather than the whole of a population. You have only taken a few out of a very large number of points which could, in theory, be plotted on a scatter diagram such as Figure 5.12. Each point (x_i, y_i) represents a possible pair of values corresponding to one observation of the bivariate population.

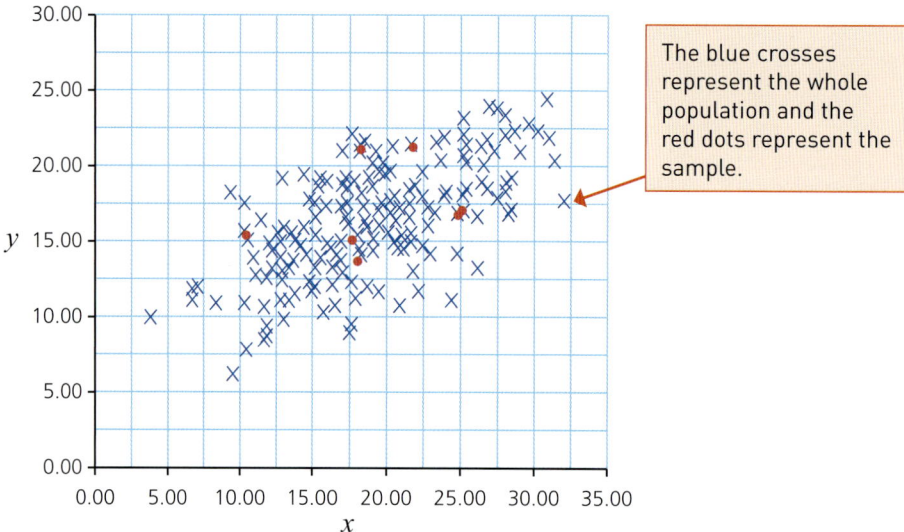

The blue crosses represent the whole population and the red dots represent the sample.

Figure 5.12 Scatter diagram showing a sample from a large bivariate population

There will be a level of correlation within the parent population and this is denoted by ρ.

The calculated value of r, which is based on the sample points, can be used as an estimate for ρ. It can also be used to carry out a hypothesis test on the value of ρ, the parent population correlation coefficient. Used in this way it is a **test statistic**.

The simplest hypothesis test which you can carry out is that there is no correlation within the parent population. This gives rise to a null hypothesis:

$H_0: \rho = 0$. There is no correlation between the two variables.

There are three possible alternative hypotheses, according to the sense of the situation you are investigating. These are:

1. $H_1: \rho \neq 0$. There is correlation between the variables (two-tailed test).
2. $H_1: \rho > 0$. There is positive correlation between the variables (one-tailed test).
3. $H_1: \rho < 0$. There is negative correlation between the variables (one-tailed test).

The test is carried out by comparing the value for r with the appropriate entry in a table of critical values. This will depend on the size of the sample, the significance level at which you are testing and whether your test is one-tailed or two-tailed.

> **Note**
>
> It is also possible to test the null hypothesis that ρ has some other value, like 0.4, but such tests are beyond the scope of this book.

Critical values

Under the null hypothesis, $H_0: \rho = 0$, i.e. when there is a complete absence of linear correlation between the variables in the **population**, it is very likely that a bivariate data set, drawn at random from the population, will produce a value of r, the **sample** product moment correlation coefficient, which is non-zero. This will be the case even if the null hypothesis is true.

The role of the significance level is that, for any sample size n, it represents the probability that the value of r will be 'further from zero' than the critical value. Three examples, using the table of critical values in Figure 5.13 (see following page) should help you to understand this concept better.

Table 5.6

Alternative hypothesis	Sample size, n	Significance level	Meaning
$H_1: \rho > 0$	11	1%	$P(r > 0.6851) = 0.01$
$H_1: \rho < 0$	6	5%	$P(r < -0.7293) = 0.05$
$H_1: \rho \neq 0$	15	10%	$P(|r| > 0.4409) = 0.1$

Example 5.2

> **Are students who are good at English also good at Mathematics?**
>
> A student believes that just because you are good at English, you are no more or less likely to be good at Mathematics. He obtains the results of an English examination and a Mathematics examination for eight students in his class. He then decides to carry out a hypothesis test to investigate whether he is correct.

The results of the two examinations for eight students are as follows.

Table 5.7

English, x	74	83	61	79	41	55	42	71
Mathematics, y	73	65	67	67	58	73	25	56

Solution

The relevant hypothesis test for this situation would be

$H_0: \rho = 0$. There is no correlation between English and Mathematics scores.

$H_1: \rho \neq 0$. There is correlation between English and Mathematics scores.

The student decides to use the 5% significance level.

The critical value for $n = 8$ at the 5% significance level for a two-tailed test is found from tables to be 0.7067.

Product moment correlation

n	5%	2½%	1%	½%	one-tailed test
	10%	5%	2%	1%	two-tailed test
1	-	-	-	-	
2	-	-	-	-	
3	0.9877	0.9969	0.9995	0.9999	
4	0.9000	0.9500	0.9800	0.9900	
5	0.8054	0.8783	0.9343	0.9587	
6	0.7293	0.8114	0.8822	0.9172	
7	0.6694	0.7545	0.8329	0.8745	
8	0.6215	0.7067	0.7887	0.8343	
9	0.5822	0.6664	0.7498	0.7977	
10	0.5494	0.6319	0.7155	0.7646	
11	0.5214	0.6021	0.6851	0.7348	
12	0.4973	0.5760	0.6581	0.7079	
13	0.4762	0.5529	0.6339	0.6835	
14	0.4575	0.5324	0.6120	0.6614	
15	0.4409	0.5140	0.5923	0.6411	

Figure 5.13 Extract from table of values for the product moment correlation coefficient, r

The calculation of the product moment correlation coefficient, r, can be set out using **Method 2** from Example 5.1.

$n = 8$

Table 5.8

x	y	x^2	y^2	xy
74	73	5476	5329	5402
83	65	6889	4225	5395
61	67	3721	4489	4087
79	67	6241	4489	5293
41	58	1681	3364	2378
55	73	3025	5329	4015
42	25	1764	625	1050
71	56	5041	3136	3976
Totals 506	484	33838	30986	31596

$$\bar{x} = \frac{\Sigma x}{n} = \frac{506}{8} = 63.25 \quad \bar{y} = \frac{\Sigma y}{n} = \frac{484}{8} = 60.5$$

$$S_{xx} = \Sigma x^2 - n\bar{x}^2 = 33838 - 8 \times 63.25^2 = 1833.5$$

$$S_{yy} = \Sigma y^2 - n\bar{y}^2 = 30986 - 8 \times 60.5^2 = 1704$$

$$S_{xy} = \Sigma xy - n\bar{x}\bar{y} = 31596 - 8 \times 63.25 \times 60.5 = 983$$

$$r = \frac{S_{xy}}{\sqrt{S_{xx}S_{yy}}} = \frac{\Sigma x_i y_i - n(\bar{x})(\bar{y})}{\sqrt{(\Sigma x_i^2 - n\bar{x}^2)(\Sigma y_i^2 - n\bar{y}^2)}} = \frac{983}{\sqrt{1833.5 \times 1704}}$$

$$= 0.5561 \quad \text{(to 4 s.f.)}$$

> **Note**
> This example is designed just to show you how to do the calculations, and no more. A sample of size 8 is smaller than you would usually use. Also there is no indication of how the student chose the eight people in his sample. An essential requirement for a hypothesis test is that the sample is representative and usually a random sample is taken.

The critical value is 0.7067. The value of r of 0.5561 is less extreme.

So there is not enough evidence to reject the null hypothesis in favour of the alternative hypothesis.

The evidence does not suggest that the student's suggestion is wrong.

USING ICT

You can use ICT to find the correlation coefficient and also to find the p-value for the test.

Figure 5.14 shows the output from a statistical package for the data in Example 5.2. You enter the correlation coefficient, the sample size and whether the test is one- or two-tailed (and the direction if one-tailed). The software then gives you the p-value as shown in Figure 5.14.

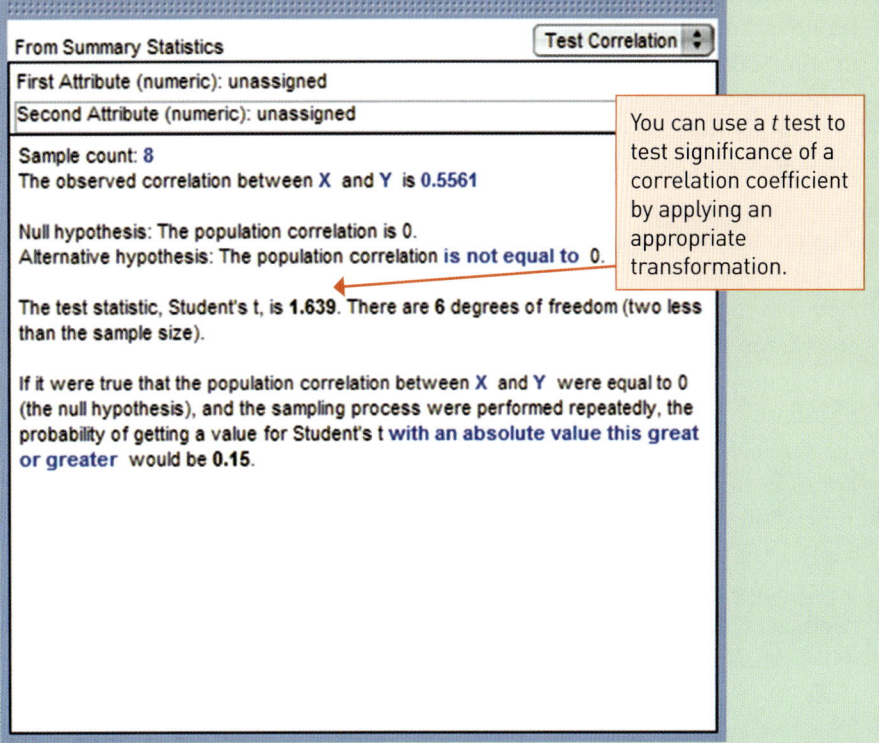

> You can use a t test to test significance of a correlation coefficient by applying an appropriate transformation.

Figure 5.14

The p-value is 0.15 which is greater than 5% so there is not enough evidence to reject the null hypothesis.

Product moment correlation

> - Hypothesis tests using Pearson's product moment correlation coefficient require modelling assumptions that both variables are **random** and that the data are drawn from a **bivariate Normal** distribution. For large data sets this is usually the case if the scatter diagram gives an approximately elliptical distribution. If one or both of the distributions is, for example, skewed or bimodal, the procedure is likely to be inaccurate.
> - The product moment correlation coefficient is a measure of correlation and so is only appropriate if the relationship between the variables is linear. For cases of non-linear association you should apply a test based on Spearman's correlation coefficient, which you will meet later in this chapter.
> - The extract from the tables gives the critical value of r for various values of the significance level and the sample size, n. You will, however, find some tables where n is replaced by ν, the **degrees of freedom**. Degrees of freedom are covered in the next section.

Degrees of freedom

Here is an example where you have just two data points.

Table 5.9

	Sean	Iain
Height of an adult man (m)	1.70	1.90
The mortgage on his house (£)	15 000	45 000

When you plot these two points on a scatter diagram it is possible to join them with a perfect straight line and you might be tempted to conclude that taller men have larger mortgages on their houses.

This conclusion would clearly be wrong. It is based on the data from only two men so you are bound to be able to join the points on the scatter diagram with a straight line and calculate r to be either $+1$ or -1 (providing their heights and/or mortgages are not the same). In order to start to carry out a test you need the data for a third man, say Dafyd (height 1.75 m and mortgage £37 000). When his data are plotted on the scatter diagram in Figure 5.15 you can see how close it lies to the line between Sean and Iain.

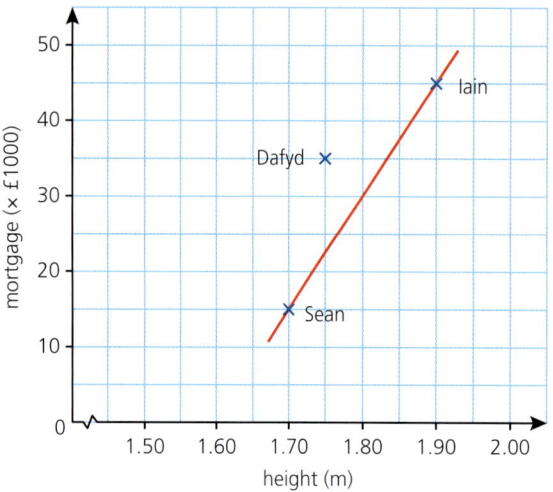

Figure 5.15

So the first two data points do not count towards a test for linear correlation. The first one to count is point number three. Similarly, if you have n points, only $n-2$ of them count towards any test. $n-2$ is called the **degrees of freedom** and denoted by ν. It is the number of free variables in the system. In this case, it is the number of points, n, less the 2 that have effectively been used to define the line of best fit.

In the case of the three men with their mortgages you would actually draw a line of best fit through all three, rather than join any particular two. So you cannot say that any two particular points have been taken out to draw the line of best fit, merely that the system as a whole has lost two.

Tables of critical values of correlation coefficients can be used without understanding the idea of degrees of freedom, but the idea is an important one throughout statistics. In general,

> degrees of freedom = sample size − number of restrictions.

Interpreting correlation

You need to be on your guard against drawing spurious conclusions from significant correlation coefficients.

Correlation does not imply causation

Figures for the years 1995–2005 show a high correlation between the sales of laptop computers and sales of microwave ovens. There is, of course, no direct connection between the two variables. You would be quite wrong to conclude that buying a laptop computer predisposes the buyer to buying a microwave oven.

Although there may be a high level of correlation between variables A and B it does not mean that A causes B or that B causes A. It may well be that a third variable C causes both A and B, or it may be that there is a more complicated set of relationships. In the case of laptop computers and microwaves, both are clearly caused by the advance of modern technology.

Non-linear association

A low value of r tells you that there is little or no correlation. There are, however, other forms of association, including non-linear as illustrated in Figure 5.16.

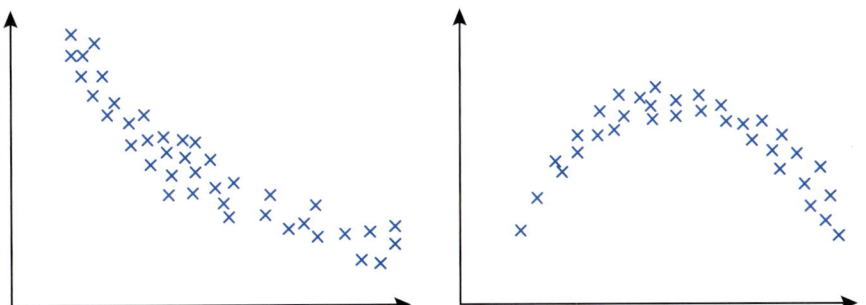

Figure 5.16 Scatter diagrams showing non-linear association

These diagrams show that there is an association between the variables, but not one that can be described as correlation.

Extrapolation

A linear relationship established over a particular domain should not be assumed to hold outside this range. For instance, there is strong correlation between the age in years and the 100 m times of female athletes between the ages of 10 and 20 years. To extend the connection, as shown in Figure 5.17, would suggest that veteran athletes are quicker than athletes who are in their prime and, if they live long enough, can even run 100 m in no time at all!

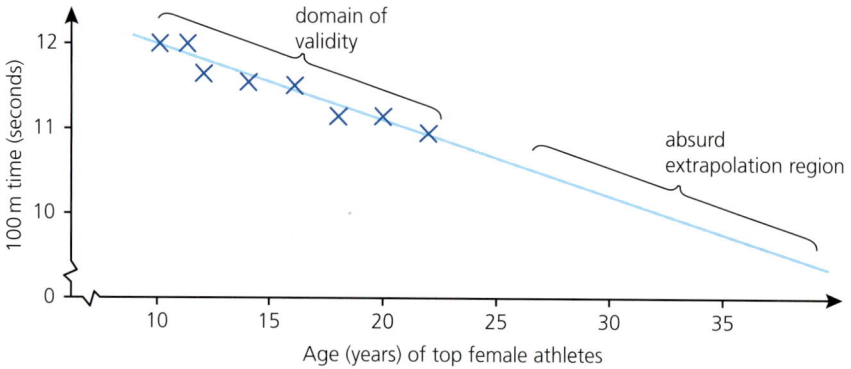

Figure 5.17

Effect size

For very large sample sizes, a very small but non-zero value of r may still be statistically significant. The null hypothesis may be rejected and the alternative hypothesis accepted. Although the result is statistically significant, there may be no real significance in terms of the use to be made of this correlation.

This is a common situation with big data where the calculation is carried out by computer and might cover millions of data items. Such data sets are often not just bivariate but multivariate covering many fields. It is quite common to find unexpected correlations but at a low level, say less than 0.1, as well as those expected at a higher level, say 0.5.

For example, suppose that there is a small but statistically significant correlation between height and intelligence. This may be of academic interest, but little use can be made of this for any practical purpose such as in selecting candidates for a job.

Effect sizes can be calculated for any significance test, but are only considered in the context of correlation in this book. One common way of judging effect size was formulated by Jacob Cohen in 1988. Cohen gave suggestions for interpreting these effect sizes, suggesting that a correlation coefficient r of 0.1 represents a small effect size, 0.3 represents a medium effect size and 0.5 represents a large effect size.

Example 5.3

The correlation matrix below shows the correlation between the four variables in the example at the start of the chapter, but for all countries in the world rather than just American countries. The four variables are population (P), life expectancy (L), birth rate (B) and GDP per capita (G).

Use Cohen's interpretation to comment on the levels of correlation.

Table 5.10

	P	L	B	G
P	1			
L	−0.017	1		
B	−0.025	−0.844	1	
G	−0.071	0.633	−0.579	1

Solution

All of the effect sizes involving population are small.

All the other effect sizes are large.

The greatest correlation is between life expectancy and birth rate. This is negative, meaning that on the whole countries with a high life expectancy tend to have a low birth rate, and vice-versa.

Exercise 5.1

①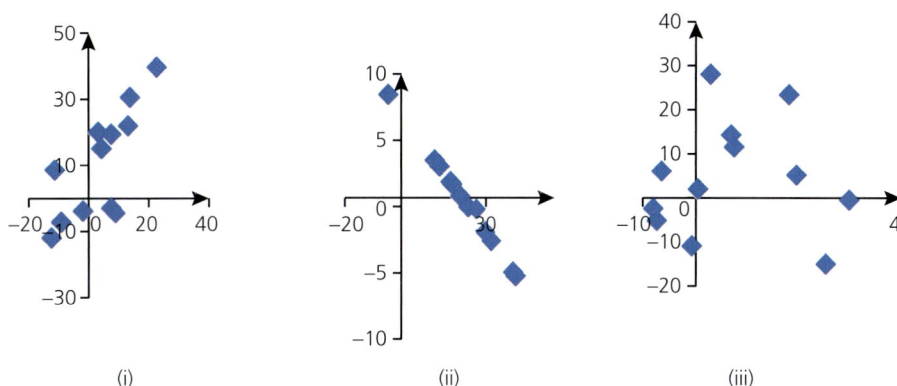

(i) (ii) (iii)

Figure 5.18

Three sets of bivariate data have been plotted on scatter diagrams, as illustrated. In each diagram the product moment correlation coefficient takes one of the values $-1, -0.8, 0, 0.8, 1$. Without doing any calculations, state the appropriate value of the correlation coefficient corresponding to the scatter diagrams (i), (ii) and (iii) in Figure 5.18.

② For each of the sets of data (i), (ii) and (iii) in Tables 5.11–5.13

   (a) draw a scatter diagram and comment on whether there appears to be any linear correlation.

   (b) calculate the product moment correlation coefficient and compare this with your assertion based on the scatter diagram.

(i) The mathematics and physics test results of 14 students.

Table 5.11

Mathematics	45	23	78	91	46	27	41	62	34	17	77	49	55	71
Physics	62	36	92	70	67	39	61	40	55	33	65	59	35	40

(ii) The wine consumption in a country in millions of litres and the years 1993 to 2000.

Table 5.12

Year	1993	1994	1995	1996	1997	1998	1999	2000
Consumption (×10⁶ litres)	35.5	37.7	41.5	46.4	44.8	45.8	53.9	62.0

(iii) The number of hours of sunshine and the monthly rainfall, in centimetres, in an eight-month period.

Table 5.13

	Jan	Feb	Mar	Apr	May	Jun	Jul	Aug
Sunshine (hours)	90	96	105	110	113	120	131	124
Rainfall (cm)	5.1	4.6	6.3	5.1	3.3	2.8	4.5	4.0

③ For each of these sets of data use one or more of the following to find the product moment correlation coefficient, r (note that the data for part (iv) are given in the form of a scatter diagram).

- a scientific calculator in two variable statistics mode
- a graphics calculator in two variable statistics mode
- a spreadsheet.

(i)

Table 5.14

x	10	11	12	13	14	15	16	17
y	19	16	28	20	31	19	32	35

(ii)

Table 5.15

x	12	14	14	15	16	17	17	19
y	86	90	78	71	77	69	80	73

(iii)

Table 5.16

x	56	78	14	80	34	78	23	61
y	45	34	67	70	42	18	25	50

(iv)

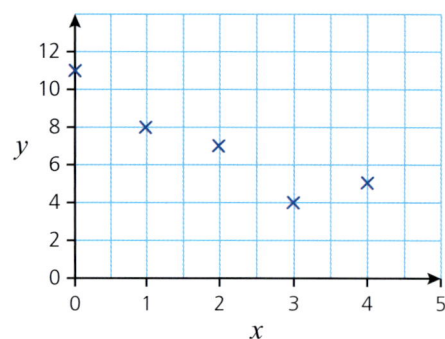

Figure 5.19

④ Find the value of r in each case below. The tables of values are not complete but, in each case, summary statistics are given for all of the data.

(i) The annual salary, in thousands of pounds, and the average number of hours worked per week by people chosen at random.

Table 5.17

Salary (× £1000)	5	7	13
Hours worked per week	18	22	35

$\Sigma x = 105$, $\Sigma y = 217$, $\Sigma x^2 = 3003$, $\Sigma y^2 = 7093$, $\Sigma xy = 3415$, $n = 7$.

(ii) The mean temperature in degrees Celsius and the amount of ice cream sold in a supermarket in hundreds of litres.

Table 5.18

	Apr	May	Jun	Jul
Mean temperature (°C)	9	13	14	17
Ice-cream sold (100 l)	11	15	17	20

$\Sigma x = 108$, $\Sigma y = 117$, $\Sigma x^2 = 1506$, $\Sigma y^2 = 1921$, $\Sigma xy = 1660$, $n = 8$.

(iii) The reaction times of women of various ages.

Table 5.19

Reaction time ($\times 10^{-3}$ s)	156	165	149	180	189
Age (years)	36	40	27	50	49

$\Sigma x = 1432$, $\Sigma y = 337$, $\Sigma x^2 = 259\,680$, $\Sigma y^2 = 15\,089$, $\Sigma xy = 61\,717$, $n = 8$.

5. A language teacher wishes to test whether students who are good at their own language are also likely to be good at a foreign language. Accordingly, she collects the marks of eight students, all native English speakers, in their end of year examinations in English and French.

Table 5.20

Candidate	A	B	C	D	E	F	G	H
English	65	34	48	72	58	63	26	80
French	74	49	45	80	63	72	12	75

(i) Calculate the product moment correlation coefficient.
(ii) State the null and alternative hypotheses.
(iii) Using the correlation coefficient as a test statistic, carry out the test at the 5% significance level.

6. 'You can't win without scoring goals.' So says the coach of a netball team. Jamila, who believes in solid defensive play, disagrees and sets out to prove that there is no correlation between scoring goals and winning matches. She collects the following data for the goals scored and the points gained by 12 teams in a netball league.

Table 5.21

Goals scored, x	41	50	54	47	47	49	52	61	50	29	47	35
Points gained, y	21	20	19	18	16	14	12	11	11	7	5	2

(i) Calculate the product moment correlation coefficient.
(ii) State suitable null and alternative hypotheses, indicating whose position each represents.
(iii) Carry out the hypothesis test at the 5% significance level and comment on the result.

⑦ A medical student is trying to estimate the birth mass of babies using prenatal scan images. The actual mass, x kg, and the estimated mass, y kg, of ten randomly selected babies are given in the table below. The data are plotted in the scatter diagram.

Table 5.22

x	2.61	2.73	2.87	2.96	3.05	3.14	3.17	3.24	3.76	4.10
y	3.2	2.6	3.5	3.1	2.8	2.7	3.4	3.3	4.4	4.1

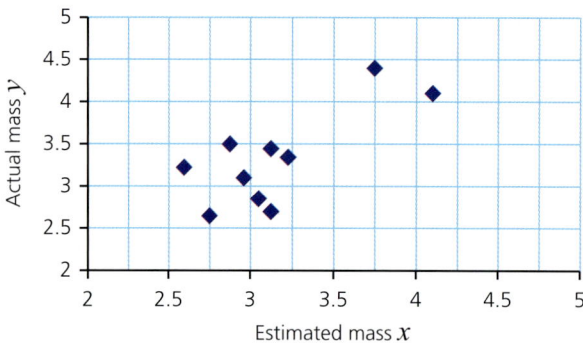

Figure 5.20

(i) The student decides to carry out a test based on the product moment correlation coefficient to investigate whether there is a positive relationship between the two variables. A friend suggests that there are two outliers so this test would not be appropriate. Explain why it may still be valid to carry out the test.

(ii) The value of the product moment correlation coefficient for these data is 0.7604. Carry out the test at the 1% significance level. [MEI]

⑧ It is widely believed that those who are good at chess are good at bridge, and vice-versa. A commentator decides to test this theory using as data the grades of a random sample of eight people who play both games.

Table 5.23

Player	A	B	C	D	E	F	G	H
Chess grade	160	187	129	162	149	151	189	158
Bridge grade	75	100	75	85	80	70	95	80

(i) Calculate the product moment correlation coefficient.
(ii) State suitable null and alternative hypotheses.
(iii) The output in Figure 5.21 comes from a statistical package. Using the p-value given at the bottom of the figure, complete the hypothesis test.

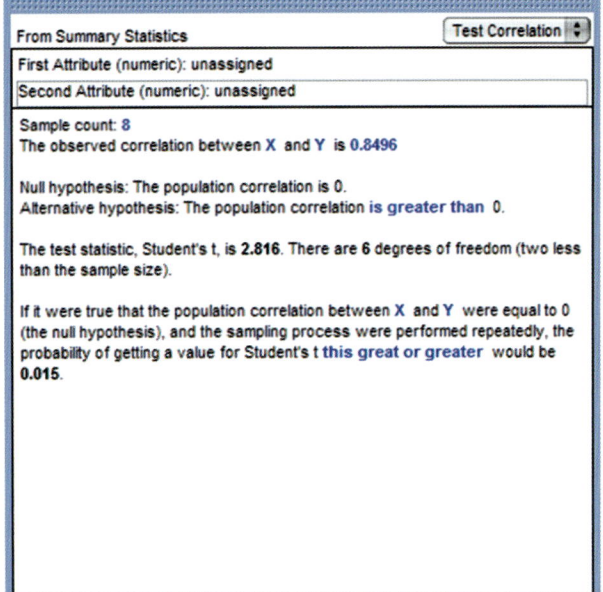

Figure 5.21

9. The scatter diagrams below were drawn by a student.

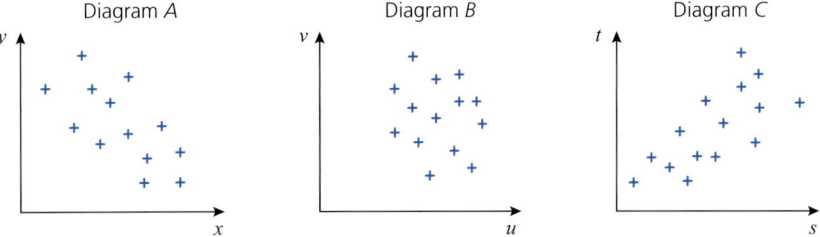

Figure 5.22

The student calculated the value of the product moment correlation coefficient for each of the sets of data.

The values were

0.68   −0.79   0.08

Write down, with a reason, which value corresponds to which scatter diagram. [Edexcel 6683 Q1 June 2005]

10. A biologist believes that a particular type of fish develops black spots on its scales in water that is polluted by certain agricultural fertilisers. She catches a number of fish; for each one she counts the number of black spots on its scales and measures the concentration of the pollutant in the water it was swimming in. She uses these data to test for positive linear correlation between the number of spots and the level of pollution.

Table 5.24

Fish	A	B	C	D	E	F	G	H	I	J
Pollutant concentration (parts per million)	124	59	78	79	150	12	23	45	91	68
Number of black spots	15	8	7	8	14	0	4	5	8	8

(i) Calculate the product moment correlation coefficient.

(ii) State suitable null and alternative hypotheses.

(iii) Carry out the hypothesis test at the 2% significance level. What can the biologist conclude?

⑪ The correlation matrix below shows the correlation between four variables in different districts of a large city. The four variables are life expectancy (L), infant mortality (M), average income (A) and a measure of income inequality (I).

Use Cohen's interpretation to comment on the levels of correlation.

Table 5.25

	L	M	A	I
L	1			
M	0.37	1		
A	−0.23	0.17	1	
I	0.06	−0.19	0.46	1

⑫ Andrew claims that the older you get, the slower is your reaction time. His mother disagrees, saying the two are unrelated. They decide that the only way to settle the discussion is to carry out a proper test. A few days later they are having a small party and so ask their 12 guests to take a test that measures their reaction times. The results are as follows.

Table 5.26

Age	Reaction time (s)	Age	Reaction time (s)
78	0.8	35	0.5
72	0.6	30	0.3
60	0.7	28	0.4
56	0.5	20	0.4
41	0.5	19	0.3
39	0.4	10	0.3

Carry out the test at the 5% significance level, stating the null and alternative hypotheses. Who won the argument, Andrew or his mother?

⑬ The teachers at a school have a discussion as to whether girls, in general, run faster or slower as they get older. They decide to collect data for a random sample of girls the next time the school cross country race is held (which everybody has to take part in). They collect the following data, with the times given in minutes and the ages in years (the conversion from months to decimal parts of a year has already been carried out).

Table 5.27

Age	Time	Age	Time	Age	Time
11.6	23.1	18.2	45.	13.9	29.1
15.0	24.0	15.4	23.2	18.1	21.2
18.8	45.0	14.4	26.1	13.4	23.9
16.0	25.2	16.1	29.4	16.2	26.0
12.8	26.4	14.6	28.1	17.5	23.4
17.6	22.9	18.7	45.0	17.0	25.0
17.4	27.1	15.4	27.0	12.5	26.3
13.2	25.2	11.8	25.4	12.7	24.2
14.5	26.8				

(i) State suitable null and alternative hypotheses and decide on an appropriate significance level for the test.

(ii) Calculate the product moment correlation coefficient and state the conclusion from the test.

(iii) Plot the data on a scatter diagram and identify any outliers. Explain how they could have arisen.

(iv) Comment on the validity of the test.

⑭ Apprentices joining a large company are given several tests. Two of the tests are 'Basic English' and 'Manual dexterity'. The results of these tests are labelled x and y respectively. A manager believes that there will be positive correlation between x and y.

The spreadsheet shows the first 3 and 60th rows of data, together with the sums of each of the 5 columns.

	A	B	C	D	E	F
1		x	y	x^2	y^2	xy
2		119	118	14161	13924	14042
3		112	116	12544	13456	12992
4		116	108	13456	11664	12528
61		115	118	13225	13924	13570
62	SUM	7105	7132	846357	853620	846371

Figure 5.23

(i) Explain how you can tell that the value of n is 60.

(ii) State the formula in cell B62.

(iii) Find the values of S_{xx}, S_{yy} and S_{xy}.

(iv) Find the value of r.

(v) State suitable null and alternative hypotheses for a test to investigate the manager's belief.

(vi) Carry out the hypothesis test at the 5% significance level.

(vii) Comment on the effect size.

(viii) Do you think that this information can be of any use to the company?

⑮ The values of x and y in the table are the marks obtained in an intelligence test and a university examination, respectively, by 20 medical students.
The data are plotted in the scatter diagram.

Table 5.28

x	98	51	71	57	44	59	75	47	39	58
y	85	40	30	25	50	40	50	35	25	90
x	77	65	58	66	79	72	45	40	49	76
y	65	25	70	45	70	50	40	20	30	60

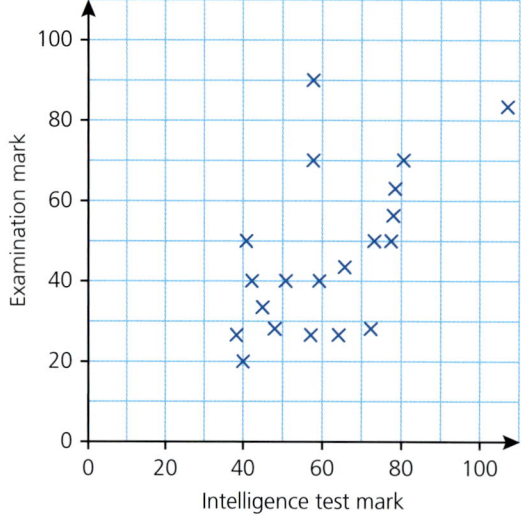

Figure 5.24

Given that $\Sigma x = 1226$, $\Sigma y = 945$, $\Sigma x^2 = 79\,732$, $\Sigma y^2 = 52\,575$ and $\Sigma xy = 61\,495$, calculate the product moment correlation coefficient, r, to 2 decimal places.

Referring to the evidence provided by the diagram and the value of r, comment briefly on the correlation between the two sets of marks.

Now eliminate from consideration those ten students whose values of x are less than 50 or more than 75. Calculate the new value of r for the marks of the remaining students. What does the comparison with the earlier value of r seem to indicate? [MEI]

16 A young family were looking for a new three-bedroom semi-detached house. A local survey recorded the price x, in £1000, and the distance y, in miles, from the station of such houses. The following summary statistics were provided.

$$S_{xx} = 113\,573, \quad S_{yy} = 8.657, \quad S_{xy} = -808.917$$

(i) Use these values to calculate the product moment correlation coefficient.

(ii) Give an interpretation of your answer to part (i).

Another family asked for the distances to be measured in km rather than miles.

(iii) State the value of the product moment correlation coefficient in this case. [Edexcel 6683 Q1 June 2007]

17 An estate agent recorded the price per square metre, p £/m², for 7 two-bedroom houses.

He then coded the data using the coding $q = \dfrac{p-a}{b}$, where a and b are positive constants.

His results are shown in the table below.

Table 5.29

p	1840	1848	1830	1824	1819	1834	1850
q	4.0	4.8	3.0	2.4	1.9	3.4	5.0

(i) Find the value of a and the value of b.

The estate agent also recorded the distance, d km, of each house from the nearest train station. The results are summarised below.

$$S_{dd} = 1.02 \qquad S_{qq} = 8.22 \qquad S_{dq} = -2.17$$

(ii) Calculate the product moment correlation coefficient between d and q.

(iii) Write down the value of the product moment correlation coefficient between d and p.

The estate agent records the price and size of 2 additional two-bedroom houses, H and J.

Table 5.30

House	Price (£)	Size (m²)
H	156 400	85
J	172 900	95

(iv) Suggest which house is most likely to be closer to a train station. Justify your answer. [Edexcel 6683 Q2 June 2015]

⑱ The table below gives the heights, h, of six male Olympic 100m sprint winners together with the times, t, they took.

Table 5.31

h	1.80	1.83	1.87	1.88	1.85	1.76
t	10.00	9.95	10.06	9.99	9.84	9.87

(i) Draw a scatter diagram to illustrate the data.

(ii) Calculate the product moment correlation coefficient.

(iii) Carry out a suitable hypothesis test, at a suitable level of significance, to determine whether or not it is reasonable to suppose that the heights and times are positively correlated.

(iv) Rewrite the table giving just the rank of each data value, using a rank of 1 for the lowest value and 6 for the highest value. For example the rank associated with a height of 1.88 m would be 6 since it is the height of the tallest person. The rank associated with a time of 9.87 s would be 2 since it is the second lowest time.

(v) Calculate the product moment correlation coefficient of the ranked data.

(vi) Comment on the difference between the two correlation coefficients.

⑲ (i) Prove algebraically that these two formulae for S_{xx} are equivalent

$$S_{xx} = \Sigma(x_i - \bar{x})^2 \qquad S_{xx} = \Sigma x^2 - n\bar{x}^2$$

(ii) Prove algebraically these two formulae for S_{xy} are equivalent

$$S_{xy} = \Sigma(x_i - \bar{x})(y_i - \bar{y}) \qquad S_{xy} = \Sigma xy - n\bar{x}\bar{y}$$

(iii) Hence prove that these two formulae for r are equivalent

$$r = \frac{\Sigma(x_i - \bar{x})(y_i - \bar{y})}{\sqrt{\Sigma(x_i - \bar{x})^2 \times \Sigma(y_i - \bar{y})^2}} \qquad r = \frac{\Sigma x_i y_i - n(\bar{x})(\bar{y})}{\sqrt{(\Sigma x_i^2 - n\bar{x}^2)(\Sigma y_i^2 - n\bar{y}^2)}}$$

2 Rank correlation

Dance competition dispute

In the dance competition at the local fete last Saturday, the two judges differed so much in their rankings of the ten competitors that initially nobody could be declared the winner. In the end, competitor C was declared the winner as the total of her ranks was the lowest (and so the best) but several of the other entrants felt that it was totally unfair.

The judgements that caused all the trouble were as follows.

Table 5.32

Competitor	A	B	C	D	E	F	G	H	I	J
Judge 1	1	9	7	2	3	10	6	5	4	8
Judge 2	8	3	1	10	9	4	7	6	5	2
Total	9	12	⑧	12	12	14	13	11	9	10

Winner

You will see that both judges ranked the ten entrants, 1st, 2nd, 3rd, ... , 10th. The winner, C, was placed 7th by one judge and 1st by the other. Their rankings look different so perhaps they were using different criteria on which to assess them. How can you use these data to decide whether that was or was not the case?

One way would be to calculate a correlation coefficient and use it to carry out a hypothesis test. However, the data you have are of a different type from any that you have used before for calculating correlation coefficients. In the point (1, 8), corresponding to competitor A, the numbers 1 and 8 are **ranks** and not scores (like marks in an examination or measurements). It is, however, possible to calculate a **rank correlation coefficient** in the same way as before. Rank correlation may be used in circumstances where ordinary correlation would not be appropriate and so is described as testing **association** rather than correlation.

The hypotheses are stated as follows:

H_0: there is no association.

H_1: there is positive association.

The null hypothesis, H_0, represents the idea that the judges are using completely different unrelated criteria to judge the competitors. The alternative hypothesis represents the idea that the judges are using similar criteria to judge the competitors.

ACTIVITY 5.1

Show that the product moment correlation coefficient of the ranks in the table above is −0.794.

Although you can carry out the calculation as in Example 5.1, it is usually done a different way.
Denoting the two sets of ranks by $x_1, x_2, ..., x_n$ and $y_1, y_2, ..., y_n$, the coefficient of association is given by

$$r_s = 1 - \frac{6\Sigma d_i^2}{n(n^2-1)}$$

where
- r_s is called Spearman's rank correlation coefficient
- d_i is the difference in the ranks for a general data item (x_i, y_i); $d_i = x_i - y_i$
- n is the number of items of data.

The calculation of $\sum d_i^2$ can then be set out in a table like this.

Table 5.33

Competitor	Judge 1, x_i	Judge 2, y_i	$d_i = x_i - y_i$	d_i^2
A	1	8	−7	49
B	9	3	6	36
C	7	1	6	36
D	2	10	−8	64
E	3	9	−6	36
F	10	4	6	36
G	6	7	−1	1
H	5	6	−1	1
I	4	5	−1	1
J	8	2	6	36
			Σd_i^2	296

The value of n is 10, so $r_s = 1 - \dfrac{6 \times 296}{10(10^2 - 1)} = -0.794$

You will see that this is the same answer as before, but the working is much shorter. It is not difficult to prove that the two methods are equivalent.

Hypothesis tests using Spearman's rank correlation coefficient

> **Note**
>
> If several items are ranked equally you give them the mean of the ranks they would have had if they had been slightly different from each other.

Spearman's rank correlation coefficient is often used as a test statistic for a hypothesis test of

H_0: there is no association between the variables

against one of three possible alternative hypotheses:

either H_1: there is association between the variables (two-tailed test)

or H_1: there is positive association between the variables (one-tailed test)

or H_1: there is negative association between the variables (one-tailed test).

The test is carried out by comparing the value of r_s with the appropriate critical value. This depends on the sample size, the significance level of the test and whether it is one-tailed or two-tailed. Critical values can be found from statistical software or tables.

Once you have found the value of r_s, this test follows the same procedure as that for correlation. However, the tables of critical values are not the same and so you need to be careful that you are using the right one.

The calculation is often carried out with the data across the page rather than in columns and this is shown in the next example.

S2 AS — Rank correlation

Example 5.4

During their course two trainee tennis coaches, Rachael and Leroy, were shown videos of seven people, A, B, C, ... , G, doing a top-spin service and were asked to rank them in order according to the quality of their style. They placed them as follows.

Table 5.34

Rank order	1	2	3	4	5	6	7
Rachael	B	G	F	D	A	C	E
Leroy	F	B	D	E	G	A	C

(i) Find Spearman's coefficient of rank correlation.

(ii) Use it to test whether there is evidence, at the 5% level, of positive association between their judgements.

Solution

(i) The rankings are as follows.

Table 5.35

Person	A	B	C	D	E	F	G
Rachael	5	1	6	4	7	3	2
Leroy	6	2	7	3	4	1	5
d_i	−1	−1	−1	1	3	2	−3
d_i^2	1	1	1	1	9	4	9

$n = 7$ $\qquad \sum d_i^2 = 26$

$$r_s = 1 - \frac{6\sum d_i^2}{n(n^2-1)} = 1 - \frac{6 \times 26}{7(7^2-1)}$$

$$= 0.54 \text{ (2 d.p.)}$$

(ii) H_0: there is no association between their rankings.
H_1: there is positive association between their rankings.
Significance level 5%.
One-tailed test.

From tables, the critical value of r_s for a one-tailed test at this significance level for $n = 7$ is 0.7143.

0.54 < 0.7143 so H_0 is accepted.

5%	2½%	1%	½%	one-tailed test
10%	5%	2%	1%	two-tailed test

n				
1	-	-	-	-
2	-	-	-	-
3	-	-	-	-
4	1.0000	-	-	-
5	0.9000	1.0000	1.0000	-
6	0.8286	0.8857	0.9429	1.0000
7	0.7143	0.7857	0.8929	0.9286
8	0.6429	0.7381	0.8333	0.8810
9	0.6000	0.7000	0.7833	0.8333
10	0.5636	0.6485	0.7455	0.7939

Figure 5.25 Extract from table of critical values for Spearman's rank correlation coefficient, rs

There is insufficient evidence to claim positive association between Rachael's and Leroy's rankings.

USING ICT

You can use ICT to find the correlation coefficient and also to find the *p*-value for the test.

Figure 5.26 shows the output from a statistical package for the data in Example 5.4. You enter the data and the software produces the output shown.

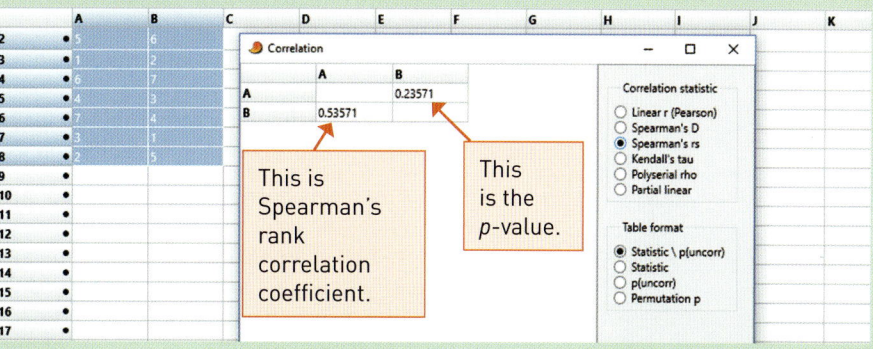

Figure 5.26

The *p*-value is 0.23571. However, because the test is one-tailed you have to divide this by 2 before comparing with the significance level. This gives approximately 0.118 which is greater than 0.05 or 5% so there is not enough evidence to reject the null hypothesis.

When to use rank correlation

Sometimes your data will be available in two forms, as values of variables or in rank order. If you have the choice you will usually work out the correlation coefficient from the variable values rather than the ranks.

It may well be the case, however, that only ranked data are available to you and, in that case, you have no choice but to use them. It may also be that, while you could collect variable values as well, it would not be worth the time, trouble or expense to do so.

Pearson's product moment correlation coefficient is a measure of correlation and so is not appropriate for non-linear data like those illustrated in the scatter diagram in Figure 5.27. You may, however, use rank correlation to investigate whether one variable generally increases (or decreases) as the other increases. The term **association** describes such a relationship.

> **Note**
>
> Spearman's rank correlation coefficient provides one among many statistical tests that can be carried out on ranks rather than variable values. Such tests are examples of **non-parametric tests**. A non-parametric test is a test on some aspect of a distribution which is not specified by its defining parameters.

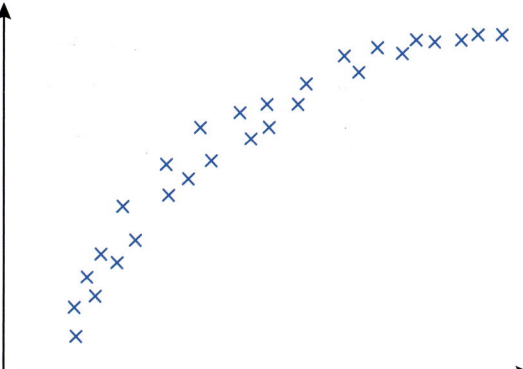

Figure 5.27 Non-linear data with a high degree of rank correlation

S2 AS — Rank correlation

You should, however, always look at the sense of your data before deciding which is the more appropriate form of correlation to use.

>  **Historical note**
>
> Karl Pearson was one of the founders of modern statistics. Born in 1857, he was a man of varied interests and practised law for three years before being appointed Professor of Applied Mathematics and Mechanics at University College, London in 1884. Pearson made contributions to various branches of Mathematics but is particularly remembered for his work on the application of statistics to biological problems in heredity and evolution. He died in 1936.
>
> Charles Spearman was born in London in 1863. After serving 14 years in the army as a cavalry officer, he went to Leipzig to study psychology. On completing his doctorate there he became a lecturer, and soon afterwards became a professor, also at University College, London. He pioneered the application of statistical techniques within psychology and developed the technique known as factor analysis in order to analyse different aspects of human ability. He died in 1945.

Exercise 5.2

① The order of merit of ten individuals at the start and finish of a training course were as follows.

Table 5.36

Individual	A	B	C	D	E	F	G	H	I	J
Order at start	1	2	3	4	5	6	7	8	9	10
Order at finish	5	3	1	9	2	6	4	7	10	8

Find Spearman's coefficient of rank correlation between the two orders.

② A sports coach obtained scores by nine athletes in two competitions (A and B). The maximum score in each competition was 12. The scores were as follows.

Table 5.37

Entrant	1	2	3	4	5	6	7	8	9
A score	9	2	12	5	3	7	10	6	8
B score	8	7	6	4	9	10	11	3	5

The spreadsheet shows this data together with partially completed working to calculate Spearman's coefficient of rank correlation.

	A	B	C	D	E	F	G
1	Entrant	A score	B score	Rank A	Rank B	$d_i = x_i - y_i$	d_i^2
2	1	9	8	7	6	1	1
3	2	2	7	1	5	-4	16
4	3	12	6	9	4	5	25
5	4	5	4				
6	5	3	9			-5	25
7	6	7	10				
8	7	10	11				
9	8	6	3			3	9
10	9	8	5				

Figure 5.28

Complete the spreadsheet and hence find Spearman's coefficient of rank correlation between the two orders. You can use the spreadsheet to find the ranks. First you highlight rows 2 to 10. You then sort on column B. You then put the numbers 1 to 9 in column D. You again highlight rows 2 to 10 and now sort on column C. You then put the numbers 1 to 9 in column E. Finally again highlight rows 2 to 10 and sort on column A.

③ Find Spearman's coefficient of rank correlation between the two variables x and y shown in the scatter diagram.

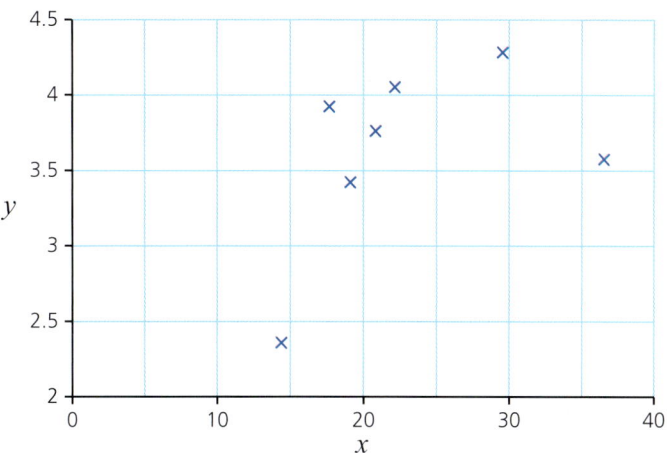

Figure 5.29

④ In a driving competition, there were eight contestants and two judges who placed them in rank order, as shown in the table below.

Table 5.38

Competitor	A	B	C	D	E	F	G	H
Judge X	2	5	6	1	8	4	7	3
Judge Y	1	6	8	3	7	2	4	5

Stating suitable null and alternative hypotheses, carry out a hypothesis test on the level of agreement of these two judges.

⑤ A coach wanted to test his theory that, although athletes have specialisms, it is still true that those who run fast at one distance are also likely to run fast at another distance. He selected six athletes at random to take part in a test and invited them to compete over 100 m and over 1500 m.

The times and places of the six athletes were as follows.

Table 5.39

Athlete	100 m time	100 m rank	1500 m time	1500 m rank
Allotey	9.8 s	1	3 m 42 s	1
Chell	10.9 s	6	4 m 11 s	2
Giles	10.4 s	2	4 m 19 s	6
Mason	10.5 s	3	4 m 18 s	5
O'Hara	10.7 s	5	4 m 12 s	3
Stuart	10.6 s	4	4 m 16 s	4

(i) Calculate the Pearson product moment and Spearman's rank correlation coefficients for these data.

(ii) State suitable null and alternative hypotheses and carry out hypothesis tests on these data.

(iii) State which you consider to be the more appropriate correlation coefficient in this situation, giving your reasons.

Rank correlation

6 During a village show, two judges, P and Q, had to award a mark out of 30 to some flower displays. The marks they awarded to a random sample of 8 displays were as follows.

Table 5.40

Display	A	B	C	D	E	F	G	H
Judge P	25	19	21	23	28	17	16	20
Judge Q	20	9	21	13	17	14	11	15

(i) Calculate Spearman's rank correlation coefficient for the marks awarded by the two judges.

After the show, one competitor complained about the judges. She claimed that there was no positive correlation between their marks.

(ii) Stating your hypotheses clearly, test at the 5% level whether or not this sample provides support for the competitor's claim.

[Edexcel 6691 Q1 June 2007]

7 A school holds an election for parent governors. Candidates are invited to write brief autobiographies and these are sent out at the same time as the voting papers.

After the election, one of the candidates, Mr Smith, says that the more words you wrote about yourself the more votes you got. He sets out to 'prove this statistically' by calculating the product moment correlation between the number of words and the number of votes.

Table 5.41

Candidate	A	B	C	D	E	F	G
Number of words	70	101	106	232	150	102	98
Number of votes	99	108	97	144	94	54	87

(i) Calculate the product moment correlation coefficient.

Mr Smith claims that this proves his point at the 5% significance level.

(ii) State his null and alternative hypotheses and show how he came to his conclusion.

(iii) Calculate Spearman's rank correlation coefficient for these data.

(iv) Explain the difference in the two correlation coefficients and criticise the procedure Mr Smith used in coming to his conclusion.

8 To test the belief that milder winters are followed by warmer summers, meteorological records are obtained for a random sample of ten years. For each year, the mean temperatures are found for January and July. The data, in °C, are given below.

Table 5.42

January	8.3	7.1	9.0	1.8	3.5	4.7	5.8	6.0	2.7	2.1
July	16.2	13.1	16.7	11.2	14.9	15.1	17.7	17.3	12.3	13.4

(i) Rank the data and calculate Spearman's rank correlation coefficient.

(ii) Test, at the 2.5% level of significance, the belief that milder winters are followed by warmer summers. State clearly the null and alternative hypotheses under test.

(iii) Would it be more appropriate, less appropriate or equally appropriate to use the product moment correlation coefficient to analyse these data? Briefly explain why.

[MEI]

⑨ A doctor is interested in the relationship between a person's Body Mass Index (BMI) and their level of fitness. She believes that a lower BMI leads to a greater level of fitness. She randomly selects 10 female 18 year-olds and calculates each individual's BMI. The females then run a race and the doctor records their finishing positions. The results are shown in the table.

Table 5.43

Individual	A	B	C	D	E	F	G	H	I	J
BMI	17.4	21.4	18.9	24.4	19.4	20.1	22.6	18.4	25.8	28.1
Finishing position	3	5	1	9	6	4	10	2	7	8

(i) Calculate Spearman's rank correlation coefficient for these data.

(ii) Stating your hypotheses clearly and using a one-tailed test with a 5% level of significance, interpret your rank correlation coefficient.

(iii) Give a reason to support the use of the rank correlation coefficient rather than the product moment correlation coefficient with these data. [Edexcel 6691 Q3 June 2009]

⑩ A county councillor is investigating the level of hardship, h, of a town and the number of calls per 100 people to the emergency services, c. He collects data for 7 randomly selected towns in the county. The results are shown in the table below.

Table 5.44

Town	A	B	C	D	E	F	G
h	14	20	16	18	37	19	24
c	52	45	43	42	61	82	55

(i) Calculate the Spearman's rank correlation coefficient between h and c.

After collecting the data, the councillor thinks there is no correlation between hardship and the number of calls to the emergency services.

(ii) Test, at the 5% level of significance, the councillor's claim. State your hypotheses clearly. [Edexcel 6691 Q2 June 2011]

⑪ In order to assess whether increased expenditure in schools produces better examination results, a survey of all the secondary schools in England was conducted. Data on a random sample of 12 of these schools are shown below. The score shown is a measure of academic performance, a higher score indicating a higher success rate in examinations; expenditure is measured in thousands of pounds per student per year.

Rank correlation

Table 5.45

Score	1.54	1.50	1.49	1.22	1.19	1.11	1.09	1.06	1.05	0.97	0.88	0.68
Expenditure	1.70	3.95	2.75	1.95	2.35	1.45	2.40	2.05	2.15	2.30	1.75	2.10

(i) Calculate the value of Spearman's rank correlation coefficient for the data.

(ii) Perform an appropriate test at the 5% level, making clear what your hypotheses are. State clearly the conclusions to be drawn from the test.

Now suppose that the value of Spearman's rank correlation coefficient, calculated for *all* the secondary schools in England, is 0.15.

(iii) What conclusion would you now reach about any association between expenditure per student and examination success, and why? [MEI]

12 In a national survey into whether low death rates are associated with greater prosperity, a random sample of 14 areas was taken. The areas, arranged in order of their prosperity, are shown in the table below together with their death rates. (The death rates are on a scale for which 100 is the national average.)

Table 5.46

most prosperous   least prosperous

Area	A	B	C	D	E	F	G	H	I	J	K	L	M	N
Death rate	66	76	84	83	102	78	100	110	105	112	122	131	165	138

(i) Calculate an appropriate correlation coefficient and use it to test, at the 5% level of significance, whether or not there is such an association. State your hypotheses and your conclusion carefully.

(ii) A newspaper carried this story under the headline 'Poverty causes increased deaths'. Explain carefully whether or not the data justify this headline.

(iii) The data include no information on the age distributions in the different areas. Explain why such additional information would be relevant. [MEI]

13 The following data, referring to the ordering of perceived risk of 25 activities and technologies and actual fatality estimates, were obtained in a study in the United States. Use these data to test at the 5% significance level for positive correlation between

(i) the League of Women Voters and college students

(ii) experts and actual fatality estimates

(iii) college students and experts.

Comment on your results and identify any outliers in the three sets of bivariate data you have just used.

Table 5.47

	League of Women Voters	College students	Experts	Actual fatalities (estimates)
Nuclear power	1	1	18	16
Motor vehicles	2	4	1	3
Handguns	3	2	4	4
Smoking	4	3	2	1
Motorcycles	5	5	6	6
Alcoholic beverages	6	6	3	2
General (private) aviation	7	12	11	11
Police work	8	7	15	18
Surgery	9	10	5	8
Fire fighting	10	9	16	17
Large construction	11	11	12	12
Hunting	12	15	20	14
Mountain climbing	13	17	24	21
Bicycles	14	19	13	13
Commercial aviation	15	13	14	20
Electric power (non-nuclear)	16	16	8	5
Swimming	17	25	9	7
Contraceptives	18	8	10	19
Skiing	19	20	25	24
X-rays	20	14	7	9
High school & college football	21	21	22	23
Railroads	22	18	17	10
Power mowers	23	23	23	22
Home appliances	24	22	19	15
Vaccinations	25	24	21	25

Source: Shwing and Albers, *Societal Risk Assessment,* Plenum

14 A population analyst wishes to test how death rates and birth rates are correlated in European countries.

(i) State appropriate null and alternative hypotheses for the test. Justify the alternative hypothesis you have given.

A random sample of ten countries from Europe was taken and their death rates (x) and birth rates (y), each per 1000 population for 1997, were noted.

Table 5.48

x	9	9	7	12	11	10	7	13	8	7
y	14	9	13	13	10	11	16	9	16	12

(ii) Represent the data graphically.

(iii) Calculate the product moment correlation coefficient.

(iv) Carry out the hypothesis test at the 5% level of significance. State clearly the conclusion reached.

Rank correlation

In fact, the value of the product moment correlation coefficient for *all* the countries in Europe in 1997 was −0.555.

(v) What does this tell you about the relationship between death rates and birth rates in European countries?

(vi) State, giving a reason, whether your conclusion in part (iv) is still valid. [MEI]

KEY POINTS

1. A scatter diagram is a graph to illustrate bivariate data.

2. Notation for n pairs of observations (x, y)

$$S_{xx} = \Sigma(x_i - \bar{x})^2 = \Sigma x^2 - n\bar{x}^2$$

$$S_{yy} = \Sigma(y_i - \bar{y})^2 = \Sigma y^2 - n\bar{y}^2$$

$$S_{xy} = \Sigma(x_i - \bar{x})(y_i - \bar{y}) = \Sigma xy - n\bar{x}\bar{y}$$

3. Pearson's product moment correlation coefficient

$$r = \frac{S_{xy}}{\sqrt{S_{xx}S_{yy}}} = \frac{\Sigma(x_i - \bar{x})(y_i - \bar{y})}{\sqrt{\Sigma(x_i - \bar{x})^2 \times \Sigma(y_i - \bar{y})^2}} = \frac{\Sigma x_i y_i - n(\bar{x})(\bar{y})}{\sqrt{(\Sigma x_i^2 - n\bar{x}^2)(\Sigma y_i^2 - n\bar{y}^2)}}$$

4. Spearman's coefficient of rank correlation

$$r_s = 1 - \frac{6\Sigma d_i^2}{n(n^2 - 1)}$$

5. Hypothesis testing based on Pearson's product moment correlation coefficient

$H_0: \rho = 0$

$H_1: \rho > 0$ or $\rho < 0$ (one-tailed test) or $\rho \neq 0$ (two-tailed test).

Test the sample value, r, against the critical value, which depends on the number of pairs in the bivariate sample, n, and the significance level.

6. Hypothesis testing based on Spearman's coefficient of rank correlation

H_0: no association

H_1: positive association or negative association (one-tailed test) or some association (two-tailed test).

Test the sample value, r_s, against the critical value, which depends on the number of pairs in the bivariate sample, n, and the significance level.

LEARNING OUTCOMES

When you have completed this chapter you should:
- understand what bivariate data are and know the conventions for choice of axis for variables in a scatter diagram
- be able to use and interpret a scatter diagram
- interpret a scatter diagram produced by software
- be able to calculate Pearson's product moment correlation coefficient from raw data or summary statistics
- know when it is appropriate to carry out a hypothesis test using Pearson's product moment correlation coefficient
- be able to carry out hypothesis tests using the Pearson's product moment correlation coefficient and tables of critical values or the p-value from software
- use the Pearson's product moment correlation coefficient as an effect size
- be able to calculate Spearman's rank correlation coefficient from raw data or summary statistics
- be able to carry out hypothesis tests using Spearman's rank correlation coefficient and tables of critical values or the output from software
- decide whether a test based on r or r_s may be more appropriate, or whether neither is appropriate.

6 Linear regression

It is utterly implausible that a mathematical formula should make the future known to us, and those who think it can would once have believed in witchcraft.

Betrand de Jouvenel

Janice grows tomatoes in her greenhouse. Every year, she uses a liquid feed to try to get a bigger yield. One year, she wonders if there is a relationship between the amount of fertiliser she uses and the yield of the tomatoes. She does an experiment where she takes seven tomato plants and gives each of them a different amount of feed. The following table shows the amount of fertiliser, x ml, and the yield, y kg, for each of the plants.

> Notice that the amount of liquid feed is not a random variable; it is a controlled variable.
>
> You can see this from the way that the values of x go up in steps of 40. These are the doses given in ml.

Table 6.1

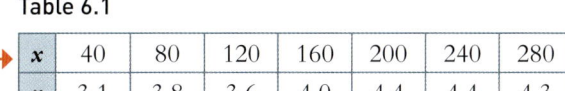

Figure 6.1 shows a scatter diagram for the data that Janice collected.

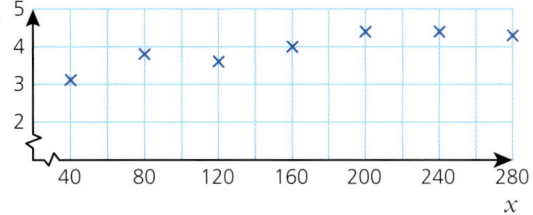

Figure 6.1

Looking at her scatter diagram, Janice thinks that there is probably a relationship between the amount of fertiliser and the yield. She considers drawing a line of best fit by eye but then a friend suggests that she should use a calculation to find the equation of a suitable line.

1 The least squares regression line

A correlation coefficient provides you with a measure of the level of linear association between the two variables in a bivariate distribution.

If this indicates that there is a relationship, your question will be 'What is it?' In the case of linear correlation, it can be expressed algebraically as a linear equation or geometrically as a straight line on the scatter diagram.

Before you do any calculations, you first need to look carefully at the two variables that give rise to your data. It is normal practice to plot the **dependent variable** on the vertical axis and the **independent variable** on the horizontal axis. In the example above, the independent variable was the dose the tomatoes were given. In many situations, the independent variable is the time at which measurements are made. Notice that these are non-random variables. The procedure that follows leads to the equation of the **regression line**, the line of best fit in these circumstances.

Look at the scatter diagram (Figure 6.2) showing the n points $A(x_1, y_1)$, $B(x_2, y_2)$, ..., $N(x_n, y_n)$. On it is marked a possible line of best fit l. If the line l passed through all the points there would be no problem since there would be perfect linear correlation. It does not, of course, pass through all the points and you would be very surprised if such a line did in any real situation.

S2 AS — The least squares regression line

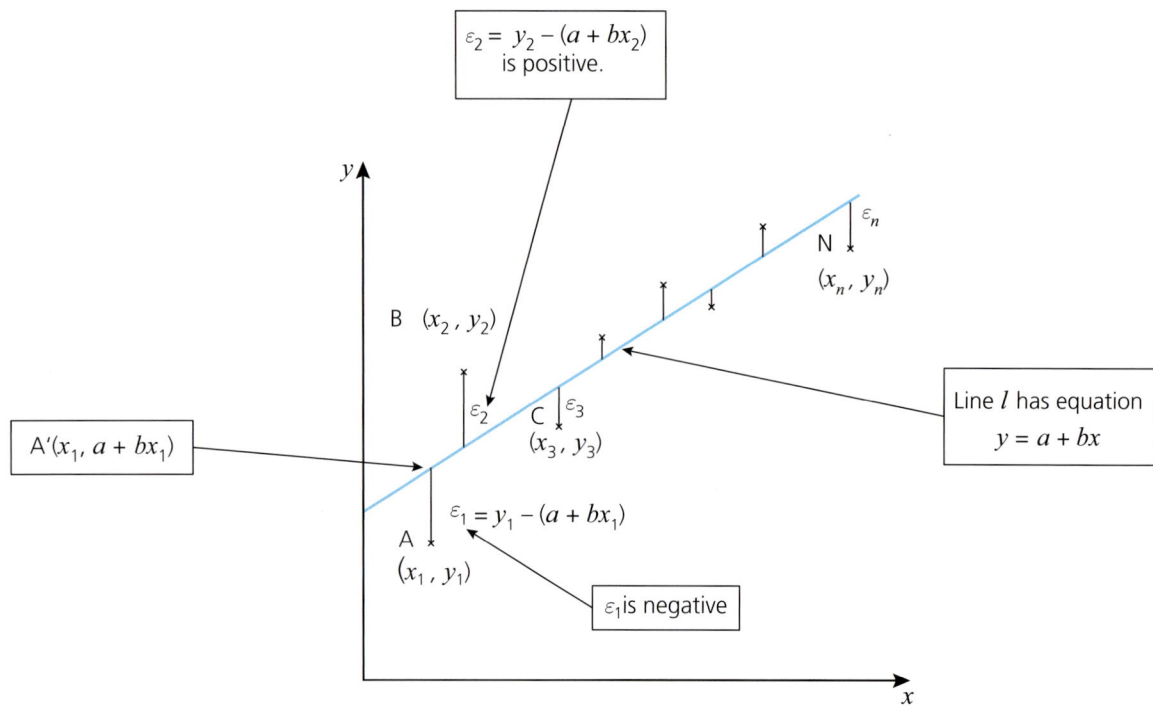

Figure 6.2 Bivariate data plotted on a scatter diagram with the regression line, l, $y = a + bx$, and the residuals $\varepsilon_1, \varepsilon_2, \cdots, \varepsilon_n$

By how much is it missing the points? The answer to that question is shown by the vertical lines from the points to the line. Their lengths $\varepsilon_1, \varepsilon_2, \cdots, \varepsilon_n$ are called the **residuals** and represent the variation which is not explained by the line l. The **least squares regression line** is the line which produces the least possible value of the sum of the squares of the residuals, $\varepsilon_1^2 + \varepsilon_2^2 + \cdots + \varepsilon_n^2$.

If the equation of the line l is $y = a + bx$, then it is easy to see that the point A′ on the diagram, directly above A, has co-ordinates $(x_1, a + bx_1)$ and so the corresponding residual, ε_1, is given by $\varepsilon_1 = y_1 - (a + bx_1)$. Similarly, for $\varepsilon_1, \varepsilon_2, \cdots, \varepsilon_n$.

> **Note**
>
> You have already met S_{xx} and S_{xy} in Chapter 5. They are defined as:
>
> $S_{xx} = \Sigma(x - \bar{x})^2$
> $\phantom{S_{xx}} = \Sigma x^2 - n\bar{x}^2$
>
> $S_{xy} = \Sigma(x - \bar{x})(y - \bar{y})$
> $\phantom{S_{xy}} = \Sigma xy - n\bar{x}\bar{y}$

The problem is to find the values of the constants a and b in the equation of the line l which make $\varepsilon_1^2 + \varepsilon_2^2 + \cdots + \varepsilon_n^2$ a minimum for any particular set of data, that is, to minimise

$$\left[y_1 - (a + bx_1)\right]^2 + \left[y_2 - (a + bx_2)\right]^2 + \cdots + \left[y_n - (a + bx_n)\right]^2$$

The mathematics involved in doing this is not particularly difficult. The resulting equation of the regression line is given below.

$$y - \bar{y} = b(x - \bar{x})$$

$$\text{where } b = \frac{S_{xy}}{S_{xx}}$$

For Janice's data on the yield from tomato plants you can find the equation of the regression line as follows.

Table 6.2

x	y	x^2	xy
40	3.1	1600	124
80	3.8	6400	304
120	3.6	14 400	432
160	4.0	25 600	640
200	4.4	40 000	880
240	4.4	57 600	1056
280	4.3	78 400	1204
Total	1120	27.6	224 000	4640

> As when calculating the correlation coefficient, the first step in calculating the equation of the regression line is to find the mean values of x and y.

$$\bar{x} = \frac{\Sigma x}{n} = \frac{1120}{7} = 160 \qquad \bar{y} = \frac{\Sigma y}{n} = \frac{27.6}{7} = 3.94\ldots$$

$$S_{xx} = \Sigma x^2 - n\bar{x}^2 = 224\,000 - 7 \times 160^2 = 44\,800$$

$$S_{xy} = \Sigma xy - n\bar{x}\,\bar{y} = 4640 - 7 \times 160 \times 3.94\ldots = 224 \qquad b = \frac{S_{xy}}{S_{xx}} = \frac{224}{44\,800} = 0.005$$

Hence the least squares regression line is given by

$$y - \bar{y} = b(x - \bar{x})$$

$$y - 3.94\ldots = 0.005(x - 160)$$

$$y = 3.14 + 0.005x$$

Notes

1. In the preceding work you will see that only variation in the y values has been considered. The reason for this is that the x values represent a non-random variable. That is why the residuals are vertical and not in any other direction. Thus $y_1, y_2, \ldots$ are values of a random variable y given by $y = a + bx + \varepsilon$, where ε is the residual variation, the variation that is not explained by the regression line.
2. This form of the regression line is often called the **regression line of y on x**.

Example 6.1

A patient is given a drip feed containing a particular chemical and its concentration in his blood is measured, in suitable units, at one hour intervals for the next five hours. The doctors believe the figures to be subject to random errors, arising both from the sampling procedure and the subsequent chemical analysis, but that a linear model is appropriate.

Table 6.3

Time, x (hours)	0	1	2	3	4	5
Concentration, y	2.4	4.3	5.2	6.8	9.1	11.8

> You can see from the intervals that x is a controlled variable.

(i) Find the equation of the regression line of y on x.
(ii) Illustrate the data and your regression line on a scatter diagram.
(iii) Estimate the concentration of the chemical in the patient's blood at
  (a) $3\frac{1}{2}$ hours,
  (b) 10 hours after treatment started.

Comment on the likely accuracy of your predictions.
(iv) Calculate the residuals for each data pair. Check that the sum of the residuals is zero and find the sum of the squares of the residuals.

The least squares regression line

Solution

(i) $n = 6$

Table 6.4

x	y	x^2	xy
0	2.4	0	0.0
1	4.3	1	4.3
2	5.2	4	10.4
3	6.8	9	20.4
4	9.1	16	36.4
5	11.8	25	59.0
15	39.6	55	130.5

$$\bar{x} = \frac{\Sigma x}{n} = \frac{15}{6} = 2.5 \quad \bar{y} = \frac{\Sigma y}{n} = \frac{39.6}{6} = 6.6$$

$$S_{xx} = \Sigma x^2 - n\bar{x}^2 = 55 - 6 \times 2.5^2 = 17.5$$

$$S_{xy} = \Sigma xy - n\bar{x}\bar{y} = 130.5 - 6 \times 2.5 \times 6.6 = 31.5$$

$$b = \frac{S_{xy}}{S_{xx}} = \frac{31.5}{17.5} = 1.8$$

Hence the least squares regression line is given by

$$y - \bar{y} = b(x - \bar{x})$$

$$y - 6.6 = 1.8(x - 2.5)$$

$$y = 2.1 + 1.8x$$

(ii)

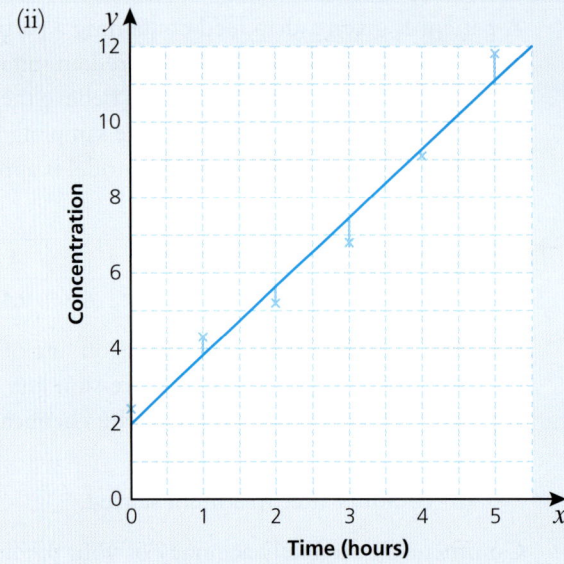

Figure 6.3

In this case, the value of x from which you want to predict the value of y lies within the range of data values.

This is an example of **interpolation** (meaning literally within the points).

In this second case, the value of x from which you want to predict the value of y lies outside the range of data values.

This is an example of **extrapolation** (meaning literally outside the points).

> **Note**
>
> So the actual data point (x, y) is in the same vertical line as $(x, \hat{y})$ on scatter diagram.

> **Note**
>
> If the same procedure were carried out using the same data with any other line the sum of the squares of the residuals would work out to be greater than 1.52.

(iii) When $x = 3.5$,
$y = 2.1 + 1.8 \times 3.5$
$= 8.4$.

When $x = 10$,
$y = 2.1 + 1.8 \times 10$
$= 20.1$.

The concentration of 8.4 lies between the measured values of the concentration at a time between 3 hours and 4 hours, so the prediction seems quite reasonable.

The time ten hours is a long way outside the set of data times; there is no indication that the linear relationship can be extrapolated to such a time, even though there seems to be a good fit, so the prediction is probably unreliable.

(iv) For each pair of data, (x, y), the corresponding point on the regression line is denoted by $(x, \hat{y})$. So the predicted value is $\hat{y} = 2.1 + 1.8x$.

Corresponding values of y and $\hat{y}$ are tabulated below, together with the residuals and their squares.

Table 6.5

x	y	$\hat{y}$	$y - \hat{y}$	$(y - \hat{y})^2$
0	2.4	2.1	0.3	0.09
1	4.3	3.9	0.4	0.16
2	5.2	5.7	−0.5	0.25
3	6.8	7.5	−0.7	0.49
4	9.1	9.3	−0.2	0.04
5	11.8	11.1	0.7	0.49
15	39.6	39.6	0.0	1.52

You can see that the sum of the residuals, $\Sigma(y - \hat{y})$ is zero, and that the sum of the squares of the residuals, $\Sigma(y - \hat{y})^2$, is 1.52.

USING ICT

A spreadsheet

Setting up a spreadsheet to work out the coefficients a and b of the least squares regression line, $y = a + bx$, is fairly straightforward; it will also display a scatter diagram. This is illustrated in Figure 6.4 using the data from the previous example. It is obtained by carrying out these steps.

1 Enter the data in two columns. ← They are columns A and B in Figure 6.4.

2 Use the formulae provided by your spreadsheet, for example SLOPE and INTERCEPT. In Figure 6.4 they are in cells B12 and B13.

S2 AS — The least squares regression line

3 To obtain a scatter diagram of your data, highlight the two columns of data and then choose the relevant options, for example Insert followed by scatter diagram. Customise it as necessary.

4 You can display the least squares regression line on the scatter diagram by highlighting the data points on the diagram, right-clicking and choosing the relevant option such as 'add trendline'. You can also display the equation by choosing the relevant option when you add the 'trendline'.

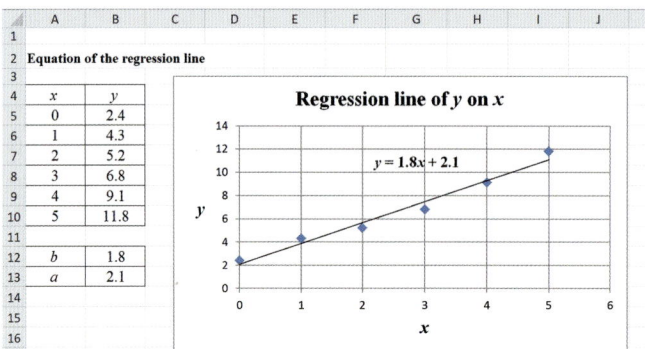

Figure 6.4

Exercise 6.1

① For the following bivariate data
  (i) Find the mean values of x and y.
  (ii) Find the values of S_{xx} and S_{xy}.
  (iii) Find the equation of the least squares regression line of y on x.

  Table 6.6

x	5	10	15	20	25
y	30	28	27	27	21

② Calculate the equation of the regression line of y on x for the following distribution and use it to estimate the value of y when $x = 42$.

  Table 6.7

x	25	30	35	40	45	50
y	78	70	65	58	48	42

③ A garage kept records of the volume, X litres, of unleaded petrol that customers bought when filling their cars.

  The data were analysed and it was suggested that a Normal distribution with a mean of 25 and a standard deviation of 10 gives a suitable model.

  (i) Determine:
    (a) $P(X < 33)$
    (b) $P(X > 18)$
    (c) $P(18 < X < 33)$.

  In October 2017, the garage was charging £1.20 per litre for unleaded petrol. Adam visited the garage once during this month and filled his car with unleaded petrol.

  (ii) Calculate the probability that the cost of Adam's petrol was more than £50.

  (iii) Give **two** reasons, in context, why the model $N(25, 10^2)$ is unlikely to be valid for a visit by **any** customer making a purchase at this garage in October 2017.

④ The speed of a car, v m s, at time t s after it starts to accelerate is shown in the table below, for $0 \leq t \leq 10$.

Table 6.8

t	0	1	2	3	4	5	6	7	8	9	10
v	0	3.0	6.8	10.2	12.9	16.4	20.0	21.4	23.0	24.6	26.1

($\Sigma t = 55$, $\Sigma v = 164.4$, $\Sigma t^2 = 385$, $\Sigma v^2 = 3267.98$, $\Sigma tv = 1117.0$.)

The relationship between t and v is initially modelled by using all the data above and calculating a single regression line.

(i) Plot a scatter diagram of the data, with t on the horizontal axis and v on the vertical axis.

(ii) Using all the data given, calculate the equation of the regression line of v on t. Give numerical coefficients in your answers correct to 3 significant figures.

(iii) Calculate the product moment correlation coefficient for the given data.

(iv) Comment on the validity of modelling the data by a single straight line and on the answer obtained in part (iii).

⑤ Maria is concerned that her phone battery is running out too quickly. Each evening she records how long her phone was on that day, x hours, and the percentage charge remaining, y %. She recharges her phone overnight each night. Her results are shown in the table.

Table 6.9

x	7	8	9	10	11	12	13	14	15
y	27	25	21	19	16	11	9	5	2

Maria models the relationship between x and y using a linear model.

(i) Calculate the equation of the least squares regression line $y = a + bx$.

(ii) Interpret the value of a in context.

(iii) Interpret the value of b in context.

(iv) Which of these values should cause Maria the greater concern, and why?

⑥ The table shows data on the rate of inflation and minimum hourly wage for a country at ten year intervals.

Table 6.10

Year	1940	1950	1960	1970	1980	1990	2000	2010
Rate of inflation (%)	0.0	3.8	1.4	5.6	12.5	6.1	3.4	1.6
Minimum hourly wage ($)	0.30	0.50	1.20	1.80	3.00	3.30	4.20	7.20

(i) Explain how you know that a linear model is not appropriate to model the minimum wage as a function of the rate of inflation.

(ii) Use the data to calculate an appropriate least squares regression line for wage against year.

A politician predicts that by 2020 the rate of inflation will be 0% and the minimum hourly wage will be $5.76.

(iii) Comment on each of these predictions.

7. In an experiment on memory, five groups of people (chosen randomly) were given varying lengths of time to memorise the same list of 40 words. Later, they were asked to recall as many words as possible from the list. The table below shows the average number of words recalled, y, and the time given, t s.

Table 6.11

t	20	40	60	80	100
y	12.1	18.5	22.8	24.6	24.0

(i) Plot the data on a scatter diagram.

(ii) Calculate the equation of the regression line for y on t.

(iii) Use your regression line to predict y when $t = 30$ and $t = 160$. Comment on the usefulness or otherwise of these results.

(iv) Discuss briefly whether the regression line provides a good model or whether there is a better way of modelling the relationship between y and t.
[MEI]

8. A farmer is investigating the relationship between the density at which a crop is planted and the quality. By using more seed per hectare (x) he can increase the yield, but he suspects that the percentage of high-quality produce (y) may fall. The farmer collects data which he enters into the spreadsheet below. He also uses the spreadsheet to produce a scatter diagram.

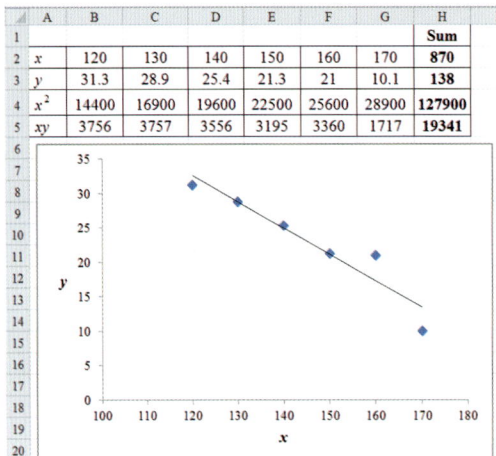

Figure 6.5

(i) Discuss how suitable a straight line model would be for the relationship between y and x.

(ii) Use the sums in the spreadsheet to help you calculate the equation of the regression line of y on x.

(iii) Obtain from your regression line the predicted values of y at $x = 145$ and $x = 180$. Comment, with reasons, on the likely accuracy of these predictions.

⑨ A car manufacturer is testing the braking distance for a new model of car. The table shows the braking distance, y m, for different speeds, x km h^{-1}, when the brakes were applied.

Table 6.12

Speed of car, x km h^{-1}	30	50	70	90	110	130
Braking distance, y m	25	50	85	155	235	350

(i) Plot a scatter diagram on graph paper.

(ii) Calculate the equation of the regression line of y on x. Draw the line on your scatter diagram, together with the residuals.

(iii) Use your regression equation to predict values of y when $x = 100$ and $x = 150$. Comment, with reasons, on the likely accuracy of these predictions.

(iv) Discuss briefly whether the regression line provides a good model or whether there is a better way of modelling the relationship between y and x. [MEI]

⑩ The authorities in a school are concerned to ensure that their students enter appropriate mathematics examinations. As part of a research project into this they wish to set up a performance-prediction model. This involves the students taking a standard mid-year test, based on the syllabus and format of the final end-of-year examination.

The school bases its model on the belief that in the final examination students will get the same things right as they did in the mid-year test and, in addition, a proportion, p, of the things they got wrong.

Consequently, a student's final mark, y%, can be predicted on the basis of his or her test mark, x%, by the relationship

$$y = x + p \times (100 - x)$$

Final mark = test mark + $p \times$ (the marks the student did not get on the test)

Investigate this model, using the following bivariate data. Start by finding the y on x regression line and then rearrange it to estimate p.

Table 6.13

x	y	x	y	x	y	x	y
40	55	27	44	60	72	46	70
22	40	32	50	50	70	70	85
10	25	26	49	90	95	33	63
46	68	68	76	30	50	40	60
66	75	54	66	64	80	56	57
8	32	68	70	100	100	45	55
48	69	88	92	44	50	78	85
58	66	48	59	58	62	68	80
50	51	82	90	54	60	78	85
80	85	66	76	24	31	89	91

2 Residual sum of squares

The residuals can be used informally to check whether the regression line is reasonable and to identify possible outliers. Any outliers will have large (positive or negative) residuals.

The residual sum of squares (RSS) can be used to help you decide if a statistical model is a good fit for the data. It measures the overall difference between the data and the values predicted by the model.

The residual sum of squares (RSS) is given by the formula

$$\text{RSS} = S_{yy} - \frac{(S_{xy})^2}{S_{xx}}$$

This is the same as squaring each of the residuals and summing the result.

The derivation of the formula for RSS is not required.

Example 6.2

The VR (verbal reasoning) and NVR (non-verbal reasoning) scores for a group of 60 students were obtained from tests given to the students. The VR score is labelled as x and the NVR score as y.

Summary figures are as follows:

$\Sigma x = 7105$, $\Sigma y = 7132$, $\Sigma x^2 = 846357$, $\Sigma y^2 = 853620$, $\Sigma xy = 846371$, $n = 60$.

The correlation coefficient between x and y is 0.337.

The following scatter diagram illustrates the data.

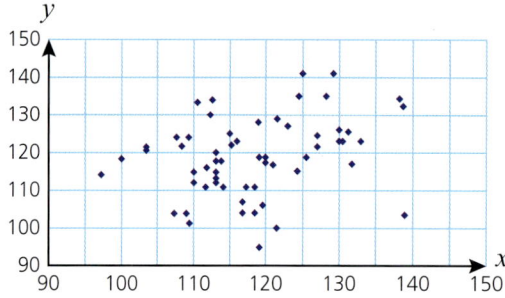

Figure 6.6

(i) Calculate the equation of the regression line of y on x.

(ii) Use the regression line to predict the NVR score for a student whose VR score was 120.

(iii) Calculate the residual sum of squares (RSS).

(iv) Comment on how reliable your estimate is likely to be.

Solution

(i) $y - \bar{y} = \dfrac{S_{xy}}{S_{xx}}(x - \bar{x})$

$S_{xx} = \Sigma x^2 - \dfrac{(\Sigma x)^2}{n} = 846\,357 - \dfrac{7105^2}{60} = 5006.583$

$S_{xy} = \Sigma xy - \dfrac{(\Sigma x)(\Sigma y)}{n} = 846\,371 - \dfrac{7105 \times 7132}{60} = 1823.333$

The equation of the regression line is

$y - \dfrac{7132}{60} = \dfrac{1823.333}{5006.583}\left(x - \dfrac{7105}{60}\right)$

So $y - 118.867 = 0.364(x - 118.417)$

So $y = 0.364x + 75.7$

(ii) The estimate is $y = 0.364 \times 120 + 75.7 = 119.4$

(iii) $S_{yy} = \Sigma y^2 - \dfrac{(\Sigma y)^2}{n} = 853\,620 - \dfrac{7132^2}{60} = 5862.933$.

$\text{RSS} = 5862.933 - \dfrac{1823.333^2}{5006.583} = 5198.899$

(iv) There are several pieces of evidence that can be used to assess the reliability of the estimate.

- The scatter diagram shows that the points do not lie close to a straight line.
- The correlation coefficient $r = 0.377$ is not particularly high.
- The value of S_{xy} is not very different from S_{xx} and S_{yy}.
- The value of RSS = 5198.899; this is the same as squaring each of the residuals and summing the result. There are $n = 60$ items of data so the average of the squared residuals is 86.65.

The evidence suggests that the linear model is not a very close fit so the estimate is likely to be unreliable.

Exercise 6.2

1. (i) Use a spreadsheet to find the equation of the line of regression for these bivariate data.

   Table 6.14

x	6.3	8.7	4.2	3.7	6.5
y	24.7	39.2	19.6	20.3	26.2

   (ii) Calculate the residual sum of squares.

2. (i) Calculate the equation of the line of regression for these bivariate data.

   Table 6.15

x	69.0	66.5	71.5	63.6	98.8	52.4	64.0	79.9
y	297.9	310.5	300.5	296.7	416.3	261.1	285.9	321.8

   (ii) Calculate the residual sum of squares.

S2 AS — Residual sum of squares

③ A long distance lorry driver recorded the distance travelled, m miles, and the amount of fuel used, f litres, each day. Summarised below are data from the driver's records for a random sample of 8 days.

The data are coded such that $x = m - 250$ and $y = f - 100$.

$\Sigma x = 130 \qquad \Sigma y = 48 \qquad \Sigma xy = 8880 \qquad S_{xx} = 20\,487.5$

(i) Find the equation of the regression line of y on x in the form $y = a + bx$.

(ii) Hence find the equation of the regression line of f on m.

(iii) Predict the amount of fuel used on a journey of 235 miles.

[Edexcel 6683 Q3 June 2005]

④ A second-hand car dealer has 10 cars for sale. She decides to investigate the link between the age of the cars, x years, and the mileage, y thousand miles. The data collected from the cars are shown in the table below.

Table 6.16

Age, x/years	2	2.5	3	4	4.5	4.5	5	3	6	6.5
Mileage, y/thousands	22	34	33	37	40	45	49	30	58	58

[You may assume that $\Sigma x = 41$, $\Sigma y = 406$, $\Sigma x^2 = 188$, $\Sigma xy = 1818.5$]

(i) Find S_{xx} and S_{xy}.

(ii) Find the equation of the least squares regression line in the form $y = a + bx$. Give the values of a and b to 2 decimal places.

(iii) Give a practical interpretation of the slope b.

(iv) Using your answer to part (ii), find the mileage predicted by the regression line for a 5-year old car.   [Edexcel 6683 Q4 Jan 2008]

⑤ A medical statistician wishes to carry out a hypothesis test to see if there is any correlation between the head circumference and body length of newly born babies.

(i) State appropriate null and alternative hypotheses for the test.

A random sample of 20 newly born babies have their head circumference, x cm, and body length, y cm, measured. This bivariate sample is illustrated in the scatter diagram.

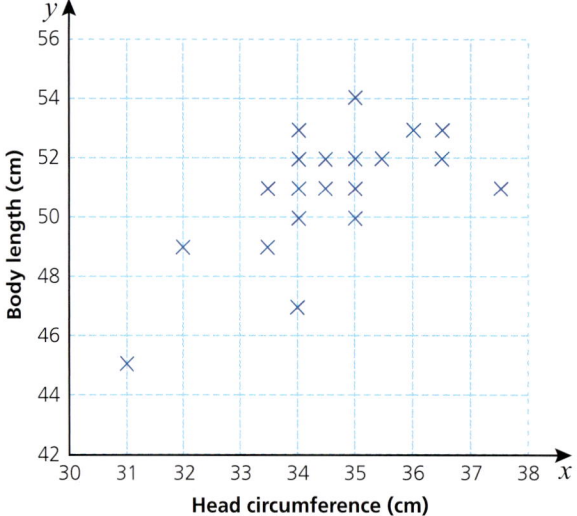

Figure 6.7

Summary statistics for this data set are as follows.

$n = 20$, $\Sigma x = 691$, $\Sigma y = 1018$, $\Sigma x^2 = 23\,917$, $\Sigma y^2 = 51\,904$, $\Sigma xy = 35\,212.5$

(ii) Calculate the product moment correlation coefficient for the data. Carry out the hypothesis test at the 1% significance level, stating the conclusion clearly. What assumption is necessary for the test to be valid?

(iii) Use a suitable regression line to estimate the body lengths of babies whose head circumferences are 36.0 cm and 29.0 cm.

(iv) Comment, with reasons, on the likely accuracy of these estimates. [MEI]

⑥ A fruit grower is investigating a crop of plums. She measures the lengths and circumferences in mm of ten randomly selected plums. She uses a spreadsheet to analyse the results. The spreadsheet below shows her analysis.

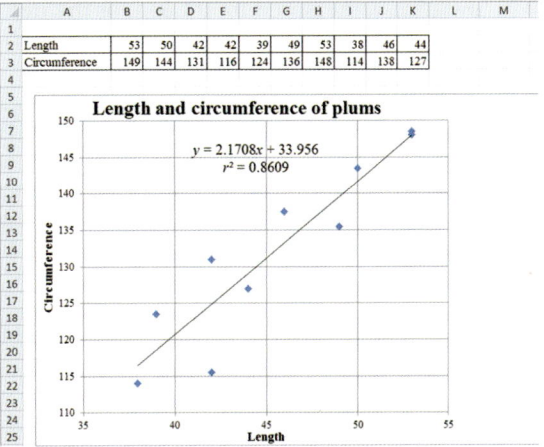

Figure 6.8

(i) The equation given is for the regression line of y on x. Estimate the circumference of a plum of length 40 mm.

(ii) Comment on the likely accuracy of this estimate.

(iii) Give a reason why it would not be sensible to use this equation to estimate the circumference of a plum of length 70 mm.

(iv) Give a reason why it would not be sensible to use this equation to estimate the length of a plum of circumference 130 mm.

⑦ A teacher is monitoring the progress of students using a computer-based revision course. The improvement in performance, y marks, is recorded for each student along with the time, x hours, that the student spent using the revision course. The results for a random sample of 10 students are recorded below.

Table 6.17

x hours	1.0	3.5	4.0	1.5	1.3	0.5	1.8	2.5	2.3	3.0
y marks	5	30	27	10	−3	−5	7	15	−10	20

[You may use $\Sigma x = 21.4$, $\Sigma y = 96$, $\Sigma x^2 = 57.22$, $\Sigma xy = 313.7$]

(i) Calculate S_{xx} and S_{xy}.

(ii) Find the equation of the least squares regression line of y on x in the form $y = a + bx$.

(iii) Give an interpretation of the gradient of your regression line.

Rosemary spends 3.3 hours using the revision course.

(iv) Predict her improvement in marks.

Lee spends 8 hours using the revision course claiming that this should give him an improvement in performance of over 60 marks.

(v) Comment on Lee's claim. [Edexcel 6683 Q1 Jan 2009]

Residual sum of squares

⑧ A farmer collected data on the annual rainfall, x cm, and the annual yield of peas, p tonnes per acre. The data for annual rainfall was coded using $v = \dfrac{x-5}{10}$ and the following statistics were found.

$S_{vv} = 5.753 \quad S_{pv} = 1.688 \quad S_{pp} = 1.168 \quad \bar{p} = 3.22 \quad \bar{v} = 4.42$

(i) Find the equation of the regression line of p on v in the form $p = a + bv$.

(ii) Using your regression line estimate the annual yield of peas per acre when the annual rainfall is 85 cm. [Edexcel 6683 Q4 Jan 2011]

⑨ A travel agent sells flights to different destinations from *Beerow* airport. The distance d, measured in 100 km, of the destination from the airport and the fare £f are recorded for a random sample of 6 destinations.

Table 6.18

Destination	A	B	C	D	E	F
d	2.2	4.0	6.0	2.5	8.0	5.0
f	18	20	25	23	32	28

[You may use $\Sigma d^2 = 152.09 \quad \Sigma f^2 = 3686 \quad \Sigma fd = 723.1$]

(i) On graph paper, draw a scatter diagram to illustrate this information.

(ii) Explain why a linear regression model may be appropriate to describe the relationship between f and d.

(iii) Calculate S_{dd} and S_{fd}.

(iv) Calculate the equation of the regression line of f on d giving your answer in the form $f = a + bd$.

(v) Give an interpretation of the value of b.

Jane is planning her holiday and wishes to fly from *Beerow* airport to a destination t km away. A rival travel agent charges 5p per km.

(vi) Find the range of values of t for which the first travel agent is cheaper than the rival. [Edexcel 6683 Q6 June 2010]

⑩ The age, t years, and weight, w grams, of each of 10 coins were recorded. These data are summarised below.

$\Sigma t^2 = 2688 \quad \Sigma tw = 1760.62 \quad \Sigma t = 158 \quad \Sigma w = 111.75 \quad S_{ww} = 0.16$

(i) Find S_{tt} and S_{tw} for these data.

(ii) Calculate, to 3 significant figures, the product moment correlation coefficient between t and w.

(iii) Find the equation of the regression line of w on t in the form $w = a + bt$.

(iv) State, with a reason, which variable is the explanatory variable.

(v) Using this model, estimate

(a) the weight of a coin which is 5 years old

(b) the effect of an increase of 4 years in age on the weight of a coin.

It was discovered that a coin in the original sample, which was 5 years old and weighed 20 grams, was a fake.

(vi) State, without any further calculations, whether the exclusion of this coin would increase or decrease the value of the product moment correlation coefficient. Give a reason for your answer.

[Edexcel 6683 Q5 Jan 2012]

⑪ A car manufacturer is introducing a new model. The car is tested for fuel economy three times at each of four different speeds. The values of the fuel economy, y miles per gallon (mpg), at each of the speeds, x miles per hour (mph), are displayed in the following table.

Table 6.19

x	40	40	40	50	50	50	60	60	60	70	70	70
y	52.7	53.8	54.5	48.1	49.7	51.3	43.3	41.1	48.0	37.5	42.0	44.7

$n = 12, \Sigma x = 660, \Sigma y = 566.7, \Sigma x^2 = 37\,800, \Sigma xy = 30\,533, \Sigma y^2 = 27\,093.81$

(i) Explain which of the variables is controlled and how you can tell that it is.

(ii) Represent the data by a scatter diagram, drawn on graph paper.

(iii) Calculate the equation of the regression line of y on x and plot it on your scatter diagram.

(iv) Hence predict the fuel economy of the car at speeds of

    (a) 45 mph

    (b) 65 mph.

(v) Use your scatter diagram to compare the reliability of your predictions in part (iv).

What do your comments suggest about the validity of a least squares regression line for this data set?

(vi) Calculate the residual sum of squares. [MEI adapted]

⑫ The approximate population of the world in billions from the year 1900 to 2000 is displayed in the following table.

Table 6.20

Year, x	1900	1910	1920	1930	1940	1950	1960	1970	1980	1990	2000
Population, y	1.67	1.75	1.86	2.07	2.30	2.56	3.04	3.71	4.45	5.28	6.08

(i) Draw a scatter diagram to illustrate the data.

(ii) Explain why it would not be sensible to find the equation of the regression line of Population on Year.

(iii) Let $z = x - 1900$. Display the values of z and of z^2, together with y in a new table.

(iv) Draw a scatter diagram with z^2 on the horizontal axis and y on the vertical axis.

(v) Calculate the equation of the regression line of y on z^2 and plot it on your scatter diagram.

(vi) Hence estimate the population of the world in

    (a) the year 1995   (b) the year 2100.

(vii) Comment on the reliability of each of your estimates.

Residual sum of squares

KEY POINTS

1. The equation of the y on x regression line $y = a + bx$ is given by

$$y - \bar{y} = b(x - \bar{x}), \text{ where } b = \frac{S_{xy}}{S_{xx}} = \frac{\Sigma(x - \bar{x})(y - \bar{y})}{\Sigma(x - \bar{x})^2} = \frac{\Sigma xy - n\bar{x}\bar{y}}{\Sigma x^2 - n\bar{x}^2}$$

$$\Rightarrow a = \bar{y} - b\bar{x}$$

2. For any data pair (x, y) the predicted value of y is

$$\hat{y} = a + bx \Rightarrow \text{ the residual is } \varepsilon = y - \hat{y}$$

3. The sum of the residuals $\Sigma\varepsilon = 0$.
4. The least squares regression line minimises the sum of the squares of the residuals, $\Sigma\varepsilon^2$.
5. When using a regression line for prediction:
   - a value within the data range (interpolation) is more likely to be predicted reliably than a value beyond the data values (extrapolation)
   - The data points should lie close to a straight line.
6. The residual sum of squares (RSS) is given by the formula

$$\text{RSS} = S_{yy} - \frac{(S_{xy})^2}{S_{xx}}$$

This is the same as squaring each of the residuals and summing the result.

LEARNING OUTCOMES

When you have completed this chapter you should be able to:
- obtain the equation of the least squares regression line for a random variable on a non-random variable, using raw data or summary statistics
- use the regression line as a model to estimate a value of the random variable and know when it is appropriate to do so
- know the meaning of the term residual and be able to calculate and interpret residuals
- check how well the model fits the data
- calculate and interpret the residual sum of squares.

7 Continuous random variables

A theory is a good theory if it satisfies two requirements: It must accurately describe a large class of observations on the basis of a model that contains only a few arbitrary elements, and it must make definite predictions about the results of future observations.

Stephen Hawking,
A Brief History of Time

Lucky escape for local fisherman

Local fisherman George Sutherland stared death in the face yesterday as he was plucked from the deck of his 56 ft boat, the *Belle Star*, by a freak wave. Only the quick thinking of his brother James, who grabbed hold of his legs, saved George from a watery grave.

'It was a bad day and suddenly this lump of water came down on us,' said George. 'It was a wave in a million, higher than the mast of the boat, and it caught me off guard'.

Hero James is a man of few words. 'All in a day's work' was his only comment.

Continuous random variables

> **Note**
> The data are historical. The last weather ship, the Polarfront, was withdrawn from service in 2010.

Freak waves do occur and they can have serious consequences in terms of damage to shipping, oil rigs and coastal defences, sometimes resulting in loss of life. It is important to estimate how often they will occur, and how high they will be. Was George Sutherland's one in a million estimate for a wave higher than the mast of the boat (11 m) at all accurate?

Before you can answer this question, you need to know the **probability density** of the heights of waves at that time of the year in the area where the *Belle Star* was fishing. The graph in Figure 7.1 shows this sort of information; it was collected in the same season of the year as the Sutherland accident by the Offshore Weather Ship *Juliet* in the North Atlantic.

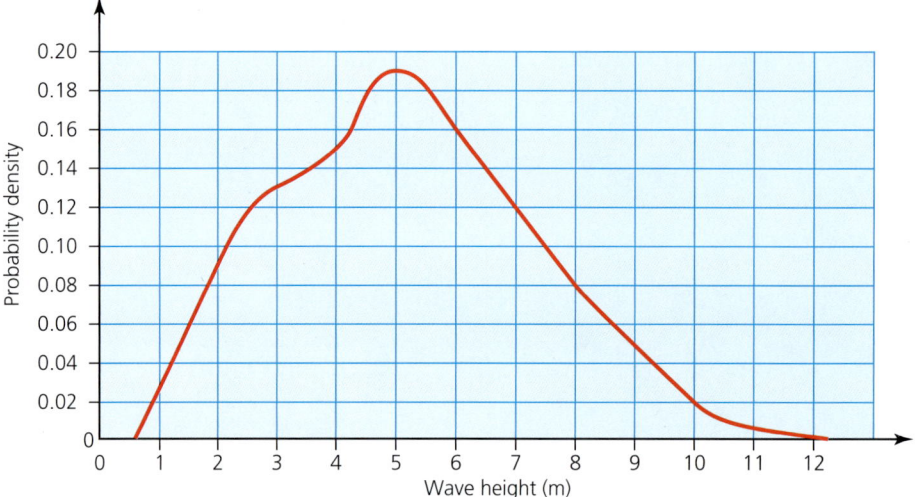

Figure 7.1

To obtain Figure 7.1 a very large amount of wave data had to be collected. This allowed the class interval widths of the wave heights to be sufficiently small for the outline of the curve to acquire this shape. It also ensured that the sample data were truly representative of the population of waves at that time of the year.

In a graph such as Figure 7.1, the vertical scale is a measure of probability density. Probabilities are found by estimating the area under the curve. The total area is 1.0, meaning that, effectively, all waves at this place have heights between 0.6 and 12.0 m (see Figure 7.2).

If this had been the place where the *Belle Star* was situated, the probability of encountering a wave at least 11 m high would have been 0.003, about 1 in 300. Clearly, George's description of it as 'a wave in a million' was not justified purely by its height. The fact that he called it a 'lump of water' suggests that perhaps it may have been more remarkable for its steep sides than its height.

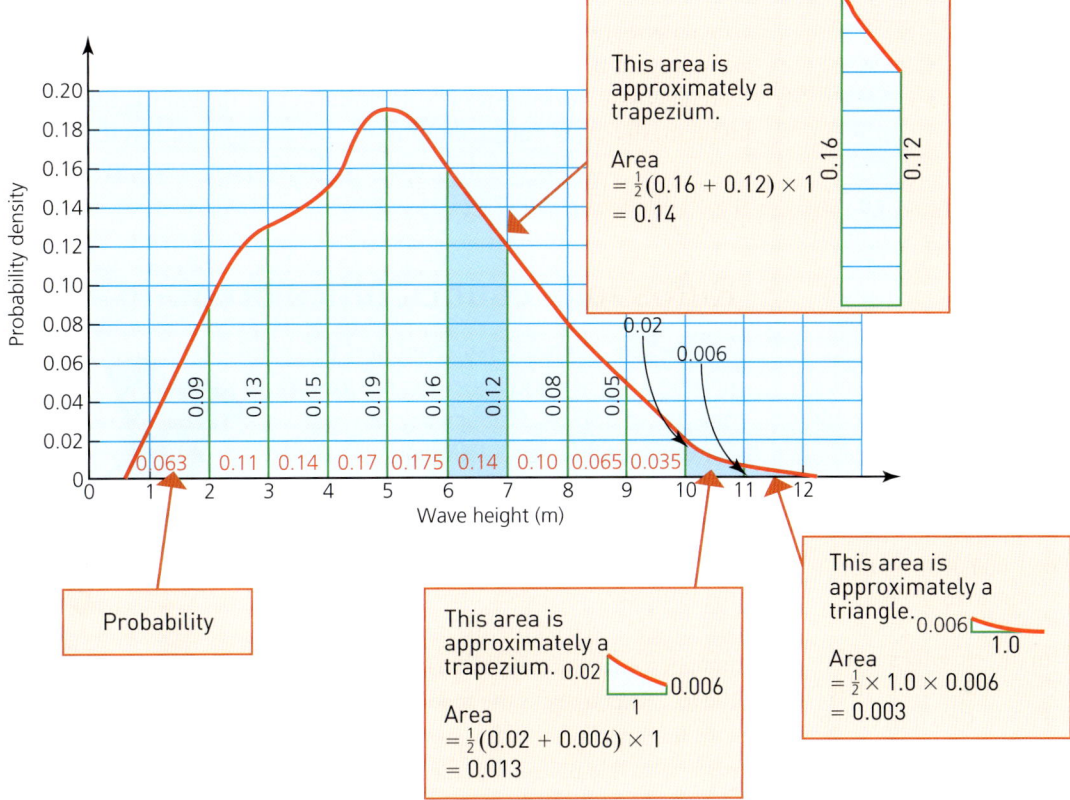

Figure 7.2

1 Probability density function

Discussion point
Is it reasonable to describe the height of a wave as **random**?

In the wave height example, the curve was determined experimentally, using equipment on board the Offshore Weather Ship *Juliet*. The curve is continuous because the random variable, the wave height, is continuous and not discrete. The possible heights of waves are not restricted to particular steps (say every 0.5 m), but may take any value within a range.

A function represented by a curve of this type is called a **probability density function**, often abbreviated to p.d.f. The probability density function of a continuous random variable, X, is usually denoted by $f(x)$. If $f(x)$ is a p.d.f. it follows that:

Note
In the wave heights example, the probability density function has quite a complicated curve and so it is not possible to find a simple algebraic expression with which to model it.

- $f(x) \geq 0$ for all x — You cannot have negative probabilities.

- $\int_{\text{All values of } x} f(x)\,dx = 1$ — The total area under the curve is 1.

For a continuous random variable with probability density function $f(x)$, the probability that X lies in the interval $[a, b]$ is given by

$$P(a \leq X \leq b) = \int_a^b f(x)\,dx$$

Probability density function

> **Note: Class boundaries**
>
> If you were to ask the question 'What is the probability of a randomly selected wave being exactly 2 m high?' the answer would be zero. If you measured a likely looking wave to enough decimal places (assuming you could do so), you would eventually come to a figure which was not zero. The wave height might be 2.01... m or 2.000 003... m but the probability of it being exactly 2 m is infinitesimally small. Consequently, in theory, it makes no difference whether you describe the class interval from 2 to 2.5 m as $2 < h < 2.5$ or as $2 \leqslant h \leqslant 2.5$.
>
> However, in practice, measurements are always rounded to some extent. The reality of measuring a wave's height means that you would probably be quite happy to record it to the nearest 0.1 m and get on with the next wave. So, in practice, measurements of 2.0 m and 2.5 m probably will be recorded, and intervals have to be defined so that it is clear which class they belong to. You would normally expect < at one end of the interval and ≤ at the other: either $2 \leqslant h < 2.5$ or $2 < h \leqslant 2.5$. In either case, the probability of the wave being within the interval would be given by $\int_2^{2.5} f(x)\,dx$.

Most of the techniques in this chapter assume that you do, in fact, have a convenient algebraic expression with which to work. However, the methods are still valid if this is not the case, but you would need to use numerical, rather than algebraic, techniques for integration and differentiation. In the high-wave incident, the areas corresponding to wave heights of less than 2 m and of at least 11 m were estimated by treating the shape as a triangle: other areas were approximated by trapezia.

Rufus foils council office break-in

Somewhere an empty-pocketed thief is nursing a sore leg and regretting the loss of a pair of trousers. Council porter Fred Lamming, and Rufus, a wiry haired Jack Russell, were doing a late-night check round the Council head office when they came upon the intruder on the ground floor. 'I didn't need to say anything,' Fred told me; 'Rufus went straight for him and grabbed him by the leg.' After a tussle, the man's trousers tore, leaving Rufus with a mouthful of material while the man made good his escape out of a window.

Following the incident, the local Council are looking at an electronic

security system. 'Rufus won't live for ever,' explained Council leader Sandra Martin.

Example 7.1

The local Council are thinking of fitting an electronic security system inside head office. They have been told by manufacturers that the lifetime, X years, of the system they have in mind has the p.d.f.:

$$f(x) = \frac{3x(20 - x)}{4000} \quad \text{for } 0 \leqslant x \leqslant 20.$$

$$\text{and } f(x) = 0 \quad \text{otherwise.}$$

(i) Show that the manufacturers' statement is consistent with f(x) being a probability density function.

(ii) Find the probability that:

(a) it fails in the first year

(b) it lasts ten years but then fails in the next year.

Solution

(i) The condition $f(x) \geqslant 0$ for all values of x between 0 and 20 is satisfied, as shown by the graph of f(x), Figure 7.3.

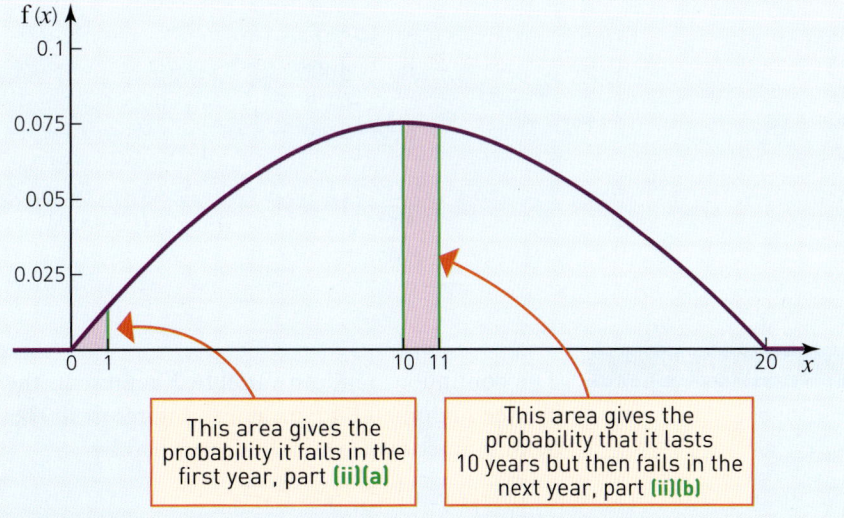

This area gives the probability it fails in the first year, part (ii)(a)

This area gives the probability that it lasts 10 years but then fails in the next year, part (ii)(b)

Figure 7.3

The other condition is that the area under the curve is 1.

$$\text{Area} = \int_{-\infty}^{\infty} f(x)dx = \int_{0}^{20} \frac{3x(20-x)}{4000}dx$$

$$= \frac{3}{4000}\int_{0}^{20}(20x - x^2)dx$$

$$= \frac{3}{4000}\left[10x^2 - \frac{x^3}{3}\right]_{0}^{20}$$

$$= \frac{3}{4000}\left[10 \times 20^2 - \frac{20^3}{3}\right]$$

$$= 1, \text{ as required.}$$

> **Note**
>
> The general rule is:
> $$F(x) = P(X \leq x) = \int_{-\infty}^{x} f(u)du$$
> However, in practice the lower limit of the integral is replaced by the lower limit of the non-zero part of the probability density function (except when this is itself equal to $-\infty$).

(a) **It fails in the first year.**

This is given by $P(X < 1) = \int_{0}^{1} \frac{3x(20-x)}{4000}dx$

$$= \frac{3}{4000}\int_{0}^{1}(20x - x^2)dx$$

$$= \frac{3}{4000}\left[10x^2 - \frac{x^3}{3}\right]_{0}^{1}$$

$$= \frac{3}{4000}\left(10 \times 1^2 - \frac{1^3}{3}\right)$$

$$= 0.0725$$

(b) **It fails in the 11th year.**

This is given by $P(10 \leq X < 11)$

$$= \int_{10}^{11} \frac{3x(20-x)}{4000} dx$$

$$= \frac{3}{4000}\left[10x^2 - \frac{1}{3}x^3\right]_{10}^{11}$$

$$= \frac{3}{4000}\left(10\times 11^2 - \frac{1}{3}\times 11^3\right) - \frac{3}{4000}\left(10\times 10^2 - \frac{1}{3}\times 10^3\right)$$

$$= 0.07475$$

Example 7.2

The continuous random variable X represents the amount of sunshine in hours between noon and 4 p.m. at a skiing resort in the high season. The probability density function, f(x), of X is modelled by

$$f(x) = \begin{cases} kx^2 & \text{for } 0 \leqslant x \leqslant 4 \\ 0 & \text{otherwise.} \end{cases}$$

(i) Find the value of k.

(ii) Find the probability that on a particular day in the high season there are more than two hours of sunshine between noon and 4 p.m.

Solution

(i) To find the value of k you must use the fact that the area under the graph of f(x) is equal to 1.

$$\int_{-\infty}^{\infty} f(x)\,dx = \int_{0}^{4} kx^2\,dx = 1$$

Therefore $\left[\dfrac{kx^3}{3}\right]_0^4 = 1$

$$\frac{64k}{3} = 1$$

So $k = \dfrac{3}{64}$

(ii)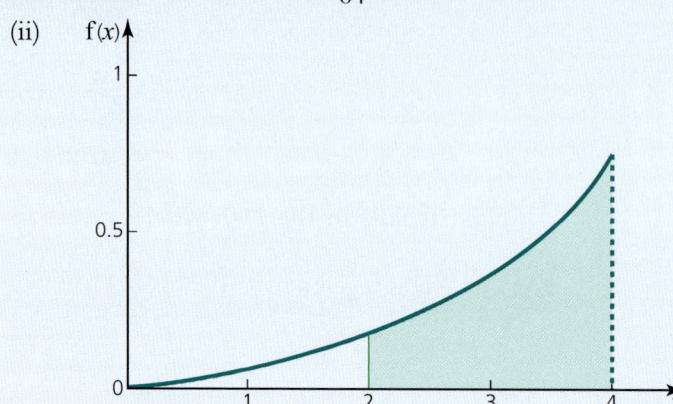

Figure 7.4

The probability of more than 2 hours of sunshine is given by

$$P(X > 2) = \int_2^\infty f(x)\,dx = \int_2^4 \frac{3x^2}{64}\,dx$$

$$= \left[\frac{x^3}{64}\right]_2^4$$

$$= \frac{64 - 8}{64}$$

$$= \frac{56}{64}$$

$$= 0.875$$

Example 7.3

The number of hours Darren spends each day working in his garden is modelled by the continuous random variable X, with p.d.f. $f(x)$ defined by

$$f(x) = \begin{cases} kx & \text{for } 0 \leq x < 3 \\ k(6 - x) & \text{for } 3 \leq x < 6 \\ 0 & \text{otherwise.} \end{cases}$$

(i) Find the value of k.

(ii) Sketch the graph of $f(x)$.

(iii) Find the probability that Darren will work between 2 and 5 hours in his garden on a randomly selected day.

Solution

(i) To find the value of k you must use the fact that the area under the graph of $f(x)$ is equal to 1. You may find the area by integration, as shown below.

$$\int_{-\infty}^{\infty} f(x)\,dx = \int_0^3 kx\,dx + \int_3^6 k(6 - x)\,dx = 1$$

$$\left[\frac{kx^2}{2}\right]_0^3 + \left[6kx - \frac{kx^2}{2}\right]_3^6 = 1$$

Therefore $\quad \frac{9k}{2} + (36k - 18k) - \left(18k - \frac{9k}{2}\right) = 1$

$$9k = 1$$

So $\quad k = \frac{1}{9}$

169

Probability density function

> In this case, you could have found k without integration because the graph of the p.d.f. is a triangle, with area given by $\frac{1}{2} \times$ base $\times$ height, resulting in the equation
> $$\frac{1}{2} \times 6 \times k(6-3) = 1$$
> hence $9k = 1$
> and $k = \frac{1}{9}$.

(ii) Sketch the graph of f(x).

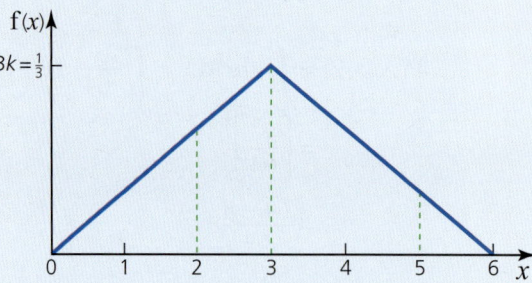

Figure 7.5

(iii) To find $P(2 \leqslant X \leqslant 5)$, you need to find both $P(2 \leqslant X < 3)$ and $P(3 \leqslant X \leqslant 5)$ because there is a different expression for each part.

$$P(2 \leqslant X \leqslant 5) = P(2 \leqslant X < 3) + P(3 \leqslant X \leqslant 5)$$

$$= \int_2^3 \frac{1}{9}x\,dx + \int_3^5 \frac{1}{9}(6-x)\,dx$$

$$= \left[\frac{x^2}{18}\right]_2^3 + \left[\frac{2x}{3} - \frac{x^2}{18}\right]_3^5$$

$$= \frac{9}{18} - \frac{4}{18} + \left(\frac{10}{3} - \frac{25}{18}\right) - \left(2 - \frac{1}{2}\right)$$

$$= 0.72 \quad \text{to two decimal places.}$$

The probability that Darren works between 2 and 5 hours in his garden on a randomly selected day is 0.72.

Discussion point

The graphs below illustrate the probability density functions for four distributions, A, B, C and D. How would you describe the shapes of these distributions?

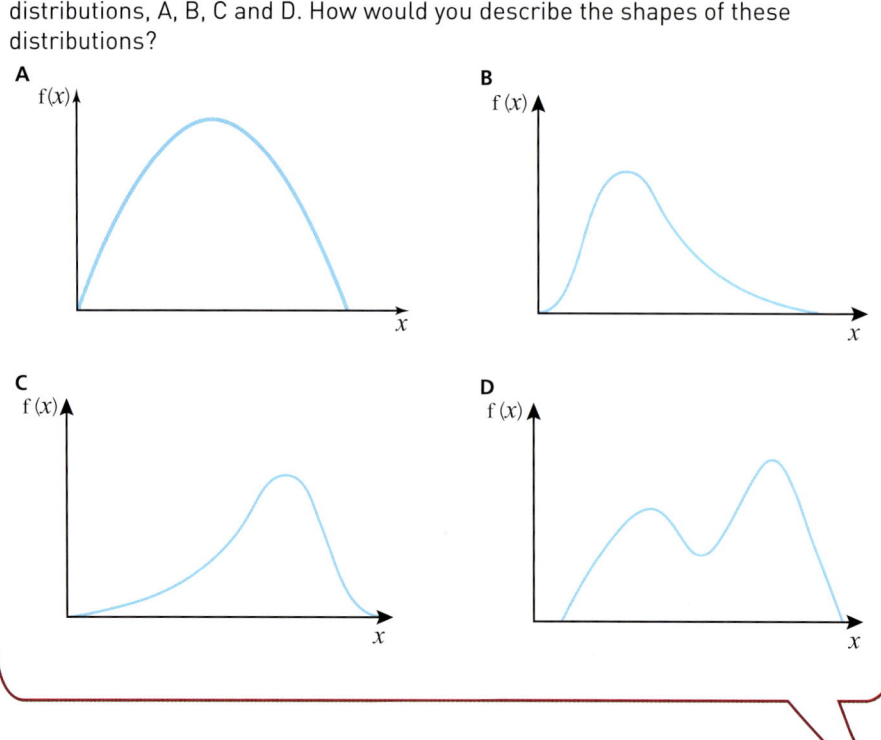

Exercise 7.1

① The continuous random variable X has probability density function $f(x)$ where

$$f(x) = \begin{cases} kx & \text{for } 1 \leq x < 6 \\ 0 & \text{otherwise.} \end{cases}$$

(i) Find the value of the constant k.
(ii) Sketch $y = f(x)$.
(iii) Find $P(X > 5)$.
(iv) Find $P(2 \leq X \leq 3)$.

② The continuous random variable X has p.d.f. $f(x)$ where

$$f(x) = \begin{cases} k(5-x) & \text{for } 0 \leq x \leq 4 \\ 0 & \text{otherwise.} \end{cases}$$

(i) Find the value of the constant k.
(ii) Sketch $y = f(x)$.
(iii) Find $P(1.5 \leq X \leq 2.3)$.

③ The continuous random variable X has p.d.f. $f(x)$ where

$$f(x) = \begin{cases} c & \text{for } -3 \leq x \leq 5 \\ 0 & \text{otherwise.} \end{cases}$$

(i) Find c.
(ii) Sketch $y = f(x)$.
(iii) Find $P(X < -1)$.
(iv) Find $P(X > 2)$.

④ The continuous random variable X has p.d.f. $f(x)$ where

$$f(x) = \begin{cases} kx & \text{for } 0 \leq x \leq 2 \\ 4k - kx & \text{for } 2 < x \leq 4 \\ 0 & \text{otherwise.} \end{cases}$$

(i) Find the value of the constant k.
(ii) Sketch $y = f(x)$.
(iii) Find $P(1 \leq X \leq 3.5)$.

⑤ The continuous random variable X has p.d.f. $f(x)$ where

$$f(x) = \begin{cases} ax^3 & \text{for } 0 \leq x \leq 3 \\ 0 & \text{otherwise.} \end{cases}$$

(i) Find the value of the constant a.
(ii) Sketch $y = f(x)$.
(iii) Find $P(X \leq 2)$.

⑥ A continuous random variable X has p.d.f.

$$f(x) = \begin{cases} k(x-1)(6-x) & \text{for } 1 \leq x \leq 6 \\ 0 & \text{otherwise.} \end{cases}$$

(i) Find the value of k.
(ii) Sketch $y = f(x)$.
(iii) Find $P(2 \leqslant X \leqslant 3)$.

7. A random variable X has p.d.f.

$$f(x) = \begin{cases} (x-1)(2-x) & \text{for } 1 \leqslant x < 2 \\ a & \text{for } 2 \leqslant x < 4 \\ 0 & \text{otherwise.} \end{cases}$$

(i) Find the value of the constant a.
(ii) Sketch $y = f(x)$.
(iii) Find $P(1.5 \leqslant X \leqslant 2.5)$.
(iv) Find $P(|X - 2| < 1)$.

8. A random variable X has p.d.f.

$$f(x) = \begin{cases} kx(3-x) & \text{for } 0 \leqslant x \leqslant 3 \\ 0 & \text{otherwise.} \end{cases}$$

(i) Find the value of k.
(ii) The lifetime (in years) of an electronic component is modelled by this distribution. Two such components are fitted in a radio which will only function if both devices are working. Find the probability that the radio will still function after two years, assuming that their failures are independent.

9. The planning officer in a council needs information about how long cars stay in the car park, and asks the attendant to do a check on the times of arrival and departure of 100 cars. The attendant provides the following data.

Table 7.1

Length of stay	Under 1 hour	1–2 hours	2–4 hours	4–10 hours	More than 10 hours
Number of cars	20	14	32	34	0

The planning officer suggests that the length of stay in hours may be modelled by the continuous random variable X with probability density function of the form

$$f(x) = \begin{cases} k(20 - 2x) & \text{for } 0 \leqslant x \leqslant 10 \\ 0 & \text{otherwise.} \end{cases}$$

(i) Find the value of k.
(ii) Sketch the graph of $f(x)$.
(iii) According to this model, how many of the 100 cars would be expected to fall into each of the four categories?
(iv) Do you think the model fits the data well?
(v) Are there any obvious weaknesses in the model? If you were the planning officer, would you be prepared to accept the model as it is, or would you want any further information?

⑩ A fish farmer has a very large number of trout in a lake. Before deciding whether to net the lake and sell the fish, she collects a sample of 100 fish and weighs them. The results (in kg) are as follows.

Table 7.2

Weight, W	Frequency
$0 < W \leqslant 0.5$	2
$0.5 < W \leqslant 1.0$	10
$1.0 < W \leqslant 1.5$	23
$1.5 < W \leqslant 2.0$	26

Weight, W	Frequency
$2.0 < W \leqslant 2.5$	27
$2.5 < W \leqslant 3.0$	12
$3.0 < W$	0

(i) Illustrate these data on a histogram, with the number of fish on the vertical scale and W on the horizontal scale. Is the distribution of the data symmetrical, positively skewed or negatively skewed?

A friend of the farmer suggests that W can be modelled as a continuous random variable and proposes four possible probability density functions.

$$f_1(w) = \frac{2}{9}w(3-w) \qquad f_2(w) = \frac{10}{81}w^2(3-w)^2$$

$$f_3(w) = \frac{4}{27}w^2(3-w) \qquad f_4(w) = \frac{4}{27}w(3-w)^2$$

in each case for $0 \leqslant w \leqslant 3$.

(ii) Using your calculator (or otherwise), sketch the curves of the four p.d.f.s and state which one matches the data most closely in general shape.

(iii) Use this p.d.f. to calculate the number of fish which that model predicts should fall within each group.

(iv) Do you think it is a good model?

⑪ During a war, the crew of an aeroplane has to destroy an enemy railway line by dropping bombs. The distance between the railway line and where the bomb hits the ground is X m, where X has the following p.d.f.

$$f(x) = \begin{cases} 10^{-4}(a+x) & \text{for } -a \leqslant x < 0 \\ 10^{-4}(a-x) & \text{for } 0 \leqslant x \leqslant a \\ 0 & \text{otherwise.} \end{cases}$$

(i) Find the value of a.
(ii) Find $P(50 \leqslant X \leqslant 60)$.
(iii) Find $P(|X| < 20)$. [MEI]

⑫ The continuous random variable X has the probability density function

$$f(x) = \begin{cases} \dfrac{1-x}{k} & 1 \leqslant x \leqslant 4 \\ 0 & \text{otherwise.} \end{cases}$$

Show that $k = \dfrac{21}{2}$.

S2 AS — Measures of average and spread for a continuous distribution

⑬ This graph shows the probability distribution function, f(x), for the heights, X, of waves at the point with Latitude 44°N Longitude 41°W.

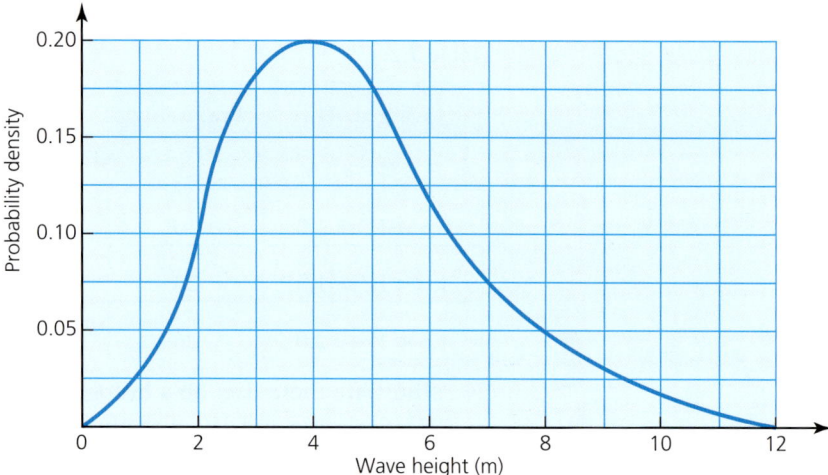

Figure 7.6

(i) Write down the values of f(x) when $x = 0, 2, 4, \ldots, 12$.

(ii) Hence estimate the probability that the height of a randomly selected wave is in the interval

(a) 0–2 m (b) 2–4 m (c) 4–6 m
(d) 6–8 m (e) 8–10 m (f) 10–12 m.

A model is proposed in which

$$f(x) = \begin{cases} kx(12-x)^2 & \text{for } 0 \leqslant x \leqslant 12 \\ 0 & \text{otherwise.} \end{cases}$$

(iii) Find the value of k.

(iv) Find, according to this model, the probability that a randomly selected wave is in the interval

(a) 0–2 m (b) 2–4 m (c) 4–6 m
(d) 6–8 m (e) 8–10 m (f) 10–12 m.

(v) By comparing the figures from the model with the real data, state whether you think it is a good model or not.

2 Measures of average and spread for a continuous distribution

You will recall that, for a discrete random variable, expectation and variance are given by:

$$E(X) = \sum_i x_i p_i$$

$$\text{Var}(X) = \sum_i (x_i - \mu)^2 p_i = \sum_i x_i^2 p_i - [E(X)]^2$$

where μ is the mean and p_i is the probability of the outcome x_i for $i = 1, 2, 3, \ldots$, with the various outcomes covering all possibilities.

> **Note**
> You met these formulae in Chapter 2.
> Notice that p_i is another way of writing $P(X = x_i)$.

The expressions for the expectation and variance of a continuous random variable are equivalent, but with summation replaced by integration.

$$E(X) = \int_{\text{All values of } x} x f(x) \, dx$$

$\int_{\text{All values of } x}$ can also be written as $\int_{-\infty}^{\infty}$

$$\text{Var}(X) = \int_{\text{All values of } x} (x-\mu)^2 f(x) \, dx = \int_{\text{All values of } x} x^2 f(x) \, dx - [E(X)]^2$$

$E(X)$ is the same as the population mean, μ, and is often called the mean of X.

Example 7.4

The response time, in seconds, for a contestant in a general knowledge quiz is modelled by a continuous random variable X whose p.d.f. is

$$f(x) = \frac{x}{50} \quad \text{for } 0 < x \leq 10.$$

The rules state that a contestant who makes no answer is disqualified from the whole competition. This has the consequence that everybody gives an answer, if only a guess, to every question. Find

(i) the mean time in seconds for a contestant to respond to a particular question

(ii) the standard deviation of the time taken.

The organiser estimates the proportion of contestants who are guessing by assuming that they are those whose time is at least one standard deviation greater than the mean.

(iii) Using this assumption, estimate the probability that a randomly selected response is a guess.

Solution

(i) Mean time:

$$E(X) = \int_0^{10} x \frac{x}{50} \, dx$$

$$= \left[\frac{x^3}{150}\right]_0^{10} = \frac{1000}{150} = \frac{20}{3}$$

$$= 6\frac{2}{3}$$

(ii) Variance:

$$\text{Var}(X) = \int_0^{10} x^2 f(x) \, dx - [E(X)]^2$$

$$= \int_0^{10} \frac{x^3}{50} \, dx - \left(6\frac{2}{3}\right)^2$$

$$= \left[\frac{x^4}{200}\right]_0^{10} - \left(6\frac{2}{3}\right)^2$$

$$= 5\frac{5}{9}$$

Standard deviation $= \sqrt{\text{variance}} = \sqrt{5\frac{5}{9}}$

The standard deviation of the times is 2.357 seconds (to 3 d.p.).

(iii) All those with response times greater than $6.667 + 2.357 = 9.024$ seconds are taken to be guessing. The longest possible time is 10 seconds.

The probability that a randomly selected response is a guess is given by

$$= \int_{9.024}^{10} \frac{x}{50} \, dx$$

$$= \left[\frac{x^2}{100} \right]_{9.024}^{10}$$

$$= 0.186$$

so just under 1 in 5 answers are deemed to be guesses.

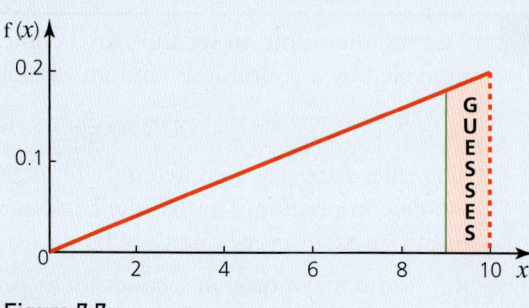

Figure 7.7

> **Note**
> Although the intermediate answers have been given rounded to three decimal places, more figures have been carried forward into subsequent calculations.

Example 7.5

This is a continuation of Example 7.3 from page 166.

The number of hours per day that Darren spends driving is modelled by the continuous random variable X, the p.d.f. of which is

$$f(x) = \begin{cases} \frac{1}{9}x & \text{for } 0 \leq x \leq 3 \\ \frac{6-x}{9} & \text{for } 3 < x \leq 6 \\ 0 & \text{otherwise.} \end{cases}$$

Find $E(X)$, the mean number of hours per day that Darren spends driving.

Solution

$$E(X) = \int_{-\infty}^{\infty} x f(x) \, dx$$

$$= \int_0^3 x \cdot \frac{1}{9} x \, dx + \int_3^6 x \cdot \frac{6-x}{9} \, dx$$

$$= \left[\frac{x^3}{27} \right]_0^3 + \left[\frac{x^2}{3} - \frac{x^3}{27} \right]_3^6$$

$$= 1 - 0 + (12 - 8) - (3 - 1)$$

$$= 3$$

> **Note**
> Notice that, in this case, E(X) can be found from the line of symmetry of the graph of f(x). This situation often arises and you should be alert to the possibility of finding E(X) by symmetry as shown in Figure 7.8.

Darren spends a mean of 3 hours per day driving.

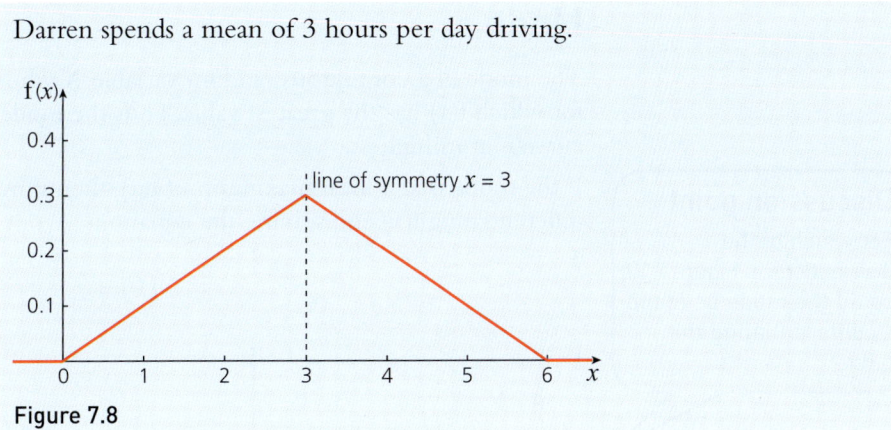

Figure 7.8

Median and percentiles

The median value of a continuous random variable X with p.d.f. $f(x)$ is the value m for which

$$P(X < m) = P(X > m) = 0.5.$$

Consequently, $\int_{-\infty}^{m} f(x)\,dx = 0.5$ and $\int_{m}^{\infty} f(x)\,dx = 0.5$.

The median is the value m such that the line $x = m$ divides the area under the curve $f(x)$ into two equal parts. In Figure 7.9, a is the smallest possible value of X, b the largest. The line $x = m$ divides the shaded region into two regions A and B, both with area 0.5.

> ⚠ In general, the **mean** does not divide the area into two equal parts but it will do so if the curve is symmetrical about it because, in that case, it is equal to the median.

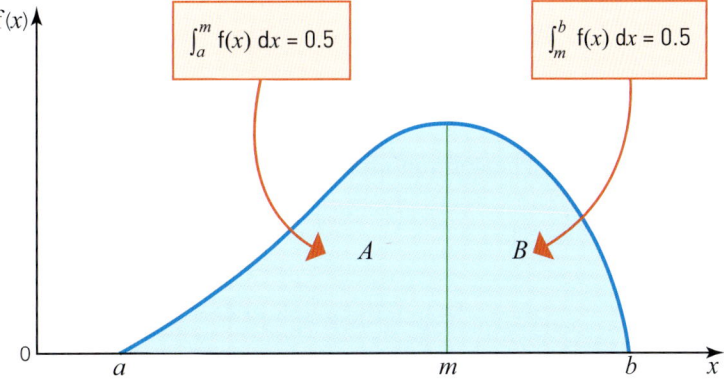

Figure 7.9

You can use a similar technique to find the percentiles of a distribution. Suppose a continuous random variable, X, has probability density function $f(x)$ and that the pth percentile is at $X = r$. Then

$$\int_{-\infty}^{r} f(x)\,dx = \frac{p}{100}.$$

Mode

The mode of a continuous random variable X whose p.d.f. is $f(x)$ is the value for which $f(x)$ has the greatest value. Thus the mode is the value of X where the curve is at its highest.

If the mode is at a local maximum of $f(x)$, then it may often be found by differentiating $f(x)$ and solving the equation

$$f'(x) = 0.$$

Discussion point

For which of the distributions in Figure 7.10 could the mode be found by differentiating the p.d.f.?

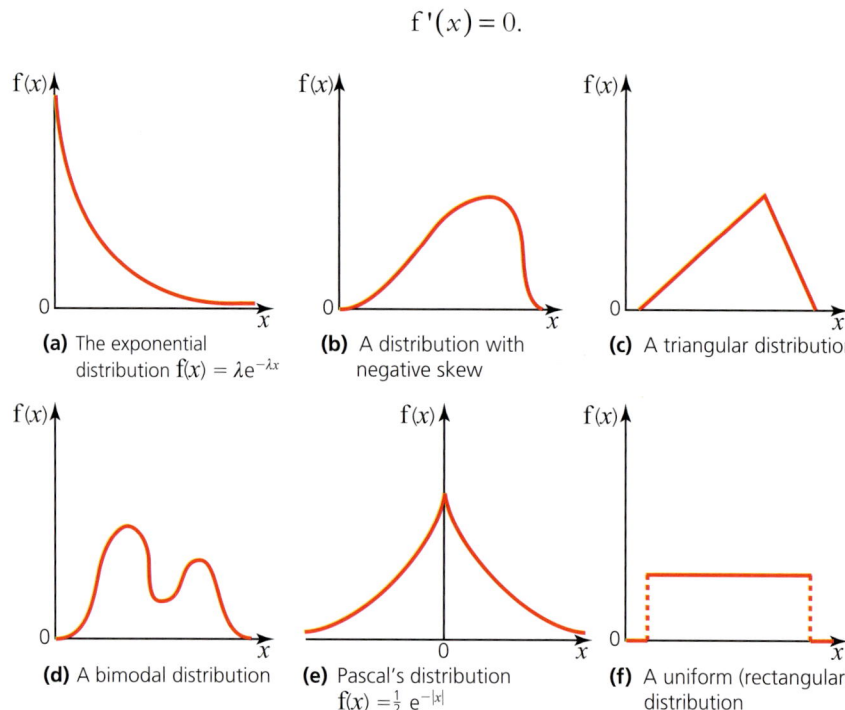

(a) The exponential distribution $f(x) = \lambda e^{-\lambda x}$

(b) A distribution with negative skew

(c) A triangular distribution

(d) A bimodal distribution

(e) Pascal's distribution $f(x) = \tfrac{1}{2} e^{-|x|}$

(f) A uniform (rectangular) distribution

Figure 7.10

Example 7.6

The continuous random variable X has p.d.f. $f(x)$ where

$$f(x) = \begin{cases} 4x(1-x^2) & \text{for } 0 \leq x \leq 1 \\ 0 & \text{otherwise.} \end{cases}$$

(i) Sketch the graph of $f(x)$ and describe the distribution.

(ii) Find the mode.

(iii) Find the median.

(iv) Verify that the value of X at the 80th percentile is approximately 0.7435.

Solution

(i)

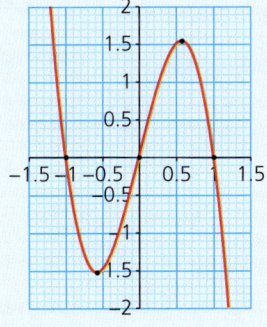

Figure 7.11 Negatively skewed

(ii) The mode is found by differentiating $f(x) = 4x - 4x^3$

$$f'(x) = 4 - 12x^2$$

Solving $f'(x) = 0$

$$x = \frac{1}{\sqrt{3}} = 0.577 \text{ to 3 decimal places.}$$

> $x = -0.577$ is also a root of $f'(x) = 0$ but is outside the range $0 \leq x \leq 1$.

It is easy to see from the shape of the graph (see Figure 7.12 below) that this must be a maximum, and so the mode is 0.577.

(iii) The median, m, is given by

$$\int_{-\infty}^{m} f(x)\,dx = 0.5$$

> Since $x \geq 0$

$$\Rightarrow \int_0^m (4x - 4x^3)\,dx = 0.5$$

$$\left[2x^2 - x^4\right]_0^m = 0.5$$

> This equation is a quadratic in m^2. You can solve it using the quadratic formula and then taking the square root of the answer.

$$2m^2 - m^4 = 0.5$$

$$2m^4 - 4m^2 + 1 = 0.$$

$$m = \pm\, 0.541 \text{ or } \pm\, 1.307$$

> Since -0.541 and ± 1.307 are all outside the domain of X, the median is 0.541.

The median is 0.541 (3 s.f.).

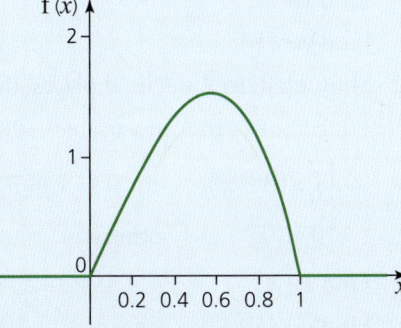

Figure 7.12

(iv) $\left[2x^2 - x^4\right]_0^{0.7435} = 0.800\,005\,228...$

$$\approx 80\%$$

Exercise 7.2

1 The continuous random variable X has p.d.f. $f(x)$, where

$$f(x) = \begin{cases} \frac{1}{8}x & \text{for } 0 \leq x \leq 4 \\ 0 & \text{otherwise.} \end{cases}$$

Find

(i) $E(X)$

(ii) $\text{Var}(X)$

(iii) the median value of X.

(iv) Find the value of X at the 25th percentile and say what is special about this value.

2 The continuous random variable T has p.d.f. $f(t)$ defined by

$$f(t) = \begin{cases} \dfrac{6-t}{18} & \text{for } 0 \leq t \leq 6 \\ 0 & \text{otherwise.} \end{cases}$$

Find

(i) $E(T)$

(ii) $\text{Var}(T)$

(iii) the median value of T.

3 The random variable X has p.d.f. $f(x)$, where

$$f(x) = \begin{cases} \dfrac{1}{6} & \text{for } -2 \leq x \leq 4 \\ 0 & \text{otherwise.} \end{cases}$$

(i) Sketch the graph of $f(x)$.

(ii) Find $P(X < 2)$.

(iii) Find $E(X)$.

(iv) Find $P(X < 1)$.

4 The continuous random variable Y has p.d.f. $f(y)$ defined by

$$f(y) = \begin{cases} 12y^2(1-y) & \text{for } 0 \leq y \leq 1 \\ 0 & \text{otherwise.} \end{cases}$$

(i) Find $E(Y)$.

(ii) Find $\text{Var}(Y)$.

(iii) Show that, to 2 decimal places, the median value of Y is 0.61.

5 The continuous random variable X has p.d.f. $f(x)$ defined by

$$f(x) = \begin{cases} \dfrac{2}{9}x(3-x) & \text{for } 0 \leq x \leq 3 \\ 0 & \text{otherwise.} \end{cases}$$

(i) Find $E(X)$.

(ii) Find $\text{Var}(X)$.

(iii) Find the mode of X.

(iv) Find the median value of X.

(v) Draw a sketch graph of $f(x)$ and comment on your answers to parts (i), (iii) and (iv) in the light of what it shows you.

⑥ The function $f(x) = \begin{cases} k(3+x) & \text{for } 0 \leq x \leq 2 \\ 0 & \text{otherwise.} \end{cases}$

is the probability density function of the random variable X.

(i) Show that $k = \frac{1}{8}$.

(ii) Find the mean and variance of X.

(iii) Find the probability that a randomly selected value of X lies between 1 and 2.

⑦ The p.d.f. of the lifetime, X hours, of a brand of electric light bulb is modelled by

$$f(x) = \begin{cases} \dfrac{1}{60\,000}x & \text{for } 0 \leq x \leq 300 \\ \dfrac{1}{50} - \dfrac{1}{20\,000}x & \text{for } 300 < x \leq 400 \\ 0 & \text{for } x > 400. \end{cases}$$

(i) Sketch the graph of $f(x)$.

(ii) Show that $f(x)$ fulfils the conditions for it to be a p.d.f.

(iii) Find the expected lifetime of a bulb.

(iv) Find the variance of the lifetimes of the bulbs.

(v) Find the probability that a randomly selected bulb will last less than 100 hours.

⑧ The marks of candidates in an examination are modelled by the continuous random variable X with p.d.f.

$f(x) = \begin{cases} kx(x-50)^2(100-x) & \text{for } 0 \leq x \leq 100 \\ 0 & \text{otherwise.} \end{cases}$

(i) Find the value of k.

(ii) Sketch the graph of $f(x)$.

(iii) Describe the shape of the graph and give an explanation of how such a graph might occur, in terms of the examination and the candidates.

(iv) Is it permissible to model a mark, which is a discrete variable going up in steps of 1, by a continuous random variable like X, as defined in this question?

⑨ The municipal tourism officer at a Mediterranean resort on the Costa Del Sol wishes to model the amount of sunshine per day during the holiday season. She denotes by X the number of hours of sunshine per day between 8 am and 8 pm and she suggests the following probability density function for X:

$f(x) = \begin{cases} k\left[(x-3)^2 + 4\right] & \text{for } 0 \leq x \leq 12 \\ 0 & \text{otherwise.} \end{cases}$

(i) Show that $k = \dfrac{1}{300}$ and sketch the graph of the p.d.f. $f(x)$.

(ii) Assuming that the model is accurate, find the mean and standard deviation of the number of hours of sunshine per day. Find also the probability of there being more than eight hours of sunshine on a randomly chosen day.

(iii) Obtain a cubic equation for m, the median number of hours of sunshine, and verify that m is about 9.74 to 2 decimal places.

10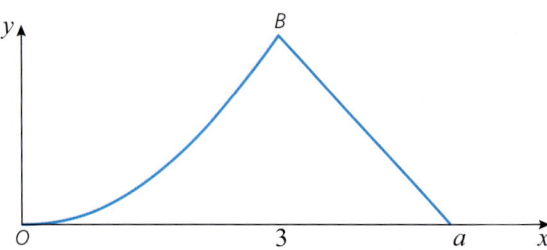

Figure 7.13

Figure 7.13 shows a sketch of the probability density function f(x) of the random variable X.
For $0 \leq x \leq 3$, f(x) is represented by a curve OB with equation $f(x) = kx^2$, where k is a constant.
For $3 \leq x \leq a$, where a is a constant, f(x) is represented by a straight line passing through B and the point $(a, 0)$.
For all other values of x, f(x) = 0.
Given that the mode of X = the median of X, find
(i)   the mode
(ii) the value of k
(iii) the value of a.
Without calculating E(X) and with reference to the skewness of the distribution
(iv) state, giving your reason, whether $E(X) < 3$, $E(X) = 3$ or $E(X) > 3$.   [Edexcel 6684 Q3 June 2011]

11 The continuous random variable X has p.d.f. f(x) defined by

$$f(x) = \begin{cases} \dfrac{a}{x} & \text{for } 1 \leq x \leq 2 \\ 0 & \text{otherwise.} \end{cases}$$

(i)   Find the value of a.
(ii) Sketch the graph of f(x).
(iii) Find the mean and variance of X.
(iv) Find the proportion of values of X between 1.5 and 2.
(v)   Find the median value of X.

12 An examination is taken by a large number of candidates. The marks scored are modelled by the continuous random variable X with probability density function

$f(x) = kx^3(120 - x)$ $0 \leq x \leq 100$.

(You should assume throughout this question that marks are on a continuous scale. Hence there is no need to consider **continuity corrections** which you will meet in Chapter 9.)

(i)   Sketch the graph of this probability density function. What does the model suggest about the abilities of the candidates in relation to this examination?
(ii) Show that $k = 10^{-9}$.
(iii) The pass mark is set at 50. Find what proportion of candidates fail the examination.
(iv) The top 20% of candidates are awarded a distinction. Determine whether a mark of 90 is sufficient for a distinction. Find the least whole number mark which is sufficient for a distinction.   [MEI]

3 The expectation and variance of a function of X

There are times when one random variable is a function of another random variable. For example:

- as part of an experiment, you are measuring temperatures in Celsius but then need to convert them to Fahrenheit: $F = 1.8C + 32$
- you are measuring the lengths of the sides of square pieces of material and deducing their areas: $A = L^2$
- you are estimating the ages, A years, of hedgerows by counting the number, n, of types of shrubs and trees in 30 m lengths: $A = 100n - 50$.

In fact, in any situation where you are entering the value of a random variable into a formula, the outcome will be another random variable which is a function of the one you entered. Under these circumstances, you may need to find the expectation and variance of such a function of a random variable.

For a discrete random variable, X, the expectation of a function $g[X]$ is given by:

$$E(g[X]) = \sum g[x_i] p_i$$

$$\text{Var}(g[X]) = \sum (g[x_i])^2 p_i - \{E(g[X])\}^2$$

The equivalent results for a continuous random variable, X, with p.d.f. $f(x)$ are:

$$E(g[X]) = \int_{\text{All values of } x} g[x] f(x) \, dx$$

$$\text{Var}(g[X]) = \int_{\text{All values of } x} (g[x])^2 f(x) \, dx - \{E(g[X])\}^2$$

Example 7.7

The continuous random variable X has p.d.f. $f(x)$ given by:

$$f(x) = \begin{cases} \dfrac{x}{50} & \text{for } 0 < x \leq 10 \\ 0 & \text{otherwise.} \end{cases}$$

(This random variable was used to model response times in Example 7.4.)

(i) Find $E(3X + 4)$.

(ii) Find $3E(X) + 4$.

(iii) Find $\text{Var}(3X + 4)$.

(iv) Verify that $\text{Var}(3X + 4) = 3^2 \text{Var}(X)$.

The expectation and variance of a function of X

Solution

Here you are using
$$E(g[X]) = \int_{\text{All values of } x} g[x]f(x)\,dx$$

(i) $E(3X+4) = \int_0^{10} (3x+4)\frac{x}{50}\,dx$

$= \int_0^{10} \frac{1}{50}(3x^2+4x)\,dx$

$= \left[\frac{x^3}{50} + \frac{x^2}{25}\right]_0^{10}$

$= 20 + 4$

$= 24$

(ii) $3E(X) + 4 = 3\int_0^{10} x\frac{x}{50}\,dx + 4$

$= \left[\frac{3}{150}x^3\right]_0^{10} + 4$

$= 20 + 4$

$= 24$

Notice here that $E(3X+4) = 24 = 3E(X) + 4$.

(iii) To find $\text{Var}(3X+4)$, use

$$\text{Var}(g[X]) = \int [g(x)]^2 f(x)\,dx - \{E(g[X])\}^2$$

with $g(x) = 3X + 4$.

$\text{Var}(3X+4) = \int_0^{10} (3x+4)^2 \frac{1}{50}x\,dx - 24^2$

$= \int_0^{10} \frac{1}{50}(9x^3 + 24x^2 + 16x)\,dx - 576$

$= \frac{1}{50}\left[\frac{9x^4}{4} + 8x^3 + 8x^2\right]_0^{10} - 576$

$= 50$

(iv) $\text{Var}(X) = E(X^2) - [E(X)]^2$

$E(X^2) = \int_0^{10} x^2 \frac{1}{50}x\,dx$ and $E(X) = \int_0^{10} x\frac{1}{50}x\,dx$

$E(X^2) = \int_0^{10} \frac{1}{50}x^3\,dx$ and $E(X) = \int_0^{10} \frac{1}{50}x^2\,dx$

$E(X^2) = \left[\frac{1}{200}x^4\right]_0^{10}$ and $E(X) = \left[\frac{1}{150}x^3\right]_0^{10}$

$E(X^2) = 50$ and $E(X) = 6.\dot{6}$

so $\text{Var}(X) = 50 - 6.\dot{6}^2 = 5.\dot{5}$

and $3^2\text{Var}(X) = 9 \times 5.\dot{5} = 50$

From part (iii), $\text{Var}(3X+4) = 50$

So $\text{Var}(3X+4) = 3^2\,\text{Var}(X)$, as required.

Note

You have of course already met these results for discrete random variables in Chapter 2.

Note

The result $\text{Var}(aX) = a^2\text{Var}(X)$ should not surprise you. It follows from the fact that if the standard deviation of a set of data is k, then the standard deviation of the set formed by multiplying all the data by a positive constant, a, is ak. That is,

standard deviation(aX)
$= a \times$ standard deviation(X)

Since variance
$=$ (standard deviation)^2
then Var(aX)
$=$ [standard deviation(aX)]^2
$=$ [$a \times$ standard deviation(X)]^2
$= a^2$[standard deviation(X)]^2
$= a^2$ Var(X).

General results

This example illustrates a number of general results for random variables, continuous or discrete.

$$\text{E}(c) = c \qquad \text{Var}(c) = 0$$
$$\text{E}(aX) = a\text{E}(X) \qquad \text{Var}(aX) = a^2\text{Var}(X)$$
$$\text{E}(aX + b) = a\text{E}(X) + b$$
$$\text{Var}(aX + b) = a^2\text{Var}(X) \quad \text{E}[g(X) \pm h(X)] = \text{E}[g(X)] \pm \text{E}[h(X)]$$

where a, b and c are constants, with a positive, and $g(X)$ and $h(X)$ are functions of X.

In Chapter 2, you met results for the expectation and variance of linear combinations of discrete independent random variables, X and Y.

$$\text{E}(aX \pm bY) = a\text{E}(X) \pm b\text{E}(Y) \text{ and } \text{Var}(aX \pm bY) = a^2\text{Var}(X) + b^2\text{E}(Y)$$

The same results hold if the variables are continuous, but they must still be independent.

Example 7.8

The continuous random variable X has p.d.f. $f(x)$ given by

$$f(x) = \begin{cases} \dfrac{3}{125}x^2 & \text{for } 0 \leqslant x \leqslant 5 \\ 0 & \text{otherwise.} \end{cases}$$

Find

(i) E(X) (ii) Var(X) (iii) E$(7X - 3)$ (iv) Var$(7X - 3)$.

Solution

(i) $\text{E}(X) = \displaystyle\int_{-\infty}^{\infty} x\text{f}(x)\,\text{d}x$

$= \displaystyle\int_0^5 x \dfrac{3}{125}x^2 \,\text{d}x$

$= \left[\dfrac{3}{500}x^4\right]_0^5$

$= 3.75$

(ii) $\text{Var}(X) = \displaystyle\int_{-\infty}^{\infty} x^2\text{f}(x)\,\text{d}x - [\text{E}(X)]^2$

$= \displaystyle\int_0^5 x^2 \dfrac{3}{125}x^2 \,\text{d}x - 3.75^2$

$= \left[\dfrac{3}{625}x^5\right]_0^5 - 14.0625$

$= 15 - 14.0625$

$= 0.9375$

(iii) $\text{E}(7X - 3) = 7\text{E}(X) - 3$

$= 7 \times 3.75 - 3$

$= 23.25$

(iv) $\text{Var}(7X - 3) = 7^2 \text{Var}(X)$

$= 49 \times 0.9375$

$= 45.94$ to 2 decimal places

S2 AS — The expectation and variance of a function of X

Exercise 7.3

① A continuous random variable X has the p.d.f.
$$f(x) = \begin{cases} k & \text{for } 0 \leq x \leq 5 \\ 0 & \text{otherwise.} \end{cases}$$

(i) Find the value of k.
(ii) Sketch the graph of $f(x)$.
(iii) Find $E(X)$.
(iv) Find $E(4X - 3)$ and show that your answer is the same as $4E(X) - 3$.

② The continuous random variable X has p.d.f.
$$f(x) = \begin{cases} 4x^3 & \text{for } 0 \leq x \leq 1 \\ 0 & \text{otherwise.} \end{cases}$$

(i) Find $E(X)$.
(ii) Find $E(X^2)$.
(iii) Find Var(X).
(iv) Verify that $E(5X + 1) = 5E(X) + 1$.

③ The number of kilograms of metal extracted from 10 kg of ore from a certain mine is modelled by a continuous random variable X with probability density function $f(x)$, where
$$f(x) = \begin{cases} cx(2-x)^2 & \text{for } 0 \leq x \leq 2 \\ 0 & \text{otherwise,} \end{cases}$$
where c is a constant.

Show that c is $\frac{3}{4}$, and find the mean and variance of X.

The cost of extracting the metal from 10 kg of ore is £$10x$. Find the expected cost of extracting the metal from 10 kg of ore.

④ A continuous random variable Y has p.d.f.
$$f(y) = \begin{cases} \frac{2}{9}y(3-y) & \text{for } 0 \leq y \leq 3 \\ 0 & \text{otherwise.} \end{cases}$$

(i) Find $E(Y)$.
(ii) Find $E(Y^2)$.
(iii) Find $E(Y^2) - [E(Y)]^2$.
(iv) Find $E(2Y^2 + 3Y + 4)$.
(v) Find $\int_0^3 (y - E(Y))^2 f(y)\,dy$.

Why is the answer the same as that for part (iii)?

⑤ A continuous random variable X has p.d.f. $f(x)$, where
$$f(x) = \begin{cases} 12x^2(1-x) & \text{for } 0 \leq x \leq 1 \\ 0 & \text{otherwise.} \end{cases}$$

(i) Find μ, the mean of X.
(ii) Find $E(6X - 7)$ and show that your answer is the same as $6E(X) - 7$.
(iii) Find the standard deviation of X.
(iv) What is the probability that a randomly selected value of X lies within one standard deviation of μ?

⑥ The continuous random variable X has p.d.f.

$$f(x) = \begin{cases} \frac{2}{25}(7-x) & \text{for } 2 \leq x \leq 7 \\ 0 & \text{otherwise.} \end{cases}$$

The function $g(X)$ is defined by $g(x) = 3x^2 + 4x + 7$.

(i) Find $E(X)$.

(ii) Find $E[g(X)]$.

(iii) Find $E(X^2)$ and hence find $3E(X^2) + 4E(X) + 7$.

(iv) Use your answers to parts (ii) and (iii) to verify that
$E[g(X)] = 3E(X^2) + 4E(X) + 7$.

⑦ A toy company sells packets of coloured plastic equilateral triangles. The triangles are actually offcuts from the manufacture of a totally different toy, and the length, X, of one side of a triangle may be modelled as a random variable with a uniform (rectangular) distribution for $2 \leq x \leq 8$.

(i) Find the p.d.f. of X.

(ii) An equilateral triangle of side x has area a. Find the relationship between a and x.

(iii) Find the probability that a randomly selected triangle has area greater than 15 cm². (Hint: What does this imply about x?)

(iv) Find the expectation and variance of the area of a triangle.

⑧ The continuous random variable X has probability density function

$$f(x) = \begin{cases} \frac{3}{1024}x(x-8)^2 & 0 \leq x \leq 8 \\ 0 & \text{otherwise.} \end{cases}$$

A sketch of $f(x)$ is shown in the diagram.

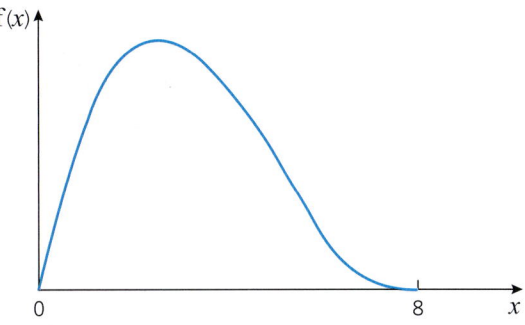

Figure 7.14

(i) Find $E(X)$ and show that $Var(X) = 2.56$.

The times, in minutes, taken by a doctor to see her patients are modelled by the continuous random variable $T = X + 2$.

(ii) Sketch the distribution of T and describe in words what this model implies about the lengths of the doctor's appointments. [MEI]

9. A continuous random variable has probability density function f defined by
$$f(x) = \begin{cases} kx(6-x) & \text{for } 0 \leq x \leq 6 \\ 0 & \text{otherwise.} \end{cases}$$

Evaluate k and the mean of the distribution.

A particle moves along a straight line in such a way that during the first six seconds of its motion its displacement at time t seconds is s m, where
$s = 2(t + 1)$.

The particle is observed at time t seconds, where t denotes a random value from a distribution whose probability density function is the function f defined above. Calculate the probability that at the time of observation the displacement of the particle is less than 5 m.

10. A continuous random variable X is uniformly distributed over the interval $[b, 4b]$ where b is a constant.

   (i) Write down $E(X)$.

   (ii) Use integration to show that $Var(X) = \dfrac{3b^2}{4}$.

   (iii) Find $Var(3 - 2X)$.

   Given that $b = 1$, find

   (iv) the cumulative distribution function of X, $F(x)$, for all values of x

   (v) the median of X.

   [Edexcel 6684 Q4 June 2013]

4 The cumulative distribution function, F

500 enter community half-marathon

There was a record entry for this year's community half-marathon, including several famous names who were treating it as a training run in preparation for the London marathon. Overall winner was Reuben Mhango in 1 hour 4 minutes and 2 seconds; the first woman home was 37-year-old Lynn Barber in 1 hour 20 minutes exactly. There were many fun runners but everybody completed the course within 4 hours.

Record numbers, but no record times this year

£150 prize to be won

Mike Harrison, chair of the Half Committee, says: 'This year we restricted entries to 500 but this meant disappointing many people. Next year we intend to allow everybody to run and expect a much bigger entry. In order to allow us to marshall the event properly we need a statistical model to predict the flow of runners, and particularly their finishing times. We are offering a prize of £150 for the best such model submitted.'

Time (hours)	Finished (%)
1¼	3
1½	15
1¾	33
2	49
2¼	57
2½	75
3	91
3½	99
4	100

An entrant for the competition proposed a model in which a runner's time, X hours, is a continuous random variable with p.d.f.

$$f(x) = \begin{cases} \frac{4}{27}(x-1)(4-x)^2 & 1 \leq x \leq 4 \\ 0 & \text{otherwise.} \end{cases}$$

According to this model, the mode is at 2 hours, and everybody finishes in between 1 hour and 4 hours; see Figure 7.15.

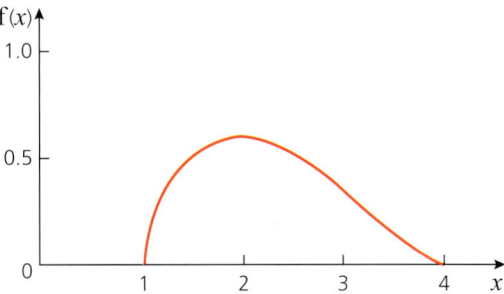

Figure 7.15

How does this model compare with the figures you were given for the actual race?

Those figures gave the **cumulative distribution**, the total numbers (expressed as percentages) of runners who had finished by certain times. To obtain the equivalent figures from the model, you want to find an expression for $P(X \leq x)$. The function giving $P(X \leq x)$ is called the **cumulative distribution function** (c.d.f.), and it is denoted by $F(x)$. The best method of obtaining the cumulative distribution function is to use indefinite integration.

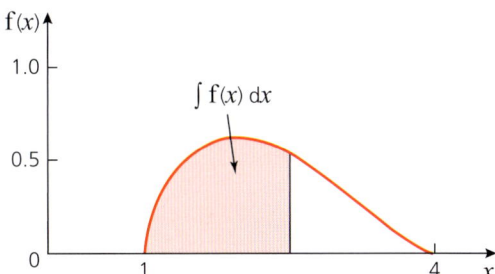

Figure 7.16

In this model, the proportion finishing by time t hours is given by

$$\int f(x) \, dx$$
$$= \int \frac{4}{27}(x-1)(4-x)^2 \, dx$$
$$= \frac{4}{27} \int \left(x^3 - 9x^2 + 24x - 16\right) dx$$
$$= \frac{4}{27}\left(\frac{1}{4}x^4 - 3x^3 + 12x^2 - 16x\right) + c$$
$$= \frac{1}{27}x^4 - \frac{4}{9}x^3 + \frac{16}{9}x^2 - \frac{64}{27}x + c$$

But $P(X \leq 1) = 0$, so $F(1) = 0$ and you can use this to find the value of c:

$$\frac{1}{27} - \frac{4}{9} + \frac{16}{9} - \frac{64}{27} + c = 0$$
$$c = 1$$

S2 AS The cumulative distribution function, F

Hence the cumulative distribution function is given by

$$F(x) = \begin{cases} 0 & \text{for } x < 1 \\ \dfrac{1}{27}x^4 - \dfrac{4}{9}x^3 + \dfrac{16}{9}x^2 - \dfrac{64}{27}x + 1 & \text{for } 1 \leq x \leq 4 \\ 1 & \text{for } x > 4. \end{cases}$$

> No runners finish in less than 1 hour, so if x is less than 1, the probability of a runner finishing in less than x hours is 0.

> All runners finish in less than 4 hours, so if x is greater than 4, the probability of a runner finishing in less than x hours is 1.

It is possible to use definite integration, but this causes a problem as you cannot use the same letter for both a limit of the integral and as the variable of integration. So you would have to change the variable of integration, which is a **dummy variable** as it does not appear in the final answer, to a different letter.

To find the proportions of runners finishing by any time, substitute that value for x; so when $x = 2$

$$F(2) = \frac{1}{27} \times 2^4 - \frac{4}{9} \times 2^3 + \frac{16}{9} \times 2^2 - \frac{64}{27} \times 2 + 1$$

$$= 0.41 \text{ to two decimal places.}$$

Here is the complete table, with all the values worked out.

Table 7.3

Time (hours)	Model	Runners
1.00	0.00	0.00
1.25	0.04	0.03
1.50	0.13	0.15
1.75	0.26	0.33
2.00	0.41	0.49
2.25	0.55	0.57
2.50	0.69	0.75
3.00	0.89	0.91
3.50	0.98	0.99
4.00	1.00	1.00

> **Discussion point**
>
> Do you think that this model is worth the £150 prize? If you were on the organising committee what more might you look for in a model?

Notice the distinctive shape of the curves of these functions (Figure 7.16), sometimes called an **ogive**.

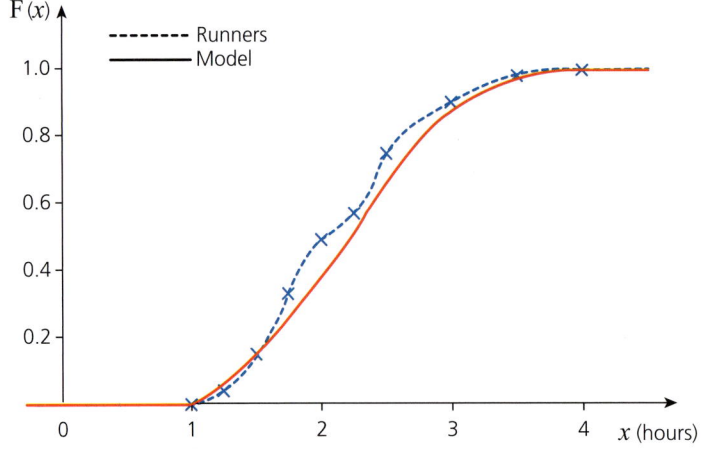

Figure 7.17

> **Note**
>
> You have probably met this shape already when drawing cumulative frequency curves.

> These graphs illustrate the probability density function of a function that can take a limited range of values. The lower limit is a and the upper limit is b. You could look at it in a different way and say, for example, that the lower limit is 0; in that case the two graphs would have line segments along the x-axis joining 0 to a.
>
> Some distributions, for example the Normal distribution, have no limits and so the range of possible values is from $-\infty$ to $+\infty$.

Properties of the cumulative distribution function, F(x)

The graphs, Figure 7.18, show the probability density function f(x) and the cumulative distribution function F(x) of a typical continuous random variable X. You will see that the values of the random variable always lie between a and b.

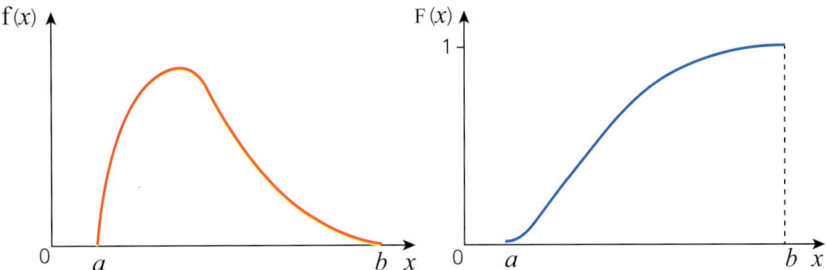

Figure 7.18

Notes

1 Notice the use of lower and upper case letters here. The probability density function is denoted by the lower case f, whereas the cumulative distribution function is given the upper case F.

2 F was derived here as a function of t rather than x, to avoid using the same variable in the expression to be integrated. In this case, t was a natural variable to use because time was involved, but that is not always the case.

It is more usual to write F as a function of x, F(x), but you would not be correct to write down an expression like

$$F(x) = \int_1^x \frac{4}{27}(x-1)(4-x)^2 \, dx$$

INCORRECT

since x would then be both a limit of the integral and the variable used within it.

To overcome this problem, a dummy variable, u, is used in the rest of this section, so that F(x) is now written,

$$F(x) = \int_1^x \frac{4}{27}(u-1)(4-u)^2 \, du$$

CORRECT

You may of course use another symbol, like y or p, rather than u; anything except x.

3 The term cumulative distribution function is often abbreviated to c.d.f.

These graphs illustrate a number of general results for cumulative distribution functions.

1   F(x) = 0   for $x \leq a$, the lower limit of x.

   The probability of X taking a value less than or equal to a is zero; the value of X must be greater than or equal to a.

2   F(x) = 1 for $x \geq b$, the upper limit of x. X cannot take values greater than b.

3   $P(c \leq X \leq d) = F(d) - F(c)$

   $P(c \leq X \leq d) = P(X \leq d) - P(X \leq c)$

   This is very useful when finding probabilities from a p.d.f. or a c.d.f.

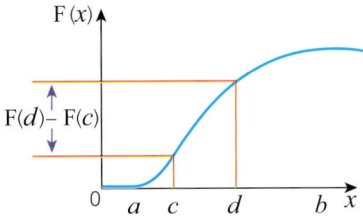

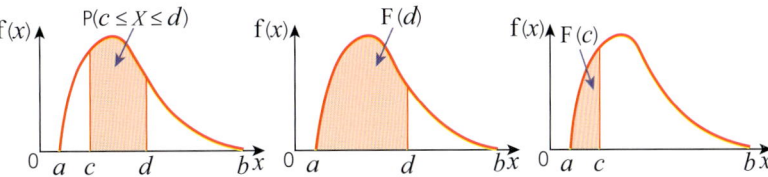

Figure 7.19

4   The median, m, satisfies the equation $F(m) = 0.5$.
   $P(X \leq m) = 0.5$ by definition of the median.

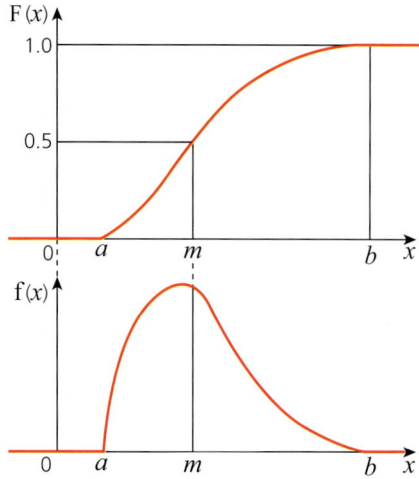

Figure 7.20

5   $f(x) = \dfrac{d}{dx} F(x) = F'(x)$

   Since you integrate $f(x)$ to obtain $F(x)$, the reverse must also be true: differentiating $F(x)$ gives $f(x)$.

6   $F(x)$ is a continuous function: the graph of $y = F(x)$ has no gaps.

Example 7.9

A machine saws planks of wood to a nominal length. The continuous random variable X represents the error in millimetres of the actual length of a plank coming off the machine. The variable X has p.d.f. $f(x)$, where

$$f(x) = \begin{cases} \dfrac{10 - x}{50} & \text{for } 0 \leq x \leq 10 \\ 0 & \text{otherwise.} \end{cases}$$

(i)   Sketch $f(x)$.

(ii) Find the cumulative distribution function $F(x)$.

(iii) Sketch $F(x)$ for $0 \leq x \leq 10$.

(iv) Find $P(2 \leq X \leq 7)$.

(v)   Find the median value of X.

A customer refuses to accept planks for which the error is greater than 8 mm.

(vi) What percentage of planks will he reject?

Solution

(i)

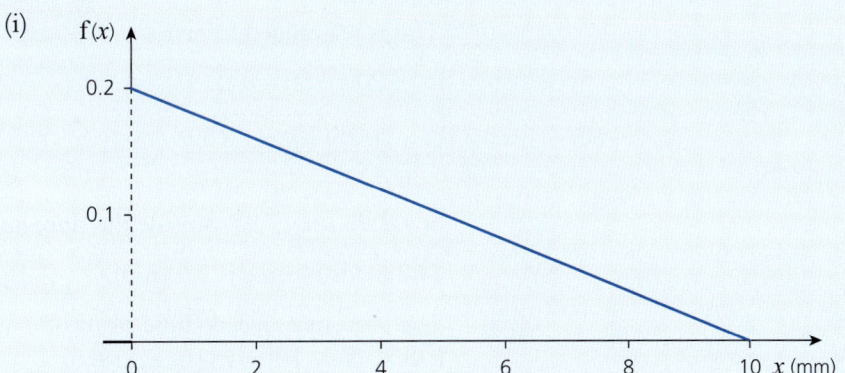

Figure 7.21

(ii) $F(x) = \int_0^x \frac{(10-u)}{50} du$

$= \frac{1}{50}\left[10u - \frac{u^2}{2}\right]_0^x$

$= \frac{1}{5}x - \frac{1}{100}x^2$

The full definition of $F(x)$ is:

$$F(x) = \begin{cases} 0 & \text{for } x < 10 \\ \frac{1}{5}x - \frac{1}{100}x^2 & \text{for } 0 \leq x \leq 10 \\ 1 & \text{for } x > 10. \end{cases}$$

(iii) The graph $F(x)$ is shown in Figure 7.22.

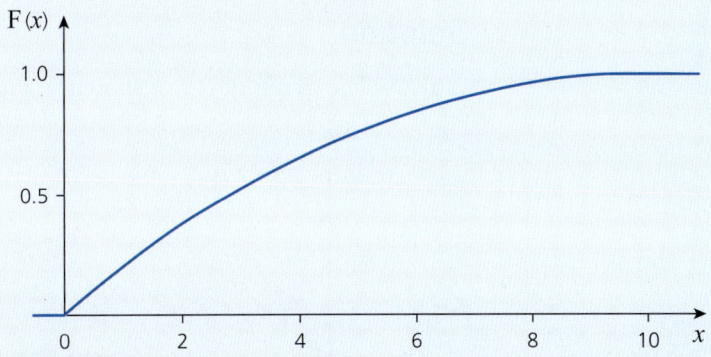

Figure 7.22

(iv) $P(2 \leq X \leq 7) = F(7) - F(2)$

$= \left[\frac{7}{5} - \frac{49}{100}\right] - \left[\frac{2}{5} - \frac{4}{100}\right]$

$= 0.91 - 0.36$

$= 0.55.$

(v) The median value of X is found by solving the equation

$F(m) = 0.5$

$$\frac{1}{5}m - \frac{1}{100}m^2 = 0.5.$$

This is rearranged to give

$$m^2 - 20m + 50 = 0$$

$$m = \frac{20 \pm \sqrt{20^2 - 4 \times 50}}{2}$$

$m = 2.93$ (or 17.07, outside the domain for X).

The median error is 2.93 mm.

(vi) The customer rejects those planks for which $8 \leqslant X \leqslant 10$.
$P(8 \leqslant X \leqslant 10) = F(10) - F(8)$
$= 1 - 0.96$

so 4% of planks are rejected.

Example 7.10

The p.d.f. of a continuous random variable X is given by:

$$f(x) = \begin{cases} \dfrac{x}{24} & \text{for } 0 \leqslant x \leqslant 4 \\ \dfrac{(12-x)}{48} & \text{for } 4 \leqslant x \leqslant 12 \\ 0 & \text{otherwise.} \end{cases}$$

(i) Sketch $f(x)$.

(ii) Find the cumulative distribution function $F(x)$.

(iii) Sketch $F(x)$.

Solution

(i) The graph of $f(x)$ is shown in Figure 7.23

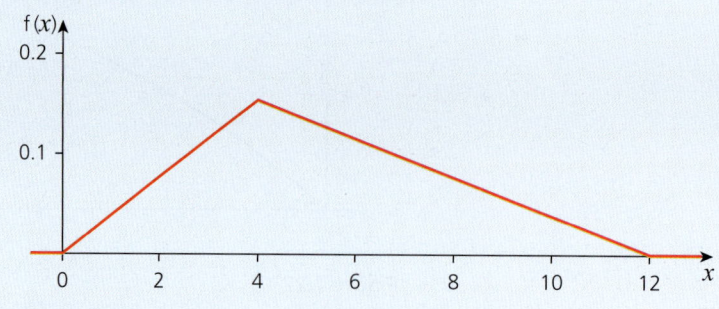

Figure 7.23

(ii) For $0 \leqslant x \leqslant 4$, $F(x) = \int \dfrac{x}{24} dx$

$$= \frac{x^2}{48} + c$$

$P(X \leqslant 0) = 0$ so $F(0) = 0$

$\Rightarrow \quad c = 0$

$\Rightarrow \quad F(4) = \tfrac{1}{3} \qquad (\star)$

For $4 \leqslant x \leqslant 12$, $F(x) = \int \dfrac{12-x}{48}\, dx$

$= \dfrac{x}{4} - \dfrac{x^2}{96} + c'$

> F(4) must take the same value in both formulae.

$F(4) = \tfrac{1}{3}$ from $(\star)$

$\Rightarrow \quad \dfrac{4}{4} - \dfrac{16}{96} + c' = \dfrac{1}{3} \quad \Rightarrow \quad c' = -\dfrac{1}{2}$

So the full definition of $F(x)$ is

$$F(x) = \begin{cases} 0 & \text{for } x < 0, \\ \dfrac{x^2}{48} & \text{for } 0 \leqslant x \leqslant 4, \\ \dfrac{x}{4} - \dfrac{x^2}{96} - \dfrac{1}{2} & \text{for } 4 \leqslant x \leqslant 12, \\ 1 & \text{for } x > 12. \end{cases}$$

> Check that this gives $F(12) = 1$ in both formulae.

(iii) The graph of $F(x)$ is shown in Figure 7.24.

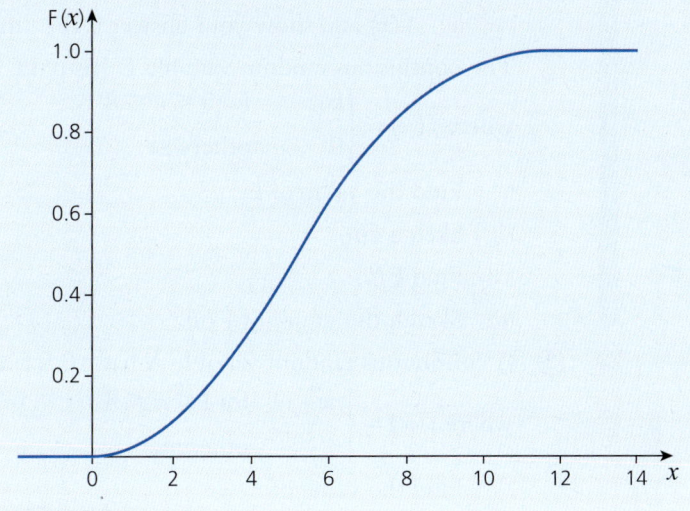

Figure 7.24

Example 7.11

The continuous random variable X has cumulative distribution function $F(x)$ given by:

$$F(x) = \begin{cases} 0 & \text{for } x < 2 \\ \dfrac{x^2}{32} - \dfrac{1}{8} & \text{for } 2 \leqslant x \leqslant 6 \\ 1 & \text{for } x > 6. \end{cases}$$

Find the p.d.f. $f(x)$.

The cumulative distribution function, F

Solution

$f(x) = \dfrac{d}{dx}F(x)$

$f(x) = \begin{cases} \dfrac{d}{dx}F(x) = 0 & \text{for } x < 2 \\ \dfrac{d}{dx}F(x) = \dfrac{x}{16} & \text{for } 2 \leqslant x \leqslant 6 \\ \dfrac{d}{dx}F(x) = 0 & \text{for } x > 6. \end{cases}$

Exercise 7.4

① The continuous random variable X has p.d.f. $f(x)$ where

$f(x) = \begin{cases} 0.2 & \text{for } 0 \leqslant x \leqslant 5 \\ 0 & \text{otherwise.} \end{cases}$

(i) Find $E(X)$.

(ii) Find the cumulative distribution function, $F(x)$.

(iii) Find $P(0 \leqslant x \leqslant 2)$ using

(a) $F(x)$

(b) $f(x)$ and show your answer is the same by each method.

② The continuous random variable U has p.d.f. $f(u)$

where $f(u) = \begin{cases} ku & \text{for } 5 \leqslant u \leqslant 8 \\ 0 & \text{otherwise.} \end{cases}$

(i) Find the value of k.

(ii) Sketch $f(u)$.

(iii) Find $F(u)$.

(iv) Sketch the graph of $F(u)$.

③ A continuous random variable X has p.d.f. $f(x)$

where $f(x) = \begin{cases} cx^2 & \text{for } 1 \leqslant x \leqslant 4 \\ 0 & \text{otherwise.} \end{cases}$

(i) Find the value of c.

(ii) Find $F(x)$.

(iii) Find the median of X.

(iv) Find the mode of X.

④ The continuous random variable X has p.d.f. $f(x)$ given by

$f(x) = \begin{cases} \dfrac{k}{(x+1)^4} & \text{for } x \geqslant 0 \\ 0 & \text{otherwise} \end{cases}$

where k is a constant.

(i) Show that $k = 3$, and find the cumulative distribution function.

(ii) Find also the value of x such that $P(X < x) = \dfrac{7}{8}$.

⑤ The continuous random variable X has c.d.f. given by

$$F(x) = \begin{cases} 0 & \text{for } x < 0 \\ 2x - x^2 & \text{for } 0 \leq x \leq 1 \\ 1 & \text{for } x > 1. \end{cases}$$

(i) Find $P(X > 0.5)$.

(ii) Find the value of q such that $P(X < q) = \frac{1}{4}$.

(iii) Find the p.d.f. $f(x)$ of X, and sketch its graph.

⑥ The continuous random variable X has p.d.f. $f(x)$ given by

$$f(x) = \begin{cases} k(9 - x^2) & \text{for } 0 \leq x \leq 3 \\ 0 & \text{otherwise} \end{cases}$$

where k is a constant.

Show that $k = \frac{1}{18}$ and find the values of $E(X)$ and $Var(X)$.

Find the cumulative distribution function for X, and verify by calculation that the median value of X is between 1.04 and 1.05.

⑦ A random variable X has p.d.f. $f(x)$ where

$$f(x) = \begin{cases} 12x^2(1 - x) & \text{for } 0 \leq x \leq 1 \\ 0 & \text{otherwise.} \end{cases}$$

Find μ, the mean of X, and show that σ, the standard deviation of X, is $\frac{1}{5}$.

Show that $F(x)$, the probability that $X \leq x$ (for any value of x between 0 and 1), satisfies

$$F(x) = \begin{cases} 0 & \text{for } x < 0 \\ 4x^3 - 3x^4 & \text{for } 0 \leq x \leq 1 \\ 1 & \text{for } x > 1. \end{cases}$$

Use this result to show that $P(|X - \mu| < \sigma) = 0.64$. [MEI]

What would this probability be if, instead, X were Normally distributed?

⑧ The temperature in degrees Celsius in a refrigerator which is operating properly has probability density function given by

$$f(t) = \begin{cases} kt^2(12 - t) & 0 < t < 12 \\ 0 & \text{otherwise.} \end{cases}$$

(i) Show that the value of k is $\frac{1}{1728}$.

(ii) Find the cumulative distribution function $F(t)$.

(iii) Show, by substitution, that the median temperature is about 7.37 °C.

(iv) The temperature in a refrigerator is too high if it is over 10 °C. Find the probability that this occurs. [MEI]

⑨ A random variable X has p.d.f. $f(x)$, where

$$f(x) = \begin{cases} k \sin 2x & \text{for } 0 \leq x \leq \frac{\pi}{2} \\ 0 & \text{otherwise.} \end{cases}$$

By integration find, in terms of x and the constant k, an expression for the cumulative distribution function of X for $0 \leq x \leq \frac{\pi}{2}$.

Hence show that $k = 1$ and find the probability that $X < \frac{\pi}{8}$. [MEI]

⑩ The time, t minutes, between customer arrivals at a country store, from Monday to Friday, can be modelled, for $t \geq 0$, by the probability density function f(t) defined by illustrating these probabilities on a sketch of the graph of f(t).

$f(t) = 0.1\, e^{-0.1t}$.

(i) Find the probability that the time between arrivals is
   (a) less than 5 minutes
   (b) more than 15 minutes.

(ii) Obtain the cumulative distribution function for T. Hence find the median time between arrivals.

The time, T minutes, between customer arrivals at the country store on Saturdays can be modelled by the probability density function g(x) defined by $g(x) = \lambda e^{-\lambda x}$, for $x \geq 0$, where λ is a positive constant.

(iii) On Saturdays, the probability of the time between customer arrivals exceeding 5 minutes is 0.4. Estimate the value of λ.

⑪ A firm has a large number of employees. The distance in miles they have to travel each day from home to work can be modelled by a continuous random variable X whose cumulative distribution function is given by

$$F(x) = \begin{cases} 0 & \text{for } x < 1 \\ k\left(1 - \frac{1}{x}\right) & \text{for } 1 \leq x \leq b \\ 1 & \text{for } x > b \end{cases}$$

where b represents the farthest distance anybody lives from work. The diagram below shows a sketch of this cumulative distribution function.

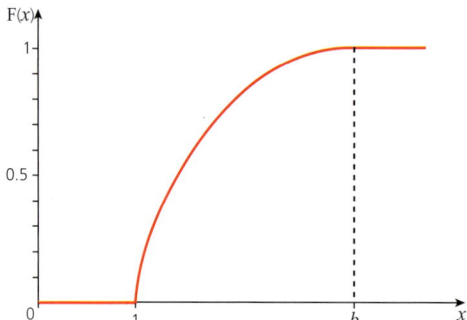

Figure 7.25

A survey suggests that $b = 5$. Use this parameter for parts (i) to (iv).

(i) Show that $k = 1.25$.

(ii) Write down and solve an equation to find the median distance travelled to work.

(iii) Find the probability that an employee lives within half a mile of the median.

(iv) Derive the probability density function for X and illustrate it with a sketch.

(v) Show that, for any value of b greater than 1, the median distance travelled does not exceed 2. [MEI]

12. The wages department of a large company models the incomes of the employees by the continuous random variable X with cumulative distribution function

$$F(x) = 1 - \left(\frac{3}{x}\right)^4 \quad 3 \leq x < \infty$$

where X is measured in an arbitrary currency unit. (X is said to have a Pareto distribution.)

(i) Find the median income. Find also the smallest income of an employee in the top 10% of incomes.

(ii) Find the probability density function of X and hence show that the mean income is 4.

(iii) Find the probability that a randomly chosen employee earns more than the mean income. [MEI]

13. A continuous random variable X has probability density function $f(x)$. The probability that $X \leq x$ is given by the function $F(x)$.

Explain why $F'(x) = f(x)$.

A rod of length $2a$ is broken into two parts at a point whose position is random. State the form of the probability distribution of the length of the smaller part, and state also the mean value of this length.

Two equal rods, each of length $2a$, are broken into two parts at points whose positions are random. X is the length of the shortest of the four parts thus obtained. Find the probability, $F(x)$, that $X \leq x$, where $0 < x \leq a$.

Hence, or otherwise, show that the probability density function of X is given by

$$f(x) = \begin{cases} \dfrac{2(a-x)}{a^2} & \text{for } 0 < x \leq a \\ 0 & \text{for } x \leq 0, \, x > a \end{cases}$$

Show that the mean value of X is $\frac{1}{3}a$.

Write down the mean value of the sum of the two smaller parts and show that the mean values of the four parts are in the proportions $1:2:4:5$. [JMB]

14. (i) Explain the significance of the results:

(a) $\displaystyle\int_{-\infty}^{\infty} \frac{1}{\sqrt{2\pi}} e^{-\frac{1}{2}z^2} \, dz = 1$ (b) $\displaystyle\int_{-\infty}^{\infty} \frac{1}{\sqrt{2\pi}} z e^{-\frac{1}{2}z^2} \, dz = 0$

The random variable Y is given by $Y = Z^2$ (where Z is the standardised Normal variable).

(ii) Using the results in part (i), find

(a) $E(Y)$ (b) $\text{Var}(Y)$.

(You will need to use integration by parts.)

5 The continuous uniform (rectangular) distribution

You will recall having studied the discrete uniform distribution in Chapter 3. You will now meet the continuous uniform distribution. Since the term uniform distribution can be applied to both discrete and continuous variables, in the continuous case it is often written as uniform (rectangular).

The **continuous uniform (rectangular) distribution** is particularly simple since its p.d.f. is constant over a range of values and zero elsewhere.

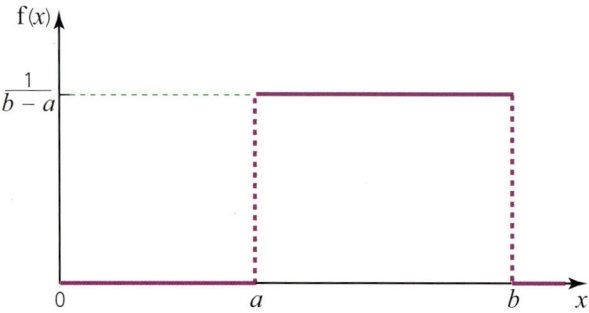

Figure 7.26

In Figure 7.26, X may take values between a and b, and is zero elsewhere. Since the area under the graph must be 1, the height is $\frac{1}{b-a}$.

> **Note**
>
> It is common to describe distributions by the shapes of the graphs of their p.d.f.s: U-shaped, J-shaped, etc.

> **Note**
>
> There are many other standard continuous distribution, including the Normal distribution with probability density function written as
> $$\phi(x) = \frac{1}{\sigma\sqrt{2\pi}} e^{-\frac{1}{2}\left(\frac{x-\mu}{\sigma}\right)^2}$$
> and the exponential distribution with probability density function $f(x) = \lambda e^{-\lambda x}$.

Example 7.12

A junior gymnastics league is open to children who are at least five years old but have not yet had their ninth birthday. The age, X years, of a member is modelled by the uniform (rectangular) distribution over the range of possible values between five and nine. Age is measured in years and decimal parts of a year, rather than just completed years. Find

(i) the p.d.f. f(x) of X

(ii) $P(6 \leq X \leq 7)$

(iii) $E(X)$

(iv) $\text{Var}(X)$

(v) the percentage of the children whose ages are within one standard deviation of the mean.

Solution

(i) The p.d.f. $f(x) = \begin{cases} \frac{1}{9-5} = \frac{1}{4} & \text{for } 5 \leq x < 9 \\ 0 & \text{otherwise.} \end{cases}$

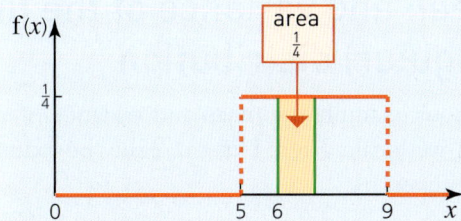

Figure 7.27

(ii) $P(6 \leq X \leq 7) = \frac{1}{4}$ by inspection of the rectangle above.

Alternatively, using integration

$$P(6 \leq X \leq 7) = \int_6^7 f(x)\,dx = \int_6^7 \frac{1}{4}\,dx$$
$$= \left[\frac{x}{4}\right]_6^7$$
$$= \frac{7}{6} - \frac{6}{4}$$
$$= \frac{1}{4}.$$

(iii) By the symmetry of the graph $E(X) = 7$. Alternatively, using integration

$$E(X) = \int_{-\infty}^{\infty} xf(x)\,dx = \int_5^9 \frac{x}{4}\,dx$$
$$= \left[\frac{x^2}{8}\right]_5^9$$
$$= \frac{81}{8} - \frac{25}{8} = 7.$$

$$Var(X) = \int_{-\infty}^{\infty} x^2 f(x)\,dx - [E(X)]^2 = \int_5^9 \frac{x^2}{4}\,dx - 7^2$$
$$= \left[\frac{x^3}{12}\right]_5^9 - 49$$
$$= \left(\frac{729}{12} - \frac{125}{12}\right) - 49$$
$$= 1.333 \text{ to 3 decimal places.}$$

(v) Standard deviation $\sqrt{\text{variance}} = \sqrt{1.333} = 1.155$.

So the percentage within 1 standard deviation of the mean is

$$\frac{2 \times 1.155}{4} \times 100\% = 57.7\%.$$

> Those within 1 standard deviation of the mean are those whose ages lie between 7 − 1.155 and 7 + 1.155. So the total range of ages is 2 × 1.155.

The mean and variance of the uniform (rectangular) distribution

In the previous example the mean and variance of a particular uniform distribution were calculated. This can easily be extended to the general uniform distribution given by:

$$f(x) = \begin{cases} \dfrac{1}{b-a} & \text{for } a \leqslant x \leqslant b \\ 0 & \text{otherwise.} \end{cases}$$

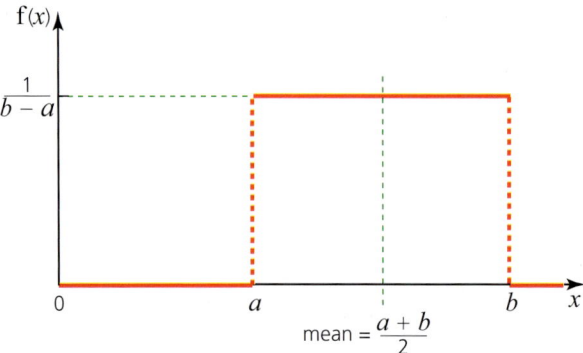

Figure 7.28

Mean: By symmetry the mean $E(X)$ is $\dfrac{a+b}{2}$.

Variance:
$$\begin{aligned}
\text{Var}(X) &= \int_{-\infty}^{\infty} x^2 f(x)\,dx - [E(X)]^2 \\
&= \int_a^b x^2 f(x)\,dx - [E(X)]^2 \\
&= \int_a^b \dfrac{x^2}{b-a}\,dx - \left(\dfrac{a+b}{2}\right)^2 \\
&= \left[\dfrac{x^3}{3(b-a)}\right]_a^b - \dfrac{1}{4}\left(a^2 + 2ab + b^2\right) \\
&= \dfrac{b^3 - a^3}{3(b-a)} - \dfrac{1}{4}\left(a^2 + 2ab + b^2\right) \\
&= \dfrac{(b-a)}{3(b-a)}\left(b^2 + ab + a^2\right) - \dfrac{1}{4}\left(a^2 + 2ab + b^2\right) \\
&= \dfrac{1}{12}\left(b^2 - 2ab + a^2\right) \\
&= \dfrac{1}{12}(b-a)^2
\end{aligned}$$

The cumulative distribution function of the uniform (rectangular) distribution

Earlier in this chapter you learnt that the cumulative distribution function, $F(x)$, of a continuous random variable, X, with probability density function $f(x)$ is given by

$$F(x) = \int f(x)\,dx.$$

The lower limit of the integral is the lowest value that X can take and the upper limit is x.

So for the general uniform (rectangular) distribution with values between a and b, as shown in Figure 12.3,

$F(x)$ is given by $\int_a^x f(u)\,du$.

> Notice the use of u as a dummy variable here because x is one of the limits of the integral.

So

$$F(x) = \int_a^x \frac{1}{b-a}\,du$$

$$\Rightarrow F(x) = \frac{x-a}{b-a} \quad \text{for} \quad a \leqslant x \leqslant b.$$

Outside this range of values, $F(x) = 0$ for $x < a$ and $F(x) = 1$ for $x > b$.

This is illustrated by the graph in Figure 7.29.

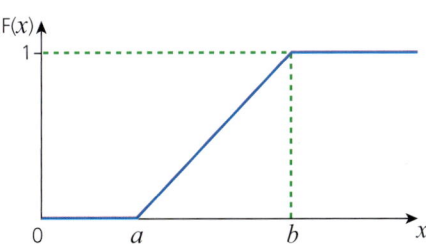

Figure 7.29

Exercise 7.5

① The random variable X has p.d.f.

$$f(x) = \begin{cases} \dfrac{1}{6} & \text{for } -2 \leqslant x \leqslant 4 \\ 0 & \text{otherwise.} \end{cases}$$

(i) Sketch the graph of $f(x)$.
(ii) Find $P(X < 2)$.
(iii) Find $E(X)$.
(iv) Find $P(X < 1)$.

② The graph shows the cumulative distribution function of a continuous random variable, X.

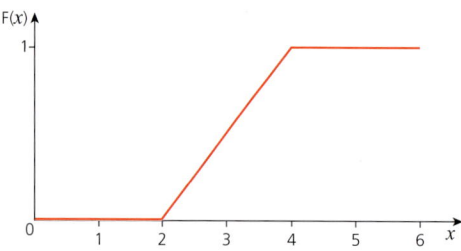

Figure 7.30

(i) The equation of the sloping line segment is $F(x) = mx + c$. Find the values of m and c.

(ii) Hence find the probability density function, $f(x)$, of X.

(iii) Describe the distribution of X.

(iv) Write down

    (a) $P(2 \leqslant X \leqslant 3)$

    (b) $P(3 \leqslant X \leqslant 4)$

    (c) $P(2 \leqslant X \leqslant 3) + P(3 \leqslant X \leqslant 4)$

③ The continuous random variable X has p.d.f. $f(x)$ where

$$f(x) = \begin{cases} 0.2 & \text{for } 0 \leqslant x \leqslant 5 \\ 0 & \text{otherwise.} \end{cases}$$

(i) Find $E(X)$.

(ii) Find the cumulative distribution function, $F(x)$.

(iii) Find $P(0 \leqslant x \leqslant 2)$ using

    (a) $F(x)$ (b) $f(x)$

and show your answer is the same by each method.

④ A continuous random variable X has a uniform (rectangular) distribution over the interval $(4, 7)$. Find

(i) the p.d.f. of X

(ii) $E(X)$

(iii) $Var(X)$

(iv) $P(4.1 \leqslant X \leqslant 4.8)$.

⑤ The probability density function of a random variable, X, is

$f(x) = 0$ for $x < a$,

$f(x) = c$ for $a \leqslant x \leqslant b$ where c is a constant,

$f(x) = 0$ for $x > b$.

Given that $E(X) = 6$ and $Var(X) = 3$, find the values of a, b and c.

⑥ The distribution of the lengths of adult Martian lizards is uniform between 10 cm and 20 cm. There are no adult lizards outside this range.

(i) Write down the p.d.f. of the lengths of the lizards.

(ii) Find the mean and variance of the lengths of the lizards.

(iii) What proportion of the lizards have lengths within

    (a) one standard deviation of the mean

    (b) two standard deviations of the mean?

⑦ A continuous random variable X has the p.d.f.

$$f(x) = \begin{cases} k & \text{for } 0 \leq x \leq 5 \\ 0 & \text{otherwise.} \end{cases}$$

(i) Find the value of k.
(ii) Sketch the graph of f(x).
(iii) Find E(X).
(iv) Find E($4X - 3$) and show that your answer is the same as $4E(X) - 3$.

⑧ A toy company sells packets of coloured plastic equilateral triangles. The triangles are actually offcuts from the manufacture of a totally different toy, and the length, X, of one side of a triangle may be modelled as a random variable with a uniform (rectangular) distribution for $2 \leq x \leq 8$.

(i) Find the p.d.f. of X.
(ii) An equilateral triangle of side x has area a. Find the relationship between a and x.
(iii) Find the probability that a randomly selected triangle has area greater than 15 cm^2. (Hint: What does this imply about x?)
(iv) Find the expectation and variance of the area of a triangle.

⑨ The random variable X has a rectangular distribution whose p.d.f. f(x) is given by $f(x) = \begin{cases} \dfrac{1}{6} & \text{for } -2 \leq x \leq 4 \\ 0 & \text{otherwise.} \end{cases}$

Sketch the p.d.f. of X and hence, or otherwise, find the probabilities that
(i) $X \leq 2$
(ii) $|X| \leq 2$
(iii) $|X| \leq x$ for $0 \leq x \leq 2$
(iv) $|X| \leq x$ for $2 \leq x \leq 4$
(v) $|X| \leq x$ for $x > 4$

Hence obtain, and sketch, the p.d.f. of $|X|$. Thus determine the mean of $|X|$.

⑩ The continuous random variable X has the rectangular distribution on the interval $a \leq x \leq b$ where a, b are unknown. Let the expectation and variance of X be μ and σ^2 respectively.

(i) State, in terms of a and b, the values of μ and σ^2.

Now let $\overline{X}$ denote the mean of a random sample of 25 values of X.

(ii) Write down, in terms of a and b, the mean and variance of $\overline{X}$.

In a random sample of 25 values of X the following totals were obtained.

$$\sum x = 1242.5 \quad \sum x^2 = 61982.89$$

(iii) Use these values to estimate μ and σ^2. Hence estimate a and b.

⑪ A computer displays coloured squares of random size. The length of the side of a square is X cm. The random variable X has a rectangular distribution on the interval $1 \leq x \leq 5$, as shown in Figure 12.6.

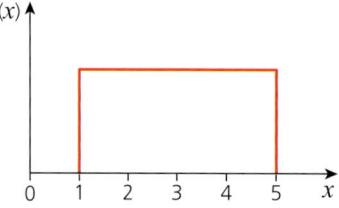

Figure 7.31

(i) Write down the probability density function for X and state its median.

(ii) Find the cumulative distribution function $F(x) = P(X \leq x)$.

The area of a square of side X is denoted by Y.

(iii) Show that the cumulative distribution function for Y is given by $P(Y \leq y) = \frac{1}{4}(\sqrt{y} - 1)$. Find $g(y)$, the probability density function for Y, and state the interval of values of y for which it applies.

(iv) Find the median area of a square.

(v) Sketch the graph of $g(y)$, and indicate the median and mean on your sketch.

> **Note**
>
> There are many other standard continuous distribution, including the Normal distribution with probability density function written as $\phi(x) = \dfrac{1}{\sigma\sqrt{2\pi}} e^{-\frac{1}{2}\left(\frac{x-\mu}{\sigma}\right)^2}$ and the exponential distribution with probability density function $f(x) = \lambda e^{-\lambda x}$.

Discussion point

Jessie organises the emergency response to incidents on an island. The time between incidents, t hours, is sometimes quite long. It is denoted by the random variable T. It is believed that the probability density function $f(t) = 0.2e^{-0.2t}$ provides a good model for T; it is shown in Figure 7.32(a).

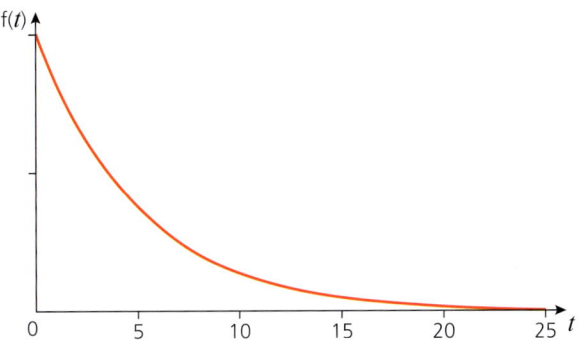

Figure 7.32 (a) Probability density function

To check this model, records are kept of the lengths of the next 50 intervals. The data are given below.

6.1	16.0	11.2	0.4	8.2	2.8	2.7	7.7	2.3	10.1
4.8	9.0	8.5	18.2	14.6	19.8	13.3	3.6	24.1	2.8
1.7	0.1	21.3	8.8	1.9	9.1	4.7	12.5	2.5	3.7
2.9	1.8	17.4	4.2	12.1	13.9	8.0	6.3	8.4	17.4
3.5	2.2	1.5	9.7	0.6	1.6	5.9	4.5	4.5	3.0

Figures 7.32 (b), (c), (d) and (e) show 4 different ways of displaying the data. In each case the labels or the scales (or both) are missing from the vertical axis.

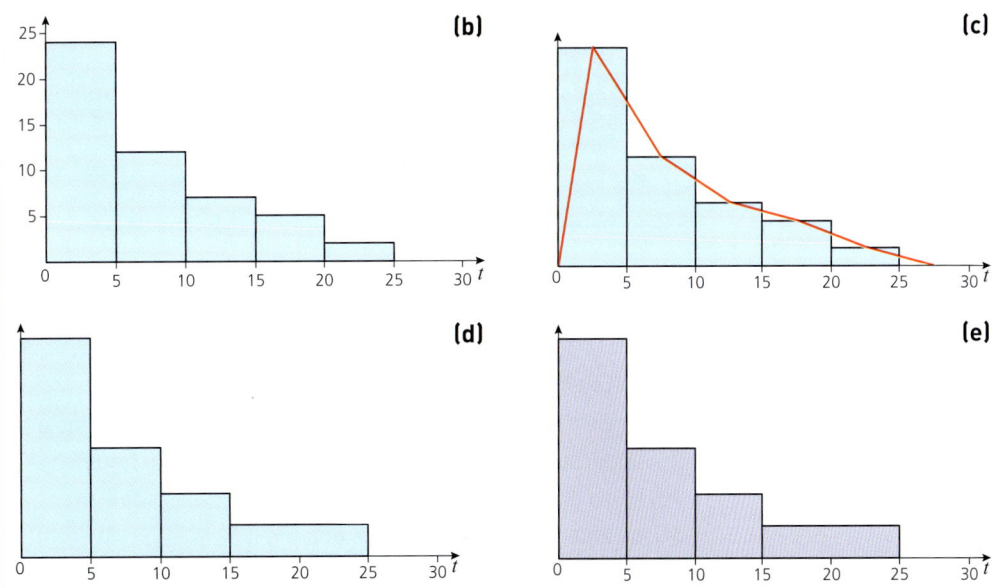

Figure 7.32 (b) Frequency chart (c) Frequency polygon (d) Histogram (e) Probability distribution

What information is missing from the vertical axes of the 5 graphs, (a) to (e)?

Which out of (b), (c), (d) and (e) do you think is the most informative way of displaying the data?

S2 AS — The continuous uniform (rectangular) distribution

KEY POINTS

1. **A continuous random variable**
   If X is a continuous random variable with probability density function $f(x)$
   - $\int_{-\infty}^{\infty} f(x)\,dx = 1$
   - $f(x) \geq 0$ for all x
   - $P(c \leq x \leq d) = \int_c^d f(x)\,dx$
   - $E(X) = \int_{-\infty}^{\infty} x f(x)\,dx$
   - $\mathrm{Var}(X) = \int_{-\infty}^{\infty} x^2 f(x)\,dx - [E(X)]^2$
   - The mode of X is the value of x for which $f(x)$ is greatest

2. **A function of a continuous random variable**
   If $g(X)$ is a function of X then
   - $E[g(X)] = \int_{-\infty}^{\infty} g(x) f(x)\,dx$
   - $\mathrm{Var}[g(X)] = \int_{-\infty}^{\infty} [g(x)]^2 f(x)\,dx - [E[g(X)]]^2$

3. **The cumulative distribution function**
   - $F(x) = \int_a^x f(u)\,du$ where the constant a is the lower limit of X.
   - $f(x) = \dfrac{d}{dx} F(x)$
   - For the median, m, $F(m) = 0.5$

4. **The uniform (rectangular) distribution over the interval (a, b)**
   - $f(x) = \dfrac{1}{b-a}$
   - $E(x) = \dfrac{1}{2}(a+b)$
   - $\mathrm{Var}(X) = \dfrac{(b-a)^2}{12}$

LEARNING OUTCOMES

When you have completed this chapter you should:
- be able to use a simple continuous random variable as a model
- understand the meaning of a p.d.f. and be able to use one to find probabilities
- know and use the properties of a p.d.f.
- be able to sketch the graph of a p.d.f.
- be able to find the mean and variance from a given p.d.f.
- understand the meaning of a c.d.f. and be able to obtain one from a given p.d.f.
- be able to sketch a c.d.f.
- be able to obtain a p.d.f. from a given c.d.f.
- use a c.d.f. to calculate the median and other percentiles

8 Conditional probability

Failure saves lives. In the airline industry, every time a plane crashes the probability of the next crash is lowered by that. The Titanic saved lives because we are building bigger and bigger ships. So these people died but we have effectively improved the safety of the system, and nothing failed in vain.

Nassim Nicholas Taleb

This picture shows a tropical lightning storm.

Typically, about two people die per year in the UK from being struck by lightning.

→ How would you estimate the probability that a particular person will be killed by lightning strike in the UK on Wednesday next week?

S2 AS — Screening tests

You might give an argument along these lines.

There are about 60 million people in the UK.

So the probability of any individual being killed by lightning strike in the next year is
$$\frac{2}{60\,000\,000}$$

There are 365 days in a year so the probability of this happening on any particular day is
$$\frac{2}{60\,000\,000} \times \frac{1}{365} \approx 9 \times 10^{-11}$$

or a little under 1 in 10 billion.

The problem with this argument is that it assumes that all people are equally likely to be victims of lightning strike and that it is equally likely to happen on any day of the year. Neither of these is true.

- Lightning is much more common over the summer months.
- Relevant data show that men are much more likely to be the victims than women.

So the figure of 1 in 10 billion is actually meaningless. The probability is conditional upon other circumstances such as the time of year and the gender and lifestyle of the person.

This chapter is about **conditional probability**.

Notation

In standard notation

- A and B are events.
- The event *not-A* is denoted by A'. $P(A') = 1 - P(A)$.
- $P(A|B)$ means the probability of event A occurring, given that event B has occurred.

You can see the meaning of $P(A|B)$ in this Venn diagram. The fact that event B has already occurred restricts the possibilities to the red region. If event A also occurs the event must be in the overlap region $A \cap B$.

So the probability is given by

$$P(A|B) = \frac{P(A \cap B)}{P(B)}$$

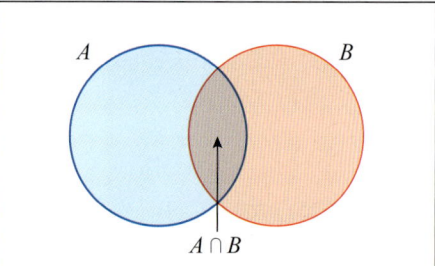

Figure 8.1 Venn diagram

This is a fundamental result in conditional probability.

> **Note**
> A false positive occurs when a person tests positive but has not got the condition. A false negative occurs when a person tests negative but has actually got the condition.

1 Screening tests

Another example of conditional probability arises with medical screening tests.

For many medical conditions, a **screening test** is given to large parts of the population to check if they have the condition. These tests are never 100% reliable and it is often the case that the test suggests that a person has the condition being screened for when in fact they do not. Such a case is called a **'false positive'**. On other occasions, the test suggests that a person does not have the condition when in fact they do have it. This type of case is known as a **'false negative'**. It may at first seem strange that screening tests do not give the correct result 100% of the time, but the human body is a very complex system and screening tests are useful. They provide early diagnosis of conditions that would later become difficult or impossible to treat.

Example 8.1

This contingency table shows the probabilities for the outcomes of a screening test for a medical condition on a randomly selected person.

The following letters describe the sets involved.

C Has the condition $\quad$ C' Does not have the condition

Y Tests positive $\quad$ N Tests negative

Table 8.1

	C	C'
Y	0.20	0.15
N	0.05	0.60

(i) Copy the table and add in the marginal totals.

(ii) Find the probabilities that a randomly selected person

(a) has the condition and tests positive, (b) has the condition, (c) tests positive.

Use set notation as well as giving the values of your answers.

(iii) Find the probability that a person who has the condition tests positive.

(iv) Find the percentages of false positives and false negatives given by the test. Hence state the percentage of incorrect results it gives.

Discussion point

In this example, set notation was used to describe most of the events. You will find it helpful to relate these to a Venn diagram below.

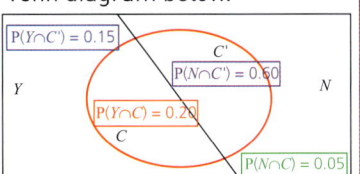

Figure 8.2

What sets do the various regions represent? What probabilities are associated with them?

Solution

(i) The marginal totals are given in **red**.

Table 8.2

	C	C'	
Y	0.20	0.15	**0.35**
N	0.05	0.60	**0.65**
	0.25	**0.75**	**1**

(ii) (a) $P(C \cap Y) = 0.20$ (b) $P(C) = 0.25$ (c) $P(Y) = 0.35$.

(iii) $P(Y|C) = \dfrac{P(Y \cap C)}{P(C)} = \dfrac{0.2}{0.25} = 0.8$.

(iv) False positives $P(C' \cap Y) = 0.15 = 15\%$

False negatives $P(C \cap N) = 0.05 = 5\%$

Incorrect results $15\% + 5\% = 20\%$

Screening tests

Another way to represent conditional probabilities is to use a tree diagram, as in the next example.

Example 8.2

This contingency table, together with the marginal totals, shows the probabilities for the outcomes of a screening test for a medical condition on a randomly selected person.

Table 8.3

	C	C'	
Y	0.20	0.15	**0.35**
N	0.05	0.60	**0.65**
	0.25	**0.75**	**1**

This is the same table as in Example 8.1.

(i) Draw a tree diagram to illustrate the same information, marking in all the relevant probabilities.

(ii) Using appropriate notation, describe the numbers that appear vertically above and below each other in the tree diagram.

(iii) Show that $P(Y) = P(Y|C) \times P(C) + P(Y|C') \times P(C')$.

Discussion point

Identify which of the numbers in the table in the question feature in the tree diagram, and where they appear.
State which numbers in the table do not feature in the tree diagram.
Would your answers be the same if you drew the tree diagram with Y and N for the left-hand branches and C and C' for the right-hand branches?

Solution

(i)
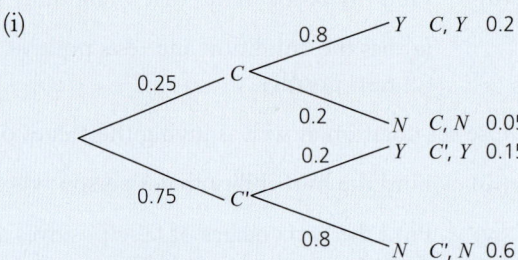

Figure 8.3

(ii) Left-hand branches $0.25 = P(C), 0.75 = P(C')$

Right-hand branches $0.8 = P(Y|C), 0.2 = P(N|C)$
$0.2 = P(Y|C'), 0.8 = P(N|C')$

Right-hand line $0.2 = P(C \cap Y), 0.05 = P(C \cap N),$
$0.15 = P(C' \cap Y), 0.6 = P(C' \cap N)$

Multiplying along the branches and adding the two cases where Y occurs.

(iii) In numbers $P(Y) = 0.25 \times 0.8 + 0.75 \times 0.2$

In symbols $P(Y) = P(C) \times P(Y|C) + P(C') \times P(Y|C')$
$= P(Y|C) \times P(C) + P(Y|C') \times P(C')$

as required.

Note

You will meet this result later in the chapter in connection with Bayes' theorem.
It will be given for events A and B in the form
$P(B) = P(B|A) \times P(A) + P(B|A') \times P(A')$.

The next example shows how important it is to specify the prior conditions correctly when using conditional probability.

Figure 8.4

Example 8.3

A robbery has taken place on one of the UK's islands. When carrying out the crime, the robber suffered from a cut and left a sample of blood at the scene. A test shows that it is group AB−: this is the rarest group in the UK, occurring in about 1% of people.

The police have the blood groups of a few of the island's 800 residents on record, including Tom who is AB−. On this evidence alone, he is arrested for the crime.

- The arresting officer says 'The national figures for this blood group mean that the probability that you are innocent is 1% and so that you are guilty is 99%'.
- Tom's solicitor says 'About eight people on the island will have this blood group, so, in the absence of any other evidence, the probability that my client is guilty is only $12\frac{1}{2}$% and that he is innocent is $87\frac{1}{2}$%'.

(i) Which one of these two statements is correct?

(ii) How would you explain the fault in the other statement?

Note

Explaining why the other statement is wrong requires some thought. The arresting officer has, in fact, made a well-known error called the **prosecutor's fallacy**.

In order to use probability in the analysis of a situation like this, you have to be very careful to ask the right question.

Solution

(i) The solicitor's statement is correct.

The expected number of people on the island with AB− blood is $800 \times 1\% = 8$.

So the probability that the blood is from Tom is $\frac{1}{8} = 0.125$ or 12.5%.

S2 AS Screening tests

> **Discussion point**
> This example was set on an island with a small population. This allowed the solicitor's statement to be fairly obviously correct. If it had been set in, say, somewhere in London it would not have been possible to quantify the probability that the accused person was innocent and so the prosecutor's fallacy might have seemed more plausible, even though it was still completely false.
>
> Is there a danger of the prosecutor's fallacy being used in cases where identification is carried out on the basis of DNA?

(ii) There are three relevant events.

- R A blood sample selected at random is AB-,
- T The blood sample came from Tom,
- I Tom is innocent.

The arresting officer asked

'What is the probability that a random blood sample is AB-?'

i.e. 'What is the value of $P(R)$?'

The officer has made the reasonable assumption that people's criminal tendencies and their blood groups are independent. So the question he asked is equivalent to

'What is the probability that an innocent person has blood group AB-?'

i.e. 'What is the value of $P(R\,|\,I)$?'

However, this is the wrong question because it is known that event R has occurred but it is not known that event I has occurred.

The correct question is

'Given that an AB- sample has been found, what is the probability that it came from Tom?'

i.e. 'What is the value of $P(T\,|\,R)$?'

If the sample did come from Tom, then he is guilty. So the probability that he is innocent is given by

$$P(I) = 1 - P(T\,|\,R).$$

EXTENSION

Bayes' theorem

The last example involved the prosecutor's fallacy. It showed that, without clear thinking, it is all too easy to go wrong with conditional probability, with potentially serious consequences. The rest of this chapter shows how Bayes' theorem can be used to place the analysis of such situations on a more formal basis.

Example 8.4

A screening test for a particular condition is not 100% reliable. In fact, the probability that a person who has the condition tests positive is 0.97. The probability that a person who does not have the condition tests positive is 0.01. The proportion of the population which has the condition is 1.5%.

A person is selected at random and tested for the condition.

(i) Using A for the event that the person does not have the condition and B for the event that the person tests positive, illustrate this situation on a tree diagram.

(ii) Find the probability that the person tests positive.

> **Note**
> You will not be tested on Bayes' theorem, but it is useful extra information.

(iii) The person tests positive. Find the probability that the person does not have the condition.

(iv) In light of your answer to part (iii), comment briefly on the effectiveness of the test.

Solution

(i)

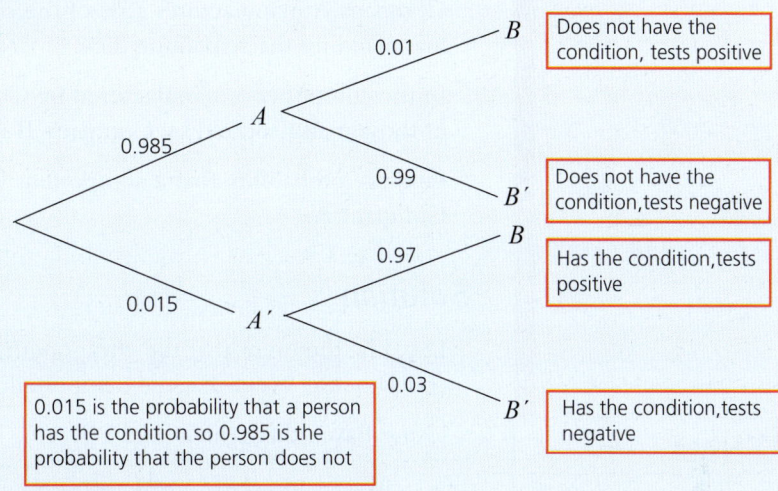

0.015 is the probability that a person has the condition so 0.985 is the probability that the person does not

Figure 8.5

To find the probability that the person tests positive, you need to include both those who have the condition and those who do not.

(ii) P(B) = 0.985 × 0.01 + 0.015 × 0.97

= 0.00985 + 0.01455

= 0.0244

(iii) To find the probability that someone who tested positive (event B) does not have the condition (event A), use the conditional probability formula

$$P(A \mid B) = \frac{P(A \cap B)}{P(B)}.$$

From part (ii) you know that P(B) = 0.0244.

From the tree diagram P($A \cap B$) = 0.985 × 0.01 = 0.00985

So $P(A' \mid B) = \frac{0.00985}{0.0244} = 0.404$.

P($A \cap B$) is the probability that the person does not have the condition and tests positive.

You can find this by multiplying along the A and B branches of the tree diagram.

(iv) The test has a false positive rate of just over 40%. This does seem rather high. The false positives would need further tests to determine whether they actually have the condition and would be unnecessarily worried. On the other hand, under 1% of the population would be in this situation so the screening test may still be worthwhile. Also over half of the positives would actually have the condition.

Note

The event $A \mid B$ is an example of a false positive

This example illustrates Bayes' theorem which states that:

$$P(A \mid B) = \frac{P(A \cap B)}{P(B \mid A)P(A) + P(B \mid A')P(A')}.$$

Substituting the numbers for the previous example, gives

$$P(A \mid B) = \frac{0.01 \times 0.985}{0.01 \times 0.985 + 0.97 \times 0.015} = 0.403\,688\ldots$$

S2 AS Screening tests

When you are solving problems like this, you can either work out the probability from first principles, as in the example above, or you can use Bayes' theorem, as in the next example.

Example 8.5

Two different companies manufacture a particular type of electrical component.

Company A manufactures 75% of the components and Company B manufactures the remaining 25%.

In the components manufactured by Company A, 1% fail early, whereas 2% of those manufactured by Company B fail early.

Find the probability that a component which fails early was manufactured by Company A.

Solution

Let A be the event that the component was manufactured by Company A and D be the event that the component is defective.

Using Bayes' theorem to find $P(A|D)$,

$$P(A|D) = \frac{P(A \cap D)}{P(D|A)P(A) + P(D|A')P(A')}$$

$P(A) = 0.75, P(A') = 0.25, P(D|A) = 0.01, P(D|A') = 0.02$

$P(A \cap D) = 0.75 \times 0.01 = 0.0075$

$$P(A|D) = \frac{0.0075}{0.01 \times 0.75 + 0.02 \times 0.25} = 0.6$$

Note

Note that A' is the event that the component was not manufactured by Company A.

So A' is the event that it was manufactured by Company B.

Exercise 8.1

① It is given that $P(A) = 0.25$, $P(B) = 0.13$ and $P(A \cap B) = 0.05$.
  (i) Find $P(A \cup B)$.
  (ii) Find $P(A|B)$.
  (iii) Find $P(B|A)$.

② Steve is going on holiday. The probability that he is delayed on his outward flight is 0.3. The probability that he is delayed on his return flight is 0.2, independently of whether or not he is delayed on the outward flight.
  (i) Find the probability that Steve is delayed on his outward flight but not on his return flight.
  (ii) Find the probability that he is delayed on at least one of the two flights.
  (iii) Given that he is delayed on at least one flight, find the probability that he is delayed on both flights. [MEI]

③ When it is reasonably dry, I cycle to work. Otherwise I take the bus. On one day in ten, on average, it is too wet for me to cycle. If I cycle, the probability that I am late for work is 0.02. If I take the bus, the probability that I am late for work is 0.15.
  (i) Find the probability that I am late on a randomly chosen day.
  (ii) Given that I am late, find the probability that I cycled.

216

④ A screening test for a particular disease has a probability of 0.99 of giving a positive result for somebody who has the disease and of 0.05 for somebody who does not have the disease. It is thought that 0.35% of the population have the disease.

  (i)   Find the probability that a randomly selected person tests positive.
  (ii) Find the probability that a randomly selected person has the disease given that they test positive.
  (iii) Write down the probability that a randomly selected person does not have the disease given that they test positive.
  (iv) In light of your answers to parts (ii) and (iii), comment briefly on the effectiveness of the test.

⑤ A machine which makes gear wheels has two settings, fast and slow. On the fast setting, 10% of the gear wheels are faulty. On the slow setting, 1% of them are faulty. One day, 1000 gear wheels are made using the fast setting and 500 using the slow setting.

  (i) Find the probability that a randomly selected gear wheel from the day's production is faulty.
  (ii) Given that a gear wheel is faulty, find the probability that the machine was on the slow setting when it was manufactured.

⑥ Two thirds of the trains which I catch on my local railway line have four coaches and the rest have eight coaches (the eight-coach trains are run during busy periods). If I catch a four-coach train, the probability of me getting a seat is 0.7. If I catch an eight-coach train, the probability of me getting a seat is 0.9. Given that I get a seat on a train, find the probability that it is a train with four coaches.

⑦ A company which runs a large fleet of HGVs (heavy goods vehicles) regularly tests its employees for consumption of illegal drugs. The test is not 100% reliable, so if somebody is found to be positive, further testing is carried out. The false positive rate is 0.5% and the false negative rate is 3%. You should assume that 1% of employees have taken drugs preceding the test.

  (i) Find the probability that a randomly selected person tests positive.
  (ii) Find the probability that a randomly selected person has consumed illegal drugs, given that they test positive.

⑧ Three machines, A, B and C, are used to make dress fabric. Machine A makes 50% of the fabric, Machine B makes 30% and Machine C makes the remainder. Each piece of fabric is checked for defects. The percentages of defective fabric for Machines A, B and C are 2, 4 and 1.5, respectively.

  (i) Find the probability that a randomly selected piece of fabric is defective.
  (ii) Given that a piece of fabric is defective, find the probability that it was produced on Machine A.

⑨ Onions often suffer from a disease called white rot. This disease stays in the soil for several years, and when onions are planted in infected ground, they can get the disease. A gardener grows onions in four separate plots Q, R, S and T, so that there is less risk of most of her onions getting white rot. They grow an equal number of onions in each plot.

S2 AS — Screening tests

The percentages of onions in each plot which get white rot are as follows
- Plot Q 2%
- Plot R 3%
- Plot S 60%
- Plot T 8%.

(i) Find the probability that a randomly selected onion has white rot.

(ii) Given that a randomly selected onion has white rot, find the probability that it came from plot S.

KEY POINTS

1. The conditional probability of event A given the event B has occurred is given by
$$P(A|B) = \frac{P(A \cap B)}{P(B)}$$

2. It is often helpful to use a tree diagram or a contingency table to illustrate a situation involving conditional probability.

3. Bayes' theorem states that
$$P(A|B) = \frac{P(A \cap B)}{P(B|A)P(A) + P(B|A')P(A')}$$

4. Bayes' theorem is used to find conditional probabilities.

5. In contexts such as a medical screening test:
   - a false positive occurs when a person tests positive but has not got the condition
   - a false negative occurs when a person tests negative but actually has got the condition.

LEARNING OUTCOMES

When you have completed this chapter you should:
- understand and be able to use the notation associated with conditional probability
- know how to use tree diagrams and contingency tables to illustrate situations involving conditional probability
- understand and be able to use Bayes' theorem to solve problems involving conditional probability.

Practice questions: set 2

1. The time, T minutes, between customer arrivals at a country store, from Monday to Friday, can be modelled, for $t \geq 0$, by the probability density function f(t).

    $$f(t) = 0.1\, e^{-0.1t}$$

    (i) Find the probability that the time between arrivals is

       (a) less than 5 minutes [1 mark]

       (b) more than 15 minutes. [1 mark]

    (ii) Obtain the cumulative distribution function for T. Hence find the median time between arrivals [3 marks]

    The time, X minutes, between customer arrivals at the country store on Saturdays can be modelled by the probability density function g(x) defined by g(x) = $\lambda e^{-\lambda t}$, where λ is a positive constant.

    (iii) On Saturdays, the probability of the time between customer arrivals exceeding 5 minutes is 0.4. Estimate the value of λ. [3 marks]

2. The wages department of a large company models the incomes of the employees by the continuous random variable X with cumulative distribution function

    $$F(x) = 1 - \left(\frac{3}{x}\right)^4 \quad x \geq 3 \text{ and } 0 \text{ otherwise,}$$

    where X is measured in an arbitrary currency unit. (X is said to have a Pareto distribution.)

    (i) Find the median income. Find also the smallest income of an employee in the top 10% of incomes. [3 marks]

    (ii) Find the probability density function of X and hence show that the mean income is 4. [3 marks]

    (iii) Find the probability that a randomly chosen employee earns more than the mean income. [2 marks]

3. Mollie is carrying out an experiment to investigate the rate at which nitrogen dioxide (NO_2) breaks down to form nitric oxide (NO) and oxygen (O_2) in a chemical reaction. The percentage of NO_2 remaining at time t seconds after the experiment begins is denoted by y.

    Mollie determines values of y at 60-second intervals and records the results on a spreadsheet. She uses the spreadsheet to find the product moment correlation coefficient and the least squares regression line for y on t. The screenshot below is an extract from Mollie's spreadsheet.

	A	B	C	D	E	F	G	H
1	t	0	60	120	180	240	300	360
2	y	100	78	62	52	45	40	36
3								
4		pmcc:	−0.9306		regression: $y = -0.169643\,t + 89.5357$			

    Figure 1

Mollie uses the regression line to estimate the percentage of nitrogen dioxide 150 seconds after the experiment begins. She states that this estimate will be accurate because the correlation coefficient is close to −1.

(i) Use the given regression line to find Mollie's estimate. Comment on the value obtained. [2 marks]

(ii) Draw a graph of the data. Comment on the value of the pmcc. [4 marks]

(iii) State two pieces of advice that you would give to Mollie. [2 marks]

4 The scatter diagrams below were constructed from data on 248 adult patients at a clinic in the United States. Weights are in kg, heights are in cm and ages are in completed years.

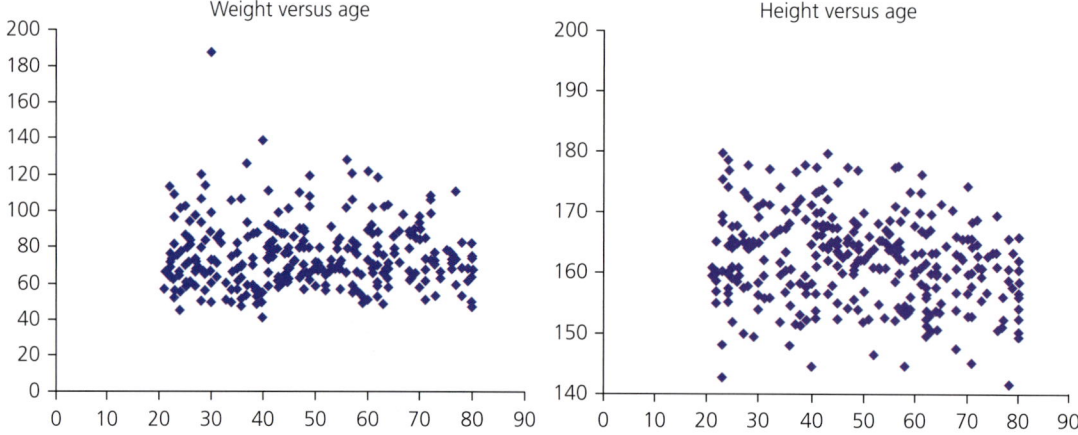

Figure 2

(i) Aster says that the average weight is much the same for patients of all ages, but the spread of weights is less for older patients. State what features of the weight versus age scatter diagram support these conclusions.

Comment in similar terms on the relationship between height and age for these patients. [4 marks]

(ii) Describe the skewness, if any, in
(a) the distributions of these patients' weights, and
(b) the distribution of their heights. [2 marks]

(iii) Estimate the mean weight and the mean height for these patients. Hence, given that all the patients are of the same sex, state with a reason whether they are men or women. [4 marks]

(iv) For these data it is possible to calculate three product moment correlation coefficients: weight and age, height and age, weight and height. The values of these three coefficients, in random order, are −0.17, −0.05, and 0.31. State, with justification, which correlation has the value 0.31. [2 marks]

(v) Explain why, without further information, it would not be advisable to regard these data as representative of the population in the area where the clinic is situated. [2 marks]

5 A primatologist was investigating whether there is any relationship between the ages of the female and the male in chimpanzee mating pairs.

For a random sample of 10 chimpanzee births, she estimated the ages, in years, of the parents. The raw data and a scatter diagram are shown on the following page.

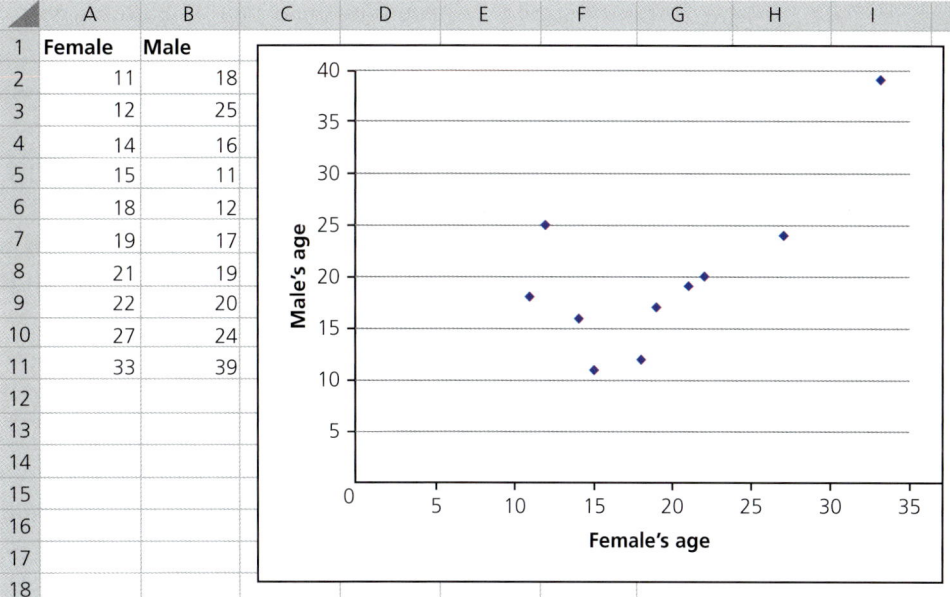

Figure 3

(i) Explain why it would not be appropriate to use the product moment correlation coefficient with these data. [1 mark]

(ii) Calculate Spearman rank correlation coefficient for the data. [2 marks]

(iii) Complete the hypothesis test that the primatologist should carry out, using a 5% significance level. [4 marks]

6 In human beings, body fat has a lower density than muscle, bone and other tissues. Overall body density can therefore be used to calculate the proportion of fat in a person's body. Body mass index (BMI) is calculated as $\frac{M}{H^2}$, for a person of mass M kg and height H m.

The following scatter diagram shows, for a sample of 240 adults, BMI (x) and body density (y). The scatter diagram also shows the linear regression line of y on x. The value shown as R^2 is the square of the product moment correlation coefficient for these data.

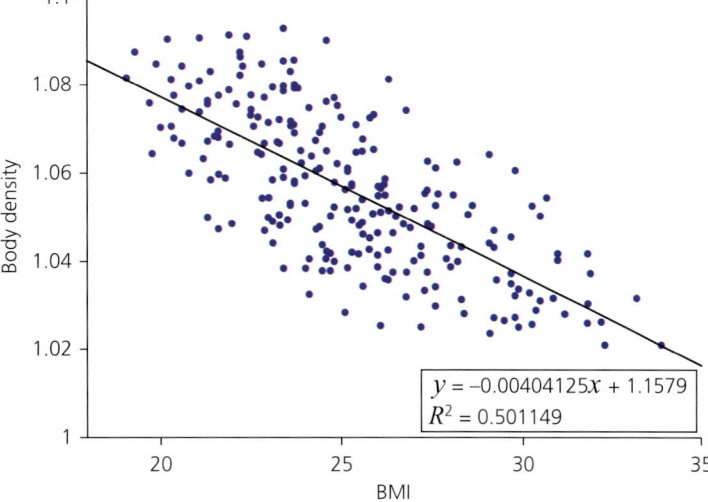

Figure 4

(i) Explain how the scatter diagram suggests that the analysis of the data carried out here is appropriate. [2 marks]

(ii) Calculate the product moment correlation coefficient for the data. [2 marks]

(iii) Use the given regression line to calculate the value of y when $x = 25$. Explain carefully what this value of y represents. [2 marks]

(iv) Explain why it would not be appropriate to use the given regression line to calculate the value of x when $y = 1.08$ [2 marks]

7 In a court of law, a defendant is put on trial for a crime. The defendant may or may not be guilty of the crime. The defendant may be or may not be convicted of committing the crime by the court. Let events be defined as follows.

G is the event that the defendant is guilty of the crime.

C is the event that the defendant is convicted of the crime.

(i) State what the events $C \mid G$ and $C \cap G$ are, making clear the distinction between them. [2 marks]

In trials for a particular type of fraud, it is estimated that 90% of defendants are actually guilty of committing the crime. Of those who are guilty, $\frac{4}{5}$ are convicted. Of those that are not guilty, 1 in 20 are convicted.

(ii) Find the probability that, for a randomly chosen defendant, the trial reaches the wrong conclusion. [2 marks]

(iii) Find the probability that a randomly chosen defendant who is convicted is actually guilty of the crime. [3 marks]

9 Further discrete distributions

The journey of a thousand miles begins with one step.

— Lao Tzu

Discussion point

What would you expect for the distributions in the following situations?

(i) There are two entrances to a hall, one to the left for the foyer and the other to the right. What can be said about the number of people using either door?

(ii) People are told to choose a whole number between 1 and 10 (inclusive). What can be said about the number of people choosing the various possibilities: 1, 2, 3 ... 10?

1 Repeating trials until success is achieved

The geometric distribution

Asha is playing a game where you need to throw a 6 on a die in order to start. She takes six throws to get a 6, and says that she has always been unlucky. One of the other players was successful on her first attempt.

- The probability that Asha is successful on any attempt is $\frac{1}{6}$.
- The probability that Asha is unsuccessful on any particular attempt is therefore $\frac{5}{6}$.
- Asha has five failures followed by one success. So the probability that Asha is successful on her sixth attempt is $\left(\frac{5}{6}\right)^5 \times \frac{1}{6} = 0.0670$.
- So, in fact, it is about half as likely for Asha to succeed on her sixth attempt as she is to succeed for the first time.

> Asha is successful on her sixth attempt so her results are F F F F F S with probabilities $\frac{5}{6} \frac{5}{6} \frac{5}{6} \frac{5}{6} \frac{5}{6} \frac{1}{6}$

The number of attempts that Asha takes to succeed is an example of a **geometric** random variable. The probability distribution is called the **geometric distribution**.

Example 9.1

Gina likes having an attempt at the coconut shy whenever she goes to the fair. From experience, she knows that the probability of her knocking over a coconut at any throw is $\frac{1}{3}$.

(i) Find the probability that Gina knocks over a coconut for the first time on her fifth attempt.

(ii) Find the probability that it takes Gina at most three attempts to knock over a coconut.

(iii) Given that Gina has already had four unsuccessful attempts, find the probability that it takes her another three attempts to succeed.

Solution

(i) Gina has 4 failures followed by 1 success. The probability of this is
$$\left(\frac{2}{3}\right)^4 \times \frac{1}{3} = \frac{16}{243} = 0.0658 \text{ (to 3 s. f.)}$$

> If Gina is successful on her fifth attempt her results are F F F F S with probabilities $\frac{2}{3} \frac{2}{3} \frac{2}{3} \frac{2}{3} \frac{1}{3}$

(ii) **Method 1**

This is the probability that she is successful on her first or second or third attempt. You can do similar calculations to the above:
$$\frac{1}{3} + \left(\frac{2}{3}\right)^1 \times \frac{1}{3} + \left(\frac{2}{3}\right)^2 \times \frac{1}{3} = \frac{1}{3} + \frac{2}{9} + \frac{4}{27} = \frac{19}{27} = 0.704 \text{ (to 3 s. f.)}.$$

First attempt — Second attempt — Third attempt

> **Note**
> There is not much difference in the difficulty of these two methods if you want the probability of at most three attempts, but if you, for example, wanted the probability of at most 20 attempts, Method 2 would be far better as the calculation would take no more effort than in this case.

Method 2

You can, instead, first work out the probability that Gina fails on all of her first three attempts and then subtract this from 1.

P(at most three attempts) = 1 − P(Fails on all of first three attempts)

$$= 1 - \left(\frac{2}{3}\right)^3 = 1 - \frac{8}{27} = \frac{19}{27} = 0.704 \text{ (to 3 s.f.)}$$

(iii) Unfortunately for Gina, however many unsuccessful attempts she has had makes no difference to how many more attempts she will need.

Thus the required probability is $\left(\frac{2}{3}\right)^2 \times \frac{1}{3} = \frac{4}{27} = 0.148$ (to 3 s.f.).

This is another example of a geometric random variable. Part (iii) illustrates the fact that the geometric distribution has 'no memory'.

For a geometric distribution to be appropriate, the following conditions must apply.

- You are finding the number of trials it takes for the first success to occur.
- On each trial, the outcomes can be classified as either success or failure.

> The probability of success is usually denoted by p and that of failure by q, so $p + q = 1$.

In addition, the following assumptions are needed if the geometric distribution is to be a good model and give reliable answers.

- The outcome of each trial is independent of the outcome of any other trial.
- The probability of success is the same on each trial.

In general, the geometric probability distribution, $X \sim \text{Geo}(p)$ over the values $\{1, 2, 3, \ldots\}$ is defined as follows:

$$P(X = r) = (1-p)^{r-1}p \text{ or } q^{r-1}p \quad \text{for } r = 1, 2, 3 \ldots$$
$$P(X = r) = 0 \quad \text{otherwise.}$$

> **Note**
> $P(X > r)$ is the probability that you take more than r attempts to succeed. The first r tries must therefore be failures. So $P(X > r) = (1-p)^r$.

In Method 2 of part (ii) of Example 9.1, you saw that the probability of Gina failing on all of her first three attempts is $\left(\frac{2}{3}\right)^3$. More generally, a useful feature of the geometric distribution is that $P(X > r) = (1-p)^r$, since this represents the probability that all of the first r attempts are failures.

The vertical line chart in Figure 9.1 illustrates the geometric distribution for $p = 0.3$.

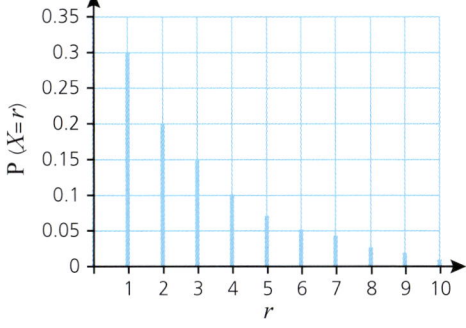

Figure 9.1

S1 AL — Repeating trials until success is achieved

General results for the geometric distribution

The general results for the total probability, mean and variance of the geometric distribution Geo(p) are given below. In this distribution the probability of success is p.

Total probability

For $X \sim \text{Geo}(p)$, $P(X = x) = p(1 - p)^{x-1}$ for $x = 1, 2, 3, \ldots$

$$\Sigma P(X = x) = p + p(1 - p) + p(1 - p)^2 \ldots$$
$$= p[1 + q + q^2 + \ldots] \quad \text{where } q = 1 - p$$

The expression $1 + q + q^2 + \ldots$ is an infinite geometric series with sum $\frac{1}{(1-q)} = \frac{1}{p}$

so $$\sum P(X = x) = p \times \frac{1}{p} = 1$$

The total probability is 1, as you would expect.

Expectation

The expectation is given by

$$E(X) = 1 \times P(X = 0) + 2 \times P(X = 2) + 3 \times P(X = 3) + \ldots$$
$$= 1 \times p + 2 \times p(1-p) + 3 \times p(1-p)^2 + \ldots$$
$$= p[1 + 2q + 3q^2 + \ldots]$$

Let $S = 1 + 2q + 3q^2 + 4q^3 \ldots$

and so $qS = q + 2q^2 + 3q^3 + \ldots$

Then $S - qS = 1 + 2q + 3q^2 + 4q^3 + \ldots - q - 2q^2 - 3q^3 - \ldots$

giving $S(1-q) = 1 + q + q^2 + q^3 + \ldots = \frac{1}{1-q}$ and so $S = \frac{1}{(1-q)^2} = \frac{1}{p^2}$.

So $E(X) = p\left(\frac{1}{p^2}\right) = \frac{1}{p}$

Variance

Variance is given by $E(X^2) - [E(X)]^2$.

So $\text{Var}(X) = (1^2 \times P(X=1) + 2^2 \times P(X=2) + 3^2 \times P(X=3) + 4^2 \times P(X=4) \ldots) - \left(\frac{1}{p}\right)^2$

$\text{Var}(X) = 1 \times p + 4 \times p(1-p) + 9 \times p(1-p)^2 + 16p(1-p)^3 \ldots - \left(\frac{1}{p}\right)^2$
and this can be written as

$\text{Var}(X) = pT - \frac{1}{p^2}$ where $T = 1 + 4q + 9q^2 + 16q^3 \ldots$

$pT = (1-q) \times T = 1 + 4q + 9q^2 + 16q^3 + \ldots - q - 4q^2 - 9q^3 - \ldots$
$pT = 1 + 3q + 5q^2 + 7q^3 + \ldots$

This can be written as $pT = 2 \times (1 + 2q + 3q^2 + 4q^3 + \ldots) - (1 + q + q^2 + q^3 + \ldots)$
And using results already obtained when finding $E(X)$ gives

$$pT = 2 \times \frac{1}{p^2} - \frac{1}{1-q} = \frac{2}{p^2} - \frac{1}{p}$$

and so $\text{Var}(X) = \frac{2}{p^2} - \frac{1}{p} - \frac{1}{p^2} = \frac{1}{p^2} - \frac{1}{p}$

This can be simplified to $\text{Var}(X) = \frac{1-p}{p^2}$ and this is the usual formula for the variance of a geometric distribution.

> For the geometric distribution with probability of success p,
> $E(X) = \frac{1}{p}$ and
> $\text{Var}(X) = \frac{1-p}{p^2}$

Example 9.2

In a communication network, messages are received one at a time and checked for errors. It is known that 6% of messages have an error in them. Errors in one message are independent of errors in any other message.

(i) Find the probability that the first message which contains an error is the fifth message to be checked.

(ii) Find the mean number of messages before the first that contains an error is found.

(iii) Find the probability that there are no errors in the first ten messages.

(iv) Find the most likely number of messages to be checked to find the first that contains an error.

Solution

In this situation, the distribution is geometric with probability 0.06, so Geo(0.06) so let $X \sim \text{Geo}(0.06)$.

(i) $P(X = 5) = 0.94^4 \times 0.06$
$= 0.0468$

(ii) $\text{Mean} = \dfrac{1}{0.06} = 16.7$ messages

(iii) $P(X > 10) = 0.94^{10} = 0.539$

(iv) The most likely number = 1 since $0.06 > 0.06 \times 0.94 > 0.06 \times 0.94^2$.

Example 9.3

The random variable $X \sim \text{Geo}(p)$. You are given that $\text{Var}(X) = 20$.

(i) Find the value of p.

(ii) Find $P(X = 7)$.

(iii) Find $P(X = 10 | X > 7)$.

Solution

(i) $\text{Var}(X) = \dfrac{1-p}{p^2} = 20$

$\Rightarrow 1 - p = 20p^2$

$\Rightarrow 20p^2 + p - 1 = 0$

$\Rightarrow (5p - 1)(4p + 1) = 0$

$\Rightarrow p = 0.2$ (discard $p = -0.25$)

(ii) $P(X = 7) = 0.8^6 \times 0.2$
$= 0.0524$

(iii) $P(X = 10 | X > 7) = P(X = 3)$
$= 0.8^2 \times 0.2$
$= 0.128$

S1 AL — Repeating trials until success is achieved

Exercise 9.1

① A fair 6-sided die is rolled. Find the probability that the first time a 6 comes up is
  (i) on the first roll
  (ii) on the third roll
  (iii) before the third roll
  (iv) after the third roll.

② A fair 3-sided spinner has sectors labelled 1, 2 and 3. It is spun until a 3 is scored. The random variable X is the number of spins required to get a 3.
  (i) Write down the probabilities that X takes each of the possible values up to and including 5.
  (ii) Find $P(X > 5)$.
  (iii) Find $E(X)$.
  (iv) Find $Var(X)$.

③ A five-sided spinner has sectors labelled 2, 3, 4, 5 and 6. It is spun repeatedly and all the numbers are equally likely to come up each time it is spun.
  (i) Find the probability that
    (a) on the first four spins it does not come up 2
    (b) it comes up 2 before the fifth spin
    (c) on the first five spins it does not come up 2
    (d) the first time it comes up 2 is on the fifth spin.
  (ii) Find which three of your answers to part (i) sum to 1 and explain why this must be the case.

④ Two fair 6-faced dice are rolled until a double comes up, i.e. they both show the same number.
  (i) What is the probability that on any roll of the dice a double comes up?
  (ii) Find the expected number of rolls that are needed to obtain the first double.
  (iii) What is the probability that the first double comes up on the sixth roll of the dice?

⑤ A game reserve runs safaris where people travel in a jeep and look out for wild animals. Experience has shown that on one in four safaris a tiger is seen. The probability of seeing a tiger on any safari is independent of the probability of seeing one on any other safari.
  Kamil goes on 5 safaris. Find the probability that
  (i) he does not see any tigers
  (ii) he sees at least one tiger
  (iii) the first tiger he sees is on his last safari
  (iv) he does not see any tigers on his first three safaris but does on his fourth and fifth safaris.

⑥ An archer is aiming for the bullseye on a target. The probability of the archer hitting the target on any attempt is 0.2, independent of any other attempt.
  (i)   Find the probability that the archer hits the bullseye at the fifth attempt.
  (ii) Find the probability that the archer hits the bullseye for the first time on her fifth attempt.
  (iii) Find the probability that the archer takes at least five attempts to hit the bullseye.
  (iv) Find the mean and variance of the number of attempts it takes for the archer to hit the bullseye.
  (v)   The archer takes n shots at the bullseye. Find the smallest value of n for which there is a chance of 50% or more that the archer will hit the target at least once.

⑦ Fair four-sided dice with faces labelled 1, 2, 3 and 4 are rolled, one at a time, until a 4 is scored.
  (i)   Find the probability that the first 4 is rolled on the fifth attempt.
  (ii) Find the probability that it takes at least six attempts to roll the first 4.
  (iii) Given that a 4 is not rolled in the first six attempts, find the probability that a 4 is rolled within the next two attempts.
  (iv) Write down the mean number of attempts which it takes to roll a 4.

⑧ In a game show, players are asked questions one after another until they get one wrong, after which it is the next player's turn. The probability that they get a question correct is 0.7, independent of any other question.
  (i)   Find the probability that the first player is asked a total of six questions.
  (ii) Find the probability that the first player is asked at least six questions.
  (iii) Find the average number of questions that the first player is asked.
  (iv) Find the probability that the first two players are asked a total of five questions.

⑨ Hamish goes on a five-day salmon fishing holiday. The probability of catching a salmon on any day is $\frac{1}{3}$ and that is independent of what happens on any other day.
  (i)   What is the probability that he does not catch a salmon on his holiday?
  (ii) What is the probability that he catches his first salmon on the last day of his holiday?
  (iii) What is the expectation of the number of days on which he catches a salmon?
  (iv) What is the expectation of the number of days on which he does not catch a salmon?

10. In order to start a board game, each player rolls a fair six-sided die, until a 6 is obtained. Let X be the number of goes a player takes to start the game.
    (i) Write down the distribution of X.
    (ii) Find (a) $P(X = 4)$, (b) $P(X < 4)$, (c) $P(X > 4)$.
    (iii) Given that $X = 3$, find the probability that the total score on all three throws of the die is less than 10.
    (iv) The game has two players and they each take turns in rolling the die. Find the probability that
        (a) neither has started within four rolls (two rolls each)
        (b) both have started within four rolls (up to two rolls each).

11. A random variable X has a geometric distribution with $\text{Var}(X) = 30$. Find $P(X = 4)$.

12. The random variables X and Y both have geometric distributions. Suggest situations which could be modelling given that
    (i) $\text{Var}(X) = E(X)$
    (ii) $\text{Var}(Y) = 5E(Y)$.

2 The negative binomial distribution

A die is thrown repeatedly. How many throws would you expect until the third time that six appears?

There is a sequence of independent attempts, each with probability $\frac{1}{6}$ of success. The random variable X is the number of attempts up to and including the third success.

$P(X = x) = P(\text{two successes and } (x - 3) \text{ failures in the first } (x - 1) \text{ attempts})$
$\qquad \times P(x^{\text{th}} \text{ attempt is a success})$

$$= {}^{x-1}C_2 \left(\frac{1}{6}\right)^2 \left(\frac{5}{6}\right)^{x-3} \times \left(\frac{1}{6}\right)$$

$$= {}^{x-1}C_2 \left(\frac{1}{6}\right)^3 \left(\frac{5}{6}\right)^{x-3}$$

> **Note**
> When $r = 1$, this is the same as a geometric distribution.

In general, if the probability of success in each attempt is p and X is the number of attempts up to and including the rth success,

$$P(X = x) = {}^{x-1}C_{r-1} \, p^r (1 - p)^{x - r} \qquad \text{for } x = r, r + 1, \ldots$$

This is the **negative binomial distribution**, $NB(p, r)$.

General results for the negative binomial distribution

The general results for the total probability, mean and variance of the negative binomial distribution $NB(p, r)$ are given below. This is the probability distribution of the number of trials needed to obtain the rth success when the probability of success on any single trial is p.

Total probability

For $X \sim \text{NB}(p, r)$, the probability distribution is given by

$P(X = x) = {}^{x-1}C_{r-1}\, p^r (1-p)^{x-r}$ for $x = r, r+1, r+2, \ldots$

So the sum of the probabilities is given by

$$\sum P(X = x) = p^r + rp^r(1-p) + \frac{(r+1)(r)}{2!} p^r (1-p)^2 + \ldots$$

$$= p^r \left(1 + rq + \frac{(r+1)r}{2!} q^2 + \ldots \right) \quad \text{where } q = 1 - p$$

$$= p^r \times (1-q)^{-r} \quad \text{using the binomial expansion for } (1-q)^{-r}$$

$$= p^r \times \frac{1}{p^r} = 1. \quad \text{and so the total probablity} = 1$$

Expectation

The expectation of X is given by

$$E(X) = r \times P(X = r) + (r+1) \times P(X = r+1) + (r+2) \times P(X = r+2) + \ldots$$

$$= rp^r + (r+1)rp^r(1-p) + (r+2)\frac{(r+1)(r)}{2!} p^r (1-p)^2 + \ldots$$

$$= rp^r \left[1 + (r+1)q + \frac{(r+2)(r+1)}{2!} q^2 + \ldots \right]$$

$$= rp^r(1-q)^{-(r+1)} = r\frac{p^r}{p^{r+1}} = \frac{r}{p}$$

Variance

The variance of X is given by $\text{Var}(X) = E(X^2) - [E(X)]^2$

To find $E(X^2)$ it is helpful to start with $E[X(X+1)]$ and to use the result that $E[X(X+1)] = E(X^2) + E(X)$

$E[X(X+1)]$

$$= r(r+1) \times P(X = r) + (r+1)(r+2) \times P(X = r+1) + (r+2)(r+3) \times P(X = r+2) + \ldots$$

$$= (r+1)(r)p^r + (r+2)(r+1)(r)p^r(1-p) + (r+3)(r+2)\frac{(r+1)r}{2!} p^r (1-p)^2 + \ldots$$

$$= (r+1)(r)p^r \left[1 + (r+2)q + \frac{(r+3)(r+2)}{2!} q^2 + \ldots \right]$$

$$= (r+1)(r)p^r(1-q)^{-(r+2)}$$

$$= (r+1)(r)\frac{p^r}{p^{r+2}}$$

$$= \frac{(r+1)(r)}{p^2}$$

So $E(X^2) + E(X) = \frac{(r+1)(r)}{p^2}$

and $E(X^2) = \frac{r^2 + r}{p^2} - \frac{r}{p}$ and $\text{Var}(X) = \frac{r}{p^2} - \frac{1}{p} = \frac{r(1-p)}{p^2}$

> For the negative binomial distribution with parameters p and r,
> $E(X) = \frac{r}{p}$ and $\text{Var}(X) = \frac{r(1-p)}{p^2}$

S1 AL The negative binomial distribution

Example 9.4

A coin is biased with P(head) = 0.25. It is tossed repeatedly.

Find the probability that the third head occurs on the fifth toss.

Solution

X = number of spins up to and including third head

X has a negative binomial distribution with $p = 0.25$ and $r = 3$.

$P(X = 5) = {}^4C_2 \, 0.25^3 \, 0.75^2 = \frac{54}{1024} = 0.0527$

Example 9.5

The random variable X follows a negative binomial distribution.

You are given that $E(X) = 10$ and $Var(X) = 40$.

Find $P(X = 10)$.

Solution

$\frac{r}{p} = 10$ and $\frac{r(1-p)}{p^2} = 40$

so $\frac{10(1-p)}{p} = 40$ which gives $p = 0.2$ and $r = 2$

$P(X = 10) = {}^9C_1 \, 0.2^2 \, 0.8^8 = 0.0604$

Exercise 9.2

1. A coin is tossed until heads comes up for the 3rd time.

   (i) Copy and complete this table giving the probabilities of different numbers of tosses.

Number of tosses	3	4	5	6	7	8
Probability						

   (ii) Add the probabilities in your table.

   Explain why your answer does not come to 1.

2. The probability of success in a trial is 0.1 and this is independent of the outcomes from any previous trials. In an experiment, trials are continued until 3 successes have been achieved.

   Find the expectation, variance and standard deviation of the number of trials needed.

3. A fair die is rolled until the number 6 has come up twice.

   (i) Find the most likely number of rolls required.

   (ii) Compare your answer with the expectation of the number of rolls required.

   (iii) What is the probability of the most likely number of rolls occurring?

   (iv) Find the probability of the expected number of rolls occurring.

④ A fair spinner has n sides, numbered 1, 2, ... n, where $n \geq 3$. The probability that the number 2 comes up for the third time at the seventh spin is 0.049152. Find the value of n.

⑤ In a game show players are asked questions until they have got three right. The probability that they get a question right is 0.7, independent of any other question.

   (i) Find the probability that the player is asked exactly six questions.

   (ii) Find the average number of questions that the player is asked.

⑥ In an experiment, the probability of success on a trial is p. The outcomes of each trial (success or failure) are independent of those of all previous trials. Trials are continued until there have been r successes. The probability that the rth success occurs on the nth trial is given by $35 \times \left(\frac{7}{11}\right)^4 \times \left(\frac{4}{11}\right)^4$.

   Write down the possible values of n, p and r.

⑦ The expectation of the random variable in the negative binomial distribution NB(p, r) is 20 and the variance is 80. Find the values of p and r.

⑧ Ramon, Selina and Tara each roll a fair die until a 6 comes up. They denote the number of rolls they take by R, S and T.

   (i) Write down the values of E(R), E(S), E(T).

   State what distribution you used.

   Write down also the value of E(R) + E(S) + E(T).

   Paul then rolls the die until it has come up 6 a total of 3 times. He denotes the number of rolls he takes by P.

   (ii) Find the expectation of P and state the distribution you used.

   (iii) Explain why E(P) = E(R) + E(S) + E(T).

   (iv) State, with a brief explanation, whether it is true that the standard deviation of P is equal to the sum of the standard deviations of R, S and T.

⑨ A fair 6-sided die is rolled. Find the probability that the fourth time a 6 comes up is

   (i) on the fourth roll

   (ii) on the 12th roll

   (iii) before the 25th roll.

⑩ The expectation of a negative binomial distribution is $\frac{1}{9}$ of the variance. The probability that the rth success occurs at the 20th trial is 0.02851798 (to 8 decimal places).

   Find the probability that the rth success occurs at the tenth trial.

S1 AL — Testing the parameter in a discrete distribution

3 Testing the parameter in a discrete distribution

You have already met hypothesis testing when testing for correlation and when testing for goodness of fit.

Hypothesis testing can be used to test the values of parameters of discrete distributions. The following examples show tests for the mean of a Poisson distribution and the values of p in binomial and geometric distributions.

Example 9.6

A researcher claims that the number of siblings (brothers or sisters) that people have can be modelled using a binomial $B(n, p)$ distribution. The researcher says that suitable values for the parameters are $n = 5$ and $p = 0.25$.

A man is selected at random. He has four siblings. Use this observation to test, at the 5% significance level, whether the researcher's value for p is too small.

Solution

The number of siblings is modelled as a binomial distribution $B(5, p)$.

Null hypothesis: $H_0: p = 0.25$

Alternative hypothesis: $H_1: p > 0.25$

Significance level: 5%

Test: one-tailed

Assuming H_0, the number of siblings will follow a $B(5, 0.25)$ distribution.

Table 9.1

Number of siblings	0	1	2	3	4	5
Probability	0.2373	0.3955	0.2637	0.0879	0.0146	0.0010
Cumulative prob.	0.2373	0.6328	0.8965	0.9844	0.9990	1.0000

If the value of p is too small then the probabilities of the larger values would be higher.

The man had four siblings. The probability of a result at least as extreme as this is $P(X \geq 4) = 0.0146 + 0.0010 = 0.0156$ or 1.56%. This result lies in the right hand tail 5% of the distribution.

If the man is typical this result is unlikely to occur by chance, so reject H_0 and conclude that the evidence supports the claim that $p > 0.25$.

> **Discussion point**
> Why is the assumption that the man is typical probably flawed?

> **Note**
> Alternatively, $P(X \geq 3) = 10.35\%$ and $P(X \geq 4) = 1.56\%$ so the critical value for a test at the 5% level is 4. An observed value in the set {0, 1, 2, 3} is consistent with H_0. The critical region is {4, 5}; an observed value in the critical region supports the alternative hypothesis.

Example 9.7

The number of accidents per month at a road junction is known to fit a Poisson distribution with mean 7.8.

In the month following a road improvement scheme there were only 5 accidents at the junction.

Test at the 5% level whether this shows that the accident rate has reduced.

Note

You should be able to find the probabilities using your calculator or tables. You do not need to find all the individual probabilities if you can find $P(X \leq 5)$ directly.

Note

The conclusion is that there is insufficient evidence to show that the mean has reduced. The observed value is consistent with H_0, but it is also consistent with other values for the parameter. The test has not shown that the mean is unchanged.

Solution

The number of accidents in a month is modelled using Poisson (λ).

Null hypothesis: $H_0: \lambda = 7.8$
Alternative hypothesis: $H_1: \lambda < 7.8$
Significance level: 5%
Test: one-tailed

Assuming H_0, $P(X \leq 5) = 0.2103$.

This is bigger than 5% so the observed value is not in the left-hand tail 5% of the distribution.

The observed value is not unlikely to have occurred by chance, so accept H_0 and conclude that there is insufficient evidence to suggest that the mean has reduced.

Note

Alternatively, $P(X \leq 3) = 0.0485$ and $P(X \leq 4) = 0.1117$ so the critical value is 3 and the critical region is {0, 1, 2, 3}. The observed value is not in the critical region so accept H_0. The critical region approach might be easier if you cannot calculate cumulative probabilities directly.

Note

When working with discrete distributions the critical region is chosen so that under H_0 the probability of a value being in the critical region is as close as possible to the significance level of 5%. This means that, strictly, you are actually testing as close as possible to 5% and your conclusion can be made with as close to 95% certainty as possible.

Example 9.8

A man flips a coin repeatedly until the coin lands on heads.

The first time he tries this he gets T T T T H.

Test at the 5% level whether this suggests that the coin is biased.

S1 AL — Testing the parameter in a discrete distribution

> **Note**
> Although the data may suggest that the value of p is smaller than 0.5 the wording of the question only asks whether the coin is biased and does not specify in which direction.

Solution

The number of attempts up to and including the first head can be modelled using a geometric distribution with parameter p, where $p = P(\text{head})$.

Null hypothesis: $H_0: p = 0.5$

Alternative hypothesis: $H_1: p \neq 0.5$

Significance level: 5%

Test: two-tailed

The observed value is 5.

Assuming H_0, $P(X \leq 5) = 1 - P(\text{first five are all T}) = 1 - (0.5)^5 = 0.96875$
left-hand side

and $P(X \geq 5) = P(\text{first four are all T}) = (0.5)^4 = 0.0625$
right-hand side

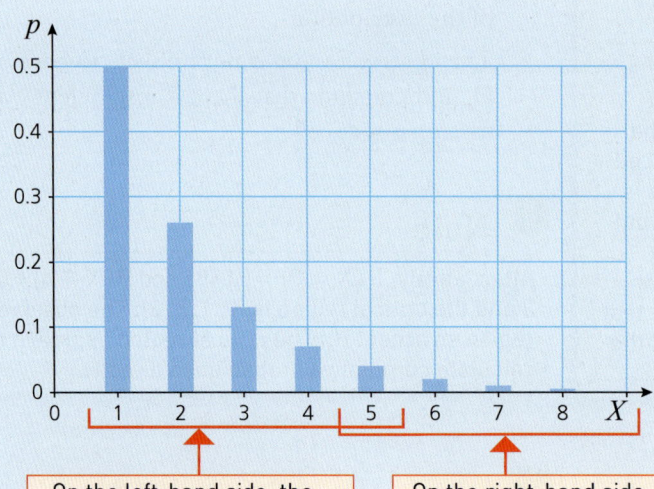

On the left-hand side, the probability of a result of 5 or fewer for the value of x is 0.96875 and this is much greater than 0.025. So there is insufficient evidence to suggest that the coin is biased towards heads, i.e., that $p > 0.5$.

On the right-hand side, the probability of a result of 5 or more for the value of X is 0.0625 and this is greater than 0.025. So there is insufficient evidence to suggest that the coin is biased towards tails, i.e., that $p < 0.5$.

So there is insufficient evidence to reject the null hypothesis that $p = 0.5$. There is no reason to claim that the coin is biased.

Alternatively, find the critical regions by finding the tail regions that have probabilities as close as possible to, but less than, 2.5%.

Left-hand tail

The smallest possible value of X is 1.

$P(X \leq 1) = P(X = 1) = P(H) = 0.5$ so there is no critical region on the left-hand side.

> **Note**
> A variation on this approach is to require the tail regions to have total probability as close as possible to 5%, but this could involve having to check several different combinations. In this example, this approach would give a critical value of 6 and a critical region of {6, 7, 8, ...}.

Right-hand tail

There is no limit to the possible value of X.

$P(X \geqslant x) = (0.5)^{x-1}$

$(0.5)^{x-1} = 0.025$ when $x - 1 = \dfrac{\log 0.025}{\log 0.5} = 5.32$

$\Rightarrow x = 6.32$

$P(X \geqslant 6) = 0.0313$ and $P(X \geqslant 7) = 0.0156$

So the critical value is 7. The observed value of 5 is not in the critical region so accept H_0 as before.

Exercise 9.3

① The number of children in a family is modelled using a Poisson distribution with mean λ. This includes families with no children. It has been claimed that $\lambda = 2.4$.

  (i) Write the null and alternative hypotheses that are used to test this claim.

  A randomly chosen family has 5 children.

  (ii) Use this information to test the claim at the 5% level.

② Chilli plants are sold in trays of 20 plants. When grown, a few of the plants are 'super hot' chilli nagas.

  (i) Write down the distribution that should be used to model the number of chilli nagas in a tray.

  The grower says that no more than 10% of the plants will grow into chilli nagas but Tom thinks that the percentage is greater than this.

  (ii) Write the null and alternative hypotheses that are used to test Tom's claim.

  A randomly chosen tray results in 5 chilli nagas.

  (iii) Test Tom's claim at the 5% level.

③ Jenny picks coloured beads randomly from a bowl that contains a very large number of beads. She has been told that the probability of a red bead is 0.02, but she thinks that it is smaller than this. She decides to test this at the 5% level.

  The first red bead that Jenny gets is her 100th bead.

  (i) Write down the distribution that should be used to model the number of beads up to and including the first red bead.

  (ii) Write the null and alternative hypotheses for Jenny's test.

  (iii) Test Jenny's claim.

④ The total number of goals scored in a football match can be modelled using a Poisson distribution. It has been claimed that the mean of the total number of goals per match is 2.5.

  Rachel coaches her local team and says that the mean of the total number of goals for their matches is more than 2.5. She wants to test this claim at the 5% level, on the basis of a match selected at random from next season's fixture.

  Find the critical region for Rachel's test.

S1 AL — Testing the parameter in a discrete distribution

5. A spinner has 100 regions, one of which is labelled 'out'. Pete counts the number of spins up to and including the first time that the spinner lands on 'out'. As a result of this, he claims that the probability that the spinner lands on 'out' is more than 0.01. Find the greatest number of spins that Pete counted, assuming that he used a 5% significance level for his test.

6. A farmer sells eggs in boxes of 6. The farmer claims that at most 2% of the eggs are cracked. Sunitra buys a box at random and finds that it contains a cracked egg. Does she have good cause to complain?

7. The number of fish caught in a day's fishing by fishermen on a certain stretch of river can be modelled using a Poisson distribution with mean 3.

   Aaron catches no fish in a day's fishing and says that this shows that the mean is less than 3.

   (i) Test Aaron's claim at the 5% level.

   (ii) What other issues should be taken into consideration?

8. In a game, a player rolls a standard six-sided die and needs to roll a six to start each of her playing pieces. She has four playing pieces and starts each of them as soon as she rolls the required six.

   The player believes that the die is biased towards six. She uses the total number of rolls until her fourth playing piece is started to test this claim at the 5% level. Her conclusion is that there is insufficient evidence to suggest that the die is biased.

   Find the least number of rolls until the fourth playing piece was started.

KEY POINTS

Geometric distribution

1. The geometric distribution may be used in situations in which:
   - there are two possible outcomes, often referred to as success and failure
   - both outcomes have fixed probabilities, p and q and $p + q = 1$
   - you are finding the number of trials which it takes for the first success to occur.

2. For the geometric distribution to be a good model:
   - the probability of success is constant
   - the probability of success in any trial is independent of the outcome of any other trial.

3. For a geometric random variable X, where $X \sim \text{Geo}(p)$
   - $P(X = r) = (1-p)^{r-1} p = q^{r-1} p$ for $r = 1, 2, 3 \ldots$

4. For Geo(p)
   - $E(X) = \dfrac{1}{p}$
   - $\text{Var}(X) = \dfrac{1-p}{p^2}$.

Negative binomial distribution

1. The negative binomial distribution may be used in situations in which
   - there are two possible outcomes, often referred to as success and failure
   - both outcomes have fixed probabilities, p and q and $p + q = 1$
   - you are finding the number of trials which it takes for the rth success to occur.

2 For the negative binomial distribution to be a good model:
   - the probability of success is constant
   - the probability of success in any trial is independent of the outcome of any other trial.
3 For a negative binomial random variable X, with parameters p and r
   - $P(X = x) = {}^{x-1}C_{r-1}\, p^r q^{x-r}$ for $x = r, r+1, \ldots$
4 For a negative binomial random variable X, with parameters p and r
   - $E(X) = \dfrac{r}{p}$
   - $Var(X) = \dfrac{r(1-p)}{p^2}$.

LEARNING OUTCOMES

When you have completed this chapter you should:

- recognise situations under which the geometric distribution is likely to be an appropriate model
- calculate the probabilities using a geometric distribution, including cumulative probabilities
- know and be able to use the mean and variance of a geometric distribution.
- calculate probabilities using a negative binomial distribution
- know and be able to use the mean and variance of a negative binomial distribution
- be able to conduct a hypothesis test on the parameter of a discrete distribution.

10 The Central Limit Theorem

> I know of scarcely anything so apt to impress the imagination as the wonderful form of cosmic order expressed by the Central Limit Theorem.
>
> Sir Francis Galton

Discussion point

In the picture the ball arrives just to the right of the middle of the skittles. The width of the bowling lane is 1.05 m and the width of the space occupied by the skittles is about 1.0 m. The point where the ball crosses the line of the skittles is x m from the left hand side. So in this instance the value of x is about 0.52.

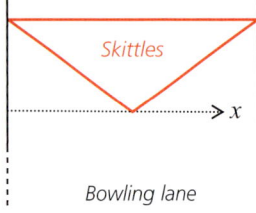

Figure 10.1

Malini is an inexperienced player; what would you expect to be the distribution of the values of x when she rolls a ball?

In the course of a typical evening she rolls 20 balls. What would you expect to be able to say about the distribution of the mean value of x per evening?

Review: The Normal distribution

This chapter involves the use of the Normal distribution. You will be familiar with this from earlier work, so what follows is a reminder of the essential points.

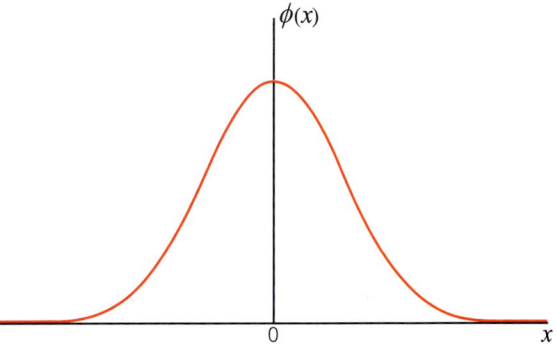

Figure 10.2

A Normal distribution, X, has a distinctive symmetrical bell shape. The probability density function is given by

$$\phi(x) = \frac{1}{\sqrt{2\pi}} e^{-\frac{1}{2}\left(\frac{x-\mu}{\sigma}\right)^2}$$

where μ is the mean and σ is the standard deviation. This is written as $N(\mu, \sigma^2)$. So, for example, $N(20, 9)$ means the Normal distribution with mean 20 and variance 9 or standard deviation 3.

All Normal distributions have the same shape, subject to stretching along the axes. It is common to represent the Normal distribution in **standardised form** as $N(0, 1)$, i.e. with mean 0 and variance 1. In this form the random variable is denoted by Z. The process of converting a Normal distribution into this form is called **standardisation** and involves the transformation

$$z = \frac{x - \mu}{\sigma}.$$

As with any other probability density function, the area under the curve represents a probability and so the cumulative distribution function, $\phi(Z)$ is particularly important. The shaded area in the diagram below represents the probability that Z lies between -0.5 and 2 and so it is given by $\phi(2) - \phi(-0.5)$.

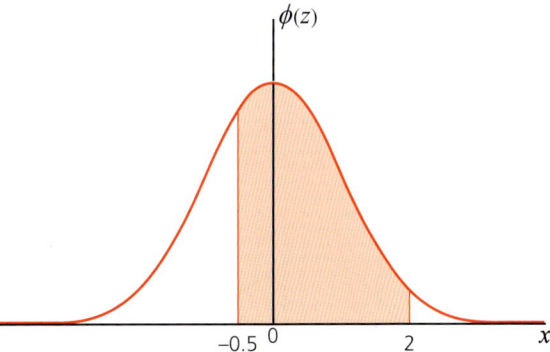

Figure 10.3

To find the value of this probability, you can use your calculator, a spreadsheet or tables. It is $0.9772... - 0.3085... = 0.669$ (to 3 s.f.)

1 The sampling distribution of the mean

> **Note**
> The idea of estimation is developed further in Chapter 16.

In many practical situations you do not know the true value of a population mean. In such situations the sample mean would seem to be a sensible value to use to estimate the population mean, but how accurate is this value likely to be?

Suppose that you could take every possible sample of size n from a population, for some fixed n, and calculate the mean of each sample. The sample means are then a random variable with a distribution, called the **sampling distribution of the mean**.

For large samples any extreme values are likely to be cancelled out by other less extreme values and the sampling distribution of the mean follows an approximately Normal-shaped distribution. This result is called the Central Limit Theorem.

> **Note**
> The Central Limit Theorem and its uses are covered more fully in Chapters 13, where most of the examples are based on continuous data. The treatment here is limited to knowledge of the results.

For samples of size n drawn from a distribution with mean μ and variance σ^2, the distribution of the sample mean is approximately $N(\mu, \frac{\sigma^2}{n})$ for sufficiently large n.

The Central Limit Theorem says that

1. The expected value of the sample mean is μ, the population mean of the original distribution. This is not a particularly surprising result but it is extremely important.

2. The standard deviation of the sample mean is $\frac{\sigma}{\sqrt{n}}$. This is often called the **standard error of the mean**. Within a sample you would expect some values above the population mean and others below it, so that overall the deviations would tend to cancel each other out, and the larger the sample the more this would be the case. Consequently, the standard deviation of the distribution of the sample mean is smaller than that of the individual values.

3. The distribution of the sample mean is approximately Normal. Even if the population distribution is not Normal, the sampling distribution of means drawn from it, for samples of a fixed size, is approximately Normal. The larger the sample size, n, the closer the sampling distribution is to a Normal distribution.

In many cases the value of n does need not be particularly large. For most population distributions you can model the distribution of the sample mean as being Normal if n is about 20 or more.

> **Note**
> The proofs of these results are not given here.

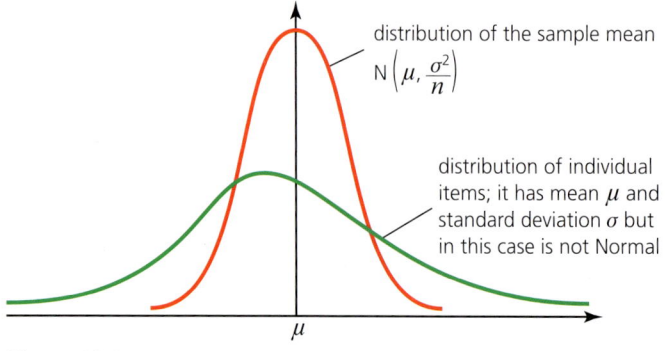

Figure 10.4

The following example is about the sampling distribution of the mean of a random variable, X, with mean μ and variance σ^2, it demonstrates the results $E(\bar{X}) = \mu$ and $\text{Var}(\bar{X}) = \dfrac{\sigma^2}{n}$.

Example 10.1

The discrete random variable X has a probability distribution as shown.

Table 10.1

X	1	2	3
Probability	0.5	0.4	0.1

A random sample of size 2 is chosen, with replacement after each selection.

(i) Find μ and σ^2.

(ii) Verify that $E(\bar{X}) = \mu$ and $\text{Var}(\bar{X}) = \dfrac{\sigma^2}{n}$

Solution

(i) $\mu = E(X) = 1 \times 0.5 + 2 \times 0.4 + 3 \times 0.1$

$= 1.6$

$E(X^2) = 1^2 \times 0.5 + 2^2 \times 0.4 + 3^2 \times 0.1$

$= 3$

$\sigma^2 = \text{Var}(X) = E(X^2) - [E(X)]^2$

$= 3 - 1.6^2 = 0.44$

(ii) The table below lists all the possible samples of size 2, their means and probabilities.

Table 10.2

Sample	1, 1	1, 2	1, 3	2, 1	2, 2	2, 3	3, 1	3, 2	3, 3
Mean	1	1.5	2	1.5	2	2.5	2	2.5	3
Probability	0.25	0.2	0.05	0.2	0.16	0.04	0.05	0.04	0.01

This gives the following probability distribution of the sample mean.

> For example P(1, 2) $= 0.5 \times 0.4 = 0.2$

Table 10.3

$\bar{X}$	1	1.5	2	2.5	3
Probability	0.25	0.4	0.26	0.08	0.01

Using this table gives

$E(\bar{X}) = 1 \times 0.25 + 1.5 \times 0.4 + 2 \times 0.26 + 2.5 \times 0.08 + 3 \times 0.01$

$= 1.6 = \mu$, as required.

$E(\bar{X}^2) = 1^2 \times 0.25 + 1.5^2 \times 0.4 + 2^2 \times 0.26 + 2.5^2 \times 0.08 + 3^2 \times 0.01$

$= 2.78$

$\text{Var}(\bar{X}) = E(\bar{X}^2) - [E(\bar{X})]^2$

$= 2.78 - (1.6)^2$

$= 0.22$

$= \dfrac{0.44}{2} = \dfrac{\sigma^2}{2}$ as required.

S1 AL — The sampling distribution of the mean

In the following examples, the Central Limit Theorem is applied to large sample distributions which are not Normal. Indeed, the variables in the first two are discrete.

Example 10.2

The discrete random variable X has the binomial distribution, $X \sim B(100, 0.4)$. A random sample of size $n = 24$ is taken. The sample mean is the random variable Y.

Estimate $P(39.98 < Y < 40.02)$.

Solution

For the binomial distribution $X \sim B(n, p)$, $E(X) = np$ and $Var(X) = npq$.

$E(X) = 40$ and $Var(X) = 24$.

$n = 24$ is large enough for the Central Limit Theorem to be used.

By the Central Limit Theorem, $Y \sim N(40, 1)$ $\qquad \dfrac{\sigma^2}{n} = \dfrac{24}{24} = 1$

Using the Normal distribution

$P(38.98 < Y < 40.02) = 0.016$ (to 2 s.f.).

Example 10.3

The random variable $\bar{X}$ is the sample mean of samples of size $n = 100$ taken from the distribution of a random variable X with mean 50 and variance 15.

Find the value of x (to 3 s.f.) for which $P(X > x)$ is closest to 0.025.

Solution

By the Central Limit Theorem, $\bar{X} \sim N\left(50, \dfrac{15}{100}\right) = N(50, 0.15)$. $\qquad \dfrac{\sigma^2}{n} = \dfrac{15}{100}$

$1 - 0.025 = 0.975$

$\phi^{-1}(0.975) = 1.960$

$x = 50 + 1.960 \times \sqrt{0.15} = 50.759$

$\Rightarrow \quad x = 50.759 = 50.8$ to 3 s.f.

Using the inverse Normal function on a calculator, $x = 50.759 = 50.8$ to 3 s.f.

In this next example the random variable (the length of a fish) is continuous but its distribution is unknown.

Example 10.4

It is known that the mean length of a species of fish is 40.5 cm with standard deviation 3.2.

(i) A random sample of five fish is taken. Explain why you cannot find the probability that the mean of this sample is at least 40.

(ii) A random sample of 50 fish is taken. Calculate an estimate of the probability that the mean of this sample is at least 40.

(iii) If the lengths of individual fish could only be measured to the nearest 1 cm, make a new estimate of the probability calculated in part (ii).

Solution

(i) You cannot find the probability that the mean is at least 40 because you are told nothing about the distribution of the parent population (the distribution of the length of this species of fish), so you cannot carry out any probability calculations.

(ii) Although you do not know the distribution of the parent population, the sample size of 50 is large enough to apply the Central Limit Theorem.

The distribution is therefore $N\left(40.5, \frac{3.2^2}{50}\right)$.

$$P(\text{sample mean} > 40) = 1 - \Phi\left(\frac{40 - 40.5}{\frac{3.2}{\sqrt{50}}}\right)$$

$$= 1 - \Phi(-1.1049)$$

$$= 0.8654$$

> Distribution of the sample means is $N\left(\mu, \frac{\sigma^2}{n}\right)$ and you are given that $\mu = 40.5$, $\sigma = 3.2$, $n = 50$.

(iii) If the lengths are only measured to the nearest 1 cm, then the sample mean can only take values … 39.96, 39.98, 40.00, 40.02, 40.04 …

You therefore need to calculate P(sample mean > 39.99) rather than P(sample mean > 40).

Using a calculator P(sample mean > 39.99) = 0.8701.

> For example, if all except one of the fish were 40 cm long and the last was 41 cm long the sample mean would be 40.02. If instead the last was 39 cm long the sample mean would be 39.98.

Exercise 10.1

① The random variable X is Normally distributed with mean 30 and variance 4.

Write down the distribution of a random sample of

(i) 2 (ii) 10 (iii) 100

observations of X.

② The random variable X has mean 200 and standard deviation of 30.

Write down if possible the approximate distribution of a random sample of

(i) 10 (ii) 30 (iii) 100

observations of X.

If you cannot write down the distribution, explain why you cannot.

3. At a men's hairdresser, haircuts take an average of 18 minutes with standard deviation 3 minutes. Find the distribution of the total time taken for 30 haircuts at the hairdresser. State any assumption which you make.

4. The distribution of a random variable, X, is shown in this table.

X	1	2	3	4
$P(X = x)$	0.4	0.1	0.2	0.3

   (i) Find $E(X)$ and $Var(X)$.

   Samples of X of size 100 are taken.

   (ii) State fully the distribution of the sample mean, $\bar{X}$.

   (iii) Find the probability that the mean of the next sample of this size will be greater than 2.5.

5. A random sample of 40 items is chosen from a population with mean 60 and variance 25. Find the probability that the mean of the sample is less than 59.

6. A company manufactures paving slabs of mean length 45.5 cm with standard deviation 0.5 cm. Find the probability that the mean length of a random sample of 100 slabs is at least 45.4 cm.

7. A certain type of low energy light bulb is claimed to have a lifetime of 8000 hours, with standard deviation 600 hours. A random sample of 25 bulbs is selected. Assuming that the claim is true:

   (i) write down the standard error of the sample mean

   (ii) find the probability that the sample mean is greater than 7900 hours.

   (iii) The mean is measured and found to be 7450 hours. Comment on this result.

8. In a certain city, the heights of women have mean 164 cm and standard deviation 8.5 cm.

   (i) Find the probability that the mean height of a random sample of 50 women will be within 1 cm of the mean.

   (ii) Find the sample size required so that the probability that a random sample of this size will be within 0.5 cm of the mean is the same as your answer to part (i).

9. A random sample of n items is chosen from a population with mean 200 and standard deviation 15. You are given that n is large. The probability that the sample mean is greater than 201 is 0.2755 to 4 significant figures. Find the value of n.

10. Wooden sleepers for use in landscape gardening are available in two lengths 1.2 m and 1.8 m. The smaller ones actually have mean length 1.23 m with standard deviation 0.05 m. The larger ones have mean length 1.84 m with standard deviation 0.10 m. Find the probability that:

    (i) 30 of the smaller sleepers have mean length less than 1.22 m

    (ii) 40 of the longer sleepers have mean length more than 1.85 m.

    (iii) The mean length of 45 smaller sleepers is greater than two thirds of the mean length of 30 longer sleepers.

⑪ Scores in an IQ test are modelled by a Normal distribution with mean 100 and standard deviation 15. The scores are reported to the nearest integer. The test is taken by a large number of children. Find the probability that

(i) the score of an individual child is greater than 110

(ii) the mean score of random sample of ten children is greater than 110.

⑫ The mean weight of the contents of a jar of jam produced in a factory is 353 g with standard deviation 8 g. The weights of the contents are recorded to the nearest gram. A random sample of 25 jars is selected.

(i) Write down the standard error of the mean.

(ii) Find the probability that the sample mean is

(a) at least 353

(b) less than 350

(c) between 352 and 354 inclusive.

KEY POINTS

1. When samples are drawn from the population of a random variable, X, the distribution of the sample means, $\bar{X}$, is called the sampling distribution of the mean.

2. The Central Limit Theorem says that for samples of size n, drawn from a population with mean μ and variance σ^2,
   - $E(\bar{X}) = \mu$
   - $\text{Var}(\bar{X}) = \dfrac{\sigma^2}{n}$
   - The distribution of $\bar{X}$ is approximately Normal whatever the distribution of X.
   - If the distribution of X is Normal, that of $\bar{X}$ is also Normal.

3. The larger the sample size, the closer the distribution of the sample means is to Normal. For large samples, it may be taken to be Normal.

4. The quantity $\dfrac{\sigma^2}{n}$ is called the standard error of the mean.

5. When the underlying distribution is discrete, it may be necessary to apply a continuity correction when working with the sampling distribution of the mean.

LEARNING OUTCOMES

When you have completed this chapter, you should:

▶ understand that the sample mean is a random variable with a probability distribution

▶ know that if the underlying distribution is Normal, then the sample mean is Normally distributed

▶ understand how and when the Central Limit Theorem may be applied to the distribution of sample means and use this result in probability calculations, using a continuity correction where appropriate.

11 Quality of tests

Errors using inadequate data are much less than those using no data at all.
Charles Babbage

1 The power of a test

You have already met the idea of errors in hypothesis testing in Chapters 3 and 6. The ideas are revisited and developed further here.

Example 11.1

A gold coin is used for the toss at a country's football matches but it is suspected of being biased. It is suggested that it shows heads more often than it should.

A test is planned in which the coin is to be tossed 19 times and the results recorded. It is decided to use a 5% significance level; so, if the coin shows heads 14 or more times, it will be declared biased.

(i) If the coin is unbiased, what is the probability of it being declared biased?

(ii) If the coin is biased, with a probability of 0.8 of coming up heads on any throw, what is the probability of it being declared biased towards heads?

Solution

Both situations can be modelled by the binomial distribution, $B(19, p)$. A 1-tail test is required.

(i) If the coin is unbiased, $p = 0.5$. This is the null hypothesis for the test.

The probabilities of outcomes near the critical value of 14 are given to 2 significant figures in Table 11.1.

Table 11.1

Number of heads	≥ 13	≥ 14	≥ 15
Probability	0.084	0.032	0.010

The table shows that the probability of getting 14 or more heads is 0.032.

So the probability of rejecting the true null hypothesis is 0.032.

(ii) Given that $p = 0.8$, the equivalent table is Table 11.2, below.

Table 11.2

Number of heads	≥ 13	≥ 14	≥ 15
Probability	0.93	0.84	0.67

So the probability of rejecting the null hypothesis is 0.84 and the probability of accepting the false null hypothesis is $1 - 0.84 = 0.16$

> When this is written as a conditional probability statement it is $P(X \leq 13 \mid p = 0.8) = 0.16$

The two parts of this example illustrate the two types of error that can occur in a hypothesis test.

- In part (i) a true null hypothesis is rejected and this is a Type I error.
- In part (ii) a false null hypothesis is accepted and this is a Type II error.

The circumstances under which these errors occur are shown below.

Table 11.3

		Decision	
		Accept H_0	Reject H_0
Reality	The null hypothesis, H_0, is true	Correct decision	H_0 wrongly rejected Type I error
	The null hypothesis, H_0, is false	H_0 wrongly accepted Type II error	Correct decision

Type I errors

The probability of a Type I error is represented by the size of the critical region. In tests involving continuous random variables this is the same as the significance level of the test. However, if the distribution is discrete, the critical region is often less than the significance level. The example above used the binomial distribution and the least number of heads in the critical region was 14, corresponding to a probability of 0.032 or 3.2%; this is the nearest probability to the significance level of 5% but below it.

So the probability of a Type I error is either equal to the significance level of the test or slightly less than it. The probability of a Type I error is sometimes called the **size** of the test.

Type II errors

In the previous example you could work out the probability of a Type II error because the value of p was given; it was 0.8. Without similar information you can't work out the probability of a Type II error and this is often the case. Indeed finding out about the population parameter, in this case p, is the object of the test so it is often not possible to calculate the probability of a Type II error. However, that does not mean such errors do not occur.

The power of a test

The power of a test is a measure of its ability to obtain the correct result by failing to accept a false null hypothesis. It is given by

Power = 1 − Probability of a Type II error.

What does the power tell you? If the alternative hypothesis, H_1, is true, then there are consequences for your data which you hope to pick up. The power of the test tells you how likely it is that you will do so, and so how powerful the test is. With this in mind you can see that as the sample size increases, the test becomes more powerful. When you ask questions like 'What is the smallest sample size that I need to have a good chance of detecting an effect in the population?', this is really a question about the power of your test.

Using the notation of conditional probability

Using the conditional probability symbol $|$ for "given that" allows you to write the probabilities more formally.

$$P(\text{Type I error}) = P(\text{rejecting } H_0 \mid H_0 \text{ is true})$$
$$P(\text{Type II error}) = P(\text{accepting } H_0 \mid H_0 \text{ is false})$$

The power of a test is $P(\text{rejecting } H_0 \mid H_0 \text{ is false})$ and so it follows that

$$= 1 - P(\text{accepting } H_0 \mid H_0 \text{ is false}) = 1 - P(\text{Type II error}).$$

So as the power of a test increases, the probability of a Type II error decreases, and vice versa.

> **Note**
>
> Notice that the probability of a Type II error changes with the value of the parameter you are testing. By contrast the probability of a Type I depends on the test you are conducting.

Example 11.2

It is known that 60% of the moths of a certain species are red; the rest are yellow.

A biologist finds a new colony of these moths and observes that more of them seem to be red than she would expect. She designs an experiment in which she will catch 10 moths at random, observe their colour and then release them. She will then carry out a hypothesis test using a 5% significance level.

(i) State the null and alternative hypotheses for this test.

(ii) Find the rejection region.

(iii) Find the probability of a Type I error.

(iv) If in fact the proportion of red moths is 80%, find the probability that the test will result in a Type II error. Hence write down the power of the test.

Solution

(i) Let p be the probability that a randomly selected moth is red.

Null hypothesis: $H_0: p = 0.6$ The proportion of red moths in this colony is 60%.

Alternative hypothesis: $H_1: p > 0.6$ The proportion of red moths is greater than 60%.

(ii) Assuming H_0 is true, you can calculate the following probabilities for the 10 moths in the sample.

All 10 moths are red: $(0.6)^{10} = 0.0060...$

9 are red and 1 yellow: $^{10}C_1 \times (0.6)^9 \times 0.4 = 0.0403...$

8 are red and 2 yellow: $^{10}C_2 \times (0.6)^8 \times (0.4)^2 = 0.1209...$

There is no need to go any further.

The probability that there are nine or ten red moths is

$$0.0403... + 0.0060... = 0.0463...$$

and this is less than the 5% significance level.

The probability that there are eight, nine or ten red moths is

$$0.1209... + 0.0403... + 0.0060... = 0.167...$$

and this is greater than 5%.

So the rejection region for this test is 9 or 10 red moths.

(iii) A Type I error occurs when a true null hypothesis is rejected.

In this case if H_0 is true, and so $p = 0.6$, the probability of it being rejected because a particular sample has 9 or 10 red moths has already been worked out to be 0.0463... in part (ii). When rounded to 3 significant figures, this gives 0.0464.

So the probability of a Type I error is 0.0464 (to 3 s.f.).

> So the size of this test is 0.0464.

(iv) If the proportion of red moths is 80%, the correct result from the test would be for the null hypothesis to be rejected in favour of the alternative hypothesis. The probability of this happening is

$$^{10}C_1 \times (0.8)^9 \times 0.2 + (0.8)^{10} = 0.376 \text{ (to 3 s.f.)}$$

A Type II error occurs when this result does not occur.

So in this situation the probability of a Type II error is $1 - 0.376 = 0.624$.

S1 AL — The power of a test

Example 11.3

A die is known to be biased in favour of a 6. One gamer believes that $p = P(6) = 0.2$ while another believes that $p = 0.4$. The die is rolled 10 times and the number of times it comes up 6 is recorded.

A hypothesis test is then carried out at the 5% significance level with the null hypothesis that $p = 0.2$.

(i) Find the critical region.

(ii) Find the probability of a Type I error.

(iii) Given that the correct value of p is 0.4, find the probability of a Type II error and the power of the test.

(iv) How could the test be improved?

Solution

Use X to denote the number of sixes thrown.

Assuming H_0 to be true, $X \sim B(10, 0.2)$.

(i) $P(X \leq 3) = 0.879... \Rightarrow P(X \geq 4) = 0.121...$ so 4 is not in the critical region

$P(X \leq 4) = 0.967... \Rightarrow P(X \geq 5) = 0.033...$ so 5 is not in the critical region

The significance level is 5% and 0.121 > 5% but 0.033 < 5%.

So the critical region is $X \geq 5$.

This is the probability that the value of X lies in the critical region.

(ii) The probability of a Type I error is 0.033.

(iii) Given that $p = 0.4$, the null hypothesis is false.

So a Type II error occurs if H_0 is accepted.

This happens if $X \leq 4$.

For $p = 0.4$, $P(X \leq 4) = 0.633$

So the probability of a Type II error is 0.633.

The power of the test is $1 - 0.633 = 0.367$.

(iv) Rolling the die more times would make the test more powerful.

The hypothesis test in the next example is based on the Poisson distribution.

Example 11.4

Those responsible for a popular website believe that it receives 25 visitors a minute between the hours of 9 a.m. and 5 p.m. each day. To test this hypothesis they count the number of visitors during a randomly selected minute.

There are 34 visitors.

(i) Carry out the test at the 5% significance level.

(ii) Draw a diagram to illustrate the probability distribution and the critical region if the mean is 25.

Solution

The situation is modelled by the Poisson distribution Po(25).

> Since the visits are likely to occur singly, independently, and at a uniform rate, it is reasonable to assume that the Poisson distribution is a good model.

(i) H_0: The mean number per minute is 25, $\lambda = 25$

H_1: The mean number per minute is not 25, $\lambda \neq 25$

2-tail test

Significance level 5%

Since it is a 2-tail test the critical region should be 2.5% or slightly less for each tail.

The observed value is 34.

Using a calculator gives the probability of a value of 34 or greater to be 0.049 78...

Since 0.049 78 > 0.025, the observed value is not in the critical region.

There is not sufficient evidence to reject the null hypothesis.

(ii)

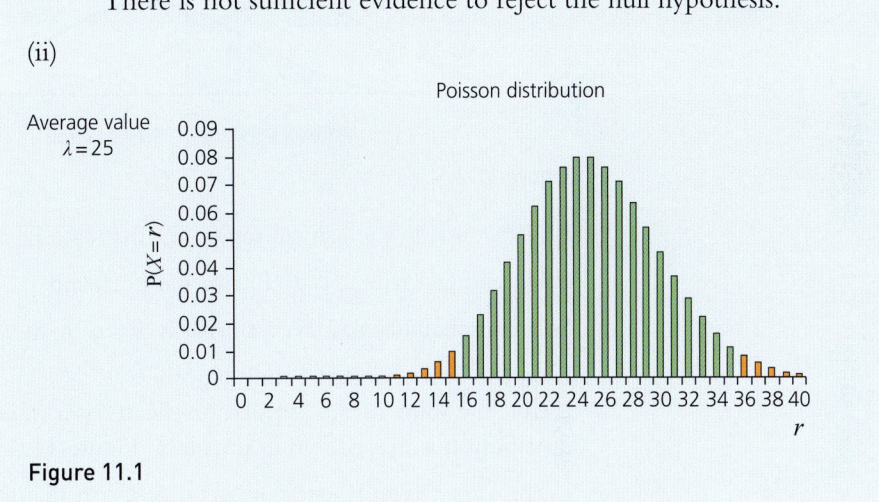

Figure 11.1

It is possible to make Type I and Type II errors in exactly the same way as for a test based on the binomial distribution.

The critical regions are $X \leq 15$ with probability 0.022 29...

> The ... tells you that this number is truncated, and that you are expected to be working with the full number on your calculator.

and $X \geq 36$ with probability 0.022 45...

So the probability of a Type I error is 0.022 29... + 0.022 45... = 0.044 75... or 4.48% (to 3 s.f.).

> Notice that this is slightly less than the significance level of 5%.

Suppose that the true value of the mean is 30 so the null hypothesis is false.

A Type II error occurs if H_0 is accepted. This occurs if X is not in the critical region and that means that $16 \leq X \leq 35$.

Using a calculator gives

$P(X \leq 15) = 0.001\,94...$

$P(X \leq 35) = 0.842\,61...$

so $P(16 \leq X \leq 35) = 0.842\,61... - 0.001\,94... = 0.840\,67...$

So the probability of a Type II error is 0.841 or 84.1% (to 3 s.f.) and the power of the test is $1 - 0.841 = 0.159$.

The power of a test

Errors for the Normal distribution

The examples of Type I and Type II errors that you have met so far have been based on hypothesis tests using the binomial and Poisson distributions and so have involved discrete random variables. The same principles apply for tests involving continuous variables, like those based on the Normal distribution.

Suppose you are working with the Normal distribution $N(\mu, \sigma^2)$, where the population variance, σ, is known but the population mean, μ, is not known. Your test statistic is denoted by X.

You are conducting a 1-tail test with

    Null hypothesis, H_0 $\mu = k$

    Alternative hypothesis, H_1 $\mu > k$.

The distribution of X is shown in red in Figure 11.2 with the critical region for the significance level of the test shaded yellow. The critical value is marked as c in the diagram.

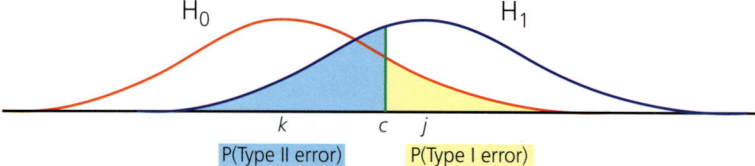

Figure 11.2

If $X \geq c$, it is in the critical region. In that case H_0 is rejected.

If H_0 is in fact true, a Type I error has occurred.

So the probability of a Type I error is given by the area of the yellow region.

If on the other hand, the null hypothesis is false and the alternative hypothesis is true, the value of the population mean has a value j where $j > k$. The distribution is that shown in purple in Figure 11.2.

In this case a Type II error occurs if $X < c$, so that the false null hypothesis is accepted. This is represented by the blue region in Figure 11.2.

Figure 11.2 allows you to see how changing the significance level of the test and the sample size affects the accuracy of the test.

- As you vary the significance level of the test, the value of c will change. If the significance level is increased, say from 5% to 10%, the value of c is decreased. So the area of the yellow region is increased and the area of the blue region is decreased. So the probability of a Type I error is increased and the probability of a Type II error is decreased. And vice versa.

> So if you decrease the significance level, say from 5% to 1%, the probability of a Type I error is decreased and that of a Type II error is increased.

- If you increase the sample size, the two curves in Figure 11.2 will get spikier as the sampling error decreases, and so you will reduce the probabilities of both Type I and Type II errors, and increase the power of the test.

The power function

The power of a test is $P(\text{rejecting } H_0 \mid H_0 \text{ is false}) = 1 - P(\text{accepting } H_0 \mid H_0 \text{ is false}) = 1 - P(\text{Type II error})$.

In Example 11.3 the alternative hypothesis took a specific value of the parameter, so the power of the test could be calculated numerically. More usually the power of a test can be written as a function of the parameter.

In the calculation of the power following Example 11.4 it was assumed that the true value of the mean was 30, so $X \sim \text{Po}(30)$. The critical region was $\{X \leq 15$ or $X \geq 36\}$. This gave the power as 0.159.

If the mean, λ, had not been known, the power could have been calculated as a function of λ.

Example 11.5

A random variable, X, has a Poisson distribution with mean λ. A test, at significance level 5%, of the hypotheses $H_0: \lambda = 25$, $H_1: \lambda \neq 25$ gave the critical region $\{X \leq 15$ or $X \geq 36\}$. Find an expression for the power of the test, in terms of λ, when $\lambda \neq 25$.

Solution

The power of the test is $1 - P(\text{accept } H_0 | H_0 \text{ is false}) = 1 - P(16 \leq X \leq 35)$

$$\text{Power} = 1 - e^{-\lambda}\left(\frac{\lambda^{16}}{16!} + \frac{\lambda^{17}}{17!} + \frac{\lambda^{18}}{18!} + \frac{\lambda^{19}}{19!} + \cdots + \frac{\lambda^{35}}{35!}\right)$$

A spreadsheet was used to tabulate the power function for Example 11.5 for various values of λ. The results are shown below, with the power given to 4 d.p.

Table 11.4

λ	5	10	15	20	30	35	40	50
power	0.9999	0.9513	0.5681	0.1573	0.1593	0.4553	0.7576	0.9838

When the true value of λ is near 25, the power is low and the test is not good at distinguishing between $\lambda = 25$ and the true value.

As the true value of λ moves further away from 25, the power increases and the test becomes stronger at distinguishing between $\lambda = 25$ and the true value.

Example 11.6

A coin is tossed 4 times. The random variable X is the number of times that it lands on heads.

It is suspected that the coin is biased towards heads, with P(heads) = p.

(i) Calculate the critical region for a test of the hypotheses $H_0: p = 0.5$, $H_1: p > 0.5$ with significance level 10%.

(ii) Calculate the power function in terms of p.

(iii) For what value of p does the power of the test equal 10%?

Solution

(i) Assuming $H_0: X \sim B(4, 0.5)$

Table 11.5

x	0	1	2	3	4
$P(X = x)$	0.0625	0.25	0.3725	0.25	0.0625

Critical region at 10% level of significance is $\{X = 4\}$

(ii) Power $= P(X = 4) = p^4$.

(iii) $p^4 = 0.10$ so $p = 0.10^{0.25} = 0.5624$

S1 AL The power of a test

The power of a test also depends on the significance level of the test. As the significance level decreases the probability of a Type I error decreases and the probability of a Type II error increases, so the power of the test decreases. Similarly, as the significance level increases the probability of a Type I error increases and the probability of a Type II error decreases, so the power of the test increases.

However, it is always more important to control the probability of a Type I error by keeping the significance level low, even if this is at the expense of the power of the test.

Exercise 11.1

① At a certain airport 20% of people take longer than an hour to check in. A new computer system is installed, and it is claimed that this will reduce the time to check in. It is decided to accept the claim if, from a random sample of 22 people, the number taking longer than an hour to check in is either 0 or 1.

(i) Calculate the significance level of the test.

(ii) State the probability that a Type I error occurs.

(iii) Calculate the probability that a Type II error occurs if the probability that a person takes longer than an hour to check in is now 0.09.

[Cambridge International AS & A Level Mathematics 9709 Paper 7 Q4 June 2007]

② A manufacturer claims that 20% of sugar-coated chocolate beans are red. George suspects that this percentage is actually less than 20% and so he takes a random sample of 15 chocolate beans and performs a hypothesis test with the null hypothesis $p = 0.2$ against the alternative hypothesis $p < 0.2$. He decides to reject the null hypothesis in favour of the alternative hypothesis if there are 0 or 1 red beans in the sample.

(i) With reference to this situation, explain what is meant by a Type I error.

(ii) Find the probability of a Type I error in George's test.

[Cambridge International AS & A Level Mathematics 9709 Paper 7 Q2 Nov 2005]

③ A particular genetic mutation occurs independently and at random in magpies. It is believed that the proportion, p, of magpies affected is 2%. To check this figure, scientists plan to collect a sample of 150 magpies. They will be checked and the number, x, with the mutation will be recorded. Then all the magpies will be released at the same time.

The random variable X represents the number of magpies with the mutations in samples of size 150. The probability distribution of X is modelled by the Poisson distribution.

(i) Write down the parameter, λ, of the Poisson distribution if $p = 0.02$.

(ii) Copy and complete this probability table.

Table 11.6

Number, x	0	1	2	3	4	5	6	7	8	9
$P(X = x)$										

The scientists conjecture that the incidence of the mutation is actually greater than 2%. They plan to use the data from their sample for a hypothesis test of this conjecture, at the 5% significance level.

(iii) Write down the null and alternative hypotheses for their test. State the critical region.

(iv) State the probability of a Type I error for this test.

Suppose that the actual value of p is 3% and not 2%.

(v) State the parameter of the Poisson distribution which would model the distribution.

By using your calculator, suitable software or a table of the Cumulative Poisson distribution, show that the probability of a Type II error is 0.831.

State the power of the scientists' test in this case.

A number of different possible values of p are considered.

(vi) Copy and complete this table for the scientists test with different values of p.

Table 11.7

p	0.025	0.03	0.04	0.05	0.06	0.08	0.10
λ							
P(Type II error)		0.831					
Power							

(vii) Draw the graph of the power function of the scientists' test.

(viii) Show that the power function is given by

$$\text{Power} = 1 - e^{-\lambda}\left(1 + \lambda + \frac{\lambda^2}{2!} + \frac{\lambda^3}{3!} + \frac{\lambda^4}{4!} + \frac{\lambda^5}{5!} + \frac{\lambda^6}{6!}\right) \text{ where } \lambda = 150p.$$

④ 350 raisins are put into a mixture which is well stirred and made into 100 small buns. Estimate how many of these buns will

(i) be without raisins

(ii) contain five or more raisins.

Raisins are put into a mixture for making buns commercially. The average number of raisins per bun, λ, should be four. A large purchaser checks each batch. As a test they select five buns at random and count the total number of raisins; if it is fewer than ten, they reject the batch.

(iii) The null hypotheses for this test is $\lambda = 4$. State the alternative hypothesis.

(iv) Find the probability of a Type I error.

(v) For one batch the true value of λ is 2. Find the probability of a Type II error for this batch.

⑤ A hypothesis test is to be carried out at the 5% level on a Normally distributed population with standard deviation 2.7. The null hypothesis is $H_0: \mu = 25$, and the alternative hypothesis is $H_1: \mu > 25$. A random sample of size 9 is selected.

(i) Find the standard error of the mean.

(ii) Calculate the test statistic.

(iii) Write down the critical value.

(iv) The test gives a sample mean of 26.71. Explain whether the null hypothesis should be rejected.

The true value of the population mean is $\mu = 27$. The test is repeated.

(v) Find (a) the probability of a Type II error (b) the power of the test.

(vi) Find the p-value associated with this test.

(6) The random variable X has a normal distribution with mean μ and variance 9.

A test is required of the null hypothesis $H_0: \mu = 20$ against the alternative hypothesis $H_1: \mu \neq 20$.

For a sample of size 25, the sample mean when H_0 is true is distributed as $N(20, 0.36)$.

The probability of a type I error is to be 1%.

Show that H_0 is to be accepted if $18.45 < \bar{x} < 21.55$ [MEI adapted]

(7) In radio-carbon dating, the number of beta particles emitted per minute by a sample of carbon from a charred bone is given by a Poisson distribution whose mean, λ, depends on the age of the sample.

A charred bone is found which, because of the deposits with which it is associated, is known to be either 1800 years old, in which case $\lambda = 7.2$, or 2600 years old, in which case $\lambda = 3.8$.

An archaeologist's strategy for deciding between these hypotheses is to count the number, B, of beta particles emitted by the sample in m minutes, accepting that $\lambda = 7.2$ if $B > c$ and that $\lambda = 3.8$ if $B < c$.

Define $\alpha = P$(accepting $\lambda = 3.8$ when $\lambda = 7.2$) and $\beta = P$(accepting $\lambda = 7.2$ when $\lambda = 3.8$).

If $m = 1$ and $c = 4.5$, determine the values of α and β. [MEI adapted]

(8) A garage uses a particular spare part at an average rate of five per week. Assuming that usage of this spare part follows a Poisson distribution, find the probability that

(i) exactly five are used in a particular week

(ii) at least five are used in a particular week

(iii) exactly ten are used in a two-week period

(iv) at least ten are used in a two-week period

(v) exactly five are used in each of two successive weeks.

(vi) If stocks are replenished weekly, determine the number of spare parts which should be in stock at the beginning of each week to ensure that, on average, the stock will be insufficient on no more than one week in a 52-week year.

(vii) A different spare part has been required 5 times a day on average, but it is thought this figure might have been reduced. The part is subsequently required 15 times in a randomly chosen five-day week. Carry out a test at the 5% level.

(viii) If the alternative hypothesis is that the part is required on average 4 times a week, find the probability of a Type I and a Type II error with a one-tailed 5% level test, and the p-value of the test.

(ix) What is the power of the test?

(9) A calculator is used to generate a random integer R, where $0 \leq R \leq 29$. The calculator is fair; there is an equal chance that each of these values for R will arise. 100 values for R are chosen.

(i) What is the distribution of X, where X is the number of times 29 is generated?

(ii) Find $P(10 < R)$ for any single reading.

(iii) The fairness of the calculator is now questioned, and a bias towards the number 29 is suspected. 100 more random numbers are selected by the calculator, and the number 29 is chosen 7 times. If you carry out a test at the 10% level, what does this suggest?

(iv) If in fact P(29) = 0.05, find the probabilities of a Type I and a Type II error for this test.

10 A one-tailed test is being undertaken to investigate a die which is thought to show a six more frequently than it should. The die is to be rolled seven times and the null hypothesis of no bias will be rejected if the die shows six more than twice.

(i) Show that the probability of a Type I error for this test is just under 10%.

(ii) Show that the power function of this test is given by
$$1 - (1-p)^5(1 + 5p + 15p^2)$$
where p is the probability of the die showing a six.

(iii) For which values of p is the probability of a Type II error below 10%? [MEI adapted]

11 The random variable, X, is distributed uniformly on the range $[0, m]$, where m is claimed to have the value 10, but is suspected to have a smaller value than this. In order to test the null hypothesis, $m = 10$, against the alternative, $m < 10$, a sample of 3 values of X is taken and the largest of the 3 values is recorded as the value of the random variable L. The null hypothesis will be rejected if this largest value is less than some critical value.

(i) Explain why P(largest of the 3 values is less than L) = $\left(\dfrac{L}{m}\right)^3$

(ii) Hence determine the critical value of L in terms of α, the significance level of the test.

(iii) Find an expression for the power of the test in terms of α and m. [MEI adapted]

KEY POINTS

1 A Type I error occurs when the null hypothesis is rejected when it is true.
2 A Type II error occurs when the null hypothesis is accepted when it is false.
3 The power of a test is 1 − the probability of a Type II error.
4 The power function of a test gives the power as a function of the parameter being tested.
5 The probability of a Type I error is sometimes called the size of the test.

LEARNING OUTCOMES

When you have completed this chapter you should be able to:
- understand what Type I and Type II errors are
- understand what the power of a test represents
- calculate the power of a test, numerically for a simple alternative hypothesis or algebraically for a compound alternative hypothesis.

12 Probability generating functions

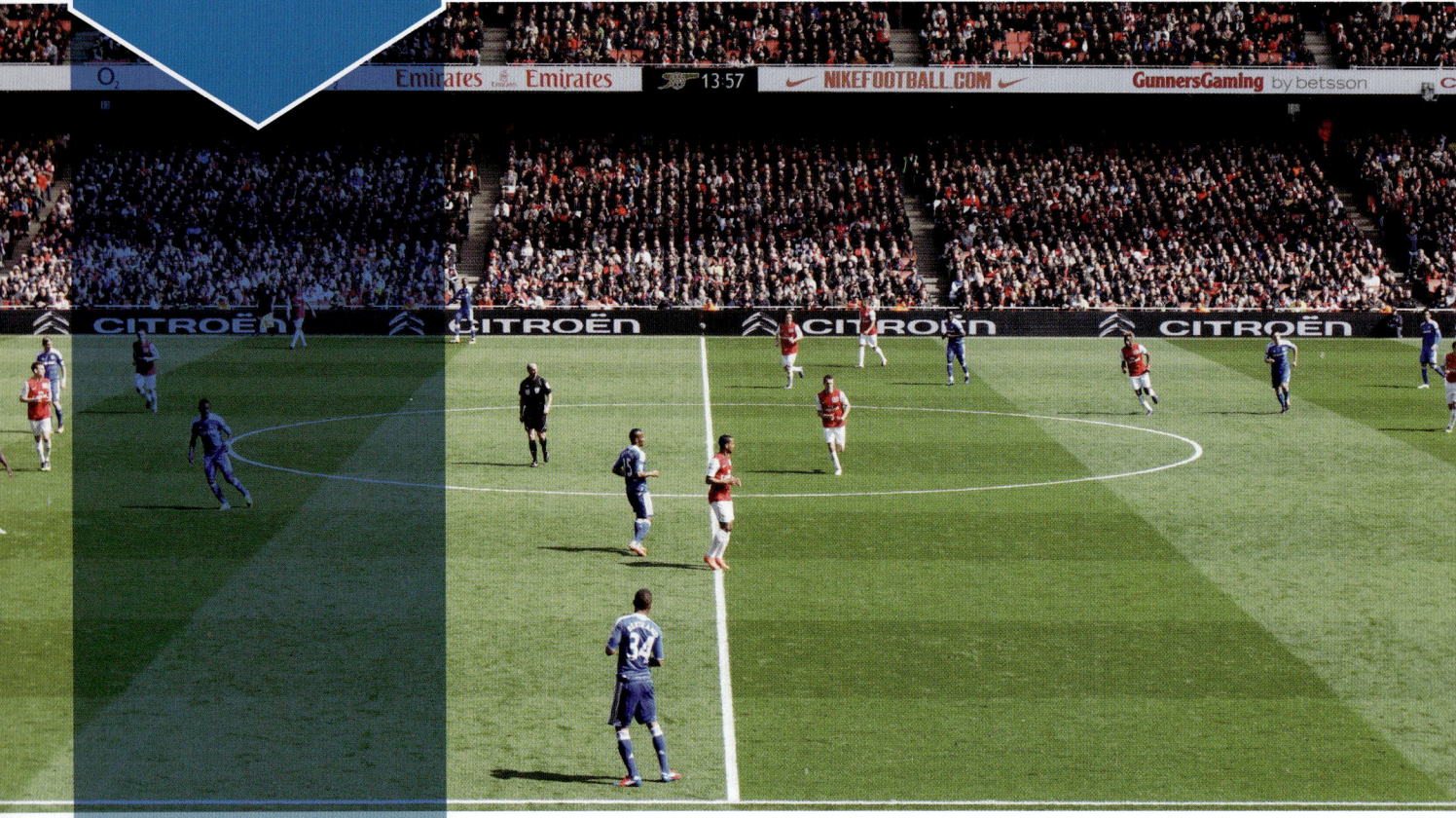

If you educate a man, you educate an individual, but if you educate a woman, you educate a whole nation.

Dr J.E.K. Aggrey (1875–1927)

During the first season of England's Premier League football competition, 462 matches were played. The frequency of the number of goals scored in each game by the **away** team, and associated relative frequencies, were recorded as follows:

Table 12.1

Number of goals	0	1	2	3	4	5	> 5
Frequency	149	179	91	37	3	3	0
Relative frequency (to 3 d.p.)	0.323	0.387	0.197	0.080	0.006	0.006	0

Assuming that these relative frequencies reflect the general pattern of scoring in Premier League matches, they may be used as **empirical** probabilities of 0, 1, 2, … goals being scored by **away** teams per match.

1 Probabilities defined by a probability generating function

Let X be the discrete random variable representing the number of goals scored per game by away teams, then one way of storing the probabilities is in the **probability generating function** (PGF).

$$G(t) = 0.323 + 0.387 \times t + 0.197 \times t^2 + 0.080 \times t^3 + 0.006 \times t^4 + 0.006 \times t^5$$

where the coefficients of t^x are the probabilities $P(X = x)$. The variable t is a **dummy** variable as, in itself, it has no significance. However, it has an important role to play in subsequent analysis, which you will appreciate later on.

Substituting $t = 1$ gives

$$G(1) = 0.323 + 0.387 + 0.197 + 0.080 + 0.006 + 0.006 = 1$$

as the sum of the probabilities has to be 1.

This can be set out in tabular form.

Table 12.2

t^x	t^0	t^1	t^2	t^3	t^4	t^5
$P(X = x)$	0.323	0.387	0.197	0.080	0.006	0.006
$t^x P(X = x)$	0.323	$0.387 \times t$	$0.197 \times t^2$	$0.080 \times t^3$	$0.006 \times t^4$	$0.006 \times t^5$

Thus the general definition of a probability generating function can be written as

$$G(t) = \sum t^x \, p(X = x)$$

and this can also be written as

$$G(t) = E(t^x).$$

The Poisson distribution

The way in which $G(t)$ generates probabilities is best seen from a **theoretical** model. From the shape of the probability distribution for goals scored by away teams, it looks as though a Poisson distribution might produce a good fit. The mean number of away goals is 1.08 (to 3 s.f.) and Figure 12.1 shows the Poisson distribution with mean 1.08.

```
Morning class            Afternoon class
(24 students)            (22 students)
                      2 | 8
                  2   3 | 1
              5 5 6   3 | 6
                      4 | 4 4 0
                  7 8 4 | 9 9 9 6 6 5
    1 1 3 3 3 3 4 4   5 | 2 2 1 1 1
        5 5 5 7 7 7 7 5 | 8 7 7 5
              1 4 4   6 | 3

                      | 6 | 3   represents 63 seconds
```

Figure 12.1

Using a suitable χ^2 test you can check that, at the 5% level of significance, a Poisson distribution with mean 1.08... (using the sample mean as an unbiased estimate of the population mean) produces a good fit. The **theoretical** probabilities for a Poisson distribution, with mean = 1.08..., are given by

$$P(X = x) = e^{-1.08} \frac{1.08^x}{x!}.$$

Therefore the PGF $G(t)$, given by $E(t^x) = \sum_x t^x P(X = x)$ may be written as

$$G(x) = \sum_{x=0}^{\infty} e^{-1.08} \frac{1.08^x}{x!} t^x.$$

When this is written as a series it is

$$G(x) = e^{-1.08}\left(1 + 1.08t + \frac{1.08^2}{2!}t^2 + \frac{1.08^3}{3!}t^3 + \ldots\right).$$

This can be written in the very neat and convenient algebraic form as

$$G(x) = e^{-1.08} e^{1.08t}$$
$$= e^{1.08(t-1)}.$$

> **Note**
> The PGF for a discrete random variable X, where $X \sim \text{Po}(\lambda)$, is given by $G(t) = e^{\lambda(t-1)}$. A formal proof is given later in this chapter.

This result shows why the PGF can be useful; using a lot of probabilities as the coefficients of powers of t in a function does not at first seem like a good idea – and does not seem to do anything that listing the probabilities in a table could not do more easily. The reason that it can be worthwhile is that the function made up by using a list of probabilities as the coefficients of powers of t can sometimes be written in an alternative, simpler way. Here, by recognising an exponential series, the function, which started out as an infinite sum, can be written in a very neat way; this is the beautiful PGF trick at work.

Basic properties of probability generating functions

The example above illustrates some basic properties of probability generating functions (PGF) for any discrete random variable X.

> Probability generating functions are used for discrete random variables. The equivalent for continuous variables are **moment generating functions** but they are beyond the scope of this book.

Definition
$$G(t) = E(t^x) = \sum_x t^x P(X = x)$$
$$G(1) = \sum_x 1^x P(X = x) = \sum_x P(X = x) = 1$$

If X takes non-negative integer values only, then the probability generating function takes the form of a polynomial in t, which is often written as

$$G(t) = \sum_x p_x t^x = p_0 + p_1 t + p_2 t^2 + \ldots + p_n t^n + \ldots$$

where p_x denotes $P(X = x)$, for example p_2 denotes $P(X = 2)$. The function $G(t)$ may be either a finite or infinite polynomial; that is, it may terminate after n terms or continue indefinitely.

The uniform distribution

Example 12.1

Let X be the discrete random variable that denotes the score when a fair die is thrown.
Find the probability generating function (PGF) for X.

Solution

The probability distribution is given by

Table 12.3

x	1	1	1	1	1	1
$P(X = x)$	$\frac{1}{6}$	$\frac{1}{6}$	$\frac{1}{6}$	$\frac{1}{6}$	$\frac{1}{6}$	$\frac{1}{6}$

The PGF for X is

$$G(t) = E(t^x) = \frac{1}{6}t + \frac{1}{6}t^2 + \frac{1}{6}t^3 + \frac{1}{6}t^4 + \frac{1}{6}t^5 + \frac{1}{6}t^6$$

$$= \frac{1}{6}t(1 + t + t^2 + t^3 + t^4 + t^5)$$

$$= \frac{1}{6}t \times \frac{1-t^6}{1-t}$$

$$= \frac{t(1-t^6)}{6(1-t)}$$

> The first of six terms of a G.P. with first term 1 and common ratio t.

The binomial distribution

Example 12.2

Dan is a keen archer. He has taken part in many competitions. When aiming for the centre of the target, called the gold, he finds that he hits the gold 70% of the time. Find the PGF for X, the number of gold hits in four consecutive shots at the target.

Solution

As $X \sim B(4, 0.7)$, the probabilities are given by $P(X = x) = {}^4C_x 0.7^x (1-0.7)^{4-x}$, which gives the following distribution:

Table 12.4

x	0	1	2	3	4
$P(X = x)$	0.3^4	$4 \times 0.7 \times 0.3^3$	$6 \times 0.7^2 \times 0.3^2$	$4 \times 0.7^3 \times 0.3$	0.7^4

The PGF for X is

$$G(t) = E(t^x) = \sum_{0}^{4} p_x t^x = \sum_{0}^{4} {}^4C_x \times 0.7^x \times 0.3^{4-x} \times t^x$$

$$= 0.3^4 + 4 \times 0.3^3 \times 0.7t + 6 \times 0.3^2 \times 0.7^2 t^2 + 4 \times 0.3 \times 0.7^3 t^3 + 0.7^4 t^4$$

$$= (0.3 + 0.7t)^4$$

> When you are working with probability generating functions, you will quite often see the binomial coefficient ${}^nC_r$ written using the alternative form $\binom{n}{r}$. In this book the ${}^nC_r$ notation is used.

This result can be generalised. For any binomial distribution B(n, p) where $P(X = x) = {}^nC_x q^{n-x}p^x$ where $q = 1 - p$, the probability generating function is $G(t) = (q + pt)^n$. A formal proof is given later in the chapter.

The geometric distribution

Example 12.3

A card is selected at random from a normal pack of 52 playing cards. If it is a heart, the experiment ends, otherwise it is replaced, the pack is shuffled and another card is selected. Let X represent the number of cards selected up to and including the first heart. Find the PGF for X.

Solution

As X follows a geometric distribution; that is, $X \sim \text{Geo}\left(\frac{1}{4}\right)$, the probabilities are given by $P(X = x) = \frac{1}{4} \times \left(1 - \frac{1}{4}\right)^{x-1}$, which gives the following distribution:

Table 12.5

x	1	2	3	4	5	...
$P(X = x)$	$\frac{1}{4}$	$\frac{1}{4} \times \frac{3}{4}$	$\frac{1}{4} \times \left(\frac{3}{4}\right)^2$	$\frac{1}{4} \times \left(\frac{3}{4}\right)^3$	$\frac{1}{4} \times \left(\frac{3}{4}\right)^4$	...

The PGF for X is

$$G(t) = E(t^x) = \sum_{x=1}^{\infty} p_x t^x = \sum_{x=1}^{\infty} \left(\frac{1}{4}\right)\left(\frac{3}{4}\right)^{x-1} t^x$$

$$= \left(\frac{1}{4}\right)t + \left(\frac{1}{4}\right)\left(\frac{3}{4}\right)t^2 + \left(\frac{1}{4}\right)\left(\frac{3}{4}\right)^2 t^3 + \left(\frac{1}{4}\right)\left(\frac{3}{4}\right)^3 t^4 + \left(\frac{1}{4}\right)\left(\frac{3}{4}\right)^4 t^5 + \ldots$$

$$= \left(\frac{1}{4}\right)t\left[1 + \left(\frac{3}{4}t\right) + \left(\frac{3}{4}t\right)^2 + \left(\frac{3}{4}t\right)^3 + \left(\frac{3}{4}t\right)^4 + \ldots\right]$$

The sum to infinity of a G.P. with first term 1 and common ratio t.

$$= \left(\frac{1}{4}\right)t\left[\frac{1}{1 - \frac{3}{4}t}\right]$$

$$= \frac{t}{4 - 3t}$$

This result can be generalised. When $X \sim \text{Geo}(p)$, i.e. $P(X = x) = pq^{x-1}$, where $q = 1 - p$, the PGF for X is given by $G(t) = \frac{pt}{1 - qt}$. A formal proof is given later in this chapter.

The negative binomial distribution

Example 12.4

A biased coin, for which P(head) = 0.6, is tossed until heads has come up 5 times. Let X represent the number of times the coin is tossed up to and including the fifth head. Find the PGF for X.

Solution

The probabilities are given by $P(X = x) = \binom{x-1}{4} \times 0.4^{x-5} \times 0.6^4 \times 0.6$, $x \geq 5$, which gives the following distribution:

Table 12.6

x	5	6	7	8	...
$P(X = x)$	$(0.6)^5$	$(5 \times 0.4 \times 0.6^4) \times 0.6$	$(15 \times 0.4^2 \times 0.6^4) \times 0.6$	$(35 \times 0.4^3 \times 0.6^4) \times 0.6$	...

The PGF for X is

$$G(t) = E(t^x) = \sum_{x=5}^{\infty} p_x t^x$$

$$= \sum_{x=5}^{\infty} {}^{x-1}C_4 (0.4)^{x-1}(0.6)^5 t^x$$

$$= \{(0.6)^5 t^5 + 5(0.4)(0.6)^5 t^6 + 15(0.4)^2(0.6)^5 t^7 + \ldots\}$$

$$= (0.6t)^5 \{1 + 5(0.4t) + 15(0.4t)^2 + \ldots\}$$

$$= \left(\frac{0.6t}{1 - 0.4t}\right)^5$$

> This is the binomial expansion for $(1 - 0.4t)^{-5}$.

This is a negative binomial distribution. The result can be generalised for the rth success in trials for which the probability of success is p and that of failure is $1 - p = q$. The probability generating function is given by $G(t) = \left(\dfrac{pt}{1 - qt}\right)^r$.

Exercise 12.1

① Let X be the discrete random variable that denotes the sum of the scores when two fair dice are thrown. Construct a table for the probability distribution and so write down the PGF as a series.

② Let X be the discrete random variable that denotes the absolute difference of the scores when two fair dice are thrown ($x = 0, 1, 2, 3, 4, 5$). Construct a table for the probability distribution and so write down the PGF as a series.

③ A fair coin is tossed three times and the number of tails appearing is noted as the discrete random variable X ($x = 0, 1, 2, 3$). Construct a table for the probability distribution and so write down the PGF as a series.

④ A box contains three red balls and two green balls. They are taken out one at a time, **without** replacement. Let X represent the number of withdrawals until a red ball is chosen. Find its PGF.

⑤ The experiment in question 4 is repeated, but this time **with** replacement. Find the PGF for the number of withdrawals until a red ball is chosen.

6. A random number generator in a computer game produces values that can be modelled by the discrete random variable X with probability distribution given by

   $P(X = r) = kr!, r = 1, 2, 3, 4, 5$.

   Determine the value of k and so write down $G(t)$, the PGF for X.

7. A student is shown three graphs. She is also given three equations, one for each graph. She matches each graph with its equation at random. Construct a table for the probability distribution of X, the number of correctly identified graphs. Deduce the PGF for X.

8. Two students are to be chosen to represent a class containing nine boys and six girls. Assuming that the students are chosen at random, find the PGF for X, the number of girls representing the group.

9. A biased coin is tossed repeatedly. The probability that the coin lands heads is p. Find the PGF for X, the number of tosses until the second occurrence of heads.

2 Expectation and variance

For any discrete probability distribution, the **expectation** (mean) and **variance** may be calculated using the following formulae. (They use the notation that $P(X = x)$ is written as p_x).

$$\mu = E(X) = \sum_x x p_x$$

$$\sigma^2 = \text{Var}(X) = E(X - \mu)^2 = \sum_x (x - \mu)^2 p_x = \sum_x x^2 p_x - \mu^2$$

$$= E(X^2) - \mu^2.$$

By successively differentiating $G(t)$, the probability generating function for X, formulae for the expectation and variance can be derived elegantly. It is here that the power of the PGF as an algebraic tool becomes apparent.

Expectation

This may be obtained from the probability generating function $G(t)$ by differentiating with respect to t and evaluating the expression at $t = 1$.

$$G(t) = E[t^X] = \sum_x p_x t^x$$

$\Rightarrow$
$$G'(t) = \sum_x x p_x t^{x-1} = E[Xt^{X-1}]$$

$\Rightarrow$
$$G'(1) = \sum_x x p_x = E[X]$$

Therefore
$$\mu = E(X) = G'(1).$$

Variance

This may be obtained by differentiating G'(t) and evaluating the expression at $t = 1$.

$$G'(t) = \sum_x x p_x t^{x-1} = E[Xt^{X-1}]$$

$\Rightarrow \quad G''(t) = \sum_x x(x-1) p_x t^{x-2} = E[X(X-1)t^{X-2}]$

$\Rightarrow \quad G''(1) = \sum_x x(x-1) p_x = E[X(X-1)]$

$\Rightarrow \quad G''(1) = \sum_x x^2 p_x - \sum_x x p_x = E[X^2] - E[X]$

$\Rightarrow \quad E[X^2] = G''(1) + E[X] = G''(1) + G'(1)$

Therefore

$$\sigma^2 = G''(1) + G'(1) - (G'(1))^2 \quad \text{or} \quad \sigma^2 = G''(1) + \mu - \mu^2.$$

Together with $\mu = G'(1)$, this is an important result that uses the power of probability generating functions.

In the following examples, the first one shows how the polynomial form for a probability generating function G(t) may be differentiated twice in order to derive the expectation and variance. You should notice that this just reproduces exactly the usual calculations for mean and variance, and there is no gain from the PGF method. However, the two further examples demonstrate the remarkable power of the PGF method when it is possible to write the function G(t) in a simpler form.

Example 12.5

Let X be the discrete random variable that denotes the absolute difference of scores when two fair dice are thrown. Using a PGF confirm that $E(X)$ is just under 2 and determine $\text{Var}(X)$.

Solution

The probability distribution of X was derived in Exercise 12.1, question 2, as

Table 12.7

x	0	1	2	3	4	5
$P(X = x)$	$\frac{3}{18}$	$\frac{5}{18}$	$\frac{4}{18}$	$\frac{3}{18}$	$\frac{2}{18}$	$\frac{1}{18}$

The PGF for X is as follows:

$$G(t) = \frac{3}{18} + \frac{5}{18}t + \frac{4}{18}t^2 + \frac{3}{18}t^3 + \frac{2}{18}t^4 + \frac{1}{18}t^5$$

$\Rightarrow \quad G'(t) = \frac{5}{18} + \frac{8}{18}t + \frac{9}{18}t^2 + \frac{8}{18}t^3 + \frac{5}{18}t^4$

$\Rightarrow \quad G''(t) = \frac{8}{18} + \frac{18}{18}t + \frac{24}{18}t^2 + \frac{20}{18}t^3$

Therefore $G'(1) = \frac{5}{18} + \frac{8}{18} + \frac{9}{18} + \frac{8}{18} + \frac{5}{18} = \frac{35}{18} = 1\frac{17}{18}$

and $G''(1) = \frac{8}{18} + \frac{18}{18} + \frac{24}{18} + \frac{20}{18} = \frac{70}{18} = 3\frac{8}{9}$.

$\Rightarrow \quad E(X) = G'(1) = 1\frac{17}{18} = 1.94$

and $\text{Var}(X) = G''(1) + G'(1) - (G'(1))^2$

$= 3\frac{8}{9} + 1\frac{17}{18} - (1\frac{17}{18})^2 = 2\frac{17}{324} \approx 2.05$

Confirm that these results may also be derived from first principles, i.e:

$$\mu = E(X) = \sum_x x p_x \quad \text{and} \quad \sigma^2 = \text{Var}(X) = \sum_x x^2 p_x - \mu^2.$$

Example 12.6

Four unbiased coins are tossed and the number of heads (X) is noted.

(i) Show that the PGF for X is given by $G(t) = \frac{1}{16}(1+t)^4$.

(ii) Hence calculate the mean of X.

(iii) Calculate the variance of X.

Solution

(i) As $X \sim B(4, \frac{1}{2})$, the probabilities are given by

$$P(X=x) = {}^4C_x \left(\frac{1}{2}\right)^x \left(\frac{1}{2}\right)^{4-x} = {}^4C_x \left(\frac{1}{2}\right)^4 = {}^4C_x \frac{1}{16}$$

The PGF for X is

$$G(t) = E(t^x) = \sum_0^4 p_x t^x = \frac{1}{16} \sum_0^4 {}^4C_x t^x$$

$$= \frac{1}{16}(1 + 4t + 6t^2 + 4t^3 + t^4)$$

$$= \frac{1}{16}(1+t)^4$$

(ii) From the PGF the mean and variance of X are found by differentiation.

$G'(t) = 4 \times \frac{1}{16}(1+t)^3 = \frac{1}{4}(1+t)^3 \quad \Rightarrow \quad G'(1) = \frac{1}{4} \times 2^3 = 2$

$G''(t) = 3 \times \frac{1}{4}(1+t)^2 = \frac{3}{4}(1+t)^2 \quad \Rightarrow \quad G''(1) = \frac{3}{4} \times 2^2 = 3$

$\Rightarrow \quad E(X) = G'(1) = 2$

(iii) $\text{Var}(X) = G''(1) + G'(1) - (G'(1))^2 = 3 + 2 - 2^2 = 1$

Example 12.7

A box contains three red balls and two green balls. They are taken out one at a time, with replacement. Let X represent the number of withdrawals until a red ball is chosen. Calculate the mean and variance of X.

Solution

In this experiment the outcomes follow a geometric distribution.

$X \sim \text{Geo}(p)$, where $p = \frac{3}{5}$ and $q = 1 - \frac{3}{5} = \frac{2}{5}$, and so the PGF is

$$G(t) = \frac{pt}{1-qt} = \frac{\frac{3}{5}t}{1-\frac{2}{5}t} = \frac{3t}{5-2t}.$$

Differentiating with respect to t, using the quotient rule:

$$G'(t) = \frac{(5-2t) \times 3 - 3t \times (-2)}{(5-2t)^2} = \frac{15}{(5-2t)^2} \quad \Rightarrow \quad G'(1) = \frac{15}{9} = 1\frac{2}{3}$$

Differentiating with respect to t, using the chain rule:

$$G''(t) = -2 \times 15 \times (5-2t)^{-3} \times (-2) = \frac{60}{(5-2t)^3} \quad \Rightarrow \quad G''(1) = \frac{60}{27} = 2\frac{2}{9}$$

$$\Rightarrow \quad E(X) = G'(1) = 1\frac{2}{3}$$

and $\text{Var}(X) = G''(1) + G'(1) - (G'(1))^2 = 2\frac{2}{9} + 1\frac{2}{3} - (1\frac{2}{3})^2 = 1\frac{1}{9}$

Example 12.8

A discrete random variable X ($x = 0, 1, 2$) has PGF given by $G(t) = a + bt + ct^2$, where a, b and c are constants. If the mean is $1\frac{1}{4}$ and the variance is $\frac{11}{16}$, find the values of a, b and c.

Solution

As $P(X = 0) = a$, $P(X = 1) = b$ and $P(X = 2) = c$:

$G(t) = a + bt + ct^2 \quad \Rightarrow \quad G(1) = a + b + c = 1$ ①

As $\mu = 1\frac{1}{4}$ and $E(X) = G'(1)$:

$G'(t) = b + 2ct \quad \Rightarrow \quad G'(1) = b + 2c = 1\frac{1}{4}$ ②

As $\sigma^2 = G''(1) + G'(1) - (G'(1))^2$, so $G''(1) = \sigma^2 - G'(1) + (G'(1))^2 = 1$:

$G''(t) = 2c \quad \Rightarrow \quad G''(1) = 2c = 1$ ③

From equation ③: $2c = 1$ $\Rightarrow$ $c = \frac{1}{2}$

From equation ②: $b + 2c = 1\frac{1}{4}$ $\Rightarrow$ $b = 1\frac{1}{4} - 2 \times \frac{1}{2} = \frac{1}{4}$

From equation ①: $a + b + c = 1$ $\Rightarrow$ $a = 1 - \frac{1}{4} - \frac{1}{2} = \frac{1}{4}$

Therefore $a = \frac{1}{4}, b = \frac{1}{4}, c = \frac{1}{2}$.

S1 AL — Expectation and variance

Exercise 12.2

① A random variable has PGF $G(t) = \frac{1}{6} + \frac{1}{3}t + \frac{1}{4}t^3 + \frac{1}{6}t^4 + \frac{1}{12}t^5$. Find its mean and variance.

② A random variable X has PGF $G(t) = \frac{1}{81}(1+2t)^4$. Calculate $E(X)$ and $Var(X)$.

③ Let X be the discrete random variable that denotes the score when a fair die is thrown. Use its PGF to find $E(X)$ and $Var(X)$.

④ A fair coin is tossed three times and the number of tails appearing is noted as the discrete random variable $X(x = 0, 1, 2, 3)$. Use the PGF

$$G(t) = \frac{1}{8}(1 + 3t + 3t^2 + t^3)$$

to calculate the mean and variance of X.

⑤ The probability generating function of a discrete random variable, X, is $G(t) = (kt + 1 - k)^n$, where k is a constant ($0 \leq k \leq 1$) and n is a positive integer. Prove that $E(X) = nk$, and find the variance of X.

⑥ An ordinary pack of playing cards is cut four times. Let X represent the number of aces appearing. Find the probability generating function for X and so find the mean and variance of X.

⑦ The King and Queen of Muldovia want a son and heir to their kingdom. Let X represent the number of children they have until a boy is born.

   (i) Assuming that each pregnancy results in a single child, and that the probability of a boy is 0.5, show that the PGF for X is given by

   $$G(t) = \frac{t}{2-t}$$

   (ii) From this show that both $E(X)$ and $Var(X) = 2$.

⑧ A random variable has PGF $G(t) = a + bt + ct^2$. It has mean $\frac{7}{6}$ and variance $\frac{29}{36}$. Find the values for a, b and c.

⑨ Two fair dice are thrown and X represents the larger of the two scores. Find the PGF for X and so find the mean and variance of X.

⑩ In a competition, the entrant has to match a number of famous landmarks (A, B, C, etc.) with the country in which it is to be found (1, 2, 3, etc.). The number of countries to choose from is the same as the number of landmarks.

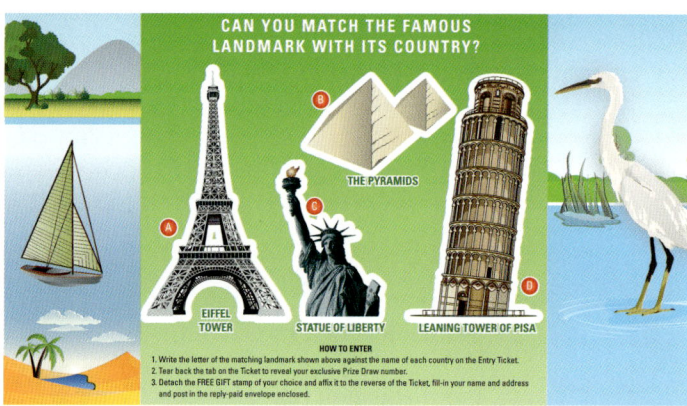

Figure 12.2

Let X represent the number of correct matches that the entrant makes. For the purposes of this investigation, assume that the matching of the landmark to its country is chosen at random.

In each case take the correct matching as $A \leftrightarrow 1, B \leftrightarrow 2, C \leftrightarrow 3$, etc.

(i) For three famous landmarks the possible matchings and values of X are as follows.

Table 12.8

A	1	1	2	2	3	3
B	2	3	1	3	1	2
C	3	2	3	1	2	1
X	3	1	1	0	0	1

Find the PGF for X and show the $E(X) = \text{Var}(X) = 1$.

(ii) (a) For **four** famous landmarks explain why there are $4! = 24$ possible matchings, only one of which is a complete matching.

(b) Show that the PGF for X is given by $G(t) = \dfrac{3}{8} + \dfrac{1}{3}t + \dfrac{1}{4}t^2 + \dfrac{1}{24}t^4$.

(c) Use the PGF to show that $E(X) = \text{Var}(X) = 1$.

(iii) (a) For n ($n > 2$) famous landmarks there are $n!$ possible matchings. Explain why $P(X = n) = \dfrac{1}{n!}$ and $P(X = n - 1) = 0$.

(b) Prove that $P(X = n - 2) = \dfrac{1}{2 \times (3 - 2)!}$ and $P(X = n - 3) = \dfrac{1}{3 \times (n - 3)!}$.

(c) Verify these results for $n = 3$ and $n = 4$.

(iv) For **five** famous landmarks, use the results from part (iii) and the assumption that $E(X) = 1$ to derive the PGF for X.

(v) For **six** famous landmarks, use the results from part (iii) and the assumptions that $E(X) = 1$ and $\text{Var}(X) = 1$ to derive the PGF for X.

(vi) (a) Compare the PGFs in parts (i), (ii), (iv) and (v). How are they related to each other?

(b) How can you use the PGF for n possible matches to derive the PGF for $n - 1$ possible matches and for $n + 1$ possible matches?

3 The PGF of the sum of independent random variables

At the beginning of the chapter you saw that, with a mean of 1.08 goals per match, the random variable X, the number of goals scored by the **away** team in each game, was distributed approximately Po(1.08). So the probability generating function for X is given by $G_X(t) = e^{1.08(t-1)}$.

Let Y represent the number of goals scored by the **home** team, then, from the data collected for the same season for **home** teams:

Table 12.9

Number of goals	Frequency	Relative frequency
0	100	0.216…
1	157	0.339…
2	110	0.238…
3	53	0.114…
4	28	0.060…
5	10	0.021…
6	3	0.006…
7	1	0.002…

The PGF of the sum of independent random variables

The sample mean is 1.56… . Again a Poisson distribution is a good fit to model the number of **home** goals; that is, $Y \sim \text{Po}(1.56\ldots)$ with PGF $G_Y(t) = e^{1.56(t-1)}$.

What do the models predict for the distribution of the total number of goals scored by both sides in a match? For instance, what do they predict for $P(X + Y = 3)$?

$P(X + Y = 3) = P(X = 0 \text{ and } Y = 3) + P(X = 1 \text{ and } Y = 2) + P(X = 2 \text{ and } Y = 1) + P(X = 3 \text{ and } Y = 0)$.

The probabilities of $P(X = a \text{ and } Y = b)$ are not simply related to the separate probabilities of away and home goals, $P(X = a)$ and $P(Y = b)$ unless you assume that the random variables X and Y are independent. You might not expect this to be the case (if one side plays a lot better and scores a lot of goals, then their opponents do not) but, surprisingly, the figures suggest that it is true to a reasonable approximation.

If X and Y are independent, $P(X = a \text{ and } Y = b) = P(X = a) \times P(Y = b)$, so

$$P(X + Y = 3) = P(X = 0) \times P(Y = 3) + P(X = 1) \times P(Y = 2)$$
$$+ P(X = 2) \times P(Y = 1) + P(X = 3) \times P(Y = 0).$$

This figure can be found very neatly from the PGFs for X and Y, because

$$G_X(t) \times G_Y(t) = (P(X = 0) + P(X = 1)t + P(X = 2)t^2 + P(X = 3)t^3 + \ldots)$$
$$\times (P(Y = 0) + P(Y = 1)t + P(Y = 2)t^2 + P(Y = 3)t^3 + \ldots).$$

Picking out from this product the terms in t^3, you get

$$P(X = 0) \times P(Y = 3)t^3 + P(X = 1)t \times P(Y = 2)t^2 + P(X = 2)t^2 \times P(Y = 1)t + P(X = 3)t^3 \times P(Y = 0)$$
$$= P(X + Y = 3) \times t^3$$

The same relationship holds for other values of $X + Y$, and so for other powers of t in the product $G_X(t) \times G_Y(t)$. This means that

$$G_X(t) \times G_Y(t) = P(X + Y = 0) + P(X + Y = 1)t +$$
$$P(X + Y = 2)t^2 + P(X + Y = 3)t^3 + \ldots,$$

but the right-hand side of this equation is just the PGF for the random variable $X + Y$, so that

$$G_X(t) \times G_Y(t) = G_{X+Y}(t)$$

and so

$$G_{X+Y}(t) = e^{1.08\ldots(t-1)} \times e^{1.56\ldots(t-1)} = e^{2.64(t-1)}.$$

This PGF is just that for a Poisson random variable with mean 2.64…, which is the sum of the means for the two independent random variables that you are adding. The result, that the sum of two independent Poisson variables with means λ and μ is a Poisson variable with mean $\lambda + \mu$, should be familiar to you from your earlier work on the Poisson distribution.

The result derived from these variables in fact holds, by exactly the same argument, for any pair of independent random variables X and Y. This can be stated as follows.

If X and Y are two **independent** discrete random variables with PGFs $G_X(t)$ and $G_Y(t)$ then the probability generating function for $X + Y$ is given by
$$G_{X+Y}(t) = G_X(t) \times G_Y(t)$$
i.e. the PGF of the sum ≡ the product of the PGFs.

> The proof of the convolution theorem is beyond the scope of this book.

This is known as the **convolution theorem**.

Example 12.10

A discrete random variable X ($x = 0, 1, 2$) has PGF $G_X(t) = \frac{1}{4} + \frac{1}{4}t + \frac{1}{2}t^2$ and another discrete random variable Y ($y = 0, 1$) has PGF $G_Y(t) = \frac{2}{3} + \frac{1}{3}t$.
Assume that X and Y are independent.
(i) Find the PGF for $X + Y$, i.e. $G_{X+Y}(t)$.
(ii) Show that $E(X + Y) = E(X) + E(Y)$ and $Var(X + Y) = Var(X) + Var(Y)$.

Solution

(i) $G_{X+Y}(t) = G_X(t) \times G_Y(t) = \left(\frac{1}{4} + \frac{1}{4}t + \frac{1}{2}t^2\right)\left(\frac{2}{3} + \frac{1}{3}t\right)$

$= \frac{1}{4} \times \frac{2}{3} + \left(\frac{1}{4} \times \frac{1}{3} + \frac{1}{4} \times \frac{2}{3}\right)t + \left(\frac{1}{4} \times \frac{1}{3} + \frac{1}{2} \times \frac{2}{3}\right)t^2 + \left(\frac{1}{2} \times \frac{1}{3}\right)t^3$

$= \frac{1}{6} + \frac{1}{4}t + \frac{5}{12}t^2 + \frac{1}{6}t^3$

(ii) By differentiation

$G'_X(t) = \frac{1}{4} + t,\; G'_Y(t) = \frac{1}{3}$ and $G'_{X+Y}(t) = \frac{1}{4} + \frac{5}{6}t + \frac{1}{2}t^2$

$\Rightarrow\quad E(X) = G'_X(1) = \frac{1}{4} + 1 = 1\frac{1}{4}$ and $E(Y) = G'_Y(1) = \frac{1}{3}$

$E(X+Y) = G'_{X+Y}(1) = \frac{1}{4} + \frac{5}{6} + \frac{1}{2} = 1\frac{7}{12}$

$\Rightarrow\quad E(X) + E(Y) = 1\frac{1}{4} + \frac{1}{3} = 1\frac{7}{12} = E(X+Y)$

By further differentiation

$G''_X(t) = 1,\; G''_Y(t) = 0$ and $G''_{X+Y}(t) = \frac{5}{6} + t$

$\Rightarrow\quad Var(X) = G''_X(1) + G'_X(1) - (G'_X(1))^2 = 1 + 1\frac{1}{4} - (1\frac{1}{4})^2 = \frac{11}{16}$

$Var(Y) = G''_Y(1) + G'_Y(1) - (G'_Y(1))^2 = 0 + \frac{1}{3} - (\frac{1}{3})^2 = \frac{2}{9}$

$Var(X+Y) = G''_{X+Y}(1) + G'_{X+Y}(1) - (G'_{X+Y}(1))^2$

$= 1\frac{5}{6} + 1\frac{7}{12} - (1\frac{7}{12})^2 = \frac{131}{144}$

$\Rightarrow\quad Var(X) + Var(Y) = \frac{11}{16} + \frac{2}{9} = \frac{131}{144} = Var(X+Y)$

S1 AL — The PGF of the sum of independent random variables

Extension to three or more random variables

The result, that the PGF of the sum of two independent random variables is the product of the PGFs, can be extended to three or more variables. For example, if X, Y and Z are three independent discrete random variables with PGFs $G_X(t)$, $G_Y(t)$ and $G_Z(t)$ then the probability generating function for $X + Y + Z$ is given by

$$G_{X+Y+Z}(t) = G_X(t) \times G_Y(t) \times G_Z(t).$$

If n independent discrete random variables all have the same PGF, $G(t)$, then the probability generating function for their sum is

$$[G(t)]^n.$$

Example 12.11

Find the probability generating function for the total number of 6s when five fair dice are thrown. Deduce the mean and variance.

Solution

When *one* die is thrown, a 6 occurs with probability $\frac{1}{6}$, therefore the PGF for the number of 6s is $G(t) = \frac{5}{6} + \frac{1}{6}t$. Therefore, when *five* dice are thrown, the PGF for X, the total number of 6s, is given by

$$G_X(t) = [G(t)]^5 = \left(\frac{5}{6} + \frac{1}{6}t\right)^5$$

Applying the results derived earlier.

$$G'_X(t) = 5 \times \left(\frac{5}{6} + \frac{1}{6}t\right)^4 \times \frac{1}{6} = \frac{5}{6}\left(\frac{5}{6} + \frac{1}{6}t\right)^4$$

$$G''_X(t) = 4 \times \frac{5}{6}\left(\frac{5}{6} + \frac{1}{6}t\right)^3 \times \frac{1}{6} = \frac{5}{9}\left(\frac{5}{6} + \frac{1}{6}t\right)^3$$

$\Rightarrow \quad E(X) = G'(1) = \frac{5}{6}\left(\frac{5}{6} + \frac{1}{6}\right)^4 = \frac{5}{6} \quad \left(\text{as } \frac{5}{6} + \frac{1}{6} = 1\right)$

and

$\text{Var}(X) = G''(1) + G'(1) - (G'(1))^2$

$= \frac{5}{9}\left(\frac{5}{6} + \frac{1}{6}\right)^3 + \frac{5}{6} - \left(\frac{5}{6}\right)^2 = \frac{5}{9} + \frac{5}{6} - \frac{25}{36} = \frac{25}{36}$

> **Note**
>
> As $X \sim B\left(5, \frac{1}{6}\right)$, the results that $E(X) = 5 \times \frac{1}{6}$ and $\text{Var}(X) = 5 \times \frac{1}{6} \times \frac{5}{6}$ have been verified. The proof for the mean and variance of $X \sim B(n, p)$ is given later in this chapter.

Example 12.12

Gina regularly practises archery. The probability of her hitting the gold (the centre of the target) with any arrow is $\frac{1}{3}$.

(i) Find the PGF $G(t)$ of the number of shots until she hits the gold for the first time.

(ii) Show that the PGF of the number of shots until she hits the gold for the second time is $[G(t)]^2$.

(iii) Explain why the PGF of the number of shots until she hits the gold for the kth time is $[G(t)]^k$.

Solution

(i) Let X represent the number of shots up to and including Gina's **first** gold, then the random variable X forms a geometric distribution with

$$P(X = x) = \tfrac{1}{3}\left(\tfrac{2}{3}\right)^{x-1}$$

The PGF for X is

$$G_X(t) = \left(\tfrac{1}{3}\right)t + \left(\tfrac{1}{3}\right)\left(\tfrac{2}{3}\right)t^2 + \left(\tfrac{1}{3}\right)\left(\tfrac{2}{3}\right)^2 t^3 + \left(\tfrac{1}{3}\right)\left(\tfrac{2}{3}\right)^3 t^4 + \left(\tfrac{1}{3}\right)\left(\tfrac{2}{3}\right)^4 t^5 + \ldots$$

$$= \left(\tfrac{1}{3}\right)t\left[1 + \left(\tfrac{2}{3}t\right) + \left(\tfrac{2}{3}t\right)^2 + \left(\tfrac{2}{3}t\right)^3 + \left(\tfrac{2}{3}t\right)^4 + \ldots\right]$$

$$= \left(\tfrac{1}{3}\right)t\left[\frac{1}{1 - \tfrac{2}{3}t}\right]$$

$$= \frac{t}{3 - 2t}$$

(ii) Let Y represent the number of shots up to and including Gina's **second** gold. This follows a **negative binomial distribution**.

$$P(Y = y) = {}^{y-1}C_1 \left(\tfrac{1}{3}\right)^2 \left(\tfrac{2}{3}\right)^{y-2} = (y-1)\left(\tfrac{1}{3}\right)^2 \left(\tfrac{2}{3}\right)^{y-2}$$

The PGF for Y is

$$G_Y(t) = \left(\tfrac{1}{3}\right)^2 t^2 + 2\left(\tfrac{1}{3}\right)^2\left(\tfrac{2}{3}\right) t^3 + 3\left(\tfrac{1}{3}\right)^2\left(\tfrac{2}{3}\right)^2 t^4 + 4\left(\tfrac{1}{3}\right)^2\left(\tfrac{2}{3}\right)^3 t^5 + \ldots$$

$$= \left(\tfrac{1}{3}t\right)^2\left[1 + 2\left(\tfrac{2}{3}t\right) + 3\left(\tfrac{2}{3}t\right)^2 + 4\left(\tfrac{2}{3}t\right)^3 + \ldots\right]$$

$$= \left(\tfrac{1}{3}\right)t^2\left[\frac{1}{(1 - \tfrac{2}{3}t)^2}\right]$$

$$= \left[\frac{t}{3 - 2t}\right]^2 = [G_X(t)]^2$$

(iii) The number of shots required until the kth success is equivalent to the sum of k geometric distributions, so the PGF for this sum is the product of the PGFs, i.e. $[G^{(t)}]^k$.

The PGFs for some standard discrete probability distributions

There are several standard discrete probability distributions that are useful for modelling statistical data. Three of these, the binomial, the Poisson and the geometric distributions, you have met and used before. In this final section you will see how the PGF is derived for each of these discrete distributions and how it is used to provide neat proofs for the expectation, $E(X)$, and variance, $\text{Var}(X)$.

The PGF of the sum of independent random variables

The binomial distribution

Let Y be the discrete random variable with probability distribution

Table 12.10

y	0	1
$P(Y = y)$	q	p

The PGF for Y is therefore $G(t) = q + pt$ and $G(1) = q + p = 1$.

Now let the discrete random variable X be defined by $X = Y_1 + Y_2 + \ldots + Y_n$, where each Y_i has PGF $G(t) = q + pt$, then X has PGF $G_X(t)$, given by

$$G_X(t) = [G(t)]^n = (q + pt)^n.$$

If '1' represents 'success' and '0' represents 'failure' then X represents the number of 'successes' in n independent trials, for which the probability of 'success' is p and the probability of failure is q.

Therefore $X \sim B(n, p)$ with PGF $G_X(t) = (q + pt)^n$.

Having established the PGF, the mean and variance may be proved.

Differentiating with respect to t:

$$G'_X(t) = n(q + pt)^{n-1} \times p = np(q + pt)^{n-1}.$$

$\Rightarrow \quad G'_X(1) = np(q + p)^{n-1} = np$

and $\quad G''_X(t) = (n-1)np(q + pt)^{n-2} \times p = n(n-1)p^2(q + pt)^{n-2}$

$\Rightarrow \quad G''_X(1) = n(n-1)p^2(q + p)^{n-2} = n(n-1)p^2$

Therefore

$$E(X) = G'_X(1) = np$$

$$\text{Var}(X) = G''_X(1) + G'_X(1) - (G'_X(1))^2 = n(n-1)p^2 + np - (np)^2$$

$$= n^2p^2 - np^2 + np - n^2p^2$$

$$= np(1 - p) = npq$$

i.e. $\quad E(X) = np$ and $\text{Var}(X) = npq$

The Poisson distribution

Let X be a Poisson random variable with parameter λ; that is, $X \sim \text{Po}(\lambda)$, then the PGF for X is

$$G_X(t) = e^{-\lambda} + e^{-\lambda} \times \lambda t + e^{-\lambda} \times \frac{\lambda^2}{2!}t^2 + \ldots + e^{-\lambda} \times \frac{\lambda^r}{r!}t^r + \ldots$$

$$= e^{-\lambda}\left(1 + \lambda t + \frac{(\lambda t)^2}{2!} + \ldots + \frac{(\lambda t)^r}{r!} + \ldots\right) = e^{-\lambda} \times e^{\lambda t} = e^{\lambda(t-1)}.$$

Having established the PGF, the mean and variance may be proved.

Differentiating with respect to t:

$$G'_X(t) = \lambda e^{\lambda(t-1)} \quad \Rightarrow \quad G'_X(1) = \lambda e^0 = \lambda$$

$$G''_X(t) = \lambda^2 e^{\lambda(t-1)} \quad \Rightarrow \quad G''_X(1) = \lambda^2 e^0 = \lambda^2$$

Therefore

$$E(X) = G'_X(1) = \lambda$$

$$\text{Var}(X) = G''_X(1) + G'_X(1) - (G'_X(1))^2 = \lambda^2 + \lambda - (\lambda)^2 = \lambda$$

i.e. $\quad E(X) = \text{Var}(X) = \lambda$

EXTENSION

The Poisson approximation to the binomial

You know that if $X \sim B(n, p)$, where n is large and p is small (for example, $n = 150, p = 0.02$), then a good approximation to the distribution of X is $X \sim Po(\lambda)$, where $\lambda = np$. The PGFs can be used to show that the Poisson distribution is the limit of the binomial distribution as $n \to \infty$, $p \to 0$ and np remains fixed.

For the binomial distribution $G(t) = (q + pt)^n$, where $q = 1 - p$

$$\Rightarrow \quad G(t) = (1 - p + pt)^n = (1 + p(t-1))^n.$$

But $\lambda = np \to p = \dfrac{\lambda}{n}$, where λ is fixed

$$\Rightarrow \quad G(t) = \left(1 + \frac{\lambda(t-1)}{n}\right)^n.$$

Using the result that $e^x = \lim\left(1 + \dfrac{x}{n}\right)^n$ as $n \to \infty$, the limit of $G(t)$ $n \to \infty$ is given by

$$G(t) = \lim\left(1 + \frac{\lambda(t-1)}{n}\right)^n \text{ as } n \to \infty = e^{\lambda(t-1)}.$$

The geometric distribution

If a sequence of independent trials is conducted, for each of which the probability of success is p and that of failure q (where $q = 1 - p$), and the random variable X is the number of the trial on which the first success occurs, then X has a geometric distribution with $P(X = x) = pq^{x-1}$, $x \geq 1$.

The corresponding PGF is given by

$$G(t) = pt + pqt^2 + pq^2 t^3 + \ldots + pq^{r-1} t^r + \ldots$$

$$= pt\left[1 + qt + (qt)^2 + \ldots + (qt)^{r-1} + \ldots\right]$$

$$= pt\left(\frac{1}{1-qt}\right) = \frac{pt}{1-qt}.$$

Differentiating with respect to t, using the quotient rule:

$$G'(t) = \frac{(1-qt) \times p - pt \times (-q)}{(1-qt)^2} = \frac{p}{(1-qt)^2} \quad \Rightarrow \quad G'(1) = \frac{p}{(1-q)^2} = \frac{1}{p}.$$

S1 AL — The PGF of the sum of independent random variables

Differentiating G'(t) with respect to t, using the chain rule:

$$G''(t) = -2p(1-qt)^3 \times (-) = \frac{2pq}{(1-qt)^3} \Rightarrow G''(1) = \frac{2pq}{(1-q)^3} = \frac{2q}{p^2}$$

$$\Rightarrow E(X) = G'(1) = \frac{1}{p}$$

$$Var(X) = G''(1) + G'(1) - (G'(1))^2$$

$$= \frac{2q}{p^2} + \frac{1}{p} - \left(\frac{1}{p}\right)^2$$

$$= \frac{2q + p - 1}{p^2}$$

$$= \frac{q}{p^2}$$

i.e. $E(X) = \frac{1}{p}$ and $Var(X) = \frac{q}{p^2}$

The negative binomial distribution

If a sequence of independent trials is conducted, for each of which the probability of success is p and that of failure q (where $q = 1 - p$), and the random variable X is the number of the trial on which the rth success occurs, then X has a negative binomial distribution with $P(X = x) = {}^{x-1}C_{r-1}p^r q^{x-r}, x \geq r$

The corresponding PGF is given by

$$G(t) = [\text{PGF for Geo}(p)]^r = \left(\frac{pt}{1-qt}\right)^r.$$

Exercise 12.3

1. (i) A fair die is thrown repeatedly until a 6 occurs. Show that the PGF for the number of throws required is given by

   $$G(t) = \frac{t}{6 - 5t}$$

   (ii) Obtain the PGF for the number of throws required to obtain two 6s (not necessarily consecutively).

   (iii) Calculate the expected value and variance of the number of throws required to obtain two 6s.

2. The probability distributions of two independent random variables X and Y are as follows.

   Table 12.11

x	1	2	
$P(X = x)$	0.6	0.4	
y	1	2	3
$P(Y = y)$	0.2	0.5	0.3

   (i) Write down the PGFs for X and Y. Hence show that the PGF for $X + Y$ is given by $G(t) = t^2(0.12 + 0.38t + 0.38t^2 + 0.12t^3)$.

   (ii) Show that $E(X + Y) = E(X) + E(Y)$ and $Var(X + Y) = Var(X) + Var(Y)$.

③ In a traffic census on a two-way stretch of road, the number of vehicles per minute, X, travelling past a checkpoint in one direction is modelled by a Poisson distribution with parameter 3.4, and the number of vehicles per minute, Y, travelling in the other direction is modelled by a Poisson distribution with parameter 4.6, i.e. $X \sim \text{Po}(3.4)$ and $Y \sim \text{Po}(4.8)$.

  (i) Find the PGFs for both X and Y.

  (ii) Show that $E(X) = \text{Var}(X) = 3.4$.

  (iii) Find the PGF for $X + Y$, the total number of vehicles passing the checkpoint per minute, assuming X and Y are independent.

  (iv) Demonstrate in two ways that $E(X + Y) = \text{Var}(X + Y) = 8.2$.

④ A game consists of rolling two dice, one six-sided and the other four-sided, and adding the scores together. Both dice are fair, the first is numbered 1 to 6 and the second is numbered 1 to 4.

Show that the PGF of Z, where Z is the sum of the scores on the two dice, is

$$G(t) = \frac{t^2}{24} \times \frac{(1-t^6)(1-t^4)}{(1-t)^2}$$

⑤ (i) The random variable X can take values 1, 2, 3, ... and its PGF is $G(t)$. Show that the probability that X is even is given by $\frac{1}{2}[1 + G(-1)]$.

  (ii) For each of the following probability distributions, state the PGF and, from this, find the probability that the outcome is even.

    (a) The total score when two ordinary dice are thrown.

    (b) The number of attempts it takes Kerry to pass her driving test, given that the probability of passing at any attempt is $\frac{2}{3}$.

⑥ Two people, A and B, fire alternately at a target, the winner of the game being the first to hit the target. The probability that A hits the target with any particular shot is $\frac{1}{4}$ and the probability that B hits the target with any particular shot is $\frac{1}{5}$.

Given that A fires first, find:

  (i) the probability that B wins the game with his first shot

  (ii) the probability that A wins the game with his second shot

  (iii) the probability that A wins the game.

  (iv) The total number of shots fired by A and B, is given by the random variable R. Show that the probability generating function of R is given by

$$G(t) = \frac{5t + 3t^2}{4(5 - 3t^2)}.$$

  Find $E(R)$.

⑦ (i) The variable X has a Poisson distribution with mean λ. Write down the value of $P(X = r)$ and the PGF for X in the form $G_X(t) = \ldots$.

  (ii) The variable Y can only take values 2, 3, 4, ... and is such that $P(Y = r) = kP(X = r)$, where k is a constant and $r > 1$. Find the value of k.

  (iii) Show that the PGF of Y is given by

$$G(t) = \frac{e^{\lambda t} - \lambda t - 1}{e^\lambda - \lambda - 1}.$$

(iv) Calculate $E(Y)$ and show that $\text{Var}(Y) = \left(\dfrac{\mu}{\gamma}\right)(\lambda^2 e^\lambda + \gamma - \mu\gamma)$ where $\mu = E(Y)$ and $\gamma = \lambda(e^\lambda - 1)$.

8. Two duellists take alternate shots at each other until one of them scores a 'hit' on the other. If the probability that the first duellist scores a 'hit' is $\frac{1}{5}$ and the probability that the second scores a 'hit' is $\frac{1}{3}$, find the expected number of shots **before** a hit.

9. Independent trials, on each of which the probability of a 'success' is p ($0 < p < 1$), are being carried out. The random variable, X, counts the number of trials up to **but not** including that on which the first 'success' is obtained.

    (i) Write down an expression for $P(X = x)$ for $x = 0, 1, 2, \ldots$ and show that the probability generating function of X is

    $$G(t) = \dfrac{p}{1 - qt}$$

    where $q = 1 - p$.

    (ii) Use $G(t)$ to find the mean and variance of X.

10. The independent random variables W, X and Y have Poisson distributions with parameters λ_1, λ_2 and λ_3 respectively.

    (i) Write down expressions for $P(W = w)$, $P(X = x)$ and $P(Y = y)$.

    (ii) Write down the PGF of the random variables W, X and Y. Hence obtain the PGF of $Z = W + X + Y$.

    (iii) Use this PGF to find:

    (a) $P(W + X + Y = k)$, where k is a non-negative integer

    (b) the mean of $W + X + Y$

    (c) the variance of $W + X + Y$.

11. Two coins are tossed. The number of heads that come up is denoted by the random variable X.

    (i) Show that the probability generating function of X is given by $G(t) = k(1 + t)^2$ where k is a constant to be determined.

    (ii) Find $E(X)$ and $\text{Var}(X)$.

    (iii) The experiment is carried out twice and the total number of heads is denoted by the random variable Y. Copy and complete this table giving the probabilities of the different possible values of Y.

    Table 12.12

y	0	1	2	3	4
$P(Y = y)$			$\frac{3}{8}$	$\frac{1}{4}$	

    (iv) Show that the probability generating function of Y is given by $G_Y(t) = \dfrac{1}{16}(1+t)^4$.

    Find $E(Y)$ and $\text{Var}(Y)$.

    (v) In a game the pair of coins is tossed once but the each head scores 2 points. The random variable P is the number of points scored.

    Find the probability generating function of P.

    Find also the mean and variance of P.

(vi) In one form of the game, players are given 3 points to start with. The random variable S is the number of points a player has after one toss of the pair of coins.

Find the probability generating function of S.

Find also the mean and variance of S.

KEY POINTS

1. For a discrete probability distribution, the **probability generating function (PGF)** is defined by

$$G(t) = E(t^x) = \sum_x t^x p_x,$$

where $p_x = P(X = x)$.

2. The sum of probabilities $= 1 \Rightarrow G(1) = 1$.
3. Expectation: $E(X) = G'(1) = \mu$.
4. Variance: $Var(X) = G''(1) + G'(1) - (G'(1))^2$.
   $= G''(1) + \mu - \mu^2$.
5. **Special probability generating functions**

   Table 12.12

Probability distribution	$P(X = r)$	PGF $G_X(t)$	Expectation $E(X)$	Variance $Var(X)$
Binomial	$^nC_x p^x q^{n-x}$	$(q + p^t)^n$	np	npq
Poisson	$e^{-\lambda} \dfrac{\lambda^x}{x!}$	$e^{\lambda(t-1)}$	λ	λ
Geometric	pq^{x-1}	$\dfrac{p^t}{1 - q^t}$	$\dfrac{1}{p}$	$\dfrac{q}{p^2}$
Negative Binomial	$^{z-1}C_{r-1} p^r q^{x-r}$	$\left(\dfrac{pt}{1 - qt}\right)^r$	$\dfrac{r}{p}$	$\dfrac{rq}{p^2}$

6. For two independent random variables X and Y

   $$G_{X+Y}(t) = G_X(t) \times G_Y(t).$$

LEARNING OUTCOMES

When you have completed this chapter you should be able to:

► understand the concept of a probability generating function (PGF)
► construct and use the PGF for given distributions (including the discrete uniform, binomial, negative binomial, geometric and Poisson distributions)
► find the mean and variance of a distribution given its PGF
► use the result that the PGF of the sum of independent random variables is the product of the PGFs of those random variables.

Practice questions: set 3

① Esperanza buys a large number of Premium Bonds. These bonds are entered into a monthly draw and may win a prize. Esperanza calculates that her probability of winning one or more prizes in any month is 0.4.

(i) Find the probability that Esperanza wins

    (a) her first prize in the third month [1 mark]

    (b) no prizes in the first six months [1 mark]

    (c) no prizes in the first six months but at least one prize in the first year [2 marks]

    (d) something in exactly two months of the first six. [2 marks]

Esperanza's brother Dan suspects that the true probability of her winning a prize in any month may be less than 0.4. He decides to carry out a hypothesis test, using a 5% significance level. Starting at a randomly chosen point in time, he asks Esperanza to count the number of months, n, up to and including her first subsequent win.

(ii) State the hypotheses for Dan's test. [1 mark]

(iii) Given that Dan accepts his null hypothesis, determine the possible values of n. [5 marks]

② A hospital serving a large town sees an average of 42 patients per year with a certain rare disease.
Occurrences are equally likely at all times of the year.

(i) State the probability distribution that should be used to model the number of patients seen with this disease in a month. [2 marks]

(ii) Find the probability that the hospital sees no occurrences of the disease in a randomly chosen month. [1 mark]

(iii) Find the probability that the hospital sees no occurrences of the disease in at least one month in a randomly chosen year. [2 marks]

A consultant in the hospital believes that the numbers of cases of this disease may have increased. She decides to carry out a hypothesis test, using a 1% significance level. In a randomly chosen month the hospital sees k occurrences of the disease.

(iv) Find the least value of k that would support the consultant's belief. [3 marks]

(v) Suppose that the true rate of occurrence of the disease has risen so that the hospital will see an average of 48 patients per year in future. Find the probability of a Type II error in the consultant's hypothesis test. Comment on your answer. [3 marks]

③ The random variable X has the following probability distribution.

$$P(X = x) = q^{x-1}p \qquad \text{for } x = 1, 2, \ldots,$$

where $q = 1 - p$.

(i) Define the probability generating function (pgf) as an expectation. Hence obtain the pgf of X. [3 marks]

(ii) Use the pgf to find $E(X)$ and $\text{Var}(X)$. [8 marks]

The random variables $X_1, X_2, \ldots, X_n$ are independent and each of them has the same distribution as X above. The random variable Y is equal to the sum $X_1, X_2, \ldots, X_n$.

(iii) Obtain the pgf of Y, explaining your reasoning. [2 marks]

(iv) State, without proof, $E(Y)$ and $Var(Y)$. [2 marks]

Now suppose that the parameter p is the probability that a randomly chosen egg has a double yolk.

(v) Explain in words what the random variable X represents. [1 mark]

(vi) Explain in words what the random variable Y represents in the case $n = 3$. [1 mark]

4 In a large office, the personnel department keeps daily records of the number of employees absent through illness. The bar chart shows the number of absences per day over a period of 5 months.

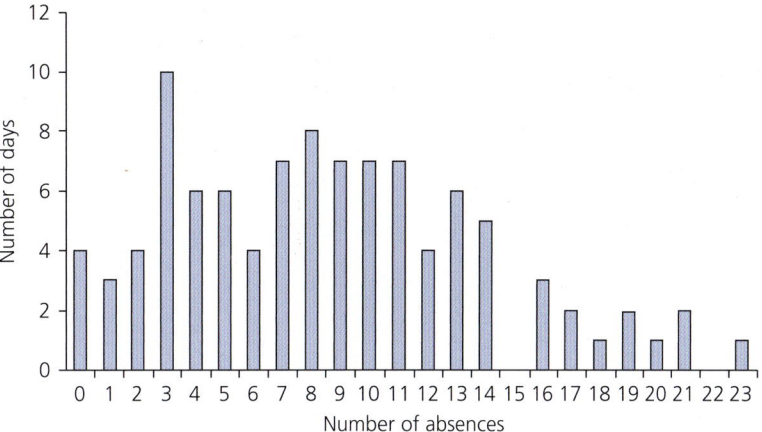

Figure 1

(i) Comment on the shape of the distribution. [1 mark]

Long term records show that the number of absences per day has mean 8.6 and standard deviation 5.3.

(ii) Explain how you can conclude from these figures that the number of absences per day does not have a Normal distribution. [2 marks]

(iii) State the approximate distribution of the mean number of absences per day in a period of 50 days. [3 marks]

(iv) Hence calculate the probability that the total number of absences in a period of 50 days exceeds 450. [3 marks]

(v) What assumption about absences have you have made in answering parts (iii) and (iv)? Discuss briefly how plausible this assumption is. [2 marks]

5 A test of significance is being conducted on the value of the parameter p in a binomial distribution. The null hypothesis is that $p = 0.2$; the alternative hypothesis is that $p > 0.2$. The sample size is 40, and the chosen significance level is 5%.

(i) Obtain the critical region for this test, and state the size of the Type I error correct to 2 significant figures. [4 marks]

(ii) Calculate the power of this test when the true value of p is 0.25. [2 marks]

(iii) Without doing any further calculations, give a sketch of the power function of this test for values of p greater than 0.2. [3 marks]

13 Expectation algebra and the Normal distribution

To approach zero defects, you must have statistical control of processes.

David Wilson

Florence is an Airedale terrier. People say that she is exceptionally heavy for a female. She weighs 26 kg.

It is true that the majority of female Airedale terriers do weigh quite a lot less than this, but how rare is it for a female to weigh this much? To answer that question you need to know the distribution of weights of female Airedale terriers. Of course, the distribution is continuous, so if you knew the probability density function for this distribution, you could work out the probability of a weight which is at least 26 kg.

> **Note**
>
> As you saw in Chapter 7 for a continuous random variable such as this, the probability that a dog's weight is exactly 26 kg is zero. Instead, you need to specify a range of values, in this case at least 26 kg.

Like many other naturally occurring variables, the weights of female Airedale terriers may be modelled by a Normal distribution, shown in Figure 13.1. You will see that this has a distinctive bell-shaped curve and is symmetrical about its middle. The curve is continuous as weight is a continuous variable.

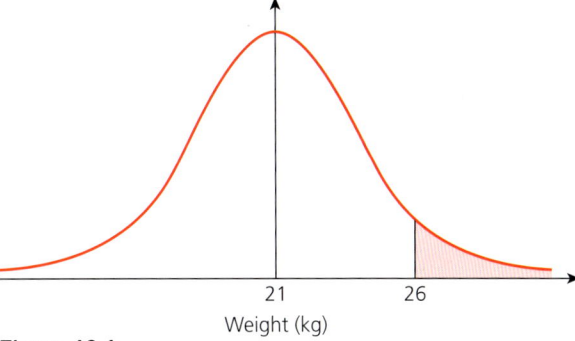

Figure 13.1

> **Note**
> The mean μ and standard deviation σ (or variance σ^2), are the two parameters used to define the distribution. Once you know their values, you know everything there is to know about the distribution.

> In this case, the lower limit is 26 kg and the upper limit is infinite. Of course, you cannot enter infinity into your calculator, so instead you have to enter a very large positive number for the upper limit.

Before you can start to find this area, you must know the mean and standard deviation of the distribution, in this case, roughly 21 kg and 2.5 kg, respectively.

You can find the probability of a weight which is at least 26 kg directly from your calculator. You need to enter the mean, standard deviation and lower and upper limits.

This probability is 0.02275, so roughly 2.3% of female Airedale terriers will weigh at least 26 kg.

Notation

- The mean of a population is denoted by μ.
- The standard deviation is denoted by σ and the variance by σ^2.
- The curve for the Normal distribution with mean μ and standard deviation σ (i.e. variance σ^2) is given by the function $\phi(x)$, where

$$\phi(x) = \frac{1}{\sigma\sqrt{2\pi}} e^{-\frac{1}{2}\left(\frac{x-\mu}{\sigma}\right)^2}$$

- The notation $X \sim N(\mu, \sigma^2)$ is used to describe this distribution.
- The number of standard deviations the value of a variable x is beyond the mean, μ, is denoted by the letter z.
- So $z = \dfrac{x - \mu}{\sigma}$. This is called standardised form.
- So the equation for the standardised Normal distribution, $Z \sim N(0,1)$ is

$$\phi(z) = \frac{1}{\sqrt{2\pi}} e^{-\frac{z^2}{2}}$$

- The function $\Phi(z)$ gives the area under the Normal distribution curve to the left of the value z, that is, the shaded area in Figure 13.2. It is the cumulative distribution function. The total area under the curve is 1, and the area given by $\Phi(z)$ represents the probability of a value smaller than z.

> **Note**
> Note that this is a model so the probability that you have calculated is unlikely to be exactly correct.
> - The distribution of the weights is unlikely to be exactly Normal.
> - The mean and standard deviation are not exactly 21 kg and 2.5 kg, respectively.

> **Note**
> Be careful using the notation $X \sim N(\mu, \sigma^2)$. The notation uses the variance not the standard deviation. So you need to take the square root the second value to find the standard deviation.

💻 USING ICT

In the past, values of $\Phi(z)$ were found from tables but advances in technology have made these tables no longer necessary.

S2 AL: Expectation algebra and the Normal distribution

> **Note**
>
> In the example on page 222, Florence's weight is 26 kg. The mean is 21 kg and the standard deviation is 2.5 kg.
>
> So $x = 26$, $\mu = 21$ and $\sigma = 2.5$, and $z = \dfrac{26-21}{2.5} = 2$.
>
> Florence's weight is 5 kg above the mean, and that is $\dfrac{5}{2.5} = 2$ standard deviations above the mean.

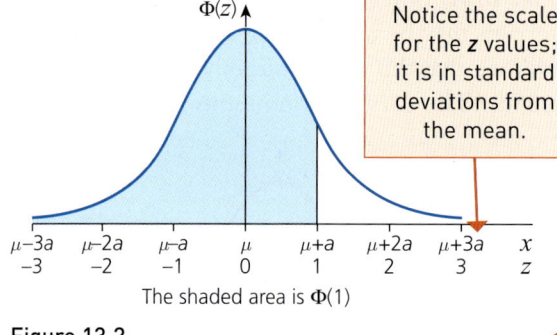

Figure 13.2

Notice the scale for the z values; it is in standard deviations from the mean.

The shaded area is $\Phi(1)$

Notice how lower case letters, x and z, are used to indicate particular values of the random variables, whereas upper case letters, X and Z, are used to describe or name those variables.

- So $P(Z < z) = \Phi(z) = \dfrac{1}{\sqrt{2\pi}} \displaystyle\int_{-\infty}^{z} e^{-\frac{1}{2}u^2}\, du$.
- Unfortunately, the integration cannot be carried out algebraically and so there is no neat expression for $\Phi(z)$. Instead, the integration can be performed numerically and, as you have seen, you can find the values directly from a calculator, both for the standardised Normal distribution and for $N(\mu, \sigma^2)$.
- The area to the right of z is $1 - \Phi(z)$.

> **Note**
>
> In the example of Florence, the Airedale terrier, you calculated
> $P(Z > 2) = 1 - P(Z < 2)$.
> $P(Z < 2) = \Phi(2)$ so the probability which you calculated is $1 - \Phi(2)$.

Example 13.1

The volume of cola in a can is modelled by a Normal distribution with mean 332 ml and standard deviation 4 ml.

(i) Find the probability that a can contains
  (a) at least 330 ml
  (b) less than 340 ml
  (c) between 325 ml and 335 ml.

(ii) 95% of cans contain at least a ml of cola. Find the value of a.

> **A word of caution**
>
> The symbols for phi Φ, ϕ (or φ) look very alike and are easily confused.
>
> - Φ is used for the cumulative distribution and is the symbol you will often use.
> - ϕ and φ are used for the equation of the actual curve; you won't actually need to use them in this book.
>
> These symbols are the Greek letter phi. Φ is upper case and ϕ is lower case.

> **USING ICT**
>
> In these calculations, 1000 is just a typically large number and 0 a typically small number.

Solution

(i) You need to enter the mean and standard deviation into your calculator and then the lower and upper limits.

(a) Lower limit 300, upper limit e.g. 1000

P(at least 330 ml) = 0.6914

(b) Lower limit e.g. 0, upper limit 340

P(less than 340 ml) = 0.9772

(c) Lower limit 325, upper limit 335

P(between 325 ml and 335 ml) = 0.7333

(ii) You can use your calculator to work out the value of a. You know that 95% of cans contain at least a ml. Therefore 5% of cans contain less than a ml. You need to use the inverse Normal function (often

The inverse Normal function on your calculator gives $P(X \leq x)$ for $X \sim N(\mu, \sigma^2)$.

labelled invNormal on a calculator). You enter the mean and standard deviation and then the maximum probability, which in this case is 5% or 0.05.

Using your calculator gives $a = 325.42$.

You can, instead, use the table of percentage points of the Normal distribution.

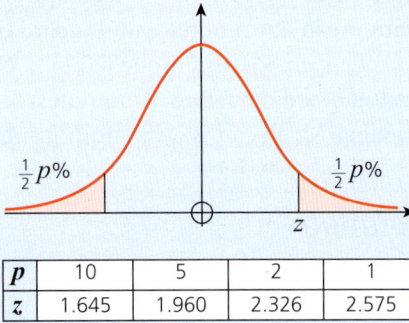

p	10	5	2	1
z	1.645	1.960	2.326	2.575

Figure 13.3

You know that 5% of cans contain less than a ml. The left-hand tail which is labelled $\frac{1}{2}p\%$ is therefore 5%. Thus $p = 10\%$. The z-value for is 1.645. However, you need the lower tail and so the z-value which you need is actually -1.645.

The standardised value $z = \dfrac{x - \mu}{\sigma}$

$$\Rightarrow -1.645 = \frac{x - 332}{4}$$

$$\Rightarrow x = 332 + 4 \times (-1.645) = 325.42$$

1 The sums and differences of Normal variables

In Chapter 2, you were given general results for discrete random variables. These results are also true for continuous random variables, as you saw in Chapter 7. Here are these results.

For any two random variables X_1 and X_2

- $E(X_1 + X_2) = E(X_1) + E(X_2)$
- $E(X_1 - X_2) = E(X_1) - E(X_2)$

If the variables X_1 and X_2 are independent then

- $\text{Var}(X_1 + X_2) = \text{Var}(X_1) + \text{Var}(X_2)$
- $\text{Var}(X_1 - X_2) = \text{Var}(X_1) + \text{Var}(X_2)$

If the variables X_1 and X_2 are Normally distributed, then the distributions of $(X_1 + X_2)$ and $(X_1 - X_2)$ are also Normal. The means of these distributions are $E(X_1) + E(X_2)$ and $E(X_1) - E(X_2)$.

You must, however, be careful when you come to their variances, since you may only use the result that $\text{Var}(X_1 \pm X_2) = \text{Var}(X_1) + \text{Var}(X_2)$ to find the variances of these distributions if the variables X_1 and X_2 are independent.

This is the situation in the next two examples.

Example 13.2

Robert Fisher, a keen chess player, visits his local club most days. The total time taken to drive to the club and back is modelled by a Normal variable with mean 25 minutes and standard deviation 3 minutes. The time spent at the chess club is also modelled by a Normal variable with mean 120 minutes and standard deviation 10 minutes. Find the probability that on a certain evening Mr Fisher is away from home for more than 2½ hours.

$E(X_1 + X_2) = E(X_1) + E(X_2) = 25 + 120 = 145$
$\text{Var}(X_1 + X_2) = \text{Var}(X_1) + \text{Var}(X_2) = 3^2 + 10^2 = 109$, so the standard deviation of $(X_1 + X_2) = \sqrt{109}$. It is **not** $3 + 10$.

Solution

Let the random variable $X_1 \sim N(25, 3^2)$ represent the driving time, and the random variable $X_2 \sim N(120, 10^2)$ represent the time spent at the chess club.

Then the random variable T, where $T = X_1 + X_2 \sim N(145, (\sqrt{109})^2)$, represents his total time away.

You can use your calculator to find the probability that Mr Fisher is away for more than $2\frac{1}{2}$ hours (150 minutes).

Alternatively it is given by

$P(T > 150) = 1 - \Phi\left(\dfrac{150 - 145}{\sqrt{109}}\right)$

$= 1 - \Phi(0.479)$

$= 0.316$

> **Note**
> You need to assume that X_1 and X_2 are independent. In other words, the time to drive to and from the chess club does not affect the time spent at the club. This assumption seems very reasonable.

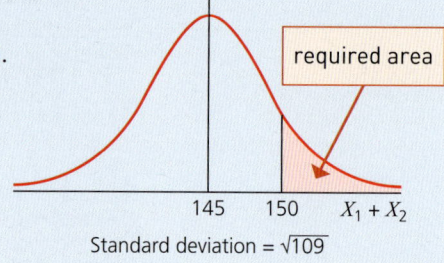

Figure 13.4
Standard deviation = $\sqrt{109}$

Example 13.3

In the manufacture of a bridge made entirely from wood, circular pegs have to fit into circular holes. The diameters of the pegs are Normally distributed with mean 1.60 cm and standard deviation 0.01 cm, while the diameters of the holes are Normally distributed with mean 1.65 cm and standard deviation of 0.02 cm. What is the probability that a randomly chosen peg will not fit into a randomly chosen hole?

Solution

Let the random variable X be the diameter of a hole:

$X \sim N(1.65, 0.02^2) = N(1.65, 0.0004)$.

Let the random variable Y be the diameter of a peg:

$Y \sim N(1.60, 0.01^2) = N(1.6, 0.0001)$.

Let $F = X - Y$. F represents the gap remaining between the peg and the hole and so the sign of F determines whether or not a peg will fit in a hole.

$E(F) = E(X) - E(Y) = 1.65 - 1.60 = 0.05$

$\text{Var}(F) = \text{Var}(X) + \text{Var}(Y) = 0.0004 + 0.0001 = 0.0005$

$F \sim N(0.05, 0.0005)$

If, for any combination of peg and hole, the value of F is negative, then the peg will not fit into the hole.

You can use your calculator to find the probability that the peg will not fit into the hole, $P(F < 0)$.

The probability that $F < 0$ is given by

$$\Phi\left(\frac{0 - 0.05}{\sqrt{0.0005}}\right) = \Phi(-2.326)$$

$= 1 - 0.9873$

$= 0.0127$.

Note that $\text{Var}(X_1 - X_2) = \text{Var}(X_1) + \text{Var}(X_2)$ and not as you might at first imagine $\text{Var}(X_1) - \text{Var}(X_2)$.

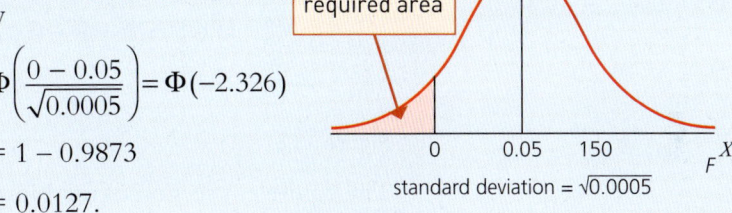

Figure 13.5

2 Modelling discrete situations

Although the Normal distribution applies strictly to a continuous variable, it is also common to use it in situations where the variable is discrete providing that:

- the distribution is approximately Normal; this requires that the steps in its possible values are small compared with its standard deviation
- **continuity corrections** are applied where appropriate.

The meaning of the term continuity correction is explained in the following example.

Example 13.4

The heights X cm of adult males living in a particular country are known to be Normally distributed with mean 174 and standard deviation 10. What proportion of adult males in this country would you expect to have heights between 176 and 180 cm inclusive (measured to the nearest cm)?

Solution

Although the random variable X is continuous, you are asked for heights measured to the nearest cm. In effect, this is a discrete variable which can be called Y. If you draw the probability distribution function for the discrete variable Y it looks like Figure 13.6. The area you require is the total of the five bars representing 176, 177, 178, 179 and 180.

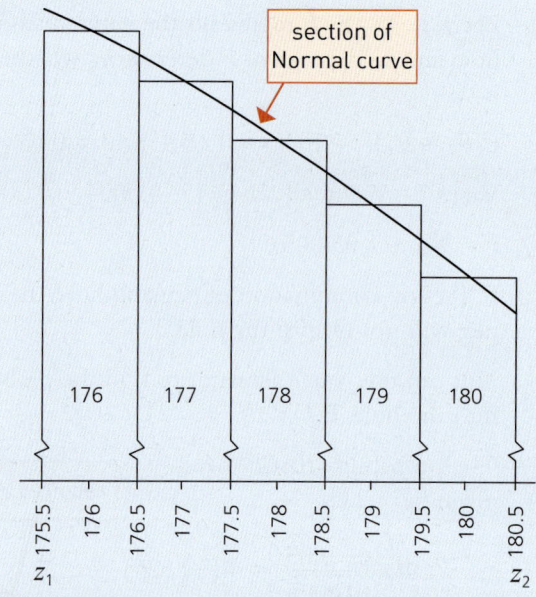

Figure 13.6

The equivalent section of the Normal curve would run not from 176 to 180 but from 175.5 to 180.5, as you can see in Figure 13.6. When you change from the discrete scale to the continuous scale, the numbers 176, 177, etc. no longer represent the whole intervals, just their centre points.

So the area you require under the Normal curve is given by $\Phi(z_2) - \Phi(z_1)$ where $z_1 = \dfrac{175.5 - 174}{10}$ and $z_2 = \dfrac{180.5 - 174}{10}$

Using a calculator this probability is 0.1825.

Answer: The proportion of adult males with heights between 176 and 180 cm (inclusive) should be approximately 18%.

In this calculation, both end values needed to be adjusted to allow for the fact that a continuous distribution was being used to approximate a discrete one. These adjustments, $176 \to 175.5$ and $180 \to 180.5$, are called continuity corrections. Whenever a discrete distribution is approximated by a continuous one, a continuity correction may need to be used.

You must always think carefully when applying a continuity correction. Should the corrections be added or subtracted? In this case, 176 and 180 are inside the required area and so any value (like 175.7 or 180.4) which would round to them must be included. It is often helpful to draw a sketch to illustrate the region you want, like the one in Figure 13.6.

If the region of interest is given in terms of inequalities, you should look carefully to see whether they are inclusive ($\leq$ or $\geq$) or exclusive ($<$ or $>$). For example, $40 \leq X \leq 50$ becomes $39.5 \leq X < 50.5$ whereas $40 < X < 50$ becomes $40.5 \leq X < 49.5$.

Exercise 13.1

① X is a random variable with distribution $N(20, 25)$. Find

(i) $P(X < 25)$ (ii) $P(X > 28)$ (iii) $P(12 \leq X \leq 24)$.

② The weights in grams of a variety of apple can be modelled as Normally distributed with mean 124 and standard deviation 15. Find the probability that the weight of a randomly selected apple will be

(i) less than 100 g, (ii) at least 120 g, (iii) between 110 and 140 grams.

③ The menu at a cafe is shown below.

Table 13.1

Main course		Dessert	
Fish and chips	£3	Ice cream	£1
Bacon and eggs	£3.50	Apple pie	£1.50
Pizza	£4	Sponge pudding	£2
Steak and chips	£5.50		

The owner of the cafe says that all the main-course dishes sell equally well, as do all the desserts, and that customers' choice of dessert is not influenced by the main course they have just eaten.

The variable M denotes the cost of the items for the main course, in pounds, and the variable D the cost of the items for the dessert. The variable T denotes the total cost of a two-course meal: $T = M + D$.

(i) Find the mean and variance of M.

(ii) Find the mean and variance of D.

(iii) List all the possible two-course meals, giving the price for each one.

(iv) Use your answer to part (iii) to find the mean and variance of T.

(v) Hence verify that for these figures

mean (T) = mean (M) + mean (D)

and

variance (T) = variance (M) + variance (D).

④ X_1 and X_2 are independent random variables with distributions $N(50, 16)$ and $N(40, 9)$, respectively. Write down the distributions of

(i) $X_1 + X_2$ (ii) $X_1 - X_2$ (iii) $X_2 - X_1$.

⑤ A play is enjoying a long run at a theatre. It is found that the performance time may be modelled as a Normal variable with mean 130 minutes and standard deviation 3 minutes, and that the length of the intermission in the middle of the performance may be modelled by a Normal variable with mean 15 minutes and standard deviation 5 minutes. Find the probability that the performance is completed in less than 140 minutes.

6. The time Melanie spends on her history assignments may be modelled as being Normally distributed, with mean 40 minutes and standard deviation 10 minutes. The times taken on assignments may be assumed to be independent. Find
   (i) the probability that a particular assignment takes longer than an hour
   (ii) the time in which 95% of all assignments can be completed
   (iii) the probability that two assignments will be completed in less than 75 minutes.

7. The weights of full cans of a particular brand of pet food may be taken to be Normally distributed, with mean 260 g and standard deviation 10 g. The weights of the empty cans may be taken to be Normally distributed, with mean 30 g and standard deviation 2 g. Find
   (i) the mean and standard deviation of the weights of the contents of the cans
   (ii) the probability that a full can weighs more than 270 g
   (iii) the probability that two full cans together weigh more than 540 g.

8. In a vending machine, the capacity of cups is Normally distributed, with mean 200 cm³ and standard deviation 4 cm³. The volume of coffee discharged per cup is Normally distributed, with mean 190 cm³ and standard deviation 5 cm³. Find the percentage of drinks which overflow.

9. On a distant island the heights of adult men and women may both be taken to be Normally distributed, with means 173 cm and 165 cm and standard deviations 10 cm and 8 cm, respectively.
   (i) Find the probability that a randomly chosen woman is taller than a randomly chosen man.
   (ii) Do you think that this is equivalent to the probability that a married woman is taller than her husband?

10. The lifetimes of a certain brand of refrigerator are approximately Normally distributed, with mean 2000 days and standard deviation 250 days.

    Mrs Chudasama and Mr Poole each buy one on the same date.

    What is the probability that Mr Poole's refrigerator is still working one year after Mrs Chudasama's refrigerator has broken down?

11. A random sample of size 2 is chosen from a Normal distribution N(100, 10).

    Find the probability that
    (i) the sum of the sample numbers exceeds 225
    (ii) the first observation is at least 12 more than the second observation.

12. An examination has a mean mark of 80 and a standard deviation of 12. The marks are assumed to be Normally distributed.
    (i) Find the probability that a randomly selected candidate scores a mark between 75 and 85 inclusive (note that you will need to include a continuity correction in your calculation).
    (ii) The bottom 5% of candidates are to be given extra help. Find the maximum mark required for extra help to be given.
    (iii) Three candidates are selected at random. Find the probability that at least one of them is given extra help (you should assume that the examination is taken by a very large group of candidates).

⑬ In a reading test for eight-year-old children, it is found that a reading score X is Normally distributed with mean 5.0 and standard deviation 2.0.

(i) What proportion of children would you expect to score between 4.5 and 6.0?

(ii) There are about 700 000 eight-year-olds in the country. How many would you expect to have a reading score of more than twice the mean?

(iii) Why might educationalists refer to the reading score X as a **score out of 10**?

The reading score is often reported, after scaling, as a value Y which is Normally distributed, with mean 100 and standard deviation 15. Values of Y are usually given to the nearest integer.

(iv) Find the probability that a randomly chosen eight-year-old gets a score, after scaling, of 103.

(v) What range of Y scores would you expect to be attained by the best 20% of readers? [MEI]

⑭ Scores on an IQ test are modelled by the Normal distribution with mean 100 and standard deviation 15. The scores are reported to the nearest integer.

(i) Find the probability that a person chosen at random scores

(a) exactly 105 (b) more than 110.

(ii) Only people with IQs in the top 2.5% are admitted to the organisation *BRAIN*. What is the minimum score for admission?

(iii) Find the probability that, in a random sample of 20 people, exactly 6 score more than 110.

(iv) Find the probability that, in a random sample of 200 people, at least 60 score more than 110.

3 More than two independent random variables

The results in Section 13.1 may be generalised to give the mean and variance of the sums and differences of n random variables, $X_1, X_2, \ldots, X_n$.

$$E(X_1 \pm X_2 \pm \ldots \pm X_n) = E(X_1) \pm E(X_2) \pm \ldots \pm E(X_n)$$

and, provided $X_1, X_2, \ldots, X_n$ are independent,

$$\text{Var}(X_1 \pm X_2 \pm \ldots \pm X_n) = \text{Var}(X_1) + \text{Var}(X_2) + \ldots + \text{Var}(X_n).$$

If $X_1, X_2, \ldots, X_n$ is a set of Normally distributed variables, then the distribution of $(X_1 \pm X_2 \pm \ldots \pm X_n)$ is also Normal.

More than two independent random variables

Example 13.5

The mass, X, of a suitcase at an airport is modelled as being Normally distributed, with mean 15 kg and standard deviation 3 kg. Find the probability that a random sample of ten suitcases weighs more than 154 kg.

Solution

The mass X of one suitcase is given by $X \sim N(15, 9)$.

Then the mass of each of the ten suitcases has the distribution of X; call them $X_1, X_2, \ldots, X_{10}$.

Let the random variable T be the total weight of ten suitcases.

$$T = X_1 + X_2 + \ldots + X_{10}.$$
$$E(T) = E(X_1) + E(X_2) + \ldots + E(X_{10})$$
$$= 15 + 15 + \ldots + 15$$
$$= 150$$

Similarly

$$Var(T) = Var(X_1) + Var(X_2) + \ldots + Var(X_{10})$$
$$= 9 + 9 + \ldots + 9$$
$$= 90$$

So $T \sim N(150, 90)$

The probability that T exceeds 154 is given by

$$1 - \Phi\left(\frac{154 - 150}{\sqrt{90}}\right)$$
$$= 1 - \Phi(0.422)$$
$$= 1 - 0.6635$$
$$= 0.3365.$$

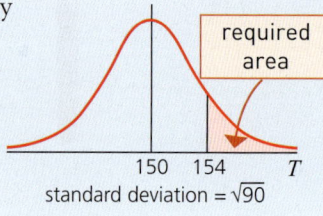

standard deviation = √90

Figure 13.7

Example 13.6

The running times of the four members of a 4 × 400 m relay race may all be taken to be Normally distributed, as follows.

Table 13.2

Member	Mean time (s)	Standard deviation (s)
Adil	52	1
Brian	53	1
Colin	55	1.5
Dexter	51	0.5

Assuming that no time is lost during changeovers, find the probability that the team finishes the race in less than 3 minutes 28 seconds.

Solution

Let the total time be T.

$$E(T) = 52 + 53 + 55 + 51 = 211$$

$$\text{Var}(T) = 1^2 + 1^2 + 1.5^2 + 0.5^2$$
$$= 1 + 1 + 2.25 + 0.25 = 4.5$$

So $T \sim N(211, 4.5)$.

The probability of a total time of less than 3 minutes 28 seconds (208 seconds) is given by

$$\Phi\left(\frac{208 - 211}{\sqrt{4.5}}\right) = \Phi(-1.414)$$

$$= 1 - 0.9213$$

$$= 0.0787$$

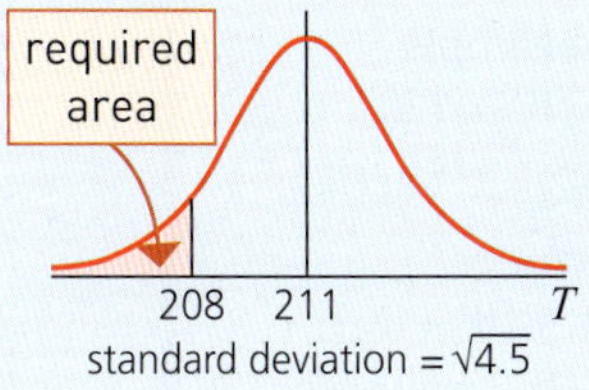

Figure 13.8

Linear combinations of two or more independent random variables

The results given in Section 13.1 can also be generalised to include linear combinations of random variables.

For any random variables X and Y,

- $E(aX + bY) = aE(X) + bE(Y)$, where a and b are constants.

If X and Y are independent

- $\text{Var}(aX + bY) = a^2\text{Var}(X) + b^2\text{Var}(Y)$.

If the distributions of X and Y are Normal, then the distribution of $(aX + bY)$ is also Normal.

These results may be extended to any number of random variables.

Example 13.7

In a workshop, joiners cut out rectangular sheets of laminated board, of length L cm and width W cm, to be made into work surfaces. Both L and W may be taken to be Normally distributed with standard deviation 1.5 cm. The mean of L is 150 cm, that of W is 60 cm, and the lengths of L and W are independent. Both of the short sides and one of the long sides have to be covered by a protective strip (the other long side is to lie against a wall and so does not need protection).

What is the probability that a protecting strip 275 cm long will be too short for a randomly selected work surface?

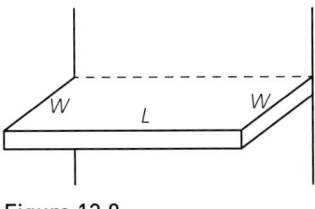

Figure 13.9

> **Note**
> You have to distinguish carefully between the random variable $2W$, which means twice the size of one observation of the random variable W, and the random variable $W_1 + W_2$, which is the sum of two independent observations of the random variable W.
> In the last example
> $E(2W) = 2E(W) = 120$
> and $Var(2W) = 2^2 Var(W)$
> $= 4 \times 2.25 = 9$.
> In contrast, $E(W_1 + W_2)$
> $= E(W_1) + E(W_2) = 60 + 60 = 120$
> and $Var(W_1 + W_2)$
> $= Var(W_1) + Var(W_2)$
> $= 2.25 + 2.25 = 4.5$

Solution

Denoting the length and width by the independent random variables L and W and the total length of strip required by T:

$$T = L + 2W$$
$$E(T) = E(L) + 2E(W)$$
$$= 150 + 2 \times 60$$
$$= 270$$
$$Var(T) = Var(L) + 2^2 Var(W)$$
$$= 1.5^2 + 4 \times 1.5^2$$
$$= 11.25$$

The probability of a strip 275 cm long being too short is given by

$$1 - \Phi\left(\frac{275 - 270}{\sqrt{11.25}}\right) = 1 - \Phi(1.491)$$
$$= 1 - 0.932$$
$$= 0.068$$

Example 13.8

A machine produces sheets of paper the thicknesses of which are Normally distributed with mean 0.1 mm and standard deviation 0.006 mm.

(i) State the distribution of the total thickness of eight randomly selected sheets of paper.

(ii) Single sheets of paper are folded three times (to give eight thicknesses). State the distribution of the total thickness.

Solution

Denote the thickness of one sheet (in mm) by the random variable W, and the total thickness of eight sheets by T.

(i) **Eight separate sheets**

In this situation $T = W_1 + W_2 + W_3 + W_4 + W_5 + W_6 + W_7 + W_8$

where $W_1, W_2, \ldots, W_8$ are eight independent observations of the variable W. The distribution of W is Normal with mean 0.1 and variance 0.006^2.

So the distribution of T is Normal with

mean $= 0.1 + 0.1 + \ldots + 0.1 = 8 \times 0.1 = 0.8$

variance $= 0.006^2 + 0.006^2 + \ldots + 0.006^2 = 8 \times 0.006^2$
$= 0.000\,288$

standard deviation = $\sqrt{0.000288} = 0.017$.

The distribution is $N(0.8, 0.017^2)$.

(ii) **Eight thicknesses of the same sheet**

In this situation $T = W_1 + W_2 + W_3 + W_4 + W_5 + W_6 + W_7 + W_8 = 8W_1$ where W_1 is a single observation of the variable W.

So the distribution of T is Normal with

mean $= 8 \times E(W) = 0.8$

variance $= 8^2 \times \text{Var}(W) = 8^2 \times 0.006^2 = 0.002304$

standard deviation $= \sqrt{0.002304} = 0.048$.

The distribution is $N(0.8, 0.048^2)$.

> ⚠ Notice that, in both cases, the mean thickness is the same but for the folded paper the variance is greater. Why is this?

Exercise 13.2

① The random variable Y is Normally distributed with mean μ and variance σ^2.

Write down the distribution of

(i) $5Y$ (ii) the sum of five independent observations of Y.

② The distributions of two independent random variables X and Y are $N(10, 5)$ and $N(20, 10)$, respectively.

Find the distributions of

(i) $X + Y$ (ii) $10X$ (iii) $3X + 4Y$.

③ A random sample of 15 items is chosen from a Normal population with mean, 30 and variance, 9. Find the probability that the sum of the variables in the sample is less than 440.

④ A company manufactures floor tiles of mean length 20 cm with standard deviation 0.2 cm. Assuming that the distribution of the lengths of the tiles is Normal, find the probability that, when 12 randomly selected floor tiles are laid in a row, their total length exceeds 241 cm.

⑤ The masses of Christmas cakes produced at a bakery are independent and may be modelled as being Normally distributed with mean 4 kg and standard deviation 100 g. Find the probability that a set of eight Christmas cakes has a total mass between 32.3 kg and 32.7 kg.

⑥ The distributions of four independent random variables X_1, X_2, X_3 and X_4 are $N(7, 9), N(8, 16), N(9, 4)$ and $N(10, 1)$, respectively.

Find the distributions of

(i) $X_1 + X_2 + X_3 + X_4$ (ii) $X_1 + X_2 - X_3 - X_4$ (iii) $X_1 + X_2 + X_3$.

⑦ The distributions of X and Y are $N(100, 25)$ and $N(110, 36)$, and X and Y are independent. Find

(i) the probability that $8X + 2Y < 1000$

(ii) the probability that $8X - 2Y > 600$.

⑧ The distributions of the independent random variables A, B and C are $N(35, 9), N(30, 8)$ and $N(35, 9)$. Write down the distributions of

(i) $A + B + C$ (ii) $5A + 4B$ (iii) $A + 2B + 3C$ (iv) $4A - B - 5C$.

9. The distributions of the independent random variables X and Y are N(60, 4) and N(90, 9). Find the probability that
   (i) $X - Y < -35$ (ii) $3X + 5Y > 638$ (iii) $3X > 2Y$.

10. Given that $X \sim$ N(60, 4) and $Y \sim$ N(90, 9) and that X and Y are independent, find the probability that
    (i) when one item is sampled from each population, the one from the Y population is more than 35 greater than the one from the X population
    (ii) the sum of a sample consisting of three items from population X and five items from population Y exceeds 638
    (iii) the sum of a sample of three items from population X exceeds that of two items from population Y.

11. Given that $X_1 \sim$ N(600, 400) and $X_2 \sim$ N(1000, 900) and that X_1 and X_2 are independent, write down the distributions of
    (i) $4X_1 + 5X_2$
    (ii) $7X_1 - 3X_2$
    (iii) $aX_1 + bX_2$, where a and b are constants.

12. The quantity of fuel used by a coach on a return trip of 200 km is modelled as a Normal variable with mean 45 litres and standard deviation 1.5 litres.
    (i) Find the probability that in nine return journeys the coach uses between 400 and 406 litres of fuel.
    (ii) Find the volume of fuel which is 95% certain to be sufficient to cover the total fuel requirements for two return journeys.

13. Assume that the weights of men and women may be taken to be Normally distributed, men with mean 75 kg and standard deviation 4 kg, and women with mean 65 kg and standard deviation 3 kg.

    At a village fair, tug-of-war teams consisting of either five men or six women are chosen at random. The competition is then run on a knock-out basis, with teams drawn out of a hat. If in the first round a women's team is drawn against a men's team, what is the probability that the women's team is the heavier?

    State any assumptions you have made and explain how they can be justified.

14. A petrol company issues a voucher with every 12 l of petrol that a customer buys. Customers who send 50 vouchers to Head Office are entitled to a 'free gift'. After this promotion has been running for some time, the company receives several hundred bundles of vouchers in each day's post. It would take a long time, and so be costly, for somebody to count each bundle and so they weigh them instead. The weight of a single voucher is a Normal variable with mean 40 mg and standard deviation 5 mg.
    (i) What is the distribution of the weights of bundles of 50 vouchers?
    (ii) Find the weight W mg which is exceeded by 95% of bundles.

        The company decides to count only the number of vouchers in those bundles which weigh less than W mg. A man has only 48 vouchers but decides to send them in, claiming that there are 50.
    (iii) What is the probability that the man is detected?

⑮ Jim Longlegs is an athlete whose specialist event is the triple jump. This is made up of a **hop**, a **step** and a **jump**. Over a season, the lengths of the **hop**, **step** and **jump** sections, denoted by H, S and J, respectively, are measured, from which the following models are proposed:

$H \sim N(5.5, 0.5^2)$, $S \sim N(5.1, 0.6^2)$, $J \sim N(6.2, 0.8^2)$

where all distances are in metres. Assume that H, S and J are independent.

(i) In what proportion of his triple jumps will Jim's total distance exceed 18 metres?

(ii) In six successive independent attempts, what is the probability that at least one total distance will exceed 18 m?

(iii) What total distance will Jim exceed 95% of the time?

(iv) Find the probability that, in Jim's next triple jump, his step will be greater than his hop. [MEI]

⑯ A country baker makes biscuits whose masses are Normally distributed with mean 30 g and standard deviation 2.3 g. She packs them by hand into either a small carton (containing 20 biscuits) or a large carton (containing 30 biscuits).

(i) State the distribution of the total mass, S, of biscuits in a small carton and find the probability that S is greater than 615 g.

(ii) Six small and four large cartons are placed in a box. Find the probability that the total mass of biscuits in the ten cartons lies between 7150 g and 7250 g.

(iii) Find the probability that three small cartons contain at least 25 g more than two large ones.

The label on a large carton of biscuits reads 'Net mass 900 g'. A trading standards officer insists that 90% of such cartons should contain biscuits with a total mass of at least 900 g.

(iv) Assuming the standard deviation remains unchanged, find the least value of the mean mass of a biscuit consistent with this requirement. [MEI]

⑰ The weights of pamphlets are Normally distributed with mean 40 g and standard deviation 2 g. What is the distribution of the total weight of

(i) a random sample of two pamphlets?

(ii) a random sample of n pamphlets?

Pamphlets are stacked in piles nominally containing 25. To save time, the following method of counting is used. A pile of pamphlets is weighed and is accepted (i.e. assumed to contain 25 pamphlets) if its weight lies between 980 g and 1020 g. Assuming that each pile is a random sample, determine, to three decimal places, the probabilities that

(iii) a pile actually containing 24 pamphlets will be accepted

(iv) a pile actually containing 25 pamphlets will be rejected.

Justify the choice of the limits as 980 g and 1020 g. [MEI]

⑱ Bricks of a certain type are meant to be 65 mm in height, but, in fact, their heights are Normally distributed with mean 64.4 mm and standard deviation 1.4 mm. A warehouse keeps a large stock of these bricks, stored on shelves. The bricks are stacked one on top of another. The heights of the bricks may be regarded as statistically independent.

(i) Find the probability that an individual brick has height greater than 65 mm.

(ii) Find the probability that the height of a stack of six bricks is greater than 390 mm.

(iii) Find the vertical gap required between shelves to ensure that, with probability 0.99, there is room for a stack of six bricks.

(iv) The vertical gap between shelves is Normally distributed with a mean of 400 mm and standard deviation 7 mm. Find the probability that such a gap has room for a stack of six bricks.

(v) A workman buys ten bricks which can be considered to be a random sample from all the bricks in the warehouse. Find the probability that the average height of these bricks is between 64.8 mm and 65.2 mm. [MEI]

4 The distribution of the sample mean

In many practical situations you do not know the true value of the mean of a variable that you are investigating, that is, the parent population mean (usually just called the **population mean**). Indeed, that may be one of the things you are trying to establish.

In such cases, you will usually take a random sample, $x_1, x_2, \ldots, x_n$, of size n from the population and work out the sample mean $\bar{x}$, given by

$$\bar{x} = \frac{x_1 + x_2 + \cdots + x_n}{n}$$

to use as an estimate for the true population mean μ.

How accurate is this estimate likely to be and how does its reliability vary with n, the sample size?

Each of the sample values $x_1, x_2, \ldots, x_n$ can be thought of as a value of an independent random variable $X_1, X_2, \ldots, X_n$. The variables $X_1, X_2, \ldots, X_n$ have the same distribution as the population and so $E(X_1) = \mu, \text{Var}(X_1) = \sigma^2$, etc.

So the sample mean is a value of the random variable $\bar{X}$ given by

$$\bar{X} = \frac{1}{n}(X_1 + X_2 + \cdots + X_n)$$

$$= \frac{1}{n}X_1 + \frac{1}{n}X_2 + \cdots + \frac{1}{n}X_n$$

and so

$$E(\bar{X}) = \frac{1}{n}E(X_1) + \frac{1}{n}E(X_2) + \cdots + \frac{1}{n}E(X_n)$$

$$= \frac{1}{n}\mu + \frac{1}{n}\mu + \cdots + \frac{1}{n}\mu$$

> **Note**
>
> The derivation has required no assumptions about the distribution of the parent population, other than that μ and σ are finite. If, in fact, the parent distribution is Normal, then the sampling distribution will also be Normal, whatever the size of n.
>
> If the parent population is not Normal, the sampling distribution will still be approximately Normal, and will be more accurately so for larger values of n.
>
> This result is called the Central Limit Theorem and will be developed after Example 13.9.
>
> The derivation does require that the sample items are independent (otherwise the result for Var(X) would not have been valid).

$$= n\left(\frac{1}{n}\mu\right)$$
$$= \mu, \text{ the population mean.}$$

Further, using the fact that $X_1, X_2, \ldots, X_n$ are independent,

$$\text{Var}(\bar{X}) = \text{Var}\left(\frac{X_1}{n}\right) + \text{Var}\left(\frac{X_2}{n}\right) + \cdots + \text{Var}\left(\frac{X_n}{n}\right)$$

$$= \frac{1}{n^2}\text{Var}(X_1) + \frac{1}{n^2}\text{Var}(X_2) + \cdots + \frac{1}{n^2}\text{Var}(X_n)$$

$$= \frac{1}{n^2}\sigma^2 + \frac{1}{n^2}\sigma^2 + \cdots + \frac{1}{n^2}\sigma^2$$

$$= n\left(\frac{1}{n^2}\sigma^2\right)$$

$$= \frac{\sigma^2}{n}$$

Thus the distribution of the means of samples of size n, drawn from a parent population with mean μ and variance σ^2, has mean μ and variance $\frac{\sigma^2}{n}$. The distribution of the sample means is called the **sampling distribution of the means**, or often just the **sampling distribution**.

Notice that $\text{Var}(\bar{X}) = \frac{\sigma^2}{n}$ means that as n increases the variance of the sample mean decreases. In other words, the value obtained for X from a large sample is more reliable as an estimate for μ than one obtained from a smaller sample. This result, simple though it is, lies at the heart of statistics: it says that you are likely to get more accurate results if you take a larger sample.

The standard deviation of sample means of size n is $\frac{\sigma}{\sqrt{n}}$ and this is called the **standard error of the mean**, or often just the **standard error**. It gives a measure of the degree of accuracy of X as an estimate for μ.

Example 13.9

The discrete random variable X has a probability distribution as shown.

Table 13.3

X	1	2	3
Probability	0.5	0.4	0.1

A random sample of size 2 is chosen, with replacement after each selection.

(i) Find μ and σ^2.

(ii) Verify that $\text{E}(\bar{X}) = \mu$ and $\text{Var}(\bar{X}) = \frac{\sigma^2}{2}$.

Solution

(i) $\mu = E(X) = 1 \times 0.5 + 2 \times 0.4 + 3 \times 0.1$

$\qquad = 1.6$

$E(X^2) = 1^2 \times 0.5 + 2^2 \times 0.4 + 3^2 \times 0.1$

$\qquad = 3$

$\sigma^2 = \text{Var}(X) = E(X^2) - [E(X)]^2$

$\qquad = 3 - 1.6^2 = 0.44$

(ii) The table below lists all the possible samples of size 2, their means and probabilities.

Table 13.4

Sample	1, 1	1, 2	1, 3	2, 1	2, 2	2, 3	3, 1	3, 2	3, 3
Mean	1	1.5	2	1.5	2	2.5	2	2.5	3
Probability	0.25	0.2	0.05	0.2	0.16	0.04	0.05	0.04	0.01

This gives the following probability distribution of the sample mean.

For example $P(1, 2) = 0.5 \times 0.4 = 0.2$

Table 13.5

$\bar{X}$	1	1.5	2	2.5	3
Probability	0.25	0.4	0.26	0.08	0.01

Using this table gives

$E(\bar{X}) = 1 \times 0.25 + 1.5 \times 0.4 + 2 \times 0.26 + 2.5 \times 0.08 + 3 \times 0.01$

$\qquad = 1.6 = \mu$, as required.

$E(\bar{X}^2) = 1^2 \times 0.25 + 1.5^2 \times 0.4 + 2^2 \times 0.26 + 2.5^2 \times 0.08 + 3^2 \times 0.01$

$\qquad = 2.78$

$\text{Var}(\bar{X}) = E(\bar{X}^2) - [E(\bar{X})]^2$

$\qquad = 2.78 - (1.6)^2$

$\qquad = 0.22$

$\qquad = \dfrac{0.44}{2} = \dfrac{\sigma^2}{2}$ as required.

The Central Limit Theorem

The Central Limit Theorem was also introduced in Chapter 10 where several of the examples involved discrete data.

This theorem is fundamental to much of statistics and so it is worth pausing to make sure you understand just what it is saying.

It deals with the distribution of the sample mean. This is called the **sampling distribution of the mean**. There are three aspects to it.

1 The expected value of the sample mean is μ, the population mean of the original distribution. That is not a particularly surprising result but it is extremely important.

For samples of size n drawn from a distribution with mean μ and finite variance σ^2, the distribution of the sample mean is approximately $N\left(\mu, \dfrac{\sigma^2}{n}\right)$ for sufficiently large n.

2 The standard deviation of the sample mean is $\dfrac{\sigma}{\sqrt{n}}$. This is often called the **standard error of the mean**. Within a sample, you would expect some values above the population mean, others below it, so that overall the deviations would tend to cancel each other out, and the larger the sample the more this would be the case. Consequently, the standard deviation of the sample means is smaller than that of individual items, by a factor of $\sqrt{n}$.

3 The distribution of the sample mean is approximately Normal.
  This last point is the most surprising part of the theorem. Even if the underlying parent distribution is not Normal, the distribution of the mean of samples of a particular size drawn from it is approximately Normal. The larger the sample size, n, the closer this distribution is to the Normal. For any given value of n, the sampling distribution will be closest to Normal when the parent distribution is not unlike the Normal.
  In many cases, the value of n does not need to be particularly large. For most parent distributions, you can rely on the distribution of the sample mean being Normal if n is about 20 or 25 (or more).

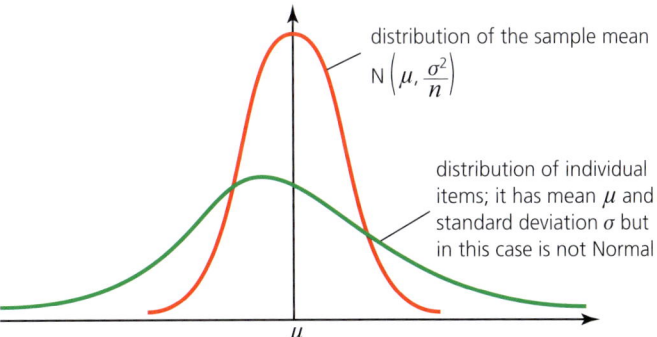

Figure 13.10

Example 13.10

It is known that the mean length of a species of fish is 40.5 cm with standard deviation 3.2.

(i) A random sample of five fish is taken. Explain why you cannot find the probability that the mean of this sample is at least 40.

(ii) A random sample of 50 fish is taken. Calculate an estimate of the probability that the mean of this sample is at least 40.

(iii) If the lengths of individual fish could only be measured to the nearest 1 cm, make a new estimate of the probability calculated in part (ii).

Solution

(i) Because you are told nothing about the distribution of the parent population (the distribution of the length of this species of fish), you cannot carry out any probability calculations.

The distribution of the sample mean

(ii) Although you do not know the distribution of the parent population, the sample size of 50 is large enough to apply the Central Limit Theorem. This states that for a large sample, the distribution of the sample mean is approximately Normal. The distribution is therefore $N(40.5, \frac{3.2^2}{50})$. Using a calculator P(sample mean > 40) = 0.8654.

(iii) If the lengths are only measured to the nearest 1 cm, then the sample mean can only take values ... 39.96, 39.98, 40.00, 40.02, 40.04 ... You therefore need to calculate P(sample mean > 39.99) rather than P(sample mean > 40). Using a calculator P(sample mean > 39.99) = 0.8701.

> Distribution of the sample means is $N\left(\mu, \frac{\sigma^2}{n}\right)$ and you are given that $\mu = 40.5$ and $\sigma = 3.2$

For example, if all except one of the fish were 40 cm long and the last was 41 cm long the sample mean would be 40.02. If instead the last was 39 cm long the sample mean would be 39.98.

> **Note**
> This is an example of a continuity correction which you met earlier in the chapter. As you can see, it requires some careful thought to work out the value needed, particularly when it involves a sample mean.

Exercise 13.3

① The random variable X is Normally distributed with mean 30 and variance 4. Write down the distribution of a random sample of
   (i) 2   (ii) 10   (iii) 100
   observations of X.

② The random variable X is has mean 200 and standard deviation of 30. Write down if possible the approximate distribution of a random sample of
   (i) 10   (ii) 30   (iii) 100
   observations of X.
   If you cannot write down the distribution, explain why you cannot.

③ At a men's hairdresser, haircuts take an average of 18 minutes with standard deviation 3 minutes. Find the distribution of the total time taken for 30 haircuts at the hairdresser. State any assumption which you make.

④ A random sample of 40 items is chosen from a population with mean 60 and variance 25. Find the probability that the mean of the sample is less than 59.

⑤ A company manufactures paving slabs of mean length 45.5 cm with standard deviation 0.5 cm. Find the probability that the mean length of a random sample of 100 slabs is at least 45.4 cm.

⑥ A certain type of low energy light bulb is claimed to have a lifetime of 8000 hours, with standard deviation 600 hours. A random sample of 25 bulbs is selected. Assuming that the claim is true:
   (i) write down the standard error of the sample mean
   (ii) find the probability that the sample mean is greater than 7900 hours.
   (iii) The mean is measured and found to be 7450 hours. Comment on this result.

⑦ In a certain city, the heights of women have mean 164 cm and standard deviation 8.5 cm.
   (i) Find the probability that the mean height of a random sample of 50 women will be within 1 cm of the mean.

(ii) Find the sample size required so that the probability that a random sample of this size will be within 0.5 cm of the mean is the same as your answer to part (i).

⑧ A random sample of n items is chosen from a population with mean 200 and standard deviation 15. You are given that n is large. The probability that the sample mean is greater than 201 is 0.2755 to 4 significant figures. Find the value of n.

⑨ Wooden sleepers for use in landscape gardening are available in two lengths 1.2 m and 1.8 m. The smaller ones actually have mean length 1.23 m with standard deviation 0.05 m. The larger ones have mean length 1.84 m with standard deviation 0.10 m. Find the probability that:

(i) 30 of the smaller sleepers have mean length less than 1.22 m
(ii) 40 of the longer sleepers have mean length more than 1.85 m.
(iii) The mean length of 45 smaller sleepers is greater than two thirds of the mean length of 30 longer sleepers.

⑩ Scores in an IQ test are modelled by a Normal distribution with mean 100 and standard deviation 15. The scores are reported to the nearest integer. The test is taken by a large number of children. Find the probability that

(i) the score of an individual child is greater than 110
(ii) the mean score of random sample of ten children is greater than 110.

⑪ The mean weight of the contents of a jar of jam produced in a factory is 353 g with standard deviation 8 g. The weights of the contents are recorded to the nearest gram. A random sample of 25 jars is selected.

(i) Write down the standard error of the mean.
(ii) Find the probability that the sample mean is
   (a) at least 353
   (b) less than 350
   (c) between 352 and 354 inclusive.

Note that a continuity correction is required for part (ii).

KEY POINTS

1. The Normal distribution with mean μ and standard deviation σ is denoted by $N(\mu, \sigma^2)$.
2. This may be given in standardised form by using the transformation
$$z = \frac{x-\mu}{\sigma}$$
3. In the standardised form, $N(0, 1)$, the mean is 0, the standard deviation and variance both 1.
4. The standardised Normal curve is given by
$$\phi(z) = \frac{1}{\sqrt{2\pi}} e^{-\frac{1}{2}z^2}.$$

5 Probabilities may be found directly from a calculator by entering the mean, standard deviation and lower and upper limits. Alternatively, the area to the left of the value z in the diagram below, representing the probability of a value less than z, is denoted by $\Phi(z)$ and can be found using tables.

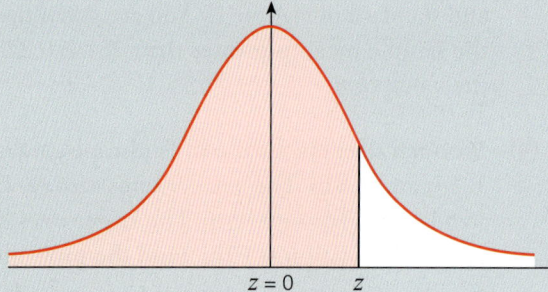

Figure 13.11

6 The Normal distribution may be used to approximate suitable discrete variables but continuity corrections are then required.

7 For two random variables X and Y, whether independent or not, and constants a and b,

- $E(X \pm Y) = E(X) \pm E(Y)$
- $E(aX + bY) = aE(X) + bE(Y)$

and, if X and Y are independent,

- $\text{Var}(X \pm Y) = \text{Var}(X) + \text{Var}(Y)$
- $\text{Var}(aX + bY) = a^2\text{Var}(X) + b^2\text{Var}(Y)$.

8 For a set of n random variables $X_1, X_2, \ldots, X_n$

- $E(X_1 \pm X_2 \pm \cdots \pm X_n) = E(X_1) \pm E(X_2) \pm \cdots \pm E(X_n)$

and, if the variables are independent,

- $\text{Var}(X_1 \pm X_2 \pm \cdots \pm X_n) = \text{Var}(X_1) + \text{Var}(X_2) + \cdots + \text{Var}(X_n)$.

9 If random variables are Normally distributed, then so are the sums, differences and other linear combinations of them.

10 For samples of size n drawn from an infinite (or large) population with mean μ and variance σ^2, or for sampling with replacement

- $E(\bar{X}) = \mu$
- $\text{Var}(\bar{X}) = \dfrac{\sigma^2}{n}$
- The standard deviation of the sample mean is $\dfrac{\sigma}{\sqrt{n}}$ called the standard error of the mean or just the standard error.

11 For samples of size n drawn from a distribution with mean μ and variance σ^2, the distribution of the sample mean is approximately $N\left(\mu, \dfrac{\sigma^2}{n}\right)$ for sufficiently large n.

Chapter 13
Expectation algebra and the Normal distribution

LEARNING OUTCOMES

When you have completed this chapter you should:

- be able to use the Normal distribution as a model, and to calculate and use probabilities from a Normal distribution
- be able to use linear combinations of independent Normal random variables in solving problems
- be able to find the mean of any linear combination of random variables and the variance of any linear combination of independent random variables
- understand that the sample mean is a random variable with a probability distribution
- be able to calculate and interpret the standard error of the mean
- know that if the underlying distribution is Normal, then the sample mean is Normally distributed
- understand how and when the Central Limit Theorem may be applied to the distribution of sample means and use this result in probability calculations, using a continuity correction where appropriate
- be able to apply the Central Limit Theorem to the sum of identically distributed independent random variables.

14 Confidence intervals for the mean

> When we spend money on testing an item, we are buying confidence in its performance.
>
> Anthony Cutler

The perfect apple grower

From our farming correspondent Tom Smith

Fruit grower, Angie Fallon, believes that, after years of trials, she has developed trees that will produce the perfect supermarket apple. 'There are two requirements', Angie told me, 'The average weight of an apple should be 100 grams and they should all be nearly the same size. I have measured hundreds of mine and the standard deviation is a mere 5 grams.'

Angie invited me to take any ten apples off the shelf and weigh them for myself. It was quite uncanny; they were all so close to the magic 100 grams: 98, 107, 105, 98, 100, 99, 104, 93, 105, 103.

> **Discussion point**
> What can you conclude from the weights of the reporter's sample of ten apples?

Before going any further, it is appropriate to question whether the reporter's sample was random. Angie invited him to 'take any ten apples off the shelf'. That is not necessarily the same as taking any ten off the tree. The apples on the shelf could all have been specially selected to impress the reporter. So what follows is based on the assumption that Angie has been honest and the ten apples really do constitute a random sample.

The sample mean is

$$\bar{x} = \frac{98 + 107 + 105 + 98 + 100 + 99 + 104 + 93 + 105 + 103}{10} = 101.2$$

> **Discussion point**
> What does that tell you about the population mean, μ?

To assess the sample, you need to know something about the distribution of the sample mean and also about the spread of the data. The usual measure of spread is the standard deviation, σ; in the article you are told that $\sigma = 5$.

You could estimate the population mean to be the same as the sample mean, namely 101.2, but you would not expect this value to be exactly right.

You can express this by saying that you estimate μ to lie within a range of values, an interval, centred on 101.2:

$101.2 -$ a bit $< \mu < 101.2 +$ a bit.

Such an interval is called a **confidence interval**.

Imagine that you take a large number of samples and use a formula to work out such an interval for each of them. Sometimes the true population mean will be in your confidence interval and sometimes it will not be. If in the long term you catch the true population mean in 90% of your intervals, the confidence interval is called a 90% confidence interval. Other percentages are also used and the confidence intervals are named accordingly.

The width of the interval is clearly twice the 'bit'.

Finding a confidence interval involves a fairly simple calculation but the reasoning behind it is somewhat subtle and requires clear thinking. It is explained in the next section.

> **Note**
> Commonly used confidence intervals are 90%, 95% and 99%.

> **Note**
> A word of advice: read through this section twice, lightly first time and then more thoroughly when you tried a few questions. That will help you to gain a good understanding of the meaning of confidence intervals.

1 The theory of confidence intervals

To understand confidence intervals you need to look not at the particular sample whose mean you have just found, but at the parent population from which it was drawn. For the data on Angie Fallon's apples this does not look very promising. All you know about it is its standard deviation σ (in this case 5). You do not know its mean, μ, which you are trying to estimate, or even its shape.

It is now that the strength of the Central Limit Theorem becomes apparent. This states that the distribution of the means of samples of size n drawn from any population is approximately Normal with mean, μ, and standard deviation $\frac{\sigma}{\sqrt{n}}$ for large values of n.

S2 AL — The theory of confidence intervals

In Figure 14.1, the central 90% region has **been shaded leaving the two 5% tails**, corresponding to z-values of ± 1.645, unshaded. So if you take a large number of samples, all of size n, and work out the sample mean $\bar{x}$ for each one, you would expect that in 90% of cases the value of $\bar{x}$ would lie in the shaded region between A and B.

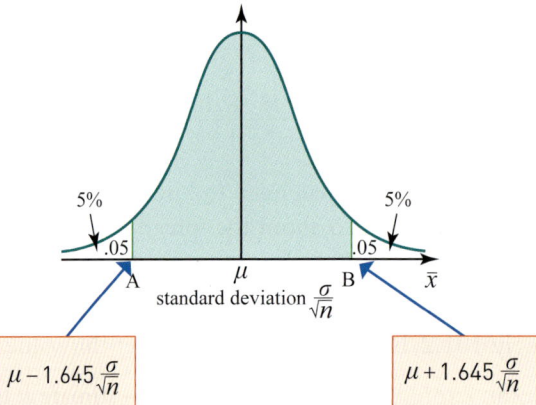

Figure 14.1

> **USING ICT**
>
> The value 1.645 is the 95% point of the cumulative Normal distribution, $\Phi(z)$. You can find this from your calculator or from a table of percentage points of the Normal distribution.

For such a value of $\bar{x}$ to be in the shaded region.

- it must be to the right of A: $\bar{x} > \mu - 1.645 \dfrac{\sigma}{\sqrt{n}}$
- it must be to the left of B: $\bar{x} < \mu + 1.645 \dfrac{\sigma}{\sqrt{n}}$

Putting these two inequalities together gives the result that, in 90% of cases,

$$\bar{x} - 1.645 \frac{\sigma}{\sqrt{n}} < \mu < \bar{x} + 1.645 \frac{\sigma}{\sqrt{n}}$$

and this is the 90% confidence interval for μ.

> The standard deviation of the sample mean, $\dfrac{\sigma}{\sqrt{n}}$, is called the **standard error**.

The boundaries of this interval, $\bar{x} - 1.645 \dfrac{\sigma}{\sqrt{n}}$ and $\bar{x} + 1.645 \dfrac{\sigma}{\sqrt{n}}$, are called the 90% **confidence limits**; 90% is the **confidence level**. If you want a different confidence level, you use a different z value from 1.645.

This number is often denoted by k; commonly used values are:

Table 14.1

Confidence level	k
90%	1.645
95%	1.96
99%	2.58

> **USING ICT**
>
> You can find all of these values of k from your calculator or from a table of percentage points of the Normal distribution.

and the confidence interval is given by

$$\bar{x} - k\frac{\sigma}{\sqrt{n}} \text{ to } \bar{x} + k\frac{\sigma}{\sqrt{n}}.$$

The P% confidence interval for the mean is an interval constructed from sample data in such a way that P% of such intervals will include the true population mean. Figure 14.2 shows a number of confidence intervals constructed from different samples, one of which fails to catch the population mean.

> **Note**
>
> Notice that this is a two-sided symmetrical confidence interval for the mean, μ. Confidence intervals do not need to be symmetrical and can be one-sided. The term confidence interval is a general one, applying not just to the mean but to other population parameters, like variance which is covered in Chapter 17.

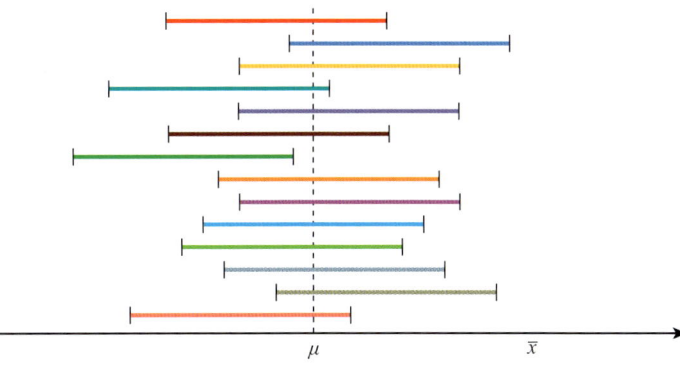

Figure 14.2

In the case of the data on Angie Fallon's apples,

$\bar{x} = 101.2$, $\sigma = 5$, $n = 10$

and so the 90% confidence interval is

$$101.2 - 1.645 \times \frac{5}{\sqrt{10}} \text{ to } 101.2 + 1.645 \times \frac{5}{\sqrt{10}}$$

98.6 to 103.8.

Since this interval includes 100, this suggests that Angie's claim that the mean weight of her apples is 100 grams may be correct.

> **Note**
>
> Note that this confidence interval assumes that the sample mean is Normally distributed.

USING ICT

Statistical software

You can use statistical software to find a confidence interval.

In order for the software to find the interval, you need to input the values of the sample mean, the variance, the value of n and the confidence level. Here is a typical output using the data on Angie Fallon's apples.

Table 14.2

Z Estimate of a Mean	
Confidence Level	0.90

	Sample
Mean	101.2
σ	5
n	10

	Result
Mean	101.2
σ	5
SE	1.5811
N	10
Lower Limit	98.6
Upper Limit	103.8
Interval	101.2 ± 2.60

USING ICT

The software gives the interval as 101.2 ± 2.60 as well as giving the lower and upper limit.

S2 AL — The theory of confidence intervals

Example 14.1

A company has two sites some distance apart and people often have to drive from one to the other. It is known from past experience that measurements of the journey time, t minutes, are Normally distributed with standard deviation 10 minutes and it is believed that the mean time is 160 minutes.

(i) Assuming that the mean journey time is indeed 160 minutes, estimate the percentage of journeys that should take at least 3 hours.

The company suspects that the mean journey is no longer 160 minutes. They time a sample of 20 drivers; in total they take 3510 minutes.

(ii) Obtain a 95% confidence interval for the mean journey time.

(iii) How might the company interpret this confidence interval?

Solution

(i) 3 hours is 180 minutes, so it is $\frac{180-160}{10} = 2$ standard deviations beyond the mean. In a Normal distribution 2.5% is at least 1.96 standard deviations above the mean and 2.5% is at least 1.96 standard deviations below the mean. For an estimate, 2 is close enough to 1.96 and so $2\frac{1}{2}\%$ of journeys should take at least 3 hours.

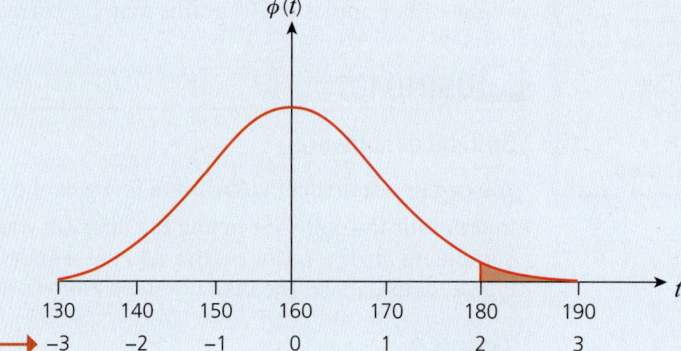

These are all z values.

Figure 14.3

(ii) The mean time of the sample is $\bar{t} = \frac{3510}{20} = 175.5$ minutes.

The 95% confidence interval is from $\bar{t} - 1.96\frac{\sigma}{\sqrt{n}}$ to $\bar{t} + 1.96\frac{\sigma}{\sqrt{n}}$ and so from $175.2 - 1.96 \times \frac{10}{\sqrt{20}} = 170.817...$ to $175.2 + 1.96 \times \frac{10}{\sqrt{20}} = 179.582...$ Rounding to the nearest minute gives the 95% confidence interval for the mean time to be 171 to 180 minutes.

(iii) The previous estimate of 160 minutes is well outside the confidence interval and so it must be regarded as unreliable. Maybe with increased traffic journeys have got slower or maybe 160 minutes never was a realistic mean time.

Known and estimated standard deviation

Notice that you can only use this procedure if you already know the value of the standard deviation of the parent population, σ. In the previous examples, Angie Fallon said that she knew from hundreds of measurements of her apples, that its value is 5 grams, and the standard deviation of the journey times was known to be 10 minutes.

It is more often the situation that you do not know the population standard deviation or variance, and have to estimate it from your sample data. If that is the case, the procedure is different in that you use the *t*-distribution rather than the Normal distribution, provided that the parent population is Normally distributed, and this results in different values of *k*. The use of the *t*-distribution will be developed later in this chapter.

However, if the sample is large, for example over 50, confidence intervals worked out using the Normal distribution will be reasonably accurate even though the standard deviation used is an estimate from the sample. So it is quite acceptable to use the Normal distribution for large samples whether the standard deviation is known or not.

Paired samples

Very often you want to find an estimate of the difference between the means of two populations, as in the following example.

Example 14.2

The senior tutor at a flying school is interested to see if there is a difference between the percentage marks achieved on Paper 1 and Paper 2 of the theory examination. She considers that her 200 students are sufficiently typical for them to be regarded as a random sample of all those taking the examination.

For each student, she calculates the difference, d = Paper 1 mark − Paper 2 mark. She then finds

- the mean of d: $\bar{d} = 5.62$
- the standard deviation of d: $s = 16.45$.

(i) Find a 95% confidence interval for the mean difference in the percentage marks on the papers.

(ii) Interpret your answer.

Solution

(i) This is a large enough sample for an assumption that the differences are distributed Normally to be unnecessary for the calculation of a 95% confidence interval for their mean value.

Note that, as the sample size of 200 is large, it is reasonable to use s as an estimate for σ.

For a 2-sided 95% confidence interval the value of k is 1.96 and so the interval is

$$5.62 - 1.96 \times \frac{16.45}{\sqrt{200}} \text{ to } 5.62 + 1.96 \times \frac{16.45}{\sqrt{200}}$$

3.34 to 7.90.

(ii) This whole interval is positive, so she concludes that Paper 1 appears to be easier.

You can see that the process here is exactly the same as for finding a confidence interval for the mean of a population with the extra initial step of finding the differences.

The situation in this example in which the data are paired (there are two marks for each student) is best for comparisons if you can obtain a suitable sample.

Such a sample is called a **paired sample** or a **matched sample**. In practice, however, it is often impossible to obtain paired data. Instead, you have to resort to taking a random sample from each population with no link between members of each sample. This is called a **two-sample** experiment; the design is described as unpaired.

EXPERIMENT 14.1

This experiment is designed to help you understand confidence intervals, rather than to teach you anything new about dice.

Imagine that you have a large number of dice. When each of them is thrown, the possible outcomes, 1, 2, 3, 4, 5, 6, are all equally likely with probability $\frac{1}{6}$.

Consequently, the expectation or mean score from throwing one of the dice is

$$\mu = 1 \times \frac{1}{6} + 2 \times \frac{1}{6} + \cdots + 6 \times \frac{1}{6} = 3.5.$$

Similarly, the standard deviation is

$$\sigma = \sqrt{\left(1^2 \times \frac{1}{6} + 2^2 \times \frac{1}{6} + \cdots + 6^2 \times \frac{1}{6}\right) - 3.5^2} = 1.708.$$

Imagine that you know σ but don't know μ and wish to construct a 90% confidence interval for it.

Converging confidence intervals

Imagine that you throw one of the dice just once. Suppose you get a 5. You have a sample of size 1, namely {5}, which you could use to work out a sort of 90% confidence interval (but see the warning below).

This confidence interval is given by

$$5 - 1.645 \times \frac{1.708}{\sqrt{1}} \text{ to } 5 + 1.645 \times \frac{1.708}{\sqrt{1}}$$

2.19 to 7.81.

> So far the procedure is not valid. The sample is small and the underlying distribution is not Normal. However, things will get better. The more dice you throw, the larger the sample size and so the more justifiable the procedure. This is because you can make use of the Central Limit Theorem which tells you that for reasonably large samples, the sample mean will be approximately Normally distributed.

Now imagine that you throw two dice. Suppose that this time you get a 5 and a 3. You now have a sample of size 2, namely {5, 3}, with mean 4, and can work out another confidence interval.

The confidence interval is given by

$$4 - 1.645 \times \frac{1.708}{\sqrt{2}} \text{ to } 4 + 1.645 \times \frac{1.708}{\sqrt{2}}$$

2.79 to 5.21.

> Again the procedure is not valid. The sample is still far too small and the underlying distribution is not Normal. You need a much larger sample.

Now consider the procedure for more dice. However instead of actually throwing more and more dice, you can use a spreadsheet to simulate throwing any number of dice.

USING ICT

A spreadsheet

You can use a spreadsheet to simulate throwing as many dice as you choose. You can then work out a 90% confidence interval for the population mean. Figure 14.3 illustrates a simulation using 25 dice.

In order to find the confidence interval, take the following steps.

1. Enter the formula provided by your spreadsheet, for example = RANDBETWEEN(1,6) in cell A1 to simulate throwing one of the dice.
2. Copy this formula into cells A2 to A25 to simulate throwing the remaining 24 dice.
3. Enter the formula provided by your spreadsheet, for example = AVERAGE(A1:A25) in cell A26 to find the sample mean.

The confidence interval will then be

$$\bar{x} - 1.645 \times \frac{1.708}{\sqrt{25}} \text{ to } \bar{x} + 1.645 \times \frac{1.708}{\sqrt{25}}$$

This can be simplified to

$$\bar{x} - 0.562 \text{ to } \bar{x} + 0.562$$

4. Enter the formula = A26 − 0.562 into cell A27 to find the lower confidence limit.
5. Enter the formula = A26 + 0.562 into cell A28 to find the upper confidence limit.

Catching the population mean

If you now copy the whole of column A into columns B to CV inclusive, you will have 100 simulated values of the sample mean. You know that the real value of μ is 3.5 and it should be that this is caught within 9 out of 10 of 90% confidence intervals. Therefore you should find that approximately 90 of your intervals catch the true value of 3.5.

> **Note**
> Of course you would not expect **exactly** 90 of your intervals catch the true value of 3.5.

	A	B	C	D	E	F	G	H	I
1		3	2	4	3	5	1	6	3
2		1	5	6	3	3	6	4	2
3		3	6	4	5	4	4	4	4
4		5	6	4	1	3	1	5	6
5		1	1	4	5	6	6	3	6
6		2	4	2	3	2	4	3	2
7		3	5	5	4	4	5	4	6
8		4	6	2	3	3	4	2	1
9		4	3	6	2	2	5	3	1
10		3	6	5	5	3	4	2	2
11		1	2	6	6	4	1	6	4
12		2	3	5	2	3	5	6	3
13		5	1	4	6	5	2	5	1
14		2	2	2	1	4	6	2	3
15		4	4	3	1	1	2	6	4
16		2	4	6	2	2	6	2	4
17		1	3	6	2	5	1	1	1
18		1	6	4	2	2	5	6	5
19		6	1	6	6	4	2	1	5
20		1	2	4	6	3	6	3	5
21		2	3	6	2	4	6	4	1
22		4	6	3	5	1	6	2	6
23		5	3	4	5	4	6	1	1
24		4	1	1	1	2	1	1	5
25		2	6	6	6	2	4	3	2
26	Sample mean	2.84	3.64	4.32	3.48	3.24	3.96	3.4	3.32
27	Lower limit	2.278	3.078	3.758	2.918	2.678	3.398	2.838	2.758
28	Upper limit	3.402	4.202	4.882	4.042	3.802	4.522	3.962	3.882

The lower limit is above 3.5.

The upper limit is below 3.5.

Figure 14.4

> For this particular simulation, you can see that the first confidence limit has not caught the true population mean, nor has the third, but all of the others shown have caught it. You can only see the first eight simulations but there are actually in fact 100 columns, each with a simulated sample mean and corresponding 90% confidence interval. When the author carried out this particular simulation, 88 of the 100 confidence intervals actually contained the true population mean, which is quite close to the 90 which you expect on average. Try it for yourself and see how many of your confidence intervals catch the population mean of 3.5.

How large a sample do you need?

You are now in a position to start to answer the question of how large a sample needs to be. The answer, as you will see in Example 14.3, depends on the precision you require, and the confidence level you are prepared to accept.

Example 14.3

A trading standards officer is investigating complaints that a coal merchant is giving short measure. Each sack should contain 25 kg but some variation will inevitably occur because of the size of the lumps of coal; the officer knows from experience that the standard deviation should be 1.5 kg.

The officer plans to take, secretly, a random sample of n sacks, find the total weight of the coal inside them and thereby estimate the mean weight of the coal per sack. He wants to present this figure correct to the nearest kilogram with 95% confidence. What value of n should he choose?

Note

This example illustrates that a larger sample size leads to a narrower confidence interval. In fact three things affect the width of the confidence interval; sample size, confidence level and population variability. Increasing the sample size results in a decrease in the width of the confidence interval. Increasing the confidence level and/or population variability results in an increase in the width of the confidence interval.

Solution

The 95% confidence interval for the mean is given by

$$\bar{x} - \frac{1.96\sigma}{\sqrt{n}} \text{ to } \bar{x} + \frac{1.96\sigma}{\sqrt{n}}$$

and so, since $\sigma = 1.5$, the inspector's requirement is that

$$\frac{1.96 \times 1.5}{\sqrt{n}} \leq 0.5$$

The officer wants the mean weight to be correct to the nearest kg, so ± 0.5 kg.

$$\Rightarrow \frac{1.96 \times 1.5}{0.5} \leq \sqrt{n}$$

$$\Rightarrow n \geq 34.57$$

So the inspector needs to take 35 sacks.

Large samples

Given that the width of a confidence interval decreases with sample size, why is it not standard practice to take very large samples?

The answer is that the cost and time involved has to be balanced against the quality of information produced. Because the width of a confidence interval depends on $\frac{1}{\sqrt{n}}$ and not on $\frac{1}{n}$, increasing the sample size does not produce a proportional reduction in the width of the interval. You have, for example, to increase the sample size by a factor of 4 to halve the width of the interval. In the previous example the inspector had to weigh 35 sacks of coal to achieve a class interval of 2 × 0.5 = 1 kg with 95% confidence. That is already quite a daunting task; does the benefit from reducing the interval to 0.5 kg justify the time, cost and trouble involved in weighing another 105 sacks?

Exercise 14.1

1. The mean of a random sample of ten observations of a random variable X is 27.2. It is known that X is Normally distributed and that the standard deviation of X is 2.9. Show that a 95% confidence interval for the mean μ of X is 25.40 to 29.00.

2. Weights in grams of eight bags of sugar are as follows.

   1023   1016   1027   1014   1023   1029   1022   1018

   It is known that the weights of such bags of sugar are Normally distributed with standard deviation 5.5 grams.

   (i) Find the sample mean.

   (ii) Write down the standard error.

   (iii) Show that a 90% confidence interval for the mean weights of bags of sugar is 1018.3 to 1024.7.

3. A biologist studying a colony of beetles selects and weighs a random sample of 20 adult males. She knows that, because of natural variability, the weights of such beetles are Normally distributed with standard deviation 0.2 g. Their weights, in grams, are as follows.

   5.2   5.4   4.9   5.0   4.8   5.7   5.2   5.2   5.4   5.1
   5.6   5.0   5.2   5.1   5.3   5.2   5.1   5.3   5.2   5.2

   (i) Find the mean weight of the beetles in this sample.

   (ii) Find 95% confidence limits for the mean weight of such beetles.

4. An aptitude test for deep-sea divers has been designed to produce scores which are approximately Normally distributed on a scale from 0 to 100 with standard deviation 25. The scores from a random sample of people taking the test were as follows.

   23   35   89   35   12   45   60   78   34   66

   (i) Find the mean score of the people in this sample.

   (ii) Construct a 90% confidence interval for the mean score of people taking the test.

   (iii) Construct a 99% confidence interval for the mean score of people taking the test. Compare this confidence interval with the 90% confidence interval.

5. In a large city the distribution of incomes per family has a standard deviation of £5200.

   (i) For a random sample of 400 families, what is the probability that the sample mean income per family is within £500 of the actual mean income per family?

   (ii) Given that the sample mean income was, in fact, £8300, calculate a 95% confidence interval for the actual mean income per family. [MEI]

⑥ A manufacturer of women's clothing wants to know the mean height of the women in a town (in order to plan what proportion of garments should be of each size). She knows that the standard deviation of their heights is 5 cm. She selects a random sample of 50 women from the town and finds their mean height to be 165.2 cm.

(i) Use the available information to estimate the proportion of women in the town who were

(a) over 170 cm tall

(b) less than 155 cm tall.

(ii) Construct a 95% confidence interval for the mean height of women in the town.

(iii) Another manufacturer in the same town wants to know the mean height of women in the town to within 0.5 cm with 95% confidence. What is the minimum sample size that would ensure this?

⑦ An examination question, marked out of 10, is answered by a very large number of candidates. A random sample of 400 scripts is taken and the marks on this question are recorded.

Table 14.3

Mark	0	1	2	3	4	5	6	7	8	9	10
Frequency	12	35	11	12	3	20	57	87	20	14	129

(i) Calculate the sample mean and the sample standard deviation.

(ii) Find 90% confidence limits for the mean score on the question.

⑧ An archaeologist discovers a short manuscript in an ancient language which he recognises but cannot read. There are 30 words in the manuscript and they contain a total of 198 letters. There are two written versions of the language. In the **early** form of the language, the mean word length is 6.2 letters with standard deviation 2.5; in the **late** form, certain words were given prefixes, raising the mean length to 7.6 letters but leaving the standard deviation unaltered. The archaeologist hopes the manuscript will help him to date the site.

(i) Construct a 95% confidence interval for the mean word length of the language in the manuscript.

(ii) What advice would you give the archaeologist?

⑨ A machine fills packets with X grams of powder where X is normally distributed with mean μ. Each packet is supposed to contain 1 kg of powder. To comply with regulations, the weight of powder in a randomly selected packet should be such that $P(X < \mu - 30) = 0.0005$.

(i) Show that this requires the standard deviation to be 9.117 g to 3 decimal places.

A random sample of 10 packets is selected from the machine. The weight, in grams, of powder in each packet is as follows:

999.8 991.6 1000.3 1006.1 1008.2 997.0 993.2 1000.0 997.1 1002.1

(ii) Assuming that the standard deviation of the population is 9.117 g, test, at the 1% significance level, whether or not the machine is delivering packets with mean weight of less than 1 kg. State your hypotheses clearly.

[Edexcel 6691 Q7 June 2014]

⑩ A woodwork teacher measures the width of a board. The measured width, X mm, is normally distributed with mean w mm and standard deviation 0.5 mm.

(i) Find the probability that X is within 0.6 mm of w.

The same board is measured 16 times and the results are recorded.

(ii) Find the probability that the mean of these results is within 0.3 mm of w.

Given that the mean of these 16 measurements is 35.6 mm,

(iii) find a 98% confidence interval for w. [Edexcel 6691 Q3 June 2010]

⑪ A delivery company has a fleet of 120 lorries. The company manager wishes to switch from conventional diesel fuel to a blend of biodiesel fuel. Before switching, he decides to check whether using biodiesel will affect the fuel consumption of the lorries. He selects 8 lorries and checks their fuel consumption using conventional diesel and then again using biodiesel. The results, measured in litres per 100 km, are as follows.

Table 14.4

Lorry	A	B	C	D	E	F	G	H
Conventional diesel	36.5	34.4	28.6	25.0	27.8	31.9	33.6	38.7
Biodiesel	38.7	36.0	29.2	25.3	27.8	32.6	33.9	39.3

The manager knows from previous data that fuel consumption is Normally distributed with standard deviation 4.61 litres per 100 km.

(i) Assuming that the standard deviation for biodiesel lorries is also 4.61, show the standard deviation of the difference in fuel consumption between lorries using each type of fuel is 6.520 to 3 decimal places.

(ii) Using the value found in part (i) for the population standard deviation of the differences, calculate 90% confidence limits for the mean difference in fuel consumption.

(iii) Does the confidence interval that you have calculated suggest that there is any difference between fuel consumption using each type of fuel? [MEI]

⑫ In a game of patience, which involves no skill, the player scores between 0 and 52 points. The standard deviation is known to be 8; the mean is unknown but thought to be about 12.

(i) Explain why players' scores cannot be Normally distributed if the mean is indeed about 12.

A casino owner wishes to make this into a gambling event but needs to know the mean score before he can set the odds profitably. He employs a student to play the game 500 times. The student's total score is 6357.

(ii) Find 99% confidence limits for the mean score.

The student recorded all her individual scores and finds, on investigation, that their standard deviation is not 8 but 6.21.

(iii) What effect would accepting this value for the standard deviation have on the 99% confidence interval?

The casino owner wants to know the mean score to the nearest 0.1 with 99% confidence.

(iv) Using the value of 6.21 for the standard deviation, find the smallest sample size that would be needed to achieve this.

⑬ An education authority decided to introduce a new P.E. programme for all 11-year-old children to try to improve the fitness of the students. In order to see whether the programme was effective, several tests were done. For one of these, the students were timed on a run of 1 km in their first week in the school and again ten weeks later. A random sample of 100 of the students did both runs. The differences of their mean times, subtracting the time of the second run from that of the first, were calculated. The mean and standard deviation were found to be 0.75 minutes and 1.62 minutes, respectively.

Calculate a 90% confidence interval for the population mean difference. You may assume that the differences are distributed Normally. What assumption have you made in finding this confidence interval?

The organiser of the programme considers that it should lead to an improvement of at least half a minute in the average times. Explain whether or not this aim has been achieved.

⑭ Each day, Stephen passes through a set of temporary traffic lights at some road works. He has been told that the time, in minutes, spent queueing at the lights each day may be modelled by a Normal distribution with mean k and standard deviation $0.25k$, for some value of k.

Stephen wants to construct a 95% confidence interval for k. He records the waiting time each day for n days.

Calculate the minimum value of n necessary for Stephen to achieve a confidence interval with width at most $0.1k$.

⑮ A football boot manufacturer did extensive testing on the wear of the front studs of its Supa range. It found that, after 30 hours use, the wear (i.e. the amount by which the length was reduced) was Normally distributed with standard deviation 1.3 mm. However, the mean wear on the studs of the boot on the dominant foot of the player was 4 mm more than on the studs of the other boot.

(i) Using the manufacturer's figure, find the standard deviation of the differences in wear between a pair of boots after 30 hours use.

The coach of a football team accepted the claim for the standard deviation but was suspicious of the claim about the mean difference. He chose ten of his squad at random. He fitted them with new boots and measured the wear after 30 hours of use with the following results.

Table 14.5

Player	1	2	3	4	5	6	7	8	9	10
Dominant foot	6.5	8.3	4.5	6.7	9.2	5.3	7.6	8.1	9.0	8.4
Other foot	4.2	4.6	2.3	3.8	7.0	4.7	1.4	3.8	8.4	5.7

(ii) Using the value found in part (i) for the population standard deviation of the differences, calculate 95% confidence limits for the mean difference in wear based on the sample data.

(iii) Use these limits to explain whether or not you consider the coach's suspicions were justified.

2 Interpreting sample data using the *t*-distribution

Students find new bat

Two students and a lecturer have found their way into the textbooks. On a recent field trip to the Seychelles they discovered a small colony of a previously unknown bat living in a cave.

'On one of the islands' is all that Shakila Mahadavan, 20, would say about its location. 'We don't want people disturbing the bats or, worse still, catching them for specimens,' she explained.

The other two members of the group, lecturer Alison Evans and 21-year-old Iain Scott, showed scores of photographs of the bats as well as pages of measurements that they had gently made on the few they had caught before releasing them back into their cave.

At a mystery location, students have pushed forward the frontiers of science

The measurements referred to in the article include the weights (in g) of eight bats which were identified as adult males.

156   132   160   142   145   138   151   144

From these figures, the team want to estimate the mean weight of an adult male bat, and 95% confidence limits for their figure.

It is clear from the newspaper report that these are the only measurements available. All that is known about the parent population is what can be inferred from these eight measurements. You know neither the mean nor the standard deviation of the parent population, but you can estimate both.

The mean is estimated to be the same as the sample mean:

$$\frac{156 + 132 + 160 + 142 + 145 + 138 + 151 + 144}{8} = 146.$$

When it comes to estimating the standard deviation, start by finding the sample variance

$$s^2 = \frac{S_{xx}}{n-1} = \sum_i \frac{(x_i - \bar{x})^2}{(n-1)}.$$

and then take the square root to find the standard deviation, *s*.

Interpreting sample data using the t-distribution

> **Note**
> The deviation is the difference (+ or −) of the value from the mean. In this example the mean is 146.

The use of $(n-1)$ as divisor illustrates the important concept of degrees of freedom.

The deviations of the eight numbers are as follows.

$156 - 146 = 10$

$132 - 146 = -14$

$160 - 146 = 14$

$142 - 146 = -4$

$145 - 146 = -1$

$138 - 146 = -8$

$151 - 146 = -5$

$144 - 146 = -2$

> **Note**
> You need to know the degrees of freedom in many situations where you are calculating confidence intervals or conducting hypothesis tests. You may recall meeting the idea in earlier chapters.

These eight deviations are not independent: they must add up to zero because of the way the mean is calculated. This means that when you have worked out the first seven deviations, it is inevitable that the final one has the value it does (in this case −2). Only seven values of the deviation are independent, and, in general, only $(n-1)$ out of the n deviations from a sample mean are independent.

Consequently, there are $n-1$ **free variables** in this situation. The number of free variables within a system is called the **degrees of freedom** and denoted by v.

> **Note**
> A particular value of the sample variance is denoted by s^2, the associated random variable by S^2.

So the sample variance is worked out using divisor $(n-1)$. The resulting value is very useful because it is an **unbiased estimate of the parent population variance**.

In the case of the bats, the estimated population variance is

> The numbers on the top line, 100, 196 and so on, are the squares of the deviations.

$$s^2 = \frac{(100 + 196 + 196 + 16 + 1 + 64 + 25 + 4)}{7} = 86$$

and the corresponding value of the standard deviation is $s = \sqrt{86} = 9.27$.

Calculating the confidence intervals

Returning to the problem of estimating the mean weight of the bats, you now know that:

$\bar{x} = 146$, $s^2 = 86$, $s = 9.27$ and $v = 8 - 1 = 7$.

Before starting on further calculations, there are some important and related points to notice.

1. This is a small sample. It would have been much better if they had managed to catch and weigh more than eight bats.
2. The true parent standard deviation, σ, is unknown and, consequently, the standard deviation of the sampling distribution given by the Central Limit Theorem, $\frac{\sigma}{\sqrt{n}}$, is also unknown.

3   In situations where the sample is small and the parent standard deviation or variance is unknown, there is little more that can be done unless you can assume that the parent population is Normal. (In this case that is a reasonable assumption, the bats being a naturally occurring population.) If you can assume Normality, then you may use the *t*-distribution, estimating the value of σ from your sample.

4   It is possible to test whether a set of data could reasonably have been taken from a Normal distribution by using Normal probability graph paper. The method involves making a cumulative frequency table and plotting points on a graph with special scales on the axes. If the graph obtained is approximately a straight line, then the data could plausibly have been drawn from a Normal population. Otherwise a Normal population is unlikely. Alternatively, you can use a spreadsheet to produce a Normal probability plot as you will see later in this chapter.

The *t*-distribution looks very like the Normal distribution, and, indeed, for large values of v is little different from it. The larger the value of v, the closer the *t*-distribution is to the Normal. Figure 14.5 shows the Normal distribution and *t*-distributions $v = 2$ and $v = 10$.

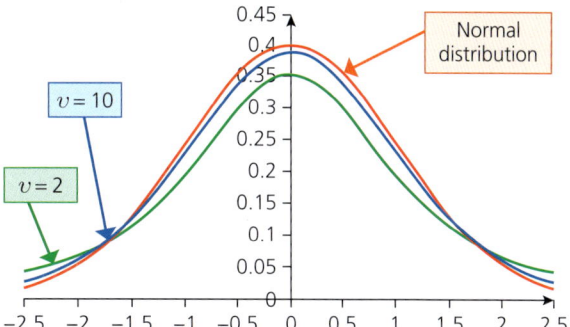

Figure 14.5

 Historical note

William S. Gosset was born in Canterbury in 1876. After studying both mathematics and chemistry at Oxford, he joined the Guinness breweries in Dublin as a scientist. He found that an immense amount of statistical data was available, relating the brewing methods and the quality of the ingredients, particularly barley and hops, to the finished product. Much of the data took the form of samples, and Gosset developed techniques to handle them, including the discovery of the *t*-distribution. Gosset published his work under the pseudonym 'Student' and so the *t*-test is often called Student's *t*-test.

Gosset's name has frequently been misspelt as Gossett (with a double t), giving rise to puns about the *t*-distribution.

S2 AL — Interpreting sample data using the *t*-distribution

Confidence intervals using the *t*-distribution are constructed in much the same way as those using the Normal, with the confidence limits given by:

$$\bar{x} \pm k \frac{s}{\sqrt{n}}$$

where the values of k are found from a spreadsheet or from tables. For instance, to find the value of k at the 5% significance level with 7 degrees of freedom, you can use the formula provided by your spreadsheet. For example $= \text{T.INV.2T}(0.05, 7)$ returns a value of 2.364624.

[2T means that you are using the two-tailed *t*-distribution]

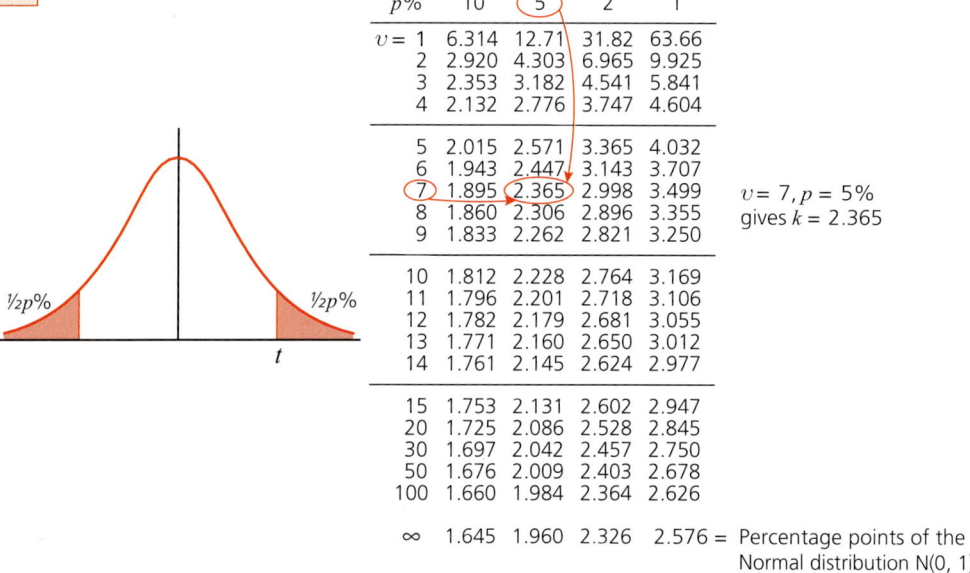

Figure 14.6

To construct a 95% confidence interval for the mean weight of the bats, you look under $p = 5\%$ and $\upsilon = 7$, to get $k = 2.365$; see Figure 14.6. This gives a 95% confidence interval of

$$146 - 2.365 \times \frac{9.27}{\sqrt{8}} \text{ to } 146 + 2.365 \times \frac{9.27}{\sqrt{8}}$$

138.2 to 153.8.

Another bat expert suggests that these bats are not, in fact, a new species, but from a known species. The average weight of adult males of this species is 160 grams. However, because the maximum value in the confidence interval is less than 160, in fact, only 153.8, this suggests that the expert may not be correct. Even if you use a 99% confidence interval, the upper limit is $146 + 3.499 \times \frac{9.27}{\sqrt{8}} = 157.5$. Therefore, it seems very unlikely that these bats are of the same species, based simply on their weights.

Example 14.4

A bus company is about to start a scheduled service between two towns some distance apart. Before deciding on an appropriate timetable, they do nine trial runs to see how long the journey takes. The times, in minutes, are:

89   92   95   94   88   90   92   93   91

(i) Use these data to set up a 95% confidence interval for the mean journey time. You should assume that the journey times are Normally distributed.

The company regards its main competition as the railway service, which takes 85 minutes.

(ii) Does your confidence interval provide evidence that the journey time by bus is different from that by train?

Solution

(i) For the given data,
$$n = 9, \quad v = 9 - 1 = 8, \quad \bar{x} = 91.56, \quad s = 2.297.$$

For a 95% confidence interval, with $v = 8$, $k = 2.306$.
The confidence limits are given by
$$\bar{x} \pm k \frac{s}{\sqrt{n}} = 91.56 \pm 2.306 \times \frac{2.297}{\sqrt{9}}.$$

So the 95% confidence interval for μ is 89.79 to 93.33.

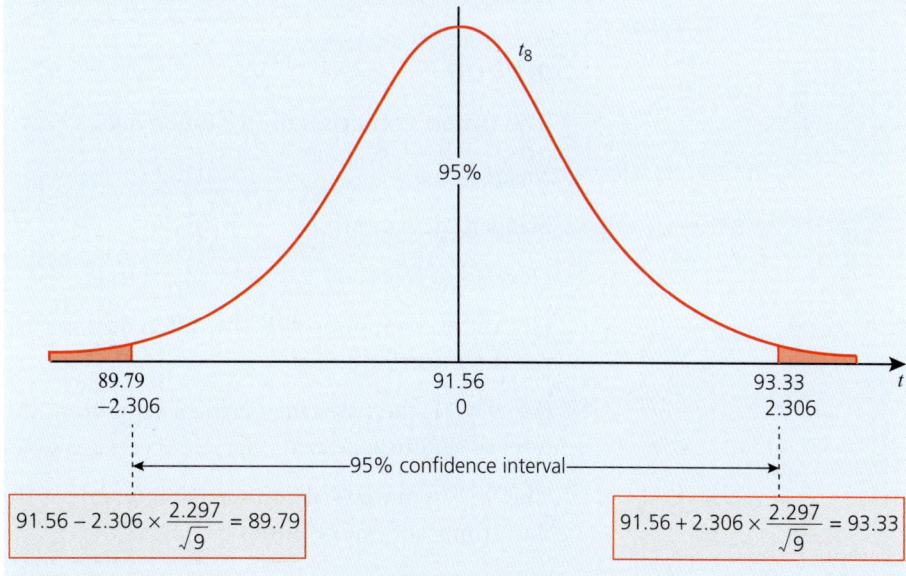

Figure 14.7

(ii) The confidence interval does not contain 85 minutes (the time taken by the train). Therefore there is sufficient evidence to suggest that the journey time by bus is different from that by train, and that it is, in fact, greater.

Interpreting sample data using the *t*-distribution

Using the *t*-distribution for paired samples

The ideas developed in the last few pages can also be used in constructing confidence intervals for the difference in the means of paired data. This is shown in the next example.

Example 14.5

In an experiment on group behaviour, 12 subjects were each asked to hold one arm out horizontally while supporting a 2 kg weight, under two conditions:

- while together in a group
- while alone with the experimenter.

The times, in seconds, for which they were able to support the weight under the two conditions were recorded as follows.

Table 14.6

Subject	A	B	C	D	E	F	G	H	I	J	K	L
'Group' time	61	71	72	53	71	43	85	72	82	54	70	73
'Alone' time	43	72	81	35	56	39	63	66	38	60	74	52
Difference, d	18	−1	−9	18	15	4	22	6	44	−6	−4	21

Find a 90% confidence interval for the true difference between 'group' times and 'alone' times. You may assume that the differences are Normally distributed. Does your result provide evidence that there is any difference in the times in the population as a whole?

Solution

The sample comprises the 12 differences.

Mean $\bar{d} = 10.67$

Standard deviation $s = 15.24$.

Degrees of freedom $v = 12 - 1 = 11$

Given the assumption that the differences are Normally distributed, you may use the *t*-distribution.

For $v = 11$, the two-tailed critical value from the *t*-distribution at the 10% level of significance is 1.796.

The 90% symmetrical confidence interval for the mean difference between the 'group' and 'alone' times is

$$\bar{d} - 1.796 \times \frac{s}{\sqrt{12}} \text{ to } \bar{d} + 1.796 \times \frac{s}{\sqrt{12}}$$

2.77 to 8.57.

Since the confidence interval does not contain zero, there is evidence that there is a difference in the times in the population as a whole.

Checking the goodness of fit of a Normal distribution

When you use a *t*-distribution, a basic assumption is that the data come from a Normal distribution. This assumption can be checked in a number of ways. In an earlier chapter you looked at chi-squared tests of goodness of fit for discrete distributions, but you did not look at them for any continuous distributions. You can, in fact, do these tests for continuous distributions, including the Normal distribution, but this is beyond the scope of this book. There are, however, other tests for Normality which you can use. These include:

- drawing a histogram
- Normal probability plots
- various hypothesis tests which are usually carried out using software.

> The theory behind these tests is beyond the scope of this book.

If the data set is fairly large, then a histogram can be used to look for a symmetrical bell shaped distribution which would suggest that the Normal distribution is a good fit to the data. However, with smaller data sets, a histogram is not very useful. Other methods can be used in this case.

Normal probability plots

Normal probability plots can be used to check whether a Normal distribution model is appropriate for a data set. They can be drawn on Normal probability graph paper or by using software such as a spreadsheet.

On Normal probability graph paper, the horizontal scale is a standard linear scale and the data values are plotted on this in order from lowest to highest. The vertical axis has a probability scale, transformed to a theoretical cumulative Normal distribution. If a Normal model fits the data perfectly, the data points lie on a straight line. The closer the points lie to a straight line, the better the model.

When using software, the vertical axis usually consists of z-values. Some software has the axes the other way round but the interpretation is the same: the straighter the line, the more likely it is that a Normal model is appropriate.

Example 14.6

The lengths in cm of a random sample of ten fish of a particular species are given below. Use a Normal probability plot to investigate whether the distribution of the population may be Normal.

Table 14.7

| 49.6 | 47.8 | 52.4 | 65.3 | 61.8 | 41.6 | 75.1 | 57.6 | 62.0 | 59.1 |

The output from two different spreadsheets is shown in Figure 14.8.

Solution

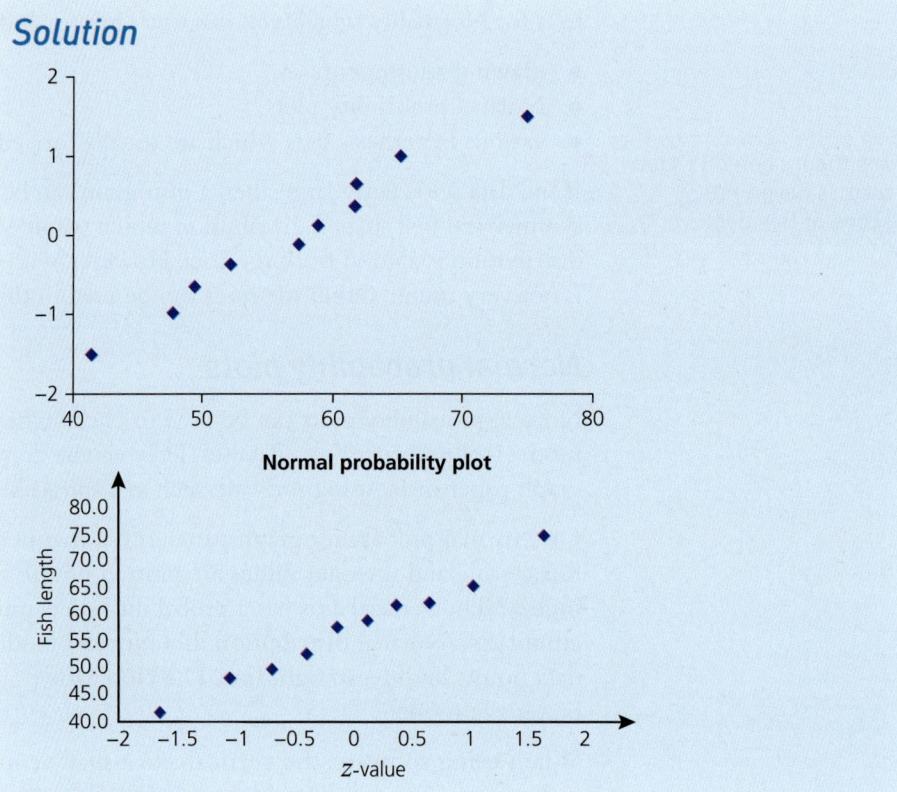

Figure 14.8

The first of these plots has the z-values on the vertical axis and the data values on the horizontal axis. The second is the other way round. Both plots suggest that a Normal model may be appropriate for these data.

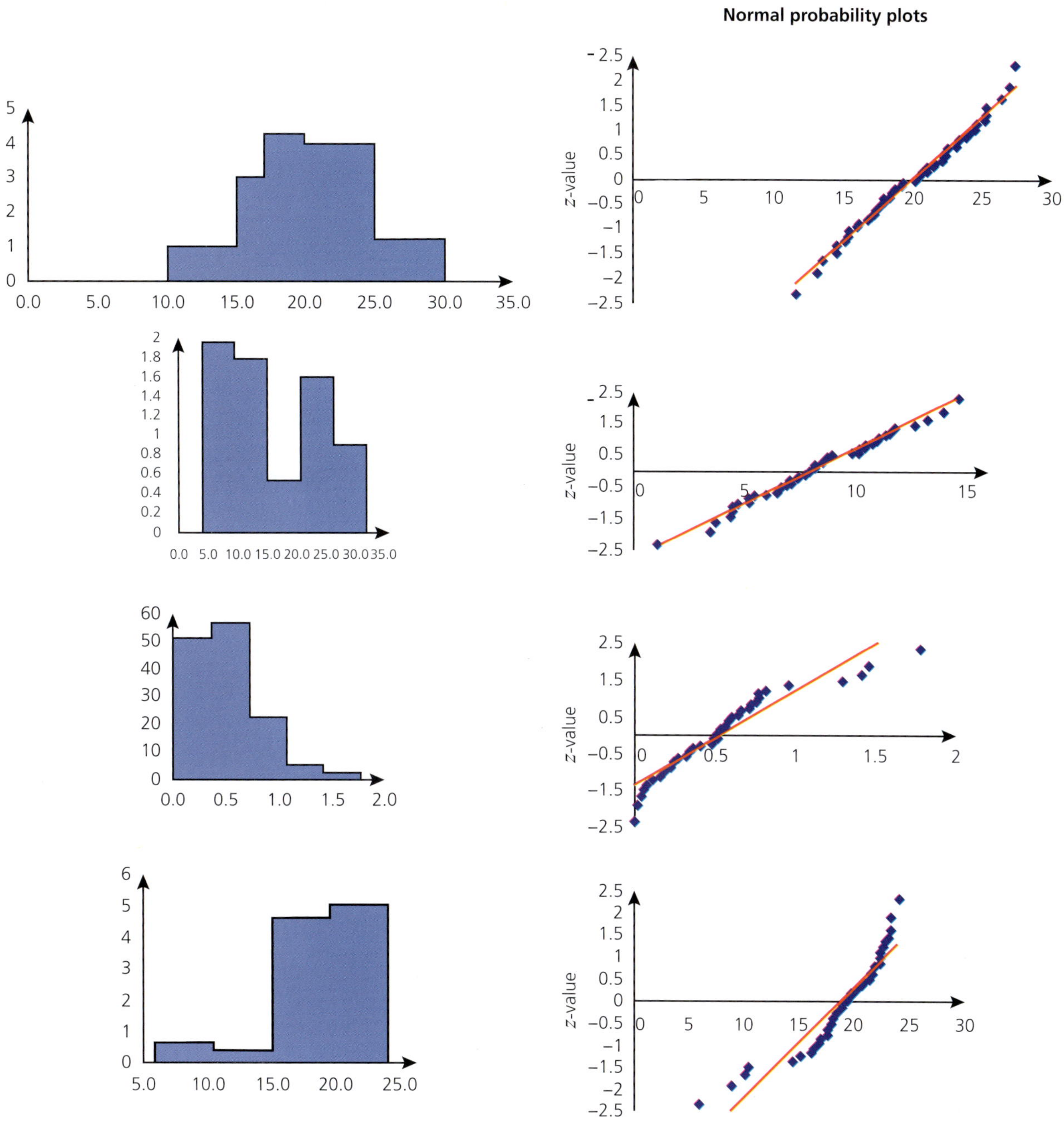

Figure 14.9

Figure 14.9 shows four histograms and their related Normal probability plots. The first shows a Normal distribution, the second shows a bimodal distribution, and the third and fourth show positively and negatively skewed distributions, respectively.

S2 AL — Interpreting sample data using the *t*-distribution

Exercise 14.2

1. The mean of a random sample of seven observations of a Normally distributed random variable X is 132.6. Based on these seven observations, an unbiased estimate of the parent population variance s^2 is 148.84.
   (i) Explain why an estimate of the standard error is given by 4.61.
   (ii) Show that a 95% confidence interval for the mean μ of X is 121.3 to 143.9.

2. The weights in grams of six beetles of a particular species are as follows
   12.3 9.7 11.8 10.1 11.2 12.4
   It may be assumed that the weights of the species are Normally distributed.
   (i) Calculate the sample mean and show that an estimate of the sample variance is 1.291.
   (ii) Show that a 90% confidence interval for the mean μ of X is 10.32 to 12.18.

3. An aptitude test for entrance to university is designed to produce scores which may be modelled by the Normal distribution. In early testing, 15 students from the appropriate age group are given the test. Their scores (out of 500) are as follows.
   321 445 219 378 317 407 289 345
   276 463 265 165 340 298 315
   (i) Use these data to estimate the mean and standard deviation to be expected for students taking this test.
   (ii) Construct a 95% confidence interval for the mean.

4. A fruit farmer has a large number of almond trees, all of the same variety and of the same age. One year, he wishes to estimate the mean yield of his trees. He collects all the almonds from eight trees and records the following weights (in kg).
   36 53 78 67 92 77 59 66
   (i) Use these data to estimate the mean and standard deviation of the yields of all the farmer's trees.
   (ii) Construct a 95% confidence interval for the mean yield.
   (iii) What statistical assumption is required for your procedure to be valid?
   (iv) How might you select a sample of eight trees from those growing in a large field?

5. A forensic scientist is trying to decide whether a man accused of fraud could have written a particular letter. As part of the investigation, she looks at the lengths of sentences used in the letter. She finds them to have the following numbers of words.
   17 18 25 14 18 16 14 16 16 21 25 19
   (i) Use these data to estimate the mean and standard deviation of the lengths of sentences used by the letter writer.
   (ii) Construct a 90% confidence interval for the mean length of the letter writer's sentences.
   (iii) What assumptions have you made to obtain your answer?
   (iv) A sample of sentences written by the accused has mean length 26 words. Does this mean he is in the clear?

⑥ A large company is investigating the number of incoming telephone calls at its exchange, in order to determine how many telephone lines it should have. During March one year, the number of calls received each day was recorded and written down, across the page, as follows.

Table 14.8

623	584	598	701	656	210	23	655	661	599
634	681	197	25	592	643	642	698	659	201
19	588	672	612	706	650	212	29	681	642
677									

(i) What day of the week was 1 March?
(ii) Which of the data do you consider relevant to the company's research and why?
(iii) Construct a 95% confidence interval for the number of incoming calls per weekday.
(iv) Your calculation is criticised on the grounds that your data are discrete and so the underlying distribution cannot possibly be Normal. How would you respond to this criticism?

⑦ A tyre company is trying out a new tread pattern which it is hoped will result in the tyres giving greater distance. In a pilot experiment, 12 tyres are tested; the mileages (×1000 miles) at which they are condemned are as follows.

65   63   71   78   65   69   59   81   72   66   63   62

(i) Construct a 95% confidence interval for the mean distance that a tyre travels before being condemned.
(ii) What assumptions, statistical and practical, are required for your answer to part (i) to be valid?

⑧ A history student wishes to estimate the life expectancy of people in Lincolnshire villages around 1750. She looks at the parish registers for five villages at that time and writes down the ages of the first ten people buried after the start of 1750. Those less than one-year-old were recorded as 0. The data were as follows.

Table 14.9

2	6	72	0	0	18	45	91	6	2
0	12	56	4	25	1	1	5	0	7
8	65	12	63	2	76	70	0	1	0
9	15	3	49	54	0	2	71	6	8
6	0	67	55	2	0	1	54	1	5

(i) Use these data to estimate the mean life expectancy at that time.
(ii) Explain why it is not possible to use these data to construct a confidence interval for the mean life expectancy.
(iii) Is a confidence interval a useful measure in this situation anyway?

A friend tells the student that she could construct a confidence interval for the mean life expectancy of those who survive childhood (age ≥ 15).

(iv) Construct a 95% confidence interval for the mean life expectancy of this group, and comment on whether you think your procedure is valid.

⑨ A large fishing-boat made a catch of 500 mackerel from a shoal. The total mass of the catch was 320 kg. The standard deviation of the mass of individual mackerel is known to be 0.06 kg.

(i) Find a 99% confidence interval for the mean mass of a mackerel in the shoal.

An individual fisherman caught ten mackerel from the same shoal. These had masses (in kg) of

1.04 0.94 0.92 0.85 0.85 0.70 0.68 0.62 0.61 0.59

(ii) From these data only, use your calculator to estimate the mean and standard deviation of the masses of mackerel in the shoal.

(iii) Assuming that the masses of mackerel are Normally distributed, use your results from part (ii) to find another 99% confidence interval for the mean mass of a mackerel in the shoal.

(iv) Give two statistical reasons why you would use the first limits you calculated in preference to the second limits.

⑩ A farmer has a large field of sunflowers. He wishes to know the average height of these sunflowers. The output from statistical software below shows the calculations for a 95% confidence interval for the heights. All figures are in metres.

Table 14.10

t estimate of a mean

Confidence Level	0.95

		Sample
	Mean	2.153
	s	0.263
	n	12
		Result
	Mean	2.153
	σ	0.263
	SE	0.07592
	N	12
	Lower limit	1.9859
	Upper limit	2.32018
	Interval	2.153 ± 0.1671

(i) Write down the confidence interval in the form $a < \mu < b$.

(ii) State the sample size.

(iii) State any assumptions required for the construction of this confidence interval.

(iv) Check that this is the same confidence interval as you obtain using the relevant value of k given in the table in Figure 14.6.

⑪ A youth club has a large number of members (referred to as the *population* in the remainder of the question). In order to find the distribution of weekly allowances of the members, a random sample of ten is questioned.

(i) Describe a method of producing the random sample.

Such a random sample produced the following weekly allowances:

£5.20, £4.40, £3.00, £2.00, £3.30,
£7.50, £5.00, £6.50, £4.80, £5.70

(ii) Estimate the population mean and variance.

(iii) Find a 95% confidence interval for the population mean. State any assumptions on which your method is based.

(iv) Explain how the width of the confidence interval may be reduced. Assuming the same variance as in part (ii) what must the sample size be to reduce the width to £2?

⑫ An aggregate material used for road building contains gravel and stones. The average size of the stones is supposed to be 55 mm. Each batch of this material is checked to ensure that the stones are of the correct size. For each batch, a random sample of eight stones is selected and a 95% confidence interval is found for the size of the stones.

(i) Explain why it would not be sensible simply to discard a batch if the confidence interval does not contain 55 mm.

(ii) Suggest what should be done instead if the confidence interval for a particular batch does not contain 55 mm.

For a particular batch, the eight observations are:

46.21   51.67   48.60   47.34   50.93   49.60   60.97   55.17

A Normal probability plot for these data is shown below:

Figure 14.10

(iii) Explain why it seems that the assumption necessary for a *t*-test may not be justified?

(iv) If, in fact, the data do come from a Normal distribution, construct a 95% confidence interval for the population mean.

⑬ A new computerised job-matching system has been developed which finds suitably-skilled applicants to fit notified vacancies. It is hoped that this will reduce unemployment rates, and a trial of the system is conducted in seven areas.

The unemployment rates in each area just before the introduction of the system and after one month of its operation are recorded in the table below.

Table 14.11

Area	1	2	3	4	5	6	7
Rate before new system (%)	10.3	3.6	17.8	5.1	4.6	11.2	7.7
Rate after new system (%)	9.3	4.1	15.2	5.0	3.3	10.3	8.1

(i) Find a 90% confidence interval for the true difference between the two rates of unemployment.

(ii) Does your confidence interval provide evidence that there is a difference in the rates after the new system is introduced?

(iii) Do you think that the assumptions required to construct the confidence interval are justified here?

14. Two timekeepers at an athletics track are being compared. They each time the nine sprints one afternoon.

(i) Find a 99% confidence interval for the true difference between the times recorded by the two timers. The times they record are listed below.

Table 14.12

Race	1	2	3	4	5	6	7	8	9
Timer 1	9.65	10.01	9.62	21.90	20.70	20.90	42.30	43.91	43.96
Timer 2	9.66	9.99	9.44	22.00	20.82	20.58	42.39	44.27	44.22

(ii) Do you think that the two timers are equivalent on average?

(iii) Are the assumptions appropriate for a confidence interval based on the t-test justified in this case?

15. Fourteen marked rats were timed twice as they ran through a maze. In one condition, they had just been fed; in the other they were hungry.

(i) Find a 95% confidence interval for the true difference between the rats' times when they are fed and when they are hungry. The data below give the rats' times in each condition.

Table 14.13

Rat	A	B	C	D	E	F	G	H	I	J	K	L	M	N
Fed time (seconds)	30	31	25	23	50	26	14	27	31	39	38	39	44	30
Hungry time (seconds)	29	18	14	27	37	34	15	22	29	18	20	10	30	32

(ii) Do you think that the assumptions required to construct the confidence interval are justified here?

(iii) Half of the rats were made to run the maze first when hungry and half ran it first when fed. Why did the experimenter do this?

16. An experiment to determine the acceleration due to gravity, g m s^{-2}, involves measuring the time, T seconds, taken by a pendulum of length 1m to perform complete swings. T is regarded as a random variable. Thirty measurements are made on T, and they are summarised by

$$\Sigma t = 59.8, \quad \Sigma t^2 = 119.7.$$

Construct a two-sided 98% confidence interval for μ, the mean value of T. Determine the corresponding range of values of g, using the formula

$$g = \frac{4\pi^2}{\mu^2}$$

This result for g is not precise enough, so a longer series of measurements of T is made. Assuming that the sample mean and standard deviation remain about the same, how many measurements will be required in total to halve the width of the 98% confidence interval for μ? What will be the corresponding effect on the range of values for g? [MEI]

3 Confidence intervals for the difference between two means

This section is about two unpaired random samples, one from one population and the other from a second population that may or may not be the same as the first population. A confidence interval for the difference between the population means is required.

Three cases are considered. These situations will be investigated further in Chapter 15.

Case 1: Small samples from Normal populations with known variances.

$X_1 \sim N(\mu_1, \sigma_1^2)$ and $X_2 \sim N(\mu_2, \sigma_2^2)$ with σ_1^2 and σ_2^2 known.

Take a random sample of size n_1 from the first population and a random sample of size n_2 from the second population.

Because the population distributions are known to be Normal and the population variances are known, the distribution of the sample mean for each will also be Normal, irrespective of the sample sizes.

$$\bar{X}_1 \sim N\left(\mu_1, \frac{\sigma_1^2}{n_1}\right) \text{ and } \bar{X}_2 \sim N\left(\mu_2, \frac{\sigma_2^2}{n_2}\right) \quad \text{So } \bar{X}_1 - \bar{X}_2 \sim N\left(\mu_1 - \mu_2, \frac{\sigma_1^2}{n_1} + \frac{\sigma_2^2}{n_2}\right)$$

The two-sided confidence interval for $\mu_1 - \mu_2$ is $\bar{x}_1 - \bar{x}_2 \pm k\sqrt{\frac{\sigma_1^2}{n_1} + \frac{\sigma_2^2}{n_2}}$

where the value of k for any confidence level can be found using Normal distribution tables.

Example 14.7

A machine fills bottles of flavoured water. The random variable X represents the amount (in ml) put into each bottle. It is known that X can be modelled using a Normal distribution with variance 5.

Random samples are taken from the production on Mondays and Fridays.

| Monday | 500 | 508 | 501 | 497 | |
| Friday | 495 | 498 | 502 | 497 | 495 |

The mean amount per bottle for the whole production on Monday is μ_1 and the mean amount per bottle for the whole production on Friday is μ_2.

(i) Use these samples to construct a 95% confidence interval for $\mu_1 - \mu_2$.

(ii) What does the confidence interval suggest?

S2 AL — Confidence intervals for the difference between two means

Solution

(i)

$\bar{X}_1 \sim N\left(\mu_1, \frac{5}{4}\right)$ and $\bar{X}_2 \sim N\left(\mu_2, \frac{5}{5}\right)$ so $\bar{X}_1 - \bar{X}_2 \sim N(\mu_1 - \mu_2, 2.25)$

$\frac{5}{4} + \frac{5}{5} = 2.25$

$\bar{x}_1 = 501.5$ and $\bar{x}_2 = 497.4$ so $\bar{x}_1 - \bar{x}_2 = 4.1$

The two-sided 95% confidence intervals for $\mu_1 - \mu_2$ is given by

$$\bar{x}_1 - \bar{x}_2 \pm k\sqrt{\frac{\sigma_1^2}{n_1} + \frac{\sigma_2^2}{n_2}}$$

$= 4.1 \pm 1.96\sqrt{2.25}$

The 95% confidence limits for $\mu_1 - \mu_2$ are

$$4.1 \pm 1.96\sqrt{2.25}$$
$$= 4.1 \pm 2.94$$
$$= (1.16, 7.04)$$

The difference between the means can be expected to fall in this interval for 95% of such samples.

(ii) The value 0 is outside the confidence interval so the mean amount on Monday would seem to be significantly greater than on Friday.

Case 2: Large samples from populations with unknown variances.

When the sample sizes are large the Central Limit Theorem can be used

$$\bar{X}_1 \sim N\left(\mu_1, \frac{\sigma_1^2}{n_1}\right) \text{ and } \bar{X}_2 \sim N\left(\mu_2, \frac{\sigma_2^2}{n_2}\right) \text{ with } \sigma_1^2 \text{ and } \sigma_2^2 \text{ unknown.}$$

so $\bar{X}_1 - \bar{X}_2 \sim N\left(\mu_1 - \mu_2, \frac{\sigma_1^2}{n_1} + \frac{\sigma_2^2}{n_2}\right)$

The variances are unknown and are estimated by s_1^2 and s_2^2, where

$$s_1^2 = \frac{1}{n_1 - 1}\sum_{1}^{n_1}(x_i - \bar{x}_1)^2 \text{ and } s_2^2 = \frac{1}{n_2 - 1}\sum_{1}^{n_2}(x_i - \bar{x}_2)^2.$$

The two-sided confidence interval for $\mu_1 - \mu_2$ is $\bar{x}_1 - \bar{x}_2 \pm k\sqrt{\frac{s_1^2}{n_1} + \frac{s_2^2}{n_2}}$.

where the value of k for any confidence level can be found using Normal distribution tables.

> **Note**
> The unbiased estimator of a population variance is $s^2 = \frac{1}{n-1}\sum_{1}^{n}(x_i - \bar{x})^2$.
> This result is proved in Chapter 16.

Example 14.8

Two different teaching methods are compared by testing a large number of students who have been taught using each method. The students all take the same test and their test scores are recorded.

	Sample size	Sample mean	Sample standard deviation
Method 1	50	65.3	14.7
Method 2	20	68.6	28.5

Use these samples to construct a 95% confidence interval for the differences in the mean scores of students taught by the two methods.

Solution

Teaching method 1

Standard deviation $= 14.7 \Rightarrow$ variance $= 14.7^2$

$$\frac{1}{50}\sum_{1}^{50}(x_i - 65.3)^2 = 14.7^2$$

$$\Rightarrow \sum_{1}^{50}(x_i - 65.3)^2 = 50 \times 14.7^2$$

$$\Rightarrow s_1^2 = \frac{50}{49} \times 14.7^2$$

$$s_1^2 = 220.5$$

Teaching method 2

Standard deviation $= 28.5 \Rightarrow$ variance $= 28.5^2$

Using the same argument as above, $s_2^2 = \frac{20}{19} \times 28.5^2 = 855$

$$\bar{X}_1 - \bar{X}_2 \sim N\left(\mu_1 - \mu_2, \frac{220.5}{50} + \frac{855}{20}\right) = N(\mu_1 - \mu_2, 47.16)$$

The two-sided confidence limits for $\mu_1 - \mu_2$ are given by

$$\bar{x}_1 - \bar{x}_2 \pm k\sqrt{\frac{\sigma_1^2}{n_1} + \frac{\sigma_2^2}{n_2}}$$

$\bar{x}_1 - \bar{x}_2 = -3.5$, $k = 1.96$

So the 95% CI for $\mu_1 - \mu_2$ is $-3.3 \pm 1.96\sqrt{47.16}$

$= -3.3 \pm 13.46$

$= (-16.76, 10.16)$

Case 3: Samples from Normal populations with equal but unknown variances.

$X_1 \sim N(\mu_1, \sigma^2)$ and $X_2 \sim N(\mu_2, \sigma^2)$ with σ^2 unknown.

A random sample of size n_1 is taken from the first population and a random sample of size n_2 from the second population.

When the Central Limit Theorem is used with small samples and the population variance is estimated a t-distribution is needed.

$$\bar{X}_1 \sim N\left(\mu_1, \frac{\sigma^2}{n_1}\right) \text{ approximately and } \bar{X}_2 \sim N\left(\mu_2, \frac{\sigma^2}{n_2}\right) \text{ approximately}$$

so $\bar{X}_1 - \bar{X}_2 \sim N\left(\mu_1 - \mu_2, \sigma^2\left(\frac{1}{n_1} + \frac{1}{n_2}\right)\right)$ approximately.

The variance is estimated by the pooled estimator

$$s^2 = \frac{(n_1 - 1)s_1^2 + (n_2 - 1)s_2^2}{(n_1 + n_2 - 2)}$$

where $s_1^2 = \frac{1}{(n_1 - 1)}\sum_{1}^{n_1}(x_i - \bar{x}_1)^2$ and $s_2^2 = \frac{1}{n_2 - 1}\sum_{1}^{n_2}(x_i - \bar{x}_1)$

So $\sigma^2\left(\frac{1}{n_1} + \frac{1}{n_2}\right)$ is estimated by $s^2\left(\frac{1}{n_1} + \frac{1}{n_2}\right)$ and a t-distribution with $n_1 + n_2 - 2$ degrees of freedom is used.

The two-sided confidence interval for $\mu_1 - \mu_2$ is $\bar{x}_1 - \bar{x}_2 \pm ks\sqrt{\left(\frac{1}{n_1} + \frac{1}{n_2}\right)}$

where the value of k for any confidence level can be found using a t-distribution with $n_1 + n_2 - 2$ degrees of freedom.

Confidence intervals for the difference between two means

Example 14.9

A machine serves tennis balls. The machine is tested by using it to serve tennis balls and measuring how far they travel. The random variable X represents the distance, in metres, that each ball travels before landing. It is known that X can be modelled using a Normal distribution.

The machine is adjusted so the mean is reduced without changing the variance. A Normal model is still appropriate.

Table 14.14

	Sample size	Sample mean	Sample variance
Before	10	22.2	9.6
After	5	15.7	9.2

The population mean before the adjustment is μ_1 and the population mean after the adjustment is μ_2.

Use these samples to construct a 95% confidence interval for $\mu_1 - \mu_2$.

Solution

Before
$$n_1 = 10, \bar{x}_1 = 22.2$$

$$\frac{1}{10}\sum_{1}^{10}(x_i - 22.2)^2 = 9.6$$

$$\Rightarrow \sum_{1}^{10}(x_i - 22.2)^2 = 96$$

$$\Rightarrow s_1^2 = \frac{96}{10-1} = 10.666\ldots$$

After
$$n_2 = 5, \bar{x}_1 = 15.7$$

$$\frac{1}{5}\sum_{1}^{5}(x_i - 15.7)^2 = 9.2$$

$$\Rightarrow s_2^2 = \frac{5}{4} \times 9.2 = 11.5$$

A t-distribution is used.

The two-sided confidence interval for $\mu_1 - \mu_2$ is $\bar{x}_1 - \bar{x}_2 \pm ks\sqrt{\frac{1}{n_1} + \frac{1}{n_2}}$

Pooled estimate for variance $s^2 = \dfrac{(n_1 - 1)s_1^2 + (n_2 - 1)s_2^2}{n_1 + n_2 - 2}$

$$= \frac{9 \times 10.666\ldots + 4 \times 11.5}{10 + 5 - 2}$$

$$= 10.923$$

Estimated standard deviation, $s = \sqrt{10.923} = 3.305$

Degrees of freedom $v = n_1 + n_2 - 2 = 13$

For a 95% confidence interval, $k = 2.160$

So the confidence interval is given by

$$(22.2 - 15.7) \pm 2.160 \times 3.305 \times \sqrt{\frac{1}{10} + \frac{1}{5}}$$

$$= 6.5 \pm 3.91$$

$$= (2.59, 10.41).$$

KEY POINTS

1. When the population standard deviation, σ, is known and the distribution is Normal, confidence intervals for μ are found using the Normal distribution.

2. Two-sided confidence intervals based on the Normal distribution are given by

    $$\bar{x} - k\frac{\sigma}{\sqrt{n}} \text{ to } \bar{x} + k\frac{\sigma}{\sqrt{n}}$$ where n is the sample size.

3. The value of k for any confidence level can be found using Normal distribution tables. Commonly used values are given in Table 14.15.

    Table 14.15

Confidence level	k
90%	1.645
95%	1.96
99%	2.58

4. For large values of n, the requirement that the distribution is Normal may be relaxed.

5. Confidence intervals for paired samples are formed in the same way but the variable is now the difference between the paired values.

6. When the population standard deviation, σ, is not known and is estimated as being the sample standard deviation, s, and the distribution is Normal, confidence intervals for μ are found using the t-distribution.

7. Two-sided confidence intervals for μ based on the t-distribution are given by

    $$\bar{x} - k\frac{s}{\sqrt{n}} \text{ to } \bar{x} + k\frac{s}{\sqrt{n}}.$$

8. The value of k for any confidence level can be found using t-distribution tables for appropriate degrees of freedom.

9. The value of s can be found using the formula

    $$s^2 = \frac{S_{xx}}{n-1} = \sum_i \frac{(x_i - \bar{x})^2}{(n-1)}.$$

10. As with confidence intervals based on the Normal distribution, confidence intervals for paired samples are formed in the same way but the variable is now the difference between the paired values.

11. The goodness of fit of a Normal distribution can be checked using a Normal probability plot.

12. Small samples from Normal populations with known variances.

    The two-sided confidence interval for $\mu_1 - \mu_2$ is $\bar{x}_1 - \bar{x}_2 \pm k\sqrt{\frac{\sigma_1^2}{n_1} + \frac{\sigma_2^2}{n_2}}$

    where the value of k for any confidence level can be found using the Normal distribution.

13. Large samples from populations with unknown variances.

    The two-sided confidence interval for $\mu_1 - \mu_2$ is $\bar{x}_1 - \bar{x}_2 \pm k\sqrt{\frac{s_1^2}{n_1} + \frac{s_2^2}{n_2}}$

    where $s_1^2 = \frac{1}{n_1-1}\sum_1^{n_1}(x_i - \bar{x}_1)^2$ and $s_2^2 = \frac{1}{n_2-1}\sum_1^{n_2}(x_i - \bar{x}_2)^2$

    and the value of k for any confidence level can be found using the Normal distribution.

S2 AL Confidence intervals for the difference between two means

14 Samples from Normal populations with equal but unknown variances.

The two-sided confidence interval for $\mu_1 - \mu_2$ is $\bar{x}_1 - \bar{x}_2 \pm ks\sqrt{\left(\dfrac{1}{n_1} + \dfrac{1}{n_2}\right)}$

where $s^2 = \dfrac{(n_1 - 1)s_1^2 + (n_2 - 1)s_2^2}{n_1 + n_2 - 2}$ and the value of k for any confidence level can be found using a t-distribution with $n_1 + n_2 - 2$ degrees of freedom.

LEARNING OUTCOMES

When you have completed this chapter you should:

- know the meaning of the term confidence interval for a parameter and associated language
- understand the factors which affect the width of a confidence interval
- be able to construct and interpret a confidence interval for a single population mean using the Normal or the t-distributions and know when it is appropriate to do so
- know when samples from two populations should be considered as paired
- be able to construct and interpret a confidence interval for the difference in means of two paired populations using a paired sample and a Normal or t-distribution and know when it is appropriate to do so
- know when samples from two populations should be considered as unpaired
- be able to construct and interpret a confidence interval for the difference in the means of two populations using unpaired samples and a Normal or t-distribution and know when it is appropriate to do so
- interpret confidence intervals given by software
- use a confidence interval for a population parameter to interpret a hypothesised value of that parameter.

15 Hypothesis tests on the mean

Every experiment may be said to exist only to give the facts a chance of disproving the null hypothesis.

R. A. Fisher

Herring gulls getting heavier?

In the last few years, the numbers of herring gulls living on roofs in towns has been increasing. It is known that in the past, before the advent of scavenging, the mean weight of adult herring gulls of a particular species was 1015 grams, with standard deviation 110. Many of the gulls scavenge food discarded by patrons of takeaway restaurants. A reporter for a local paper says that as a result of the 'easy pickings' from these restaurants, the gulls in the town where she lives are heavier now than they were in the past, before they discovered free takeaway food. She enlists the help of an ornithologist to capture and weigh ten adult females of that species in the town, before releasing them so that they can continue to take advantage of the free food.

The weights of the ten herring gulls measured in grams are

1050   1145   1205   985   1015   1100   890   940   1290   1080

1 Testing the mean and variance of a single sample

You can carry out a hypothesis test based on the Normal distribution to investigate whether it is likely that the scavenging seagulls are heavier than seagulls used to be. Two requirements for this test to be valid are that:

- the population from which the sample is drawn has a Normal distribution (as the sample size is too small for the Central Limit Theorem to apply)
- the sample is representative of the population; this is usually ensured by having a random sample.

> **Note**
> The null hypothesis for a test of the mean always has the form $H_0: \mu = a$, where a is the 'original' value of the population mean – the value of the mean if nothing has changed.

> Since the weight of a herring gulls is a naturally occurring variable, it is reasonable to assume its distribution is Normal.

> You have no information about how the sample was collected but it is reasonable to assume that it was random.

To use the Normal distribution, you also have to know the standard deviation of the parent population. You do not know this for the scavenging seagulls, but you do know that in the past it was 110 g. You have to assume that it is still 110 g.

To carry out the test, you need to decide the significance level; in this case 5% is chosen. You also need to state the null and alternative hypotheses; the alternative hypothesis will tell you whether it is a one- or two-tailed test. In this case,

Null hypothesis: $H_0: \mu = 1015$
Alternative hypothesis: $H_1: \mu > 1015$
Significance level: 5%
Test: one-tailed.

> **Note**
> You are investigating whether the weights of herring gulls have increased. The test is therefore one-tailed because the alternative hypothesis is $\mu > 1015$ and so is one-sided.
>
> If you had instead been investigating whether the weights of herring gulls had changed, then it would have been a two-tailed test and the alternative hypothesis would have been $\mu \neq 1015$.

There are three slightly different methods used to carry out a hypothesis test based on the Normal distribution. They all use the fact that the distribution of the sample means is given by

$$\overline{X} \sim N\left(\mu, \frac{\sigma^2}{n}\right)$$

for large values of the sample size, n.

Start by calculating the sample mean.

$$\left(\frac{1050 + 1145 + 1205 + 985 + 1015 + 1100 + 890 + 940 + 1290 + 1080}{10}\right)$$

$$= \frac{10700}{10} = 1070$$

Method 1: Using critical regions

Since the distribution of sample means is $N\left(\mu, \frac{\sigma^2}{n}\right)$, critical values for a test on the sample mean are given by

$$\mu \pm k \times \frac{\sigma}{\sqrt{n}}.$$

In this case, if H_0 is true: $\mu = 1015$; $\sigma = 110$; $n = 10$.
The test is one-tailed, for $\mu > 1015$, so only the right-hand tail applies.
k is the critical value for the standardised value, z.
Using the inverse Normal function on a calculator, the value of k is 1.645.

> **Note**
> For a one-tailed test rather than the symbol $\pm$ in the critical values, you either have $+$ or $-$ according to whether you are interested in the upper or lower tail.

> Make sure that you know how to get this value from your calculator
>
> For some calculators, you may need to calculate $1 - 0.05 = 0.95$.

For a one-tailed test at the 5% significance level, the critical value is $1015 + 1.645 \times \dfrac{110}{\sqrt{10}} = 1072.2$, as shown in the diagram below.

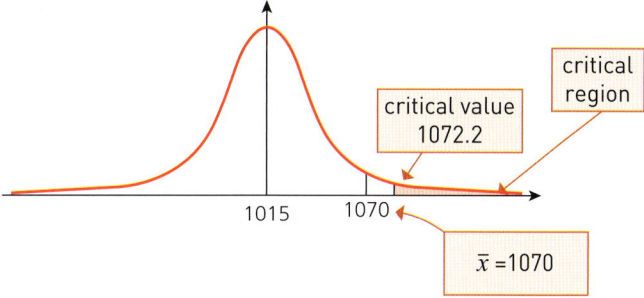

Figure 15.1

However, the sample mean $\bar{x} = 1070$, and $1070 < 1072.2$.

Therefore the sample mean lies outside the critical region and so there is insufficient evidence to reject the null hypothesis.

There is insufficient evidence to suggest that the mean weight of scavenging gulls is greater than before the advent of scavenging.

This does not mean that the null hypothesis is definitely true but that there is not enough evidence to show that it is false.

Method 2: Using probabilities

The distribution of sample means, $\bar{x}$, is $N\left(\mu, \dfrac{\sigma^2}{n}\right)$.

According to the null hypothesis, $\mu = 1015$ and it is known that $\sigma = 110$ and $n = 10$.

So this distribution is $N\left(1015, \dfrac{110^2}{10}\right)$, as shown in the following diagram.

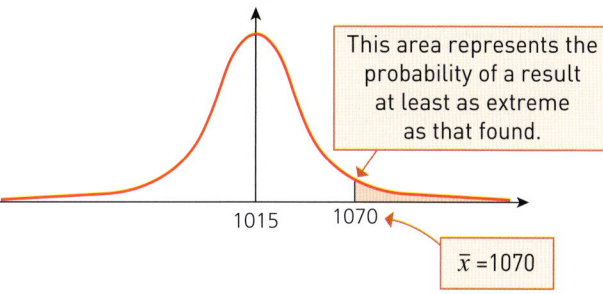

Figure 15.2

The probability of the mean, $\bar{X}$, of a randomly chosen sample being greater than the value found, i.e. 1070, is given by

$$P(\bar{X} \geqslant 1070) = 1 - \Phi\left(\dfrac{1070 - 1015}{\dfrac{110}{\sqrt{10}}}\right)$$

$$= 1 - \Phi(1.581)$$

$$= 1 - 0.9431$$

$$= 0.0569.$$

Notes

1 A hypothesis test should be formulated before the data are collected and not after. In this case, the reporter thought that the gulls might be larger and so data was collected to test this hypothesis. If sample data had led the reporter to form a hypothesis, then a new set of data would have been required in order to carry out a suitable test.

2 If the data were not collected properly, any test carried out on them may be worthless. So if the seagulls did not form a random sample from the population, then the test would be invalid.

S2 AL — Testing the mean and variance of a single sample

> 0.05 is the significance level.

Since $0.0569 > 0.05$, there is insufficient evidence to reject the null hypothesis.

There is insufficient evidence to suggest that the mean weight of scavenging gulls is greater than before the advent of scavenging.

Method 3: Using the test statistics (or critical ratios)

The **test statistic** (or **critical ratio**) is given by $z = \left(\dfrac{\bar{x} - \mu}{\frac{\sigma}{\sqrt{n}}} \right)$

In this case, $z = \left(\dfrac{1070 - 1015}{\frac{110}{\sqrt{10}}} \right) = 1.581$.

This is now compared with the critical value for z, in this case, $z = 1.645$.

Since $1.581 < 1.645$, H_0 is not rejected. There is insufficient evidence to reject the null hypothesis.

There is insufficient evidence to suggest that the mean weight of scavenging gulls is greater than before the advent of scavenging.

> This is the value of k in Method 1.

> **Note**
> The calculated probability is often called the *p*-value.
> If the *p*-value is smaller than the significance level, then you reject H_0 and accept H_1. Otherwise, as in this case, you do not reject H_0.

⚠ You need to be careful when interpreting the *p*-value. In many tests, like this one, rejecting the null hypothesis, H_0, means that you have found an interesting result, expressed as the alternative hypothesis, H_1; it is sometimes summarised as 'The test is significant'. So a *p*-value that is less than the significance level might be seen as desirable.

By contrast, in goodness of fit tests, the interesting result, that you have found a suitable distribution to use as a model, is given by the null hypothesis. So, in such tests, a *p*-value that is higher than the significance level can be seen as desirable.

💻 USING ICT

Statistical software

You can use statistical software to do all of the calculations for this test. In order for the software to process the test, you need to input the following information:

- the population mean if H_0 is true
- the form of the alternative hypothesis ($<$, $>$ or $\neq$)
- the population standard deviation
- the sample mean
- the sample size.

Figure 15.3 shows the output for the test above.

	A	B	C
1	Z Test of a mean		
2			
3	Null hypothesis	1015	
4	Alternative hypothesis	>	
5	Mean	1070	
6	σ	110	
7	n	10	
8			
9	Mean	1070	
10	σ	110	
11	SE	34.7851	
12	n	10	
13	z	1.5811	
14	p	0.0569	
15			

Rows 3–7: INPUTS. Rows 9–14: OUTPUTS.

The standard error is $\dfrac{\sigma}{\sqrt{n}}$.

Standard error → row 11 SE
Test statistic → row 13 z
p-value → row 14 p

Figure 15.3

You can see that the output from the software gives both the *p*-value of 0.0569 and the test statistic of 1.5811.

Large samples

If you have a large sample, then you do not need to know the population variance in order to carry out a hypothesis test for the mean. You can use an unbiased estimate of the population variance, and for a large sample this should be good enough. Nor do you need to know that the parent population has a Normal distribution; if you have a large sample, then the Central Limit Theorem implies that the distribution of the sample mean will be approximately Normal even though the parent population distribution is not. Therefore, if your sample is large, you can carry out a test for the mean based on the Normal distribution when you do not know either the distribution or the standard deviation of the parent population. The example below illustrates a situation where you know neither of these.

> This is just the same as with confidence intervals.

Example 15.1

A psychologist is investigating reaction times. From many previous experiments, she knows that the mean reaction time for an adult to respond to a stimulus provided by a computer is 0.650 seconds. She thinks that the reaction time of 12-year-old children will be quicker. She selects a sample of 60 of these children and measures their reaction times, x seconds, to the stimulus. The results are summarised by

$$\Sigma x = 38.24 \qquad \Sigma x^2 = 25.18$$

(i) State any conditions necessary for a hypothesis test to investigate whether the reaction times of 12-year olds are lower.

(ii) Carry out the test described in part (i) at the 10% significance level.

Solution

(i) The sample of children must be random.

(ii) You can use any of the three methods given above. *Method 3* is used here.

First, you must state the null and alternative hypotheses.

Null hypothesis: $H_0: \mu = 0.650$
Alternative hypothesis: $H_1: \mu < 0.650$
Significance level: 10%
Test: one-tailed

> You are investigating whether the reaction times of children are quicker. The test is therefore one-tailed.

To estimate the population standard deviation σ, find the sample standard deviation, s. The general form of this is

$$s = \sqrt{\text{Variance}} = \sqrt{\frac{S_{xx}}{n-1}}, \text{ where } S_{xx} = \Sigma x^2 - n\bar{x}^2$$

$$\bar{x} = \frac{\Sigma x}{n} = \frac{38.24}{60} = 0.6373\ldots$$

and so $S_{xx} = \Sigma x^2 - n\bar{x}^2 = 25.18 - 60 \times 0.6373\ldots^2 = 0.80837\ldots$

$$s = \sqrt{\frac{0.80837\ldots}{59}} = 0.117\ldots$$

S2 AL — Testing the mean and variance of a single sample

> **Note**
>
> This is a large sample and so you do not need to know the standard deviation of the parent population. Instead, you can use the sample standard deviation as an estimate for it.
>
> Also, you do not need to know anything about the distribution of the parent population since the Central Limit Theorem tells you that the distribution of the sample mean will be approximately Normal.

The **test statistic** is given by

$$z = \frac{\bar{x} - \mu}{\frac{\sigma}{\sqrt{n}}} = \frac{0.6373\ldots - 0.650}{\frac{0.117\ldots}{\sqrt{60}}} = -0.838\ldots$$

This is now compared with the critical value for z, in this case $z = -1.282$.

Since $-0.838 > -1.282$, H_0 is not rejected. There is insufficient evidence to reject the null hypothesis.

There is insufficient evidence to suggest that the 12-year-old children have quicker reaction times to the stimulus compared to adults.

Small samples

In Chapter 13, you met the t-distribution. This is the distribution of a sample mean when the parent population is Normally distributed but the standard deviation of the parent population is unknown and has to be estimated using the sample standard deviation, s. In addition to finding a confidence interval, you can also carry out a hypothesis test based on the t-distribution.

Example 15.2

> **ACTIVITY 15.1**
>
> Use Methods 1 and 2 to carry out the test. You should of course come to the same conclusion.

Tests are being carried out on a new drug designed to relieve the symptoms of the common cold. One of the tests is to investigate whether the drug has any effect on the number of hours that people sleep; this may be assumed to have a Normal distribution.

The drug is given in tablet form one evening to a random sample of 16 people who have colds. The number of hours they sleep may be assumed to be Normally distributed and is recorded as follows.

| 8.1 | 6.7 | 3.3 | 7.2 | 8.1 | 9.2 | 6.0 | 7.4 |
| 6.4 | 6.9 | 7.0 | 7.8 | 6.7 | 7.2 | 7.6 | 7.9 |

There is also a large control group of people who have colds but are not given the drug. The mean number of hours they sleep is 6.6.

(i) Use these data to set up a 95% confidence interval for the mean length of time somebody with a cold sleeps after taking the tablet.

(ii) Carry out a test, at the 1% significance level, of the hypothesis that the new drug has an effect on the number of hours a person sleeps.

> **Note**
>
> The critical value in this case is negative since you are carrying out a lower tail test. You can either give the comparison as shown above $-0.838 > -1.282$ or use the absolute values (moduli) in which case the comparison would be $|-0.838| < |-1.282|$. You should **not** compare -0.838 with $+1.282$.

Solution

(i) For the given data, $n = 16$, $\upsilon = 16 - 1 = 15$, $\bar{x} = 7.094$, $s = 1.276$.

For a 95% confidence interval, with $\upsilon = 15$, $k = 2.131$ (from tables).

The confidence limits are given by

$$\bar{x} \pm k \frac{s}{\sqrt{n}} = 7.094 \pm 2.131 \times \frac{1.276}{\sqrt{16}},$$

So the 95% confidence interval for μ is 6.41 to 7.77.

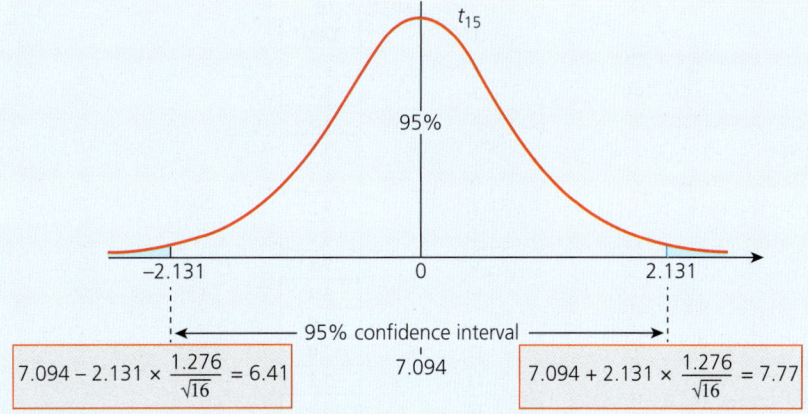

Figure 15.4

(ii) H_0: there is no change in the mean number of hours sleep. $\mu = 6.6$
H_1: there is a change in the mean number of hours sleep. $\mu \neq 6.6$
Two-tailed test at the 1% significance level.

For this sample, $n = 16$, $\upsilon = 16 - 1 = 15$, $\bar{x} = 7.094$, $s = 1.276$.

The critical value for t, for $\upsilon = 15$, at the 1% significance level, is found from tables to be 2.947.

The test statistic $t = \left(\dfrac{\bar{x} - \mu}{\frac{s}{\sqrt{n}}} \right) = \left(\dfrac{7.094 - 6.6}{\frac{1.276}{\sqrt{16}}} \right) = 1.55$.

This is to be compared with 2.947, the critical value, for the 1% significance level.

Since $1.55 < 2.947$, there is no reason at the 1% significance level to reject the null hypothesis.

There is insufficient evidence to suggest that the mean number of hours sleep is different when people take the drug.

USING ICT

Using a spreadsheet

You can use a spreadsheet (see Figure 15.5 on the following page) to do all of the calculations for this test using the following steps:

1. Enter the data (in this case into cells B2 to B17)
2. Use the spreadsheet functions provided by your spreadsheet, for example = AVERAGE and = STDEV to find the mean and sample standard deviation.
3. Calculate the t value using the formula = (B18-6.6)/(B19/SQRT(16))
4. Use the spreadsheet function provided by your spreadsheet, for example = T.INV.2T(0.01,15) to calculate the critical value.
5. You can also find the p-value using the spreadsheet function provided by your spreadsheet, for example = T.DIST.2T(1.5482,15)

S2 AL Testing the mean and variance of a single sample

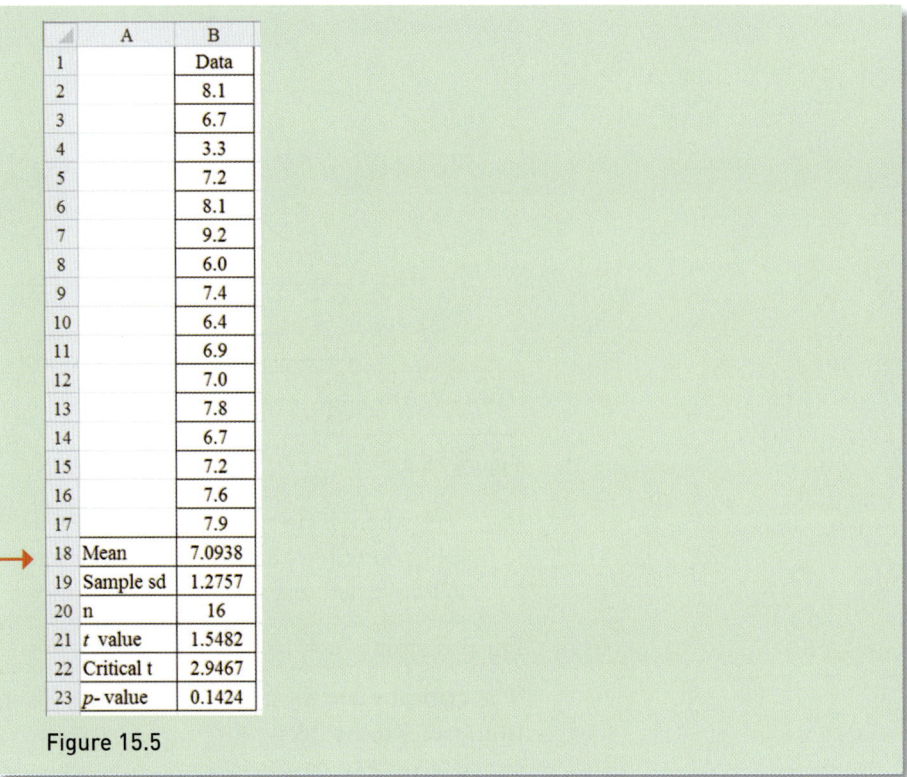

In this case, the values of $\bar{x}$ and s are given in cells B18 and B19.

Figure 15.5

Exercise 15.1

① A hypothesis test is to be carried out at the 5% level on a Normally distributed population with standard deviation 2.7. The null hypothesis is $H_0: \mu = 25$, and the alternative hypothesis is $H_1: \mu > 25$. A random sample of size 9 is selected and the sample mean is 26.71.

(i) Find the standard error of the mean.

(ii) Calculate the test statistic.

(iii) Write down the critical value.

(iv) Explain whether the null hypothesis should be rejected.

② Bags of flour are supposed to weigh at least 1.5 kg. Sampling is carried out at the factory where the bags are filled to check that the bags are not underweight. On a particular day, the sample mean of a random sample of 40 bags is 1.4972 kg with sample standard deviation 0.009 kg. A hypothesis test is carried out at the 1% significance level to check whether the bags are satisfactory.

(i) Explain why you do not need to know the distribution of the parent population in order to carry out the test.

(ii) Write down suitable null and alternative hypotheses.

(iii) Find the standard error of the mean.

(iv) Calculate the test statistic.

(v) Write down the critical value.

(vi) Explain whether the null hypothesis should be rejected.

③ The spreadsheet below shows the data and the output for a *t*-test to investigate the hypotheses

$H_0: \mu = 180$ and $H_1: \mu \neq 180$.

(i) State whether the test is one-tailed or two-tailed and use the 'Critical *t*' in cell B13 to find the significance level of the test.

(ii) State the result of the test relating it to the content of cells B12 and B13.

(iii) State the result of the test, this time relating it to the content of cell B14.

	A	B
1		Data
2		172.24
3		198.79
4		167.81
5		192.23
6		183.81
7		178.49
8		183.27
9	Mean	182.37
10	Sample sd	10.80
11	n	7
12	*t* value	0.8796
13	Critical *t*	2.3646
14	*p*-value	0.4082

Figure 15.6

④ Some biologists were studying a large group of wading birds. A random sample of 36 were measured and the wing length, x mm, of each wading bird was recorded. The results are summarised as follows.

$$\Sigma x = 6046, \quad \Sigma x^2 = 1\,016\,338.$$

(i) Calculate unbiased estimates of the mean and the variance of the wing lengths of these birds.

Given that the standard deviation of the wing lengths of this particular type of bird is actually 5.1 mm,

(ii) find a 99 % confidence interval for the mean wing length of the birds from this group. [Edexcel 6691 Q1 June 2008]

⑤ A report states that employees spend, on average, 80 minutes every working day on personal use of the internet. A company takes a random sample of 100 employees and finds their mean personal internet use is 83 minutes with a standard deviation of 15 minutes. The company's managing director claims that his employees spend more time on average on personal use of the internet than the report states.

Test, at the 5% level of significance, the managing director's claim. State your hypotheses clearly. [Edexcel 6691 Q1 June 2010]

⑥ A council is investigating the weight of rubbish in domestic dustbins. It has recently started a recycling initiative and wishes to determine whether there has been a reduction in the weight of rubbish. Before the initiative, the mean weight was 33.5 kg. A random sample of 60 domestic dustbins is selected and the weight x kg of rubbish in each bin is recorded. The results are summarised as follows.

$$\Sigma x = 1801.2 \quad \Sigma x^2 = 56963$$

Carry out a hypothesis test at the 1% level to investigate whether the mean weight has been reduced.

7. Over a long period, it has been found that the time that it takes for an underground train to complete a particular journey is 18.6 minutes. New signalling equipment is introduced which might affect the journey time. The times m minutes taken for a random sample of 30 journeys after the change are summarised as follows.

$$\Sigma m = 548.0 \qquad \Sigma m^2 = 10055$$

   (i) Carry out a test at the 10% significance level using the Normal distribution to investigate whether there has been any change in journey time since the introduction of the new signalling equipment. You should use an estimate of the population standard deviation.

   (ii) It is suggested that, because the sample size is only 30, the estimate of population standard deviation may not be very reliable. Carry out an alternative test and state an assumption necessary for this test.

8. Freeze drying is often used in the production of coffee. For best results, the drying rate (measured in suitable units) should be 72.0. It is thought that the drying rate in a particular batch may be higher than this. In order to test this, a sample of 12 observations was selected and the drying rates were as follows.

   75   73   81   88   75   79   69   91   82   76   73   72

   (i) Carry out a test at the 0.5% significance level to determine whether the drying rates in this batch are greater than the ideal. State clearly your null and alternative hypotheses and your conclusion.

   (ii) What assumptions are required for your answer to part (i) to be valid?

9. A fisherman claims that pollack are not as big as they used to be. 'They used to average three quarters of a kilogram each', he says. When challenged to prove his point, he catches 20 pollack from the same shoal. Their masses (in kg) are as follows.

   0.65   0.68   0.77   0.71   0.67   0.75   0.69   0.72   0.73   0.69

   0.70   0.70   0.72   0.76   0.73   0.78   0.75   0.69   0.70   0.71

   (i) State the null and alternative hypotheses for a test to investigate whether the fisherman's claim is true.

   (ii) Carry out the test at the 5% significance level and state the conclusion.

   (iii) State any assumptions underlying your procedure and comment on their validity.

10. In the game of bridge, a standard pack of 52 playing cards is dealt into four hands of 13 cards each. Players usually assess the value of their hands by counting 4 points for an ace, 3 for a King, 2 for a Queen, 1 for a Jack and nothing for any other card. The total points available from the four suits are $(4 + 3 + 2 + 1) \times 4 = 40$. So the mean number of points per hand is $\frac{40}{4} = 10$. Helene claims that she never gets good cards. One day, she is challenged to prove this and agrees to keep a record of the number of points she gets on each hand next time she plays, with the following results.

5	16	7	1	11	2	8	9	14	12
21	10	0	7	12	7	6	8	13	4

(i) What assumption underlies the use of the *t*-test in this situation? To what extent do you think the assumption is justified?

(ii) State null and alternative hypotheses relating to Helene's claim.

(iii) Carry out the test at the 5% significance level and comment on the result.

(This question is set in memory of a lady called Helene who claimed that bridge hands had not been the same since the Second World War.)

⑪ (i) Explain what you understand by the Central Limit Theorem.

A garage services hire cars on behalf of a hire company. The garage knows that the lifetime of the brake pads has a standard deviation of 5000 miles. The garage records the lifetimes, x miles, of the brake pads it has replaced. The garage takes a random sample of 100 brake pads and finds that $\Sigma x = 1\,740\,000$.

(ii) Find a 95% confidence interval for the mean lifetime of a brake pad.

(iii) Explain the relevance of the Central Limit Theorem in part (ii).

Brake pads are made to be changed every 20 000 miles on average. The hire car company complain that the garage is changing the brake pads too soon.

(iv) Comment on the hire company's complaint. Give a reason for your answer. [Edexcel 6691 Q3 June 2012]

⑫ Sugar is automatically packed by a machine into bags of nominal weight 1000 g. Due to random fluctuations and the set up of the machine, the weights of bags are, in fact, Normally distributed with mean 1020 g and standard deviation 25 g. Two bags are selected at random.

(i) Find the probability that the total weight of the two bags is less than 2000 g.

(ii) Find the probability that the weights of the two bags differ by less than 20 g.

Another machine is also in use for packing sugar into bags of nominal weight 1000 g. It is assumed that the distribution of the weights for this machine is also Normal. A random sample of nine bags packed by this machine is found to have the following weights (in grams).

1012   996   984   1005   1008   994   1003   1017   1002

(iii) Test at the 5% level of significance whether it may be assumed that the mean weight for this machine is 1000 g. [MEI]

⑬ A trial is being made of a new diet for feeding pigs. Ten pigs are selected and their increases in weight (in kilograms) are measured, over a certain period, using the new diet. The data are as follows.

15.2   13.8   14.6   15.8   13.1   14.9   17.2   15.1   14.9   15.2

The underlying population can be assumed to be Normally distributed.

(i) Using an established diet, the mean increase in weight of pigs over the period is known to be 14.0 kg. Test at the 5% level of significance whether the new diet is an improvement, stating carefully your null and alternative hypotheses and your conclusion.

(ii) Find a 95% confidence interval for the mean increase in weight using the new diet.

(iii) Little information about the conduct of the trial is given in the opening paragraph of the question. Comment on **two** aspects of how the trial should have been conducted. [MEI]

14. A notional allowance of 9 minutes has been given for the completion of a routine task on a production line. The operatives have complained that it appears usually to be taking slightly longer.

An inspector took a sample of 12 measurements of the time required to undertake this task. The results (in minutes) were as follows.

9.4   8.8   9.3   9.1   9.4   8.9   9.3   9.2   9.6   9.3   9.3   9.1

Stating carefully your null and alternative hypotheses and the assumptions underlying your analysis, test at the 1% level of significance whether the task is indeed taking on average longer than 9 minutes. [MEI]

15. Archaeologists have discovered that all skulls found in excavated sites in a certain country belong either to racial Group A or to racial Group B. The mean lengths of skulls from Group A and Group B are 190 mm and 196 mm, respectively. The standard deviation for each group is 8 mm, and skull lengths are distributed Normally and independently.

A new excavation produced 12 skulls of mean length x and there is reason to believe that all these skulls belong to Group A. It is required to test this belief statistically with the null hypothesis (H_0) that all the skulls belong to Group A and the alternative hypothesis (H_1) that all the skulls belong to Group B.

(i) State the distribution of the mean length of 12 skulls when H_0 is true.

(ii) Explain why a test of H_0 versus H_1 should take the form: 'Reject H_0 if $\bar{x} > c$', where c is some critical value.

(iii) Calculate this critical value c to the nearest 0.1 mm when the probability of rejecting H_0 when it is, in fact, true is chosen to be 0.05.

(iv) Perform the test, given that the lengths (in mm) of the 12 skulls are as follows.

204.1   201.1   187.4   196.4   202.5   185.0
192.6   181.6   194.5   183.2   200.3   202.9

[MEI]

2 Testing the difference of the means of two populations

Hypothesis testing is also used to compare the means of two populations which may be paired or unpaired.

For paired data, a hypothesis test can be carried out on the differences, using the same idea as for a confidence interval.

Unpaired samples

Figure 15.7

> When he returned to the office at two o'clock, his desk was clear. Until the afternoon post arrived, he had nothing to do. A ray of sunlight shone through a high window and he watched the specks of dust fall gently through it. With one finger, he pushed a paperclip around his blotter. The only sound was the distant whirring of a printer in a room down the ccorridor. For what seemed like hours he sat, sunk in a deep reverie. Then he lowered his eyes and glanced at his watch. It was five past two.

Have you noticed how time often seems to pass more slowly after lunch?

If time passes more slowly, one minute of real time should seem longer, so if you ask people to estimate when a minute appears to have elapsed, the real time elapsed will be less.

You could ask the question: 'Will the mean real time elapsed when one minute appears to have elapsed be less after lunch than before?'

In this example you are interested, not in what the mean value of a random variable is, but in the difference between the mean values in two different situations. Statistical problems giving rise to different versions of this general question often require either paired sample tests or two sample tests.

> **Note**
>
> What are the advantages and disadvantages of using separate groups of people for the before-lunch and after-lunch times? What are the advantages and disadvantages of instead conducting an experiment in which the same people are asked before and after lunch and only the difference in their real times is recorded.

EXPERIMENT 15.1

Find a group of volunteers and approach them before lunch. Give each of them a starting signal and ask them to say when one minute has elapsed. Record the real time elapsed. You will then need to find a second group of volunteers to approach after lunch. You will need reasonably large groups to get useful results.

The volunteers in research projects are called **subjects** and 'before lunch' and 'after lunch' are the two **conditions** in which testing occurs. An experiment such as the one described above, where a different group of subjects is tested in each of the two conditions is called an **unpaired design**; this is in contrast to **paired design**, where the same set of subjects is tested in both conditions.

S2 AL — Testing the difference of the means of two populations

> Three samples are neither very small nor very large. They are used to illustrate both small-sample and large-sample techniques in this section.

The members of a maths class were asked one morning to check the time shown by their watches, then look away, and when they had estimated that a minute had elapsed, they were asked to check their watches again to see how long had in fact elapsed.

The same procedure was followed with another class, from the same year group, that afternoon. The back-to-back stem-and-leaf diagram below shows the results.

```
         Morning class          Afternoon class
         (24 students)          (22 students)
                         2 | 8
                       2 | 3 | 1
                   5 5 6 | 3 | 6
                         | 4 | 4 4 0
                     7 8 | 4 | 9 9 9 6 6 5
       1 1 3 3 3 3 4 4   | 5 | 2 2 1 1 1
           5 5 5 7 7 7 7 | 5 | 8 7 7 5
  Mean 51.542     1 4 4  | 6 | 3          Mean 47.909
                         | 6 | 3     represents 63 seconds
```

Figure 15.8

> You cannot look here at the difference between a before-lunch and an after-lunch time for a particular person but you can look at the difference between the mean before lunch time and the mean after lunch time.

This experiment gives a set of data with which you could investigate the question. This experiment has an **unpaired design**: two separate groups of subjects are used in the two conditions.

In this section the data in Figure 15.8 are used to work through the process of hypothesis testing in the context of an unpaired design.

> If you have carried out your own investigation, you might find it helpful to repeat the calculations using your data.

The measurements come from two independent random variables.

- B is the before-lunch time estimate, with population mean μ_B and variance σ_B^2.
- A is the after-lunch time estimate, with population mean μ_A and variance σ_A^2.

H_0: there is no difference between the mean of people's estimates of one minute before and after lunch.

$$\mu_B - \mu_A = 0$$

H_1: after lunch, the mean of people's estimates of one minute tends to be shorter after lunch than before lunch.

$$\mu_B - \mu_A > 0$$

For these particular samples, the observed means are $\bar{b} = 51.542$ and $\bar{a} = 47.909$, so $\bar{b} - \bar{a} = 3.633$. This is the test statistic.

You need to calculate the distribution of this sample statistic on the assumption that the null hypothesis is true. This is done in the following worked example in which three different sets of assumptions are considered.

Example 15.3

Using the data from the experiment described above, carry out hypothesis tests of whether the after-lunch time estimates for one minute are less than the before-lunch estimates in the following cases.

Case 1

The populations are known to be Normal with variances $\sigma_A^2 = 66$ and $\sigma_B^2 = 75$. The samples are treated as small.

> Notice that these are the same three cases that you met in Chapter 14.

Case 2

The population variance are unknown. The samples are treated as large.

Case 3

The populations are known to be Normal with equal but unknown variance. The samples are treated as small.

Solution

Case 1: Small samples from Normal populations with known variances.

Null hypothesis $\quad\quad\quad H_0: \mu_B = \mu_A \quad$ or $\quad \mu_B - \mu_A = 0$

Alternative hypothesis $\quad H_1: \mu_B > \mu_A \quad$ or $\quad \mu_B - \mu_A < 0$

In this case,

$B \sim N(\mu_B, 75)$ and $A \sim N(\mu_A, 66)$

Because the population distributions are known to be Normal and the population variances are known, the distribution of the sample mean for each will also be Normal, irrespective of the sample sizes. Using the Central Limit Theorem:

$\bar{B} \sim N\left(\mu_B, \frac{75}{24}\right)$ and $\bar{A} \sim N\left(\mu_A, \frac{66}{22}\right)$

so $\bar{B} - \bar{A} \sim N\left(\mu_B - \mu_A, \frac{75}{24} + \frac{66}{22}\right) = N(\mu_B - \mu_A, 6.125)$

If the null hypothesis is true then $\bar{B} - \bar{A} \sim N(0, 6.125)$

The observed value of $\bar{b} - \bar{a}$ is $51.542 - 47.909 = 3.633$

This gives the test statistic $z = \dfrac{3.633 - 0}{\sqrt{6.125}} = 1.468$

For a one-tailed test at the 5% level of significance the critical value is $z = 1.645$.

Since $1.468 < 1.645$, H_0 is accepted.

There is insufficient evidence to suggest that the mean of the population of after-lunch estimates is smaller than the mean of the before-lunch estimates.

Note: Case 1 in general

- A random sample of size n_x is taken from $X \sim N(\mu_x, \sigma_x^2)$ and
- a random and sample of size n_y is taken from $Y \sim N(\mu_y, \sigma_y^2)$, where σ_x^2 and σ_y^2 are known.

Then

$$\bar{X} - \bar{Y} \sim N\left(\mu_x - \mu_y, \frac{\sigma_x^2}{n_x} + \frac{\sigma_y^2}{n_y}\right).$$

$H_0: \mu_x = \mu_y$

Assuming H_0 it follows that $\bar{X} - \bar{Y} \sim N\left(0, \dfrac{\sigma_x^2}{n_x} + \dfrac{\sigma_y^2}{n_y}\right)$.

So $z = \dfrac{\bar{x} - \bar{y}}{\sqrt{\dfrac{\sigma_x^2}{n_x} + \dfrac{\sigma_y^2}{n_y}}}$ is an observation from $N(0, 1)$.

S2 AL — Testing the difference of the means of two populations

Solution

Case 2: Large samples from populations with unknown variances.

The population variance σ_b^2 and σ_a^2 are unknown.

Null hypothesis $\qquad H_0: \mu_b = \mu_a \qquad$ or $\qquad \mu_b - \mu_a = 0$

Alternative hypothesis $\qquad H_1: \mu_b > \mu_a \qquad$ or $\qquad \mu_b - \mu_a > 0$

If the sample sizes are treated as being large then the Central Limit Theorem says that

$$\overline{B} \sim N\left(\mu_b, \frac{\sigma_b^2}{24}\right) \text{ and } \overline{A} \sim N\left(\mu_a, \frac{\sigma_a^2}{22}\right)$$

so $\overline{B} - \overline{A} \sim N\left(\mu_b - \mu_a, \frac{\sigma_b^2}{24} + \frac{\sigma_a^2}{22}\right).$

The population variances are unknown so they are estimated.

For the before-lunch estimates:

$n_b = 24, \overline{b} = 51.542$ and $\sum_{i}^{24}(b_i - \overline{b})^2 = 1780$ so $s_b^2 = 77.389$

$\boxed{\dfrac{1780}{24} - 1 = 77.389}$

For the after-lunch estimates:

$n_a = 22, \overline{a} = 47.909$ and $\sum_{i}^{22}(a_i - \overline{a})^2 = 1543.8$ so $s_a^2 = 73.515$

So $\overline{B} - \overline{A} \sim N\left(\mu_b - \mu_a, \frac{77.389}{24} + \frac{73.515}{22}\right) = N(\mu_b - \mu_a, 6.566)$

If the null hypothesis is true then $\overline{B} - \overline{A} \sim N(0, 6.566)$

The observed value of $\overline{b} - \overline{a}$ is $51.542 - 47.909 = 3.633$

This gives the test statistic $z = \dfrac{3.633 - 0}{\sqrt{6.566}} = 1.418$

For a one-tailed test at the 5% level of significance, the critical value is $z = 1.645$.

Since $1.418 < 1.645$ H_0 is accepted.

There is insufficient evidence to suggest that the mean of the population of after-lunch estimates is smaller than the mean of the before-lunch estimates.

> **Note**
> The unbiased estimator of a population variance is $s^2 = \dfrac{1}{n-1}\sum_{i=1}^{n}(x_i - \overline{x})^2$.

Note: Case 2 in general

In general, a random sample of size n_x is taken from a population with mean μ_x and variance σ_x^2 and a random sample of size n_y is taken from a population with mean μ_y and variance σ_y^2, with σ_x^2 and σ_y^2 unknown.

Estimate σ_x^2 by $s_x^2 = \dfrac{1}{n_x - 1}\sum_{i=1}^{n_x}(x_i - \overline{x})^2$ and σ_y^2 by $s_y^2 = \dfrac{1}{n_y - 1}\sum_{i=1}^{n_y}(y_i - \overline{y})^2$

$H_0: \mu_x = \mu_y$

Assuming H_0 to be true, it follows that $\overline{X} - \overline{Y} \sim N\left(0, \dfrac{s_x^2}{n_x} + \dfrac{s_y^2}{n_y}\right).$

so $z = \dfrac{\overline{x} - \overline{y}}{\sqrt{\dfrac{s_x^2}{n_x} + \dfrac{s_y^2}{n_y}}}$ is an observation from $N(0, 1)$.

> **Note**
>
> The unbiased estimator of a population variance is
> $s^2 = \frac{1}{n-1}\sum_{i=1}^{n}(x_i - \bar{x})^2$

Solution

Case 3: Samples from Normal populations with equal but unknown variances.

Null hypothesis $\quad H_0: \mu_b = \mu_a \quad$ or $\quad \mu_b - \mu_a = 0$

Alternative hypothesis $\quad H_1: \mu_b > \mu_a \quad$ or $\quad \mu_b - \mu_a > 0$

In this case σ^2 is unknown but is the same for both populations.

It is estimated using the pooled estimator for variance

$$s_b^2 = \frac{1}{(n_b - 1) + (n_a - 1)}\left\{\sum_{1}^{n_b}(b_i - \bar{b})^2\right\}$$

$$s^2 = \frac{(n_b - 1)s_b^2 + (n_a - 1)s_a^2}{(n_b + n_a - 2)}$$

Before lunch:
$n_b = 24$, $\bar{b} = 51.542$ and $\sum_{1}^{24}(b_i - \bar{b})^2 = 1780$ so $s_b^2 = 77.389$

After lunch:
$n_a = 22$, $\bar{a} = 47.909$ and $\sum_{1}^{22}(a_i - \bar{a})^2 = 1543.8$ so $s_a^2 = 73.515$

> The values $s_b = 8.797$ and $s_a = 8.574$ can be found directly using a calculator.

The pooled estimator of σ^2 is $s^2 = \frac{(23 \times 77.389) + (21 \times 73.515)}{24 + 22 - 2} = 75.540$

The sampling distribution of $\bar{B} - \bar{A}$ has mean $\mu_b - \mu_a$ and variance $\sigma^2\left(\frac{1}{n_b} + \frac{1}{n_a}\right)$, which is estimated by $75.540\left(\frac{1}{24} + \frac{1}{22}\right) = 6.581$

If the null hypothesis is true then $T = \frac{\bar{B} - \bar{A}}{\sqrt{6.581}}$ follows a t-distribution with $24 + 22 - 2 = 44$ degrees of freedom.

The observed value of $\bar{b} - \bar{a}$ is $51.542 - 47.909 = 3.633$

This gives the test statistic $t = \frac{3.633}{\sqrt{6.5816}} = 1.416$

For a one-tailed test at the 5% level of significance, the critical value is $t = 1.680$.

Since $1.416 < 1.680$ H_0 is accepted.

There is insufficient evidence to suggest that the mean of the population of after-lunch estimates is smaller than the mean of the before-lunch estimates.

> ### Note: Case 3 in general
>
> In general, a random sample of size n_x is taken from a population with mean μ_x and variance σ^2 and a random sample of size n_y is taken from a population with mean μ_y and variance σ^2, with σ^2 unknown.
>
> Estimate σ^2 by $s^2 = \frac{(n_x - 1)s_x^2 + (n_y - 1)s_y^2}{(n_x + n_y - 2)}$, where $s_x^2 = \frac{1}{n_x - 1}\sum_{i=1}^{n_x}(x_i - \bar{x})^2$ and $s_y^2 = \frac{1}{n_y - 1}\sum_{i=1}^{n_y}(y_i - \bar{y})^2$
>
> $H_0: \mu_x = \mu_y$
>
> Assuming H_0 to be true, it follows that $t = \frac{\bar{x} - \bar{y}}{s\sqrt{\frac{1}{n_x} + \frac{1}{n_y}}}$ is an observation from a t-distribution with $n_x + n_y - 2$ degrees of freedom.

S2 AL — Testing the difference of the means of two populations

Exercise 15.2

1. A college careers officer investigates the income at age 24 of a group of students who left school at 16, and a group who stayed on to take A-levels. The results he finds (measuring incomes in £ per week) are summarised by this table.

   Table 15.1

	16-year-old leavers	A-levels leavers
Mean	156	164
Sample estimate of variance	673	593
Sample size	37	28

   Show, using an unpaired t-test, that the hypothesis that those staying on at school have higher incomes at age 24 is rejected, on the evidence of this sample. What assumptions are you making for an unpaired t-test to be appropriate? How plausible are they?

2. The reaction times (in milliseconds) of ten workers were measured at the start of their working day and again at the end of their working day. Reaction times can be modelled using Normal distributions.

   The results are given below.

   Table 15.2

Worker	A	B	C	D	E	F	G	H	I	J
Start	200	180	150	120	120	140	150	190	170	210
End	220	250	160	170	120	130	150	180	200	250
Difference	20	70	10	50	0	−10	0	−10	30	40

   The differences have a mean of 20 and a variance of 660.

   Test whether the workers' reactions are significantly slower at the end of their working day.

3. Ten people take part in a two-week weight loss program. Their weights, in kg, are recorded at the start and end of the program. The weights may be assumed to follow Normal distributions.

   The results are given below.

   Table 15.3

Person	A	B	C	D	E	F	G	H	I	J
Start	84	92	104	97	77	89	95	102	98	83
End	80	84	88	104	82	82	85	98	90	86

   Test, at the 5% level of significance, whether the weights at the end of the program are significantly lower than at the start of the program.

4. The heights, in cm, of adults aged over 50 are modelled as $N(\mu_1, 100)$ and the heights of adults aged under thirty are modelled as $N(\mu_2, 225)$. The heights of a random sample of seven adults aged over 50 has mean 170 cm, and the heights of an independent random sample of 16 adults aged under 30 has mean 180 cm.

   Test, at the 10% level, whether $\mu_2 > \mu_1$.

5. Some personality theorists classify people into 'introverts', or quiet personalities, and 'extroverts', or outgoing personalities. In an attempt to find physical correlates of these personality types, the heights of a group of 17 extroverts and a group of 17 introverts were measured.

   (i) Show that an unpaired t-test with the data below accepts, at the 5% level, the hypothesis that extroverts are taller on average than introverts, stating carefully the assumptions you are making.

   Table 15.4

Heights of extroverts (cm)					
179	174	166	170	170	168
173	170	173	173	168	174
163	156	157	162	169	

Heights of introverts (cm)					
159	172	167	168	163	155
164	163	166	164	175	155
165	168	164	166	162	

   (ii) The scientist who made these measurements concluded that taller people tend to be more confident and hence extroverted. Is this conclusion justified?

6. A species of finch has subspecies on two different Galapagos Islands. The weights of a sample of finches from each island are listed below.

   Table 15.5

Weights from Daphne Major (grams)											
64	67	64	61	68	61	61	67	70	62	66	63

Weights from Daphne Minor (grams)						
63	61	65	65	61	63	63

   (i) Is there evidence that the finches on Daphne Major are heavier on average than those on Daphne Minor?

   (ii) What assumptions do you need to make? Are they reasonable?

7. Two different varieties of sprout being grown around the country, on a variety of different plots with different soil types and weather conditions, are monitored by a crop-testing station.

   (i) Test, using an unpaired t-test, the hypothesis that the two varieties have the same average yield. The yields of sprouts per square metre in kilograms are as follows.

   Table 15.6

Yields of 'Old Cobby' (kg m^{-2})								
18	15	14	21	21	19	9	20	26	27
21	12	15	14	27	22	23	18	10	19

Yields of 'Early Yellow' (kg m^{-2})								
12	20	18	14	15	13	17	15	20	26
17	14	8	29	24	16	20	20	18	23
20	23	23	18	9					

(ii) Plot the data on a back-to-back stem-and-leaf diagram. State and comment on the assumptions you are making in carrying out the test. Do you think the assumptions are justified?

(iii) Devise an experiment to test this hypothesis which would use a paired *t*-test. Would this be an improvement, do you think?

8 A machine fills jars with jam. The amount of jam dispensed into a jar is assumed to follow a Normal distribution. After a power cut it is suspected that the mean amount dispensed may have changed.

Data is available on the amount dispensed, in grams, into 60 jars before the power cut and 90 jars after the power cut. The samples are independent.

Before: $n = 60$ $\bar{x} = 458.6$ $s = 25$

After: $n = 90$ $\bar{x} = 452$ $s = 12$

(i) Test at the 5% level whether the mean has changed significantly.

(ii) If the means and standard deviations are rounded to the nearest integer before the test is carried out, would this alter the conclusion that is reached?

9 Two teachers, Mr Adams and Ms Brown, record the test marks for their classes. It may be assumed that the test marks for each teacher come from independent Normal distributions with equal variances. It has been suggested that Mr Adams marks more generously than Ms Brown, so his classes will have a higher mean mark.

Independent random samples are taken from the marking of each teacher.

Mr Adams: $n = 4$ $\bar{x} = 88$ $s = 14.2$

Ms Brown: $n = 8$ $\bar{x} = 76$ $s = 9.9$

Test the claim at the 5% level.

10 The travel times, in minutes, for two bus routes are recorded over several journeys.

Table 15.7

Route X100	20	25	25	18	
Route X202	56	60	51	52	54

It is known that the travel times for any bus route can be modelled as a Normal random variable and the variance of the times is the same for each route. It is also assumed that the times for different routes are independent of one another.

It has been suggested that, because it is a longer route, the travel times for route X202 should be, on average, 30 minutes greater than for route X100. Test this claim at the 10% level.

KEY POINTS

1. **Hypothesis test for the mean using a Normal distribution**
   - Sample data may be used to carry out a hypothesis test on the null hypothesis that the population mean has some particular value μ_0, i.e. $H_0: \mu = \mu_0$.
   - The alternative hypothesis takes one of the three forms.
     - $H_1: \mu \neq \mu_0$
     - $H_1: \mu < \mu_0$
     - $H_1: \mu > \mu_0$
   - The test statistic $z = \dfrac{\bar{x} - \mu_0}{\frac{\sigma}{\sqrt{n}}}$ is used, based on:

     (a) A sample drawn from a normal population of known, given or assumed variance.

     (b) A large sample drawn from any population with known, given or assumed variance, using the Central Limit Theorem.

     (c) A large sample drawn from any population with unknown variance, using an unbiased estimate of the population variance.
   - If the sample is large, you can use the sample standard deviation s as an estimate of the population standard deviation σ if the latter is not known.
   - If the sample is large, then using the Central Limit Theorem, you can carry out a test using a Normal distribution even if the distribution of the parent population is not known.

2. **Hypothesis test for the mean using a t-distribution**
   - If the population is Normally distributed but the population standard deviation σ is not known then, whatever the sample size, you can carry out a hypothesis test using a t-distribution.
   - The null hypothesis is the same as for the Normal distribution $H_0: \mu = \mu_0$.
   - You estimate the population standard deviation using the sample standard deviation s.
   - The test statistic $t = \dfrac{\bar{x} - \mu_0}{\frac{s}{\sqrt{n}}}$ is used.

3. To test the value of the difference between the means of paired samples from Normal distributions, the test statistic is
$$\dfrac{\overline{X} - \mu}{\frac{s}{\sqrt{n}}} \sim t_{(n-1)}$$

4. To test for a difference between the means of unpaired samples from independent Normal distributions with known variances, the test statistic under H_0 is
$$\dfrac{(\overline{X} - \overline{Y}) - (\mu_x - \mu_y)}{\sqrt{\dfrac{\sigma_x^2}{n_x} + \dfrac{\sigma_y^2}{n_y}}} \sim N(0,1)$$

5 To test for a difference between the means of unpaired samples from independent Normal distributions with unknown variances, the test statistic under H_0 for large samples is

$$\frac{(\overline{X}-\overline{Y})-(\mu_x-\mu_y)}{\sqrt{\frac{s_x^2}{n_x}+\frac{s_y^2}{n_y}}} \sim N(0,1)$$

6 To test for a difference between the means of unpaired samples from independent Normal distributions with equal but unknown variances, the test statistic under H_0 is

$$\frac{(\overline{X}-\overline{Y})-(\mu_x-\mu_y)}{s\sqrt{\frac{1}{n_x}+\frac{1}{n_y}}} \sim t_{(n_x+n_y-2)}$$

where s^2 is the pooled estimator of the population variance, given by

$$s^2 = \frac{(n_x-1)s_x^2+(n_y-1)s_y^2}{n_x+n_y-2}$$

LEARNING OUTCOMES

When you have completed this chapter, you should be able to:

- carry out a hypothesis test for a single population mean using the Normal or t-distributions and know when it is appropriate to do so
- test the difference between the means of paired samples from Normal distributions
- test for a difference between the means of unpaired samples from independent Normal distributions with known variances
- test for a difference between the means of unpaired samples from independent Normal distributions with unknown variances
- test for a difference between the means of unpaired samples from independent Normal distributions with equal but unknown variances.

16 Estimation

Depend upon it, a lucky guess is never merely luck – there's always some talent in it.
Jane Austen

The fossil record provides invaluable information on the evolution of life on Earth.

Recently, the remains of a primitive human were found in the Great Rift Valley of Africa. It was the third specimen of its particular subspecies of human. The three specimens found so far are calculated to have been aged 9, 16 and 35 at the times of their deaths.

What can be said about the distribution of ages at death in the population of this species of human, from the sample available?

The obvious answer to this question is 'not much': there is a very small sample to work with, and a very large amount of information is needed to determine the details of a distribution. You need to start by making a heroic assumption about the general shape of the distribution of ages at death. The easiest assumption to start with is that it is a uniform (rectangular) distribution: humans from this species were equally likely to die at every age up to some maximum. Formally, if X is the random variable 'age at death', then you are assuming that X is uniformly distributed on $[0, m]$, where m is the maximum age to which these humans lived.

The task is now dramatically simplified. You need to use the sample merely to **estimate the parameter** m of the distribution of ages at death.

What estimate would you make of the parameter m, the maximum age at death, from the sample given?

Two **possible** methods for estimating m are given below, but if you have thought of others, you could follow the work through with your method.

Method 1: Find the mean of the sample and double it.

Rationale: The mean age at death ought to be about half of the maximum.

This estimator is given by $M_1 = 2\left(\dfrac{X_1 + X_2 + X_3}{3}\right)$.

For this sample it gives an estimate of $m_1 = 2 \times \left(\dfrac{9 + 16 + 35}{3}\right) = 40$

Method 2: Take the largest age in the sample and multiply by $\dfrac{6}{5}$.

Rationale: The largest age in the sample ought to be in the middle of the highest third of the distribution, i.e. at about $\dfrac{5}{6}$ of the maximum age.

This estimator is given by $M_2 = \dfrac{6}{5}\max\{X_1, X_2, X_3\}$.

For this sample it gives an estimate of $\dfrac{6}{5} \times 35 = 42$

Notice the vocabulary. An **estimator** is a function which converts a particular set of sample values into a numerical value, an **estimate**. In this example M_1 and M_2 are estimators for m.

When you make an estimate you obviously want to know how good it is. This raises two important questions about the estimator you are using.

- Is the estimator biased?
- How accurate is the estimate likely to be?

1 Bias in estimators

The estimator for Method 1 involves the sum of three independent random variables. If you assume that the three sets of remains found were a random sample from the entire population of this subspecies of human, then the ages at death of the three individuals are the values of the three random variables X_1, X_2 and X_3, each of which comes from a population with a uniform (rectangular) distribution on $[0, m]$.

The variables M_1 and M_2, which are the estimators described by Methods 1 and 2, respectively, are then given by

$$M_1 = \dfrac{2}{3}(X_1 + X_2 + X_3)$$

and

$$M_2 = \dfrac{6}{5}\max\{X_1, X_2, X_3\}.$$

As a random variable, an estimator has a distribution. If it is a discrete random variable, the distribution will be the set of possible values which it might take, together with the probability with which each value arises. If it is a continuous random variable, the distribution is given by a density function. The uncertainty involved arises from the process of taking samples at random from the population as a whole, and therefore this distribution is called the **sampling distribution** of the estimator.

If the mean of the sampling distribution of the estimator is equal to the population parameter in question, the estimator is **unbiased**. If they are not equal the estimator is **biased** and the difference between the values is the **bias**.

Is M_1 biased?

You can see that M_1 is unbiased. Each of the X_i has a uniform distribution on $[0, m]$ and hence $E[X_1] = \frac{1}{2}m$.

Therefore

$$E[M_1] = E\left[\frac{2}{3}(X_1 + X_2 + X_3)\right]$$
$$= E\left[\frac{2}{3}(X_1)\right] + E\left[\frac{2}{3}(X_2)\right] + E\left[\frac{2}{3}(X_3)\right]$$
$$= \frac{2}{3}\left(E[X_1] + E[X_2] + E[X_3]\right)$$
$$= \frac{2}{3}\left(\frac{m}{2} + \frac{m}{2} + \frac{m}{2}\right) = m.$$

Since $E(M_1) = m$, M_1 is an unbiased estimator.

Is M_2 biased?

Before you can decide whether M_2 is biased you need to know its sampling distribution.

The probability density function for X is

$$f(x) = \begin{cases} \frac{1}{m} & 0 \leq x \leq m \\ 0 & \text{otherwise} \end{cases}$$

The graph of this function for $m = 40$ is sketched in Figure 16.1.

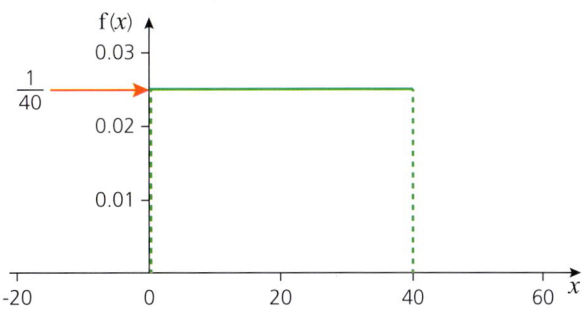

Figure 16.1

The cumulative distribution function for X is obtained by integration.

$$F(x) = P(X \leq x) = \int_0^x f(t)\,dt = \begin{cases} 0 & x \leq 0 \\ \frac{x}{m} & 0 \leq x \leq m \\ 1 & x \geq m \end{cases}$$

> In this integral t is a dummy variable.

The cumulative distribution function H, the maximum value of X_1, X_2 and X_3 is found by noting that the largest of three numbers will be less than a value x if and only if all three of them are less than x. Thus the probability that the largest of the three sample values is less than x is the same as the probability that all three sample values are less than x.

So you can find the cumulative distribution function of H from the cumulative distribution function of X as follows.

$$F(x) = P(H \leq x) = P(\max\{X_1, X_2, X_3\} \leq x)$$
$$= P(X_1 \leq x \text{ and } X_2 \leq x \text{ and } X_3 \leq x)$$
$$= P(X_1 \leq x) \times P(X_2 \leq x) \times P(X_3 \leq x)$$
$$= \begin{cases} 0 & x \leq 0 \\ \left(\dfrac{x}{m}\right)^3 & 0 \leq x \leq m \\ 1 & x \geq m \end{cases}$$

Then the probability density function of H, which is found by differentiating the cumulative distribution function, is

$$f(x) = \frac{d}{dx} F(x) \begin{cases} \dfrac{3x^2}{m} & 0 \leq x \leq m \\ 0 & \text{otherwise} \end{cases}$$

The estimator $M_2 = \dfrac{6}{5} H$. A graph of the probability density function of M_2 is sketched in Figure 16.2.

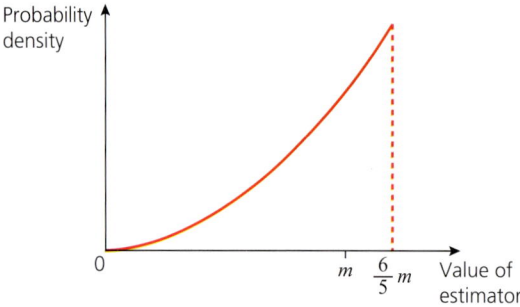

Figure 16.2

This graph illustrates the following points.

- Since each of the X_i is restricted to the range 0 to m, so is the value of H. Thus the estimator M_2 cannot take values greater than $\dfrac{6}{5} m$.

- The distribution is strongly negatively skewed: this should not be surprising, since the largest of the X_i is relatively unlikely to be in the lower part of the range 0 to m.

ACTIVITY 16.1

Using a spreadsheet (or a calculator which produces random numbers in the range 0 to 1) you can simulate, for the case $m = 1$, the sampling process used here.

1. Find three random numbers between 0 and 1.
2. Choose the largest.
3. Multiply this by $\frac{6}{5}$.
4. Repeat steps 1–3 two hundred times.
5. Plot a histogram showing the simulated sampling distribution of M_1.

It should resemble the probability density function shown in Figure 16.2, with $m = 1$.

The expected value of H is

$$E[H] = \int_{-\infty}^{\infty} x f(x)\, dx = \int_0^m \frac{3x^2}{m^3}\, dx = \frac{3m}{4}$$

So the expected value of M_2 is

$$E[M_2] = \frac{6}{5} E[H] = \frac{9m}{10}$$

This shows that M_2 is biased. It does not, on average over all possible samples, give the true value of the population parameter.

The bias is $\frac{m}{10}$.

Earlier on in this chapter two questions were posed:

- Is the estimator biased?
- How accurate is the estimate likely to be?

The first of these questions has now been answered for M_1 and M_2.

The answer to the second depends both on any bias in the estimator and also on the variability of the estimates it produces. A measure of the variability is provided by the variance (or by the standard deviation) of the distribution. This tells you about the spread of the estimates but not necessarily about their accuracy.

> For a uniform (rectangular) distribution for $a \leq x \leq b$, the mean is $\frac{a+b}{2}$ and the variance is $\frac{(b-a)^2}{12}$.

For $M_1 = 2\left(\frac{X_1 + X_2 + X_3}{3}\right) = \frac{2}{3}X_1 + \frac{2}{3}X_2 + \frac{2}{3}X_3$ using the formula, for independent variables, $\text{Var}(aX_1 + bX_2 + cX_3) = a^2\text{Var}(X_1) + b^2\text{Var}(X_2) + c^2\text{Var}(X_3)$ and the standard result for the variance of a uniform (rectangular) distribution, this gives

$$\text{Var}(M_1) = \frac{m^2}{9}$$

and so the standard deviation of M_1 is $\frac{m}{3}$.

For M_2, start by finding $\text{Var}(H)$ using $\text{Var}(H) = E[H^2] - (E[H])$ using $\text{Var}(X) = E(X^2) - [E(X)]^2$. It is already known that $E(H) \frac{3m}{4}$ and $E(H^2)$ is found using integration; the p.d.f. $f(x)$ has also been found already

$$E[H^2] = \int_{-\infty}^{\infty} x^2 f(x)\, dx = \int_0^m x^2 \frac{3x^2}{m^3}\, dx$$

This calculation $\text{Var}(H) = \dfrac{3m^2}{80}$ and so $\text{Var}(M_2) = \left(\dfrac{6}{5}\right)^2 . \text{Var}(H)$ gives $\text{Var}(M_2) = 0.054m^2$. So the standard deviation of M_2 is $0.23m$.

Thus M_2 has a lower standard deviation and so produces more reliable estimates.

> **Note**
>
> You have found that M_2 is biased, but the estimator
> $$M_3 = \dfrac{10}{9} M_2 = \dfrac{4}{3} \text{Max}\{X_1, X_2, X_3\}$$
> is unbiased, because
> $$\text{E}[M_3] = \text{E}\left[\dfrac{10}{9} M_2\right] = \dfrac{10}{9} \text{E}[M_2] = \dfrac{10}{9} \times \dfrac{9m}{10} = m.$$
>
> For the original sample, the value of M_3 is $m_3 = \dfrac{4}{3} \times 35 = 46\dfrac{2}{3}$.
>
> You cannot always create an unbiased estimator from a biased one by multiplying by a constant. It was only possible to do so in this case because the expected value of the biased estimator, although not equal to the population parameter, was just a constant multiple of it.

ACTIVITY 16.2

You made the assumption at the start of the analysis that the age at death had a uniform distribution, but this is not very realistic. In fact, it seems likely that primitive humans were much more likely to die younger. A better model for the distribution of the age at death in such a species might therefore be the triangular distribution with density function

$$f(x) = \begin{cases} \dfrac{2}{m^2}(m-x) & 0 \leqslant x \leqslant m \\ 0 & \text{otherwise} \end{cases}$$

The graph of this density function is sketched in Figure 16.3.

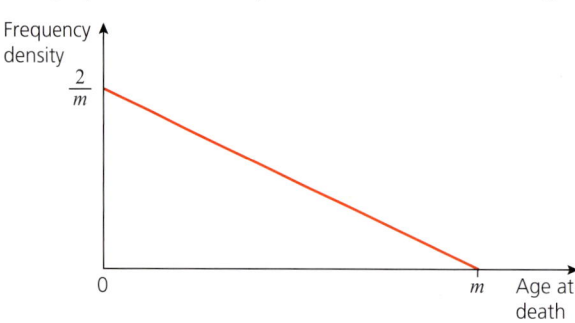

Figure 16.3

Using the same estimators M_1 and M_2 for m and assuming X to have this new distribution, find

(i) the sampling distribution of M_1

(ii) the expectation of M_1, using your sampling distribution

(iii) the expectation of M_2

and so determine whether M_1 and M_2 are biased or unbiased. If either estimator is biased, find a multiple of it which is unbiased.

> **Note**
>
> In both the original example and in this investigation, the population distribution has been assumed to have a very definite form, so that the determination of the exact population distribution has required the estimation of only one parameter. In general, you may want to make less restrictive assumptions about the form of the population distribution, in which case you may need to estimate two or more parameters. For instance, if you assume that the population distribution is Normal, you will want to estimate both its mean and its variance.

ACTIVITY 16.3

Use the original (uniform) distribution for X, but suppose that a sample of size n is available. The estimators M_4 and M_5 are extensions of M_3 and M_2 to this situation.

$$M_4 = \frac{n+1}{n}\max\{X_1, X_2, \ldots, X_n\}$$

and

$$M_5 = \frac{2}{n}(X_1 + X_2 + \ldots + X_n).$$

Determine

(i) the sampling distribution for M_4

(ii) the expected value of M_4, using your sampling distribution

(iii) the expected value for M_5

and so decide whether M_4 and M_5 are biased or unbiased.

Example 16.7

Suppose that you buy scratch cards until you have won three times. The number of cards you have purchased by then is recorded as the value of the random variable N. Calculate the sampling distribution of N and show that $P = \dfrac{2}{N-1}$ is an unbiased estimator of p, the probability that a scratch card is a winning card.

> **Note**
>
> This is a negative binomial distribution, as described in Chapter 9.

Solution

The probability that N has the value n is the probability that the third win will occur with the nth card.

That is,

the nth card is a winner

and

there have been exactly two winners in the first $n-1$ cards

so

P(third win with nth card) = P(nth card wins) × P(2 wins in first $n-1$ cards).

The nth card wins with probability p, and the binomial distribution gives the probability that two of the first $n-1$ cards are winners (and so $(n-3)$ are not winners) as

$$^{n-1}C_2 \, p^2 (1-p)^{(n-3)}.$$

Therefore

$$P(\text{third win with } n\text{th card}) = p \times {}^{n-1}C_2 p^2 (1-p)^{(n-3)}$$
$$= {}^{n-1}C_2 p^3 (1-p)^{(n-3)}.$$

> Note that ${}^{n-1}C_2 = \dfrac{(n-1)(n-2)}{2}$

Hence the sampling distribution of N is

$$P(N=n) = \tfrac{1}{2}(n-1)(n-2) p^3 (1-p)^{n-3} \quad (n \geqslant 3).$$

To determine whether $\Pi = \dfrac{2}{N-1}$ is unbiased, you must calculate $E[\Pi]$.

> Note that $n < 3$ is impossible: the third win cannot occur before the third card.

Because Π is a function of N, you can use the result

$$E[f(N)] = \sum_{\text{all } n} f(n) \times P(N=n)$$

to obtain

$$E[\Pi] = \sum_{n=3}^{\infty} \dfrac{2}{n-1} \times \tfrac{1}{2}(n-1)(n-2) p^3 (1-p)^{n-3}$$
$$= p^3 \sum_{n=3}^{\infty} (n-2)(1-p)^{n-3}.$$

Now substitute $n - 3 = r$. This gives also $n - 2 = r + 1$, and means that $n = 3$ corresponds to $r = 0$ while $n = \infty$ corresponds to $r = \infty$, so $E[\Pi]$ becomes

$$E[\Pi] = p^3 \sum_{r=0}^{\infty} (r+1)(1-p)^r.$$

But this is just a binomial series

$$(1-a)^{-2} = 1 + (-2)(-a) + \dfrac{(-2)(-3)}{2!}(-a)^2 + \cdots$$
$$+ \dfrac{(-2)(-3)\ldots(-[r+1])}{r!}(-a)^r + \cdots$$
$$= 1 + 2a + 3a^2 + \cdots + (r+1)a^r + \cdots$$
$$= \sum_{r=0}^{\infty} (r+1)a^r.$$

So

$$E[\Pi] = p^3 \left(1 - [1-p]\right)^{-2} = p^3 p^{-2}$$
$$= p$$

and Π is unbiased, as required.

2 Estimators for the population mean and variance

You have seen how to estimate the mean and variance of a population from a sample. Now that you have looked at the idea of an estimator, you can derive these results more formally.

Consider the following model. You take a random sample of size n from a population which is very large (so that you can consider sampling with or without replacement to be identical). You measure, for each element of the sample, a variable, X, for which the distribution in the population has mean μ and variance σ^2, which are unknown parameters. Taking the sample can be considered as measuring the values of the n random variables $X_1, X_2, \ldots, X_n$ each of which has mean μ and variance σ^2 and which are all independent of one another, because of the assumption that the population is very large. The problem is to use $X_1, X_2, \ldots, X_n$ to construct random variables which are estimators for μ and σ^2.

An unbiased estimator for μ

An obvious candidate for an estimator for μ is

$$\overline{X} = \frac{X_1 + X_2 + \cdots + X_n}{n}$$

the sample mean. To show that M is unbiased, calculate its expectation.

$$\begin{aligned}
\mathrm{E}[\overline{X}] &= \mathrm{E}\left[\frac{X_1 + X_2 + \cdots + X_n}{n}\right] = \mathrm{E}\left[\frac{1}{n}\{X_1 + X_2 + \cdots + X_n\}\right] \\
&= \frac{1}{n}\mathrm{E}[X_1 + X_2 + \cdots + X_n] \\
&= \frac{1}{n}\{\mathrm{E}[X_1] + \mathrm{E}[X_2] + \cdots + \mathrm{E}[X_n]\} \\
&= \frac{1}{n}\{\mu + \mu + \cdots + \mu\} = \mu.
\end{aligned}$$

So $\overline{X}$ is an unbiased estimator for μ.

An estimator for σ^2

You might naturally try, as an estimator of σ^2, the variance of the n elements of the sample.

Define $S^2 = \dfrac{\Sigma(X_i - \overline{X})^2}{n-1}$.

So the numerator of S^2 is $S = (X_1 - \overline{X})^2 + (X_2 - \overline{X})^2 + \ldots + (X_n - \overline{X})^2$.

Multiplying out all the brackets and collecting the similar terms gives

$$\begin{aligned}
S &= \left(X_1^2 - 2X_1\overline{X} + \overline{X}^2\right) + \left(X_2^2 - 2X_2\overline{X} + \overline{X}^2\right) + \cdots + \left(X_n^2 - 2X_n\overline{X} + \overline{X}^2\right) \\
&= \left(X_1^2 + X_2^2 + \cdots + X_n^2\right) - 2\overline{X}(X_1 + X_2 + \cdots + X_n) + n\overline{X}^2 \\
&= \left(X_1^2 + X_2^2 + \cdots X_n^2\right) - 2n\overline{X}^2 + n\overline{X}^2 \\
&= \left(X_1^2 + X_2^2 + \cdots + X_n^2\right) - n\overline{X}^2
\end{aligned}$$

> **Notice that**
> $X_1 + X_2 + \ldots + X_n = n\overline{X}$

Estimators for the population mean and variance

> **Using the results**
> $E[nA] = nE[A]$
> and $E[A + B +...]$
> $= E[A] + E[B] + ...$

So,

$$\Rightarrow E[S^2] = \frac{1}{n-1}\left(E[X_1^2] + E[X_2^2] + \cdots + E[X_n^2]\right) - nE[\overline{X}^2]$$

$$\Rightarrow E[S^2] = \frac{n}{n-1}\left(E[X^2] - E[\overline{X}^2]\right)$$

> You know that for any random variable, R
> $\text{Var}[R] = E[R^2] - (E[R])^2$
> and so
> $E[R^2] = \text{Var}[R] + (E[R])^2$.
> Thus,
> $E[X^2] = \text{Var}[X] + \overline{X}^2$
> and
> $E[\overline{X}^2] = \text{Var}[\overline{X}] + \overline{X}^2$

Hence

$$E[S^2] = \frac{n}{n-1}\left(\text{Var}[X] + \overline{X}^2 - \left(\text{Var}[\overline{X}] + \overline{X}^2\right)\right)$$

$$= \frac{n}{n-1}\left(\text{Var}[X] - \text{Var}[\overline{X}]\right).$$

You can also obtain the familiar result

$$\text{Var}[\overline{X}] = \text{Var}\left[\frac{1}{n}\{X_1 + X_2 + \cdots + X_n\}\right]$$

$$= \frac{1}{n^2}\text{Var}[X_1 + X_2 + \cdots + X_n]$$

$$= \frac{1}{n^2}\{\text{Var}[X_1] + \text{Var}[X_2] + \cdots + \text{Var}[X_n]\}$$

$$= \frac{1}{n^2}\{\sigma^2 + \sigma^2 + \cdots + \sigma^2\} = \frac{\sigma^2}{n}.$$

> The quantity $\frac{\sigma}{\sqrt{n}} = \sqrt{\text{Var}(X)}$
> is the standard deviation of the means of sample of size n.
> It is called the **standard error**.

Thus

$$E[S^2] = \frac{n}{n-1}\left(\text{Var}[X] - \text{Var}[\overline{X}]\right)$$

$$= \frac{n}{n-1}\left(\sigma^2 - \frac{\sigma^2}{n}\right)$$

$$= \sigma^2$$

So S^2 is an unbiased estimator of σ^2, where S^2 is $\left(\frac{n}{n-1}\right) \times$ sample variance.

However $\sqrt{S^2}$ is not an unbiased estimator of σ.

> **Note**
>
> This estimate of the variance is said to have $(n-1)$ **degrees of freedom**. Intuitively this is because one of the n independent sources of variability in the sample is 'used up' in estimating the mean. You are calculating the variance of the sample data about the sample mean, which is the value which minimises this variance, rather than about the true mean.

The sample estimate of the population variance

$$S^2 = \frac{\sum_{i=1}^{n} x_i^2 - n\overline{x}^2}{n-1}$$

can also be written in the form

$$S^2 = \frac{\sum_{i=1}^{n} x_i^2 - n \cdot \overline{x}^2}{n-1}$$

which is sometimes more convenient.

> **Discussion point**
> A local council wishes to assess how successful a scheme has been to promote car sharing for people travelling to work. The council carries out two different surveys to estimate the average number of people in a car.
>
> In the first survey, an observer watches cars arriving at a city centre car park and records the number of people in each car.
>
> In the second survey, an interviewer stops people leaving a car park on foot (having parked their cars) and asks them how many people there were in the car they had been travelling in.
>
> The first survey gives an estimate of 1.73 people per car. The second survey gives an estimate of 2.21 people per car.
>
> Explain why these two methods of estimation give such different answers. State which of them is better. (You should assume that the samples taken were random and large enough for sampling variation not to be the reason for the difference.)

Exercise 16.1

① The random variable W is the wingspan of a very rare species of bird. The distribution of W is not known; nor are the mean and variance.

Sightings of the bird in different places, A, B and C, allow three independent measurements to be made of W. They are treated as values of the random variables W_A, W_B and W_C respectively.

Three different estimators, M1, M2 and M3, are considered for the population mean, μ.

$$M_1 = \frac{1}{3}(W_A + W_B + W_C)$$

$$M_2 = \frac{1}{4}W_A + \frac{1}{2}W_B + \frac{1}{4}W_C$$

$$M_3 = \frac{1}{3}W_A + \frac{2}{3}W_B + \frac{1}{3}W_C.$$

(i) Show that two out of M_1, M_2 and M_3 are unbiased estimators of W but that one is biased.

(ii) State the bias of the biased estimator.

(iii) Find the variances of M_1, M_2 and M_3.

(iv) State, with reasons, which out of M_1, M_2 and M_3 you consider to be the most reliable estimator for μ.

The measurement of W from places A, B and C are 0.95 m, 1.2 m and 0.85 m respectively.

(v) Use your answer to part (iv) to estimate the value of the bird's wing span.

A new sighting is made at a place D. The wingspan of this bird is measured at 1.24 m. It is suggested that the estimated value of μ should be changed to 1.12 m.

(vi) Suggest an estimator, M_4, that gives this result. Do you think it is a good estimator?

S2 AL — Estimators for the population mean and variance

② A sequence of independent measurements are made of a random variable, X. They are treated as values of the random variables $X_1, X_2, X_3, \ldots, X_k, \ldots$
The random variables $M_1, M_2, M_3, M_4, \ldots$ are defined as

$$M_1 = X_1, \quad M_2 = \tfrac{1}{2}(X_1 + X_2), \quad M_3 = \tfrac{1}{3}(X_1 + X_2 + X_3),$$

$$M_4 = \tfrac{1}{4}(X_1 + X_2 + X_3 + X_4), \ldots$$

The mean, μ, of X is unknown. $M_1, M_2, M_3, M_4, \ldots$ are used as estimators for μ.

(i) Show that $E(M_2) = \mu$.
What does this tell you about M_2 as an estimator for μ?

(ii) Show that M_k is an unbiased estimator for μ.

(iii) Show that $\text{Var}(M_2) = \tfrac{1}{2}\text{Var}(X)$.

(iv) Find $\text{Var}(W_k)$ in terms of $\text{Var}(X)$ and k.

(v) Show that if $r > k$ then $\text{Var}(W_r) < \text{Var}(W_k)$.
What does this tell you about larger and smaller samples?

③ (i) Show that if $\alpha_1, \alpha_2, \ldots, \alpha_n$ are any n numbers with the property that

$$\sum_{1}^{n} \alpha_i = 1,$$

then the weighted mean

$$\overline{X}_\alpha = \sum_{1}^{n} \alpha_i X_i$$

is an unbiased estimator of μ.

(ii) The results above show that, given a random sample of size n drawn from a population of which the distribution has mean μ,

$$T_1 = X_1$$

$$T_2 = \tfrac{1}{5}(4X_1 - 2X_2 + 3X_3)$$

$$T_3 = \tfrac{1}{n}(X_1 + X_2 + \ldots + X_n)$$

are all unbiased estimators for μ.
Which of them is preferable and why? (You might try a computer simulation.)

④ X_1 and X_2 are independent random variables. They both have the same mean μ. Their variances are σ_1^2 and σ_2^2, respectively, where σ_1^2 and σ_2^2 are assumed to be known constants.

It is proposed to estimate μ by an estimator T having the form
$$T = c_1 X_1 + c_2 X_2$$
where c_1 and c_2 are constants to be determined.

(i) Show that T will be an unbiased estimator of μ if $c_1 + c_2 = 1$. For the remainder of the question, assume that this condition holds.

(ii) Find an expression for $\text{Var}(T)$ in terms of c_1, σ_1^2 and σ_2^2.
Hence obtain the value of c_1 that minimises $\text{Var}(T)$. Write down the corresponding value of c_2. Find the minimum of $\text{Var}(T)$.

(iii) Summarise the desirable properties possessed by T (with the obtained values of c_1 and c_2) as an estimator of μ. In practice, what difficulty would arise in estimating p in this way?

⑤ Sam has a bag containing n discs numbered $1, 2, \ldots, n$, where n is unknown.

(i) Show that the mean, μ, of the numbers $1, 2, \ldots, n$ is $\frac{1}{2}(n + 1)$.

Sam proposes to take a random sample of three discs, without replacement. Let the mean of the numbers on these three discs be denoted by M.

(ii) Show that $2M - 1$ is an unbiased estimator of n.

Now suppose that the three discs in Sam's sample have the numbers 1, 3 and 11.

(iii) Find the unbiased estimate of n based on Sam's sample.

(iv) Identify a problem with this estimate, and comment on what this indicates about unbiasedness.

⑥ Fluorite is a crystalline material that often occurs with square faces.

Anthea and Bilal are investigating the average size of fluorite squares. They find four pieces of fluorite that they regard as a random sample. The four pieces have square faces with sides 6, 7, 9 and 10 mm respectively. Anthea calculates the best possible estimate, a, of the mean area of a fluorite square. Bilal calculates the best possible estimate, b, of the mean length of side of a fluorite square.

(i) State the best possible estimator of a population mean from a random sample. What two important properties does such an estimator have?

(ii) Show that $\sqrt{a} \neq b$.

(iii) Who, if either, of Andrea and Bilal can claim to have the better estimate of the mean size of fluorite squares.

(iv) The working in part (ii) shows that the square root of an unbiased estimator is biased.

Name a parameter for which this is an important result, explaining exactly what it means.

⑦ (i) The random variable X takes values 1 and 2 with probabilities p and q, where $p > 0$, $q > 0$ and $p + q = 1$.

(a) Find $E(X)$ and $E(\frac{1}{X})$

(b) Determine the values of p and q, if any, for which $E\left(\frac{1}{X}\right) = \frac{1}{E(x)}$.

(ii) Umar and Violet take the same bus to school every day in a big city. The journey is 4 km long, but the journey times are very variable because traffic conditions differ from day to day. Umar and Violet want to estimate the average speed of the bus on this journey. They record the journey times, in minutes, on a random sample of 5 days.

Umar calculates his estimate of the average speed from the average of the 5 journey times. Violet calculates the 5 different speeds and uses the average of those as her estimate. An extract from the spreadsheet they use is shown below.

	Times (minutes)	Speeds (km/hour)
	22	10.9
	28	8.6
	31	7.7
	19	12.6
	21	11.4
Totals	121	51.3

Estimators for the population mean and variance

(a) Calculate Umar's and Violet's estimates of the average speed of the bus.

(b) Explain why the estimates are different.

8 The random variable X has mean μ and variance σ^2. A random sample of two values of X is taken. The two values, treated as random variables, are denoted by X_1 and X_2.

An estimator of μ is $U = aX_1 + bX_2$.

(i) Given that U is to be unbiased, write down the condition on a and b.

(ii) Find, in terms of a and σ^2, an expression for the variance of U.

(iii) Find the value of a, and hence the value b, that minimise the variance of U.

A third value of X, denoted by X_3, now becomes available. A new estimator of μ is $W = cU + dX_3$.

(iv) Given that W is to be unbiased, find the values of c and d that minimise the variance of W. Hence write W in terms of X_1, X_2, and X_3.

9 The random variable Z is normally distributed with mean 0 and variance 1. A spreadsheet was used to investigate whether or not the sample median is a good estimator of the population mean. One hundred samples of size 10 from the distribution of Z were generated. The boxplots below are for the 100 sample means and the 100 sample medians.

Boxplot of sample means

Boxplot of sample medians

(i) State one property of the median of the normal distribution which suggests that the sample median could be used as an estimate of the population mean.

(ii) What does the simulation suggest about the relative merits of the sample mean and sample median as estimators of the population mean?

10 A coin is tossed n times and the number of heads, H is counted. The probability of a head is p.

(i) Show that $P = \dfrac{H}{n}$ is an unbiased estimator of p.

(ii) The variance of the underlying binomial distribution is given by $np(1-p)$. Show that $V = nP(1-P)$ is not an unbiased estimate of the variance of the distribution. What multiple of V is unbiased?

11 A typist makes errors at a rate of r per minute. The number of errors, N, they make in t minutes is counted. Assuming that N is a Poisson variable, write $E[N]$ in terms of r and t.

Show that $R = \dfrac{N}{t}$ is an unbiased estimator of r.

> **Hint:** You may find the binomial sum
> $$\sum_{r=0}^{n} r^2 \,{}^nC_r\, p^r(1-p)^{n-r} = n(n-1)p^2 + np$$
> helpful.

12 An experimenter places six balls, numbered from 1 to 6, in a bag. Three balls are to be drawn from this bag, the largest, L, of the three numbers and the median, M, of the three numbers being noted.

(i) Assuming that the balls are drawn without replacement

  (a) write down all 20 possible samples of size three, and hence calculate the probability distributions for L and M

  (b) use these distributions to calculate the expected values of L and M.

  A subject, who knows that the bag contains n balls numbered 1 to n, but does not know the value of n, is asked to draw three balls from the bag. He uses L and $2M - 1$ as estimates for the value of n.

  (c) Explain why $2M - 1$ is unbiased as an estimator for n, but L is not.

(ii) Repeat part (i) for the case where the three balls are drawn with replacement. (There are 56 distinct samples, not all equally likely.)

13 A fairground game requires players to attempt to throw ping-pong balls into pots; the game is so difficult that balls land essentially at random. The pot that wins £1 has twice the area of the pot that wins a teddy-bear, so that, if p is the probability that a ball lands in the teddy-bear pot, $2p$ is the probability that it lands in the £1 pot and $(1 - 3p)$ is the probability that a ball misses the prize pots altogether. To obtain an estimator Π of p, throw two ping-pong balls and see where they land; then record the value of Π, as in this table.

Table 16.2

If balls land:	Value of Π
both in one of the prize pots	$\frac{1}{3}$
exactly one in one of the prize pots	$\frac{1}{6}$
both missing the prize pots	0

(i) Calculate, in terms of p, the probability of each of these outcomes.

(ii) Hence show that Π is an unbiased estimator of p.

14 A coin is tossed repeatedly until h heads have appeared, and the number of tosses flips, N, required to do this is recorded.

Hint: You will need to recognise a binomial sum.

(i) Explain why, if p is the probability of a head: $P(N = n) = P$(exactly $h - 1$ heads in the first $n - 1$ flips and a head on the nth flip) and hence show that

$$P(N = n) = {}^{n-1}C_{h-1}\, p^h (1-p)^{n-h} \quad (n = h, h+1, \ldots)$$

is the distribution of N.

(ii) Show that $\Pi = \dfrac{h-1}{N-1}$ is an unbiased estimator of p.

15 A library has been given three books belonging to a very rare set. These books carry volume numbers X_1, X_2 and X_3 (where $X_1 < X_2 < X_3$), but it is not known how many volumes there are altogether in the set.

Suppose that there are n volumes, numbered $1, 2, \ldots, n$, in the set and that the three books in the library are regarded as a random sample from this total of n. Two estimators of n are proposed.

$Y = X_3$ + average gap between the observed numbers = $\frac{1}{2}(3X_3 - X_1)$

Z = twice sample median $- 1 = 2X_2 - 1$

S2 AL — Estimators for the population mean and variance

(i) Consider the case where n is, in fact, 3. Show that the value of Y is certain to be 4 and that the value of Z is certain to be 3.

(ii) Consider the case where n is, in fact, 4. Show that the possible values of Y are 4, 5 and $5\frac{1}{2}$.

Show that the mean of Y is 5 and that the variance of Y is $\frac{3}{8}$.

(iii) Still considering the case where n is, in fact, 4, find the possible values of Z and the mean and variance of Z.

(iv) If n is in fact 5, the following results may be derived:

mean of $Y = 6$ variance of $Y = 0.9$

mean of $Z = 5$ variance of $Z = 2.4$.

(You are **not** required to verify these results.)

Using this information and the results you have obtained in parts (i), (ii) and (iii), discuss, with reasons, whether either Y or Z is preferable as an estimator of n, for the values of n considered in this question.

KEY POINTS

1. If the form of the population distribution of a random variable is known or can be assumed, but the value of some parameter, α, of the distribution is unknown, then a statistic, A, calculated from a random sample of values of the variable can often be found which gives an **estimator** of that parameter. The value of the estimator from a particular sample is an **estimate** of the parameter.

2. The estimator is itself a random variable, and the probability, or probability density, with which each of its possible values arises in random samples from the population is called the **sampling distribution** of the estimator.

3. If the expected value $E[A]$ of the estimator is equal to the population parameter α, i.e. if $E[A] = \alpha$, the estimator is said to be **unbiased**. Otherwise, if $E[A] \neq \alpha$ the estimator is **biased** and the **bias** of A is $|E[A] - \alpha|$.

4. The **standard error** of A is the standard deviation of $\bar{A} = \sqrt{\mathrm{Var}(\bar{A})}$.

5. Given a sample of n values $x_1, x_2, \ldots, x_n$ from a distribution with mean μ and variance σ^2, unbiased estimates of μ and σ^2 are given by

$$\bar{x} = \frac{\sum_{i=1}^{n} x_i}{n}$$

$$s^2 = \frac{\sum_{i=1}^{n}(x_i - \bar{x})^2}{n-1} = \frac{\sum_{i=1}^{n} x_i^2 - n\bar{x}^2}{n-1}.$$

LEARNING OUTCOMES

When you have completed this chapter you should:

- understand what an unbiased estimator is
- know that the sample mean is an unbiased estimator of the population mean
- know that $s^2 = \frac{1}{n-1}\sum_{i=1}^{n}(x_i - \bar{x})^2$ is an unbiased estimator of the population variance
- be able to use bias and variance to compare estimators.

17 Variance: confidence intervals and hypothesis tests

Climate change is no longer some far-off problem; it is happening here, it is happening now.

Barack Obama

There is currently a great deal of debate about possible changes to the climate of the Earth through 'global warming'. The climate of the planet is governed by many factors and, as a result, is extremely complicated. We may receive the first indication that the climate has changed significantly from a change in variability rather than from a change in the mean. This chapter looks at ways of testing to see if the variance has changed.

The distribution of S^2

In Chapter 15 you learnt about hypothesis tests for the mean of a Normally distributed random variable. In order to do this you had to know the sampling distribution of $\bar{X}$, the sample mean.

Because $E[\bar{X}] = \mu$ and $Var[\bar{X}] = \dfrac{\sigma^2}{n}$ when sampling from $N(\mu, \sigma^2)$,

- if the population variance is known

$$Z = \dfrac{\bar{X} - \mu}{\dfrac{\sigma}{\sqrt{n}}} \sim N(0,1)$$

- if the population variance is unknown

$$t_{n-1} = \dfrac{\bar{X} - \mu}{\dfrac{S}{\sqrt{n}}}.$$

S is the estimated population standard deviation.

This information enables you to carry out hypothesis tests for the population mean.

In this section you will learn similar tests for the population variance and, to do this, you will need to know the sampling distribution of $S^2 = \dfrac{\sum_{i=1}^{n_x}(x_i - \bar{x})^2}{n-1}$ the unbiased estimate of the population variance. Before doing this you need to take a closer look at the χ^2 distribution.

1 The distribution of S^2

In Chapter 4 you met the χ^2 test in which you used the χ^2 statistic.

If Z is a random variable with distribution $N(0,1)$, then the distribution of Z^2 is the χ^2 distribution with 1 degree of freedom. Similarly, if $Z_1, Z_2, Z_3, \ldots Z_n$, are independent random variables each with distribution $N(0, 1)$, then the distribution of $Y = Z_1^2 + Z_2^2 + Z_3^2 + \ldots + Z_n^2$ is the χ^2 distribution with n degrees of freedom and is written χ_n^2.

$S^2 = \dfrac{\sum(X_i - \bar{X})^2}{n-1}$ is an unbiased estimate of the population variance σ^2 (S^2 is often denoted by $\hat{\sigma}^2$). It can be shown that $\dfrac{(n-1)s^2}{\sigma^2} \sim \chi_{n-1}^2$.

> **Note**
> The proof of this is not difficult but it is beyond the scope of this book.

Hypothesis test for the population variance

Now that you know the sampling distribution of S^2 you are able to carry out hypothesis tests on the population variance. To see how this is done, look at the next two examples.

Example 17.1

The nominal weight of a bag of crisps is 25 grams. A machine is set to fill bags with a quantity of crisps which is distributed $N(26, \tfrac{1}{9})$. This will ensure that only about one bag per thousand will contain less than the nominal amount. After a number of complaints about underweight bags, a sample of ten bags is taken and weighed. The results, in grams, are: 26.1, 26.2, 25.4, 26.6, 26.0, 25.7, 26.4, 26.7, 26.8, 26.1. Does this sample provide evidence at the 5% level that the variance of the weights is now greater than $\tfrac{1}{9}$?

Solution

The hypotheses are: $H_0: \sigma^2 = \frac{1}{9}$

$$H_1: \sigma^2 > \frac{1}{9}$$

This is a one-tailed test at the 5% significance level.

You know that $\frac{(n-1)S^2}{\sigma^2} \sim \chi^2_{n-1}$ and, in this case, $n = 10$.

Under H_0, $\sigma^2 = \frac{1}{9}$, and so $S^2 \sim \frac{1}{81} \chi^2_9$

Looking at your tables, in the column headed 5.0 and the row $\nu = 9$ you can see that the critical value of χ^2_9 is 16.92

Table 17.1

$p\%$	99	97.5	95	90	10	5.0	2.5	1.0	0.5
$\nu = 1$	0.0001	0.0010	0.0039	0.0158	2.706	3.841	5.024	6.635	7.879
2	0.0201	0.0506	0.103	0.211	4.605	5.991	7.378	9.210	10.60
3	0.115	0.216	0.352	0.584	6.251	7.815	9.348	11.34	12.84
4	0.297	0.484	0.711	1.064	7.779	9.488	11.14	13.28	14.86
5	0.554	0.831	1.145	1.610	9.236	11.07	12.83	15.09	16.75
6	0.872	1.237	1.635	2.204	10.64	12.59	14.45	16.81	18.55
7	1.239	1.690	2.167	2.833	12.02	14.07	16.01	18.48	20.28
8	1.646	2.180	2.733	3.490	13.36	15.51	17.53	20.09	21.95
9	2.088	2.700	3.325	4.168	14.68	16.92	19.02	21.67	23.59
10	2.558	3.247	3.940	4.865	15.99	18.31	20.48	23.21	25.19
11	3.053	3.816	4.575	5.578	17.28	19.68	21.92	24.72	26.76
12	3.571	4.404	5.226	6.304	18.55	21.03	23.34	26.22	28.30
13	4.107	5.009	5.892	7.042	19.81	22.36	24.74	27.69	29.82
14	4.660	5.629	6.571	7.790	21.06	23.68	26.12	29.14	31.32

The critical value for S^2 is $\frac{16.92}{81} = 0.2089$

H_0 is rejected if S^2 exceeds this value.

From the sample above, $S^2 = 0.1956$

As $0.1956 < 0.2089$ there is insufficient evidence at the 5% level to reject H_0. The conclusion is that the machine is behaving as intended and that the variance is still $\frac{1}{9}$.

S2 AL — The distribution of S^2

Example 17.2

After an extensive overhaul, the crisp machine in the previous example continues to fill bags with a quantity of crisps which is Normally distributed with mean 26g. However, it is not clear whether or not the variance is still $\frac{1}{9}$. To test this, a sample of 15 bags were weighed and the results were:

$$26.3, 26.2, 25.6, 25.6, 26.0, 26.1, 26.9, 25.6,$$
$$25.1, 25.9, 26.4, 26.9, 26.4, 25.4, 26.9$$

Does this provide evidence at the 5% level that the variance is no longer $\frac{1}{9}$?

Solution

In this case the hypotheses are: $H_0 = \sigma^2 = \frac{1}{9}$

$$H_1 = \sigma^2 \neq \frac{1}{9}$$

This is a two-tailed test at the 5% level.

You know that $\frac{(n-1)S^2}{\sigma^2} \sim \chi^2_{n-1}$, $n = 15$ and that H_0 is $\sigma^2 = \frac{1}{9}$

So $S^2 \sim \frac{1}{126}\chi^2_{14}$

> Use the table on the previous page.

From your tables in the row $\nu = 14$ in the columns headed 2.5 and 97.5 you can see that the critical values of χ^2_{14} are 5.629 and 26.12.

Hence the critical values of S^2 are $\frac{5.629}{126}$ and $\frac{26.12}{126}$ i.e. 0.045 and 0.207.

So reject H_0 if the value of S^2 is outside the interval $(0.045, 0.207)$.

For this sample $S^2 = 0.31$.

0.317 is outside the interval $(0.045, 0.207)$ and hence there is sufficient evidence to reject H_0 and conclude that the variance is no longer $\frac{1}{9}$.

Confidence intervals for the population variance

Now that you have developed a hypothesis test for the population variance, little extra work is required to derive a confidence interval.

You know that $\frac{(n-1)S^2}{\sigma^2} \sim \chi^2_{n-1}$ and so you can use the fact that:

$$P\left(\chi^2_{n-1}(97.5) < \frac{(n-1)S^2}{\sigma^2} < \chi^2_{n-1}(2.5\%)\right) = 0.95$$

to get a 95% confidence interval.

Rearranging gives $P\left(\frac{(n-1)S^2}{\chi^2_{n-1}(2.5\%)} < \sigma^2 < \frac{(n-1)S^2}{\chi^2_{n-1}(97.5\%)}\right) = 0.95$

So the values $\frac{(n-1)S^2}{\chi^2_{n-1}(2.5\%)}$ and $\frac{(n-1)S^2}{\chi^2_{n-1}(97.5\%)}$

are the limits of a 95% confidence interval for σ^2.

Example 17.3

Assuming that the distribution of marks scored in an examination follow a Normal distribution, find a symmetrical 95% confidence interval for the population variance σ^2 if a random sample of 12 marks gave:

51, 60, 50, 65, 64, 50, 60, 66, 71, 62, 62, 68

Solution

With $\nu = 11$ the values of $\chi^2_{11}(2.5\%)$ and $\chi^2_{11}(97.5\%)$ are 21.92 and 3.82

For this sample $S^2 = 49.48$

So the limits of the confidence interval are $\frac{(11)(49.48)}{21.92}$ and $\frac{(11)(49.48)}{3.82}$

And so the 95% confidence interval for σ^2 is (24.83, 142.47)

Interpretation

It is important to realise that this does *not* mean that the probability of σ^2 lying in this interval is 0.95. The population variance, σ^2, has a fixed value which either lies in this interval or not and so the probability that σ^2 is in the interval is either 0 or 1. The correct interpretation is that each sample of size 12 will lead to a different confidence interval and that σ^2 will lie in 95% of these intervals.

In the example above it was specified that a symmetrical confidence interval was required. This means that the probability in each tail has to be equal and so you looked at $\chi^2(2.5\%)$ and $\chi^2(97.5\%)$.

Example 17.4

Assuming that the weights of newborn babies in a certain county is a Normally distributed random variable, calculate a one-sided 95% confidence interval giving an upper bound for the population variance using the two random samples below which were taken on different days (all measurements are in kg).

Table 17.3

Day 1	4.8	3.2	3.1	3.8	3.7	3.5	3.7	
Day 2	3.9	4.4	3.8	3.6	4.8	3.3	4.1	4.6	4.0

Solution

From the data $\quad n_1 = 7$ and $S_1^2 = 0.311$

$\quad n_2 = 9$ and $S_2^2 = 0.230$

So $\frac{6S_1^2 + 8S_2^2}{\sigma^2} \sim \chi^2_{14}$ and so $\frac{3.71}{\sigma^2} \sim \chi^2_{14}$

Notice the use of the pooled estimate of variance.

From your tables you can see that $\chi^2_{14}(95\%) = 6.571$

So $P\left(\frac{3.71}{\sigma^2} > 6.571\right) = 0.95$ and so $P(0.565 > \sigma^2) = 0.95$

This gives the one-sided confidence interval (0, 0.565)

S2 AL — The distribution of S^2

Exercise 17.1

① A random sample of size 16 from a Normal population gave $S^2 = 28.6$.
Test the hypothesis at the 5% level that the population variance is 14.8.

② A random sample of size 10 from a Normal population gave $S^2 = 1.08$.
Test the hypothesis at the 10% level that the population variance is 2.

③ Nine observations of a random variable X were:

10.4   11.12   10.7   10.3   9.9   10.6   11.0   11.1   10.5

Assuming that X is Normally distributed, test, at the 5% level, the hypothesis that the population variance is 0.08.

④ A random sample of size 30 from a Normal population has a variance of 11.5. Construct:

(i) a 95% confidence interval;   (ii) a 98% confidence interval;

for the population variance.

⑤ Ten bags of fertiliser supplied to a garden centre were found to have the following masses;

2.57   2.05   1.65   2.62   2.44   1.48   2.31   1.58   2.60   1.85

Find a 95% confidence interval for the population variance, assuming that the population distribution is Normal.

⑥ The number of hours per day that each of 25 children spent playing computer games was recorded. The sample mean was 2.3 and the sample standard deviation was 1.25. Assume that these times come from a Normal population, $N(\mu, \sigma^2)$.

It is claimed that σ is 1.

Test this claim at the 5% level.

⑦ The finishing time, in minutes, of each of 25 runners in a race was recorded. These times come from a Normal distribution with variance σ^2. The sample variance was 120.

It is claimed that σ^2 is 100.

Test this claim at the 5% level.

⑧ Over a long period of cycling to college, a student had found that the standard deviation of his journey time was 9.3 minutes. After the opening of a new cycle path he recorded the following journey times (in minutes) for a random sample of eight journeys:

48.6   39.9   61.2   50.3   44.7   46.6   51.2   38.3

Test at the 5% level of significance whether there has been a change in the variability of the journey time.

⑨ The management of an airline are concerned about the day-to-day variability in the number of passengers using a certain flight. Over a long period of time, the standard deviation of the number of passengers had been found to be 18.4. In an attempt to reduce this, a simplified fares structure has been introduced. Following this introduction, the numbers of passengers on this flight on ten randomly chosen days were:

87   64   52   73   81   65   73   80   49   70

(i) Does this constitute evidence at the 5% level that the variance has been reduced?

(ii) What assumptions do you make in applying the hypothesis test?

10. A sample of 14 from a certain population gives the results:
    9.1 14.3 11.2 8.4 8.5 14.0 9.9
    8.9 11.0 10.2 10.8 11.4 13.0 9.9
    (i) Test at the 5% level whether the sample could have been drawn from a parent population with variance 2.
    (ii) Test at the 5% level whether the sample could have been drawn from a Normal population with a variance greater than 2.

11. Nine observations of a random variable X were:
    10.4 11.2 10.7 10.3 9.9 10.6 11.0 11.1 10.5
    (i) Assuming that X is normally distributed, find a 90% confidence interval for the population variance.
    (ii) Deduce a 90% confidence interval for the population standard deviation.

12. Over a long period of driving to work, a man had found that his journey time was Normally distributed with a standard deviation of 9.3 minutes. After the opening of a new stretch of motorway, he recorded the following journey times (in minutes) for a random sample of ten journeys:
    37.4 48.6 39.9 61.2 50.3 44.7 46.6 51.2 38.3 45.9
    Construct a 90% confidence interval for the population variance.

13. The managers of a large company are concerned about the day-to-day variability in the number of people arriving late for work. Over a long period of time this number is found to be well modelled by a Normal distribution with standard deviation 18.4. In an attempt to reduce this, new working practices are introduced. To find out whether these are working, the number of people arriving late is recorded on ten randomly chosen working days:
    67 64 52 73 81 65 73 80 69 70
    (i) Construct a 95% confidence interval for the population standard deviation.
    (ii) What conclusion would you come to in the light of your answer to (i)?

14. Three samples are taken from a Normal population and the population variance estimated. The estimates were:
    6.03 from a sample of size 7;
    7.11 from a sample of size 6;
    6.82 from a sample of size 12.
    (i) Pool the sample variances to form an unbiased estimate of the population variance.
    (ii) Does this provide evidence at the 5% level to support the hypothesis that the population variance is:
        (a) 10
        (b) 4

15. Over a long period of driving to work, a commuter had found that the standard deviation of her journey time was 6.8 minutes. After the opening of a new stretch of motorway, the journey times (in minutes) for a random sample of ten journeys were found to be:
    34.2 41.6 56.0 32.1 52.8 44.5 45.1 68.7 48.3 45.4

(i) Test at the 5% level of significance whether there has been a change in the variability of the journey time, stating carefully your null and alternative hypotheses.

(ii) Provide a two-sided 90% confidence interval for the new variance.

(iii) What underlying distributional assumption is required for your analysis to be satisfactory?

[MEI]

2 Hypothesis test for the difference in population variances

In this section a method to determine whether or not two samples could have come from populations with equal variances is developed. For an unpaired t-test to compare two population means to be valid it is necessary for the variances of the two populations to be equal. So you need to test the two samples to see if it is reasonable to assume that they come from populations with the same variance.

Using the F-distribution to test the equality of two variances

A test for the difference between two variances is based on the F-distribution which is defined in the following way:

$$F_{v_1, v_2} = \frac{\dfrac{\chi^2_{v_1}}{v_1}}{\dfrac{\chi^2_{v_2}}{v_2}}$$

where v_1 and v_2 are the degrees of freedom of two independent χ^2 distributions.

Critical values for the F-distribution for different significance levels are given in tables or can be found on your calculator.

Table 17.4

v_2 \ v_1	1	2	3	4	5	6	7	8	10	12	24	∞
1	161.4	199.5	215.7	224.6	230.2	234.0	236.8	238.9	241.9	243.9	249.0	254.3
2	18.5	19.0	19.2	19.2	19.3	19.4	19.4	19.4	19.4	19.4	19.5	19.5
3	10.13	9.55	9.28	9.12	9.01	8.94	8.89	8.85	8.79	8.74	8.64	8.53
4	7.71	6.94	6.59	6.39	6.26	6.16	6.09	6.04	5.96	5.91	5.77	5.63
5	6.61	5.79	5.41	5.19	5.05	4.95	4.88	4.82	4.74	4.68	4.53	4.36
6	5.99	5.14	4.76	4.53	4.39	4.28	4.21	4.15	4.06	4.00	3.84	3.67
7	5.59	4.74	4.35	4.12	3.97	3.87	3.79	3.73	3.64	3.57	3.41	3.23
8	5.32	4.46	4.07	3.84	3.69	3.58	3.50	3.44	3.35	3.28	3.12	2.93
9	5.12	4.26	3.86	3.63	3.48	3.37	3.29	3.23	3.14	3.07	2.90	2.71

To test the equality of the variances, σ_1^2 and σ_2^2, of two Normal distributions, you need to take an independent random sample from each distribution, of size n_1 and n_2, and to calculate unbiased estimates, S_1^2 and S_2^2, of σ_1^2 and σ_2^2.

The hypotheses will be $H_0: \sigma_1^2 = \sigma_2^2 = \sigma^2$
$$H_1: \sigma_1^2 \neq \sigma_2^2$$

So under H_0 $\dfrac{(n_1 - 1)s_1^2}{\sigma^2} \sim \chi_{n_1-1}^2$ and $\dfrac{(n_2 - 1)s_2^2}{\sigma^2} \sim \chi_{n_2-1}^2$

So from the definition of the F-distribution $Fv_1, v_2 = \dfrac{\dfrac{\chi_{v_1}^2}{v_1}}{\dfrac{\chi_{v_2}^2}{v_2}}$

you can deduce that $F_{n_1-1, n_2-1} = \dfrac{\dfrac{(n_1 - 1)s_1^2}{\sigma^2(n_1 - 1)}}{\dfrac{(n_2 - 1)s_2^2}{\sigma^2(n_2 - 1)}}$

Which cancels nicely to give $F_{n_1-1, n_2-1} = \dfrac{s_1^2}{s_2^2}$

This ratio can now be compared with the critical values from the F-tables. In order to reduce the amount of tabulation required, this ratio is always calculated with the larger of the two values in the numerator and so you will always get a value of $F > 1$. This means that only the right-hand tail of F is of interest. A value of F which is close to 1 would suggest that the two population variances are equal. The further F differs from 1, the more likely it is that the variances are not equal. The critical values of F are given in the tables.

You are now in a position to carry out a test for the equality of two variances.

Example 17.5

Angela and Bryony are members of a Golf Club. They are both applicants for the post of Club Captain. One of the criteria involved in making the selection between the applicants is golfing ability. In order to examine whether there is any difference between Angela and Bryony in terms of golfing ability, a random sample of eight scores achieved by Angela and a random sample of 12 scores achieved by Bryony were taken. The scores are recorded below.

Table 17.5

Angela	75	77	71	73	73	79	81	74				
Bryony	74	74	79	80	72	72	75	76	74	79	74	77

(i) State formally the null and alternative hypotheses that are to be tested.

(ii) State an appropriate assumption concerning underlying Normality.

(iii) State a further necessary assumption concerning the underlying distributions.

(iv) Carry out the test, using a 5% level of significance.

Hypothesis test for the difference in population variances

Solution

(i) First define μ_A and μ_B to be the population means of the scores obtained by Angela and Bryony respectively.

The hypotheses are:
$$H_0: \mu_A = \mu_B$$
$$H_1: \mu_A \neq \mu_B$$

(ii) The underlying distributions of the scores obtained by Angela and Bryony must both be Normal.

(iii) The variances of the two distributions, σ_A^2 and σ_B^2 must be equal. You can now test this.

From the two samples you should find: $n_A = 8$ and $n_B = 12$

also $S_A^2 = 11.41$ and $S_B^2 = 7.36$

The hypotheses are:
$$H_0: \sigma_A^2 = \sigma_B^2 = \sigma^2$$
$$H_1: \sigma_A^2 \neq \sigma_B^2$$

This is a two-tailed test at the 5% level, so use the 2.5% points of the F-distribution.

As $11.41 > 7.36$ the relevant F-distribution is $F_{7,11}$.

From your tables you can see that the critical value of F is 3.76 and so you should reject H_0 if $\frac{S_A^2}{S_B^2}$ is greater than this value.

In fact, $\frac{S_A^2}{S_B^2} = \frac{11.41}{7.36} = 1.55$ and so H_0 cannot be rejected and you can assume that the population variances are equal.

> Now you can carry out the two-sample t-test with a clear conscience.

(iv) The hypotheses are:
$$H_0: \mu_A = \mu_B$$
$$H_1: \mu_A \neq \mu_B$$

This is a two-tailed test at the 5% level and the appropriate t-distribution is t_{18}

As t_{18} is not tabulated you will need to interpolate between t_{15} and t_{20}

The value you should get is $t_{18} = 2.104$

From the two saamples $\bar{x}_A = 75.375$ and $\bar{x}_B = 75.5$

The pooled, unbiased estimate of the common variance is given by

$$S^2 = \frac{7(11.41) + 11(7.36)}{18} = 8.94$$

and so $t_{calc} = \dfrac{75.375 - 75.5}{\sqrt{8.94\left(\frac{1}{8} + \frac{1}{12}\right)}} = -0.092$

$0.092 < 2.104$ and so you cannot reject H_0. Hence you conclude that the mean scores obtained by Angela and Bryony are not significantly different.

Example 17.6

The captain of the first team at Fisher Cricket Club wishes to promote one of two promising batsmen, Alan and Brian, from the second team. He wants to make his decision in a statistically meaningful way. To do this he takes a random sample from each player's previous scores for the second team. The samples were:

Table 17.6

Alan	23	27	48	13	34	64	19	22	64	43	12	20
Brian	0	7	98	66	11	17	113	1	2	9		

Their mean scores are very similar, but it seems that Alan is more consistent.

Test at the 5% level to see if the population variance of Alan's scores is less than that of Brian.

Solution

The hypotheses are: $H_0 : \sigma_A^2 = \sigma_B^2 = \sigma^2$
$H_1 : \sigma_A^2 < \sigma_B^2$

This is a one-tailed test at the 5% level, so use the 5% points of the F-distribution.

For these samples $n_A = 12$, $S_A^2 = 336.99$ and $n_B = 10$, $S_B^2 = 1864.04$

As $1864.04 > 336.99$ the appropriate F-distribution is $F_{9,11}$.

$F_{9,11}$ is not tabulated so interpolate between $F_{8,11}$ and $F_{10,11}$

From the tables you can see that the values are 2.95 and 2.85 respectively, therefore take $F_{9,11}$ to have a critical value of 2.90 and reject H_0 if the calculated value of $\frac{s_B^2}{s_A^2}$ is greater than this value.

But $\frac{s_B^2}{s_A^2} = \frac{1864.04}{336.99} = 5.53$ and so you should reject H_0 and conclude that Alan is more consistent than Brian.

Exercise 17.2

① Test, at the 5% significance level, whether it is reasonable to suppose that a random sample of size 25, which gives an unbiased population variance estimate of 5.17, could be from the Normal population as a random sample of size 20 which gives an unbiased population variance estimate of 3.92.

② A sample of size 10 gave an unbiased population variance estimate of 3.44. Another sample of size 8 gave an unbiased population variance estimate of 6.62. Test at the 2% level whether it is reasonable to assume that both samples are taken from the same Normal population.

③ The test scores of each of eight girls and four boys were recorded. The data came from independent Normal populations. The data for the girls had sample standard deviation 10.5 and the data for the boys had sample standard deviation 6.8.

Test whether the population variances are equal.

④ Two samples from independent Normal populations have

$n_1 = 10 \qquad s_1 = 82.3 \qquad n_2 = 18$

Find the least value of s_2 that is consistent with equal population ariances.

⑤ The number of hours per day that each of 12 girls and nine boys spent exercising was recorded. The data came from independent Normal populations. The data for the girls had sample standard deviation 1.25 and the data for the boys had sample standard deviation 1.8.

Test whether the population variances are equal.

⑥ Two independent samples of size 25 are taken from a Normal population. The first gives an unbiased population variance estimate of 28.36. Calculate how large the unbiased population variance estimate of the second sample would have to be before the two samples will be accepted as from different populations at the 5% level, although they are from the same population. (You may recognise this as the probability of making a type I error.)

⑦ Two apprentices were tested by being asked to make 60 identical components on the same machine. Each batch of 60 was measured and the mean and standard deviation **of the sample** calculated.

Table 17.7

	Mean (mm)	Standard deviation (mm)
Apprentice A	14.71	0.017
Apprentice B	14.74	0.037

Does this provide evidence at the 5% level that Apprentice B is less consistent?

⑧ In an experiment on the reaction time in seconds of two individuals, A and B, measured under identical conditions, the following results were obtained:

Table 17.8

Person A	0.41	0.38	0.37	0.42	0.35	0.38
Person B	0.32	0.36	0.38	0.33	0.38	

(i) Show that there is no significant difference between the samples in terms of variance.

(ii) What assumptions have you made?

⑨ The masses (in grams) of two samples of eggs, one sample from species A and one sample from a related species, B, are given below:

Table 17.9

Species A	12.3	13.7	10.4	11.4	14.9	12.6	
Species B	15.7	10.3	12.6	14.5	12.6	13.8	11.9

(i) Test whether there is a significant difference between the two samples in terms of variability and then, if appropriate, test if there is any difference in the mean masses of the two populations.

(ii) What assumptions have you made?

⑩ In order to test the effectiveness of a new rust-proofing treatment for steel, the times until rust started to appear were recorded for two samples, one being given the standard treatment and the other the new treatment. The table below gives the times (in units of 100 days):

Table 17.10

New	22.0	23.9	20.9	23.8	25.0	24.0	21.7	22.8	23.1	23.5	23.0
Standard	23.2	22.0	22.2	21.2	21.6	21.9	22.9	22.8			

(i) Test at the 5% level the hypothesis that the population variances for the new and the standard treatments are the same.

(ii) Test at the 5% level whether the mean time for the new treatment is significantly greater than the mean time for the standard treatment.

(iii) Clearly state the assumptions you needed to make in order to carry out the tests in (i) and (ii).

11 (i) The data below are the number of hours of labour per hectare required for planting winter wheat on a random sample of 15 farms:

2.2 2.5 1.9 2.3 2.6 2.7 1.8 2.0 2.5 2.6 3.4 2.9 3.1 2.7 2.3

Obtain a 95% confidence interval for the variance of the population underlying this data. What assumptions about the statistical nature of this population are necessary for your analysis to be valid?

(ii) It is thought that the variability in the number of hours of labour per hectare for planting spring wheat may be greater than that for winter wheat. A random sample of nine farms gave the following number of hours of labour per hectare for planting spring wheat:

2.7 2.8 3.1 3.8 2.1 2.3 2.9 3.7 2.2

(a) At the 5% level of significance, test whether there is greater variability in the number of hours of labour per hectare for planting spring wheat than for planting winter wheat.

(b) State carefully the null and alternative hypotheses you are testing.

(c) What further assumptions are necessary for your analysis to be valid?

[MEI]

12 In a chemical plant, a certain product is made by two processes. The product always contains a small amount of impurity and it is necessary to know whether the processes are equally variable in terms of the percentage of impurity in the product. The percentages of impurity in a random sample of ten batches of the product made by one process were:

4.4 3.6 4.8 2.8 4.4 4.1 3.9 4.2 4.5 5.1

and the percentages of impurity in a random sample of eight batches of the product made by the other process were:

4.7 4.1 5.6 4.8 3.4 5.4 4.7 3.5

(i) At the 5% level of significance, test whether it is reasonable to assume that the underlying population variances are equal. State your null and alternative hypotheses and the critical value of your test statistic at the 5% level of significance.

(ii) State the assumptions about the underlying distributions that are required for your analysis to be satisfactory.

(iii) In the light of your test, would you be prepared to use a t-test to examine whether the underlying population means are equal? Explain your answer.

[MEI]

⑬ An orchestra performs a weekly series of concerts. Some of these concerts consist wholly or mainly of works that are established as part of the normal concert repertoire; these are referred to as 'traditional' concerts. The other concerts consist mainly of unfamiliar works; these are referred to as 'innovative' concerts. The orchestra managers accept that on the whole, the audiences for the 'innovative' concerts may be smaller than those for the 'traditional' ones, but are concerned about the variability in the audience sizes for the two types of concert.

The next 18 concerts consist of eight 'innovative' and ten 'traditional' concerts. Regard these as random samples. For the eight 'innovative' concerts, the usual unbiased estimate of the population variance of audience size is found to be 818. For the ten 'traditional' concerts, the corresponding estimate is found to be 384.

(i) At the 5% level, test whether the population variances of audience size for the two types of concert may be assumed to be equal.

(ii) Produce a two-sided 80% confidence interval for the population variance of audience size for the 'traditional' concerts.

(iii) Denoting the lower and upper limits of this confidence interval by l and u respectively, a manager states that the probability that the population variance is between l and u is 80%. Explain why this interpretation is incorrect. Give the correct interpretation.

KEY POINTS

1. The distribution of S^2, the unbiased estimate of the variance of a Normal population is given by

$$\frac{(n-1)S^2}{\sigma^2} \sim \chi^2_{n-1}$$

2. The limits of a 95% confidence interval for σ^2, based on a sample of size n, are

$$\frac{(n-1)S^2}{\chi^2_{n-1}(2.5\%)} \text{ and } \frac{(n-1)S^2}{\chi^2_{n-1}(97.5\%)}$$

3. To test the equality of the variances of two Normal populations

   (i) Collect a random sample from each population of sizes n_1 and n_2

   (ii) Calculate S_1^2 and S_2^2, unbiased estimates of the variance of each population

   (iii) Calculate $\frac{S_1^2}{S_2^2}$ (assuming $S_1^2 > S_2^2$)

   (iv) Compare your value of $\frac{S_1^2}{S_2^2}$ with the critical value of F_{n_1-1,n_2-1}

LEARNING OUTCOMES

When you have completed this chapter you should be able to:

▶ carry out a hypothesis test on the variance of a Normal population

▶ construct a confidence interval for the variance of a Normal population

▶ construct a hypothesis test for the equality of variance of two Normal populations

Practice questions: set 4

1. A group of 12 patients were given an experimental treatment for calcium deficiency. One side-effect of the treatment was thought to be reduced appetite and hence weight loss. Over a three-month period of treatment, the mean weight lost by the 12 patients was 2.4 kg with a standard deviation of 1.3 kg.

   (i) Construct a 95% confidence interval from these data. [4 marks]

   (ii) State the assumptions made in constructing the confidence interval. [2 marks]

   (iii) Explain carefully how to interpret the confidence interval. [1 mark]

2. A website allows people to give approval ratings to restaurants by moving a slider on a scale from 0 to 10.

   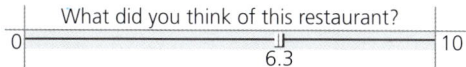

   Figure 1

   Having eaten there, I think that Paulo's Pizza Parlour should get an average rating of 6, and I want to test that hypothesis. The 10 most recent scores for Paulo's Pizza Parlour on the website are as follows.

   3.3   2.2   0.0   4.6   2.8   9.9   8.7   4.9   4.3   8.2

   (i) Identify two scores that look as though they may not be genuine. [1 mark]

   These two scores should be eliminated from the data set in carrying out further analysis. You should assume that the underlying distribution for the genuine scores is $N(\mu, \sigma^2)$.

   (ii) Obtain unbiased estimates of μ and σ^2. Explain what it means to say that an estimate is unbiased. [3 marks]

   (iii) Carry out an appropriate hypothesis test, using a 10% level, to investigate whether or not the true value for μ is 6. [5 marks]

3. A school nurse wishes to investigate the variability in heights of 17-year-old boys and girls. The table summarises the data on random samples of 17-year-old boys and girls in the school. The heights are measured in inches.

	Sample size	Sum of heights	Sum of squares of heights
Boys	24	1641	112 367
Girls	20	1310	86 046

   Table 1

   (i) Find, for boys and girls separately, estimates of the mean and variance of heights. [3 marks]

   (ii) The school nurse is well aware that boys and girls differ in mean height. However, he wishes to investigate whether the variances of the heights of boys and girls differ. Carry out an appropriate hypothesis test for this investigation, using a 5% significance level. [5 marks]

   (iii) Find a 90% confidence interval for the variance of boys' heights. [4 marks]

   (iv) State an assumption required for the procedures carried out in parts (ii) and (iii) to be valid. [1 mark]

S2 AL

Practice questions: set 4

(4) The random variables X and Y are independent and have the following distributions.
$$X \sim N(\mu, \sigma_1^2), \quad Y \sim N(\mu, \sigma_2^2).$$
The random variable $W = aX + bY$, where a and b are constants.

(i) Write down expressions for $E(W)$ and $Var(W)$. [2 marks]

You are now given that W is an unbiased estimator of μ.

(ii) Show that $a + b = 1$. [1 mark]

(iii) Show that W has the smallest possible variance when [5 marks]
$$\frac{a}{b} = \frac{\sigma_2^2}{\sigma_1^2}.$$

Two sets of scales, A and B, are used to weigh a baby. The weights given by the scales are not exact, but are subject to small random fluctuations. Both sets of scales are thought to be unbiased, but the weights given by set A have twice the standard deviation of the weights given by set B. The weight of the baby given by set A is 3.20 kg; the weight given by set B is 3.23 kg.

(iv) Find the best estimate of the weight of the baby. [3 marks]

(5) A psychology student carried out an experiment to investigate whether people's reactions are slower at midnight than at midday. She obtained the reaction times, in seconds, to a visual stimulus for 15 people at midday, and for 12 people at midnight. The data and some analysis is shown in the screenshot.

	A	B	C	D	E	F
1	Midday	Midnight			Midday	Midnight
2	0.27	0.30		Mean	0.2074	0.2876
3	0.24	0.23		Variance	0.0109	0.0116
4	0.33	0.43		Observations	15	12
5	0.18	0.35		Pooled Variance	0.0112	
6	0.19	0.31		Hypothesized Mean Difference	0	
7	0.31	0.19		Observed Mean Difference	−0.0802	
8	0.29	0.11		df	25	
9	0.20	0.37		t stat	−1.9585	
10	0.10	0.21		p value, one-tail	0.0307	
11	0.17	0.16		t crtical one-tail	1.7081	
12	0.41	0.44		p value, two-tail	0.0614	
13	0.02	0.36		t critical two-tail	2.0595	
14	0.10					
15	0.11					
16	0.19					

Figure 2

(i) Give spreadsheet formulae to show how the following cells are calculated.

(a) E5 [1 mark]

(b) E7 [1 mark]

(c) E8 [1 mark]

(d) E9 [1 mark]

(ii) Determine the significance level used to calculate the values in cells E10–E13. [2 marks]

(iii) Explain what conclusion the student should reach at this significance level. [2 marks]

(iv) Identify one important feature that the student's two samples should have. [1 mark]

(v) Identify two important assumptions about the underlying distributions. [2 marks]

6 A study was carried out on the effectiveness of two different pain-killers, tablets P and Q. Ten patients who suffered from frequent headaches were told to use tablet P to treat all their headaches for the next month, and to use tablet Q for all their headaches in the month after that. They were asked to record the times, in minutes, from taking the tablet to gaining relief from the headache. The averages of these times, for each patient with each tablet, are shown in columns B and C in the screenshot. The units are minutes.

	A	B	C	D
1	Patient	Tablet P	Tablet Q	Difference
2	1	73.4	71.8	−1.6
3	2	65.3	73.0	7.7
4	3	86.3	87.7	1.4
5	4	49.3	56.0	6.7
6	5	78.7	85.6	6.9
7	6	34.8	31.8	−3.0
8	7	30.2	26.9	−3.3
9	8	64.1	68.8	4.7
10	9	60.6	68.8	8.2
11	10	38.4	35.8	−2.6
12	mean:	58.11	60.62	2.51
13	variance:	370.51	486.05	23.25

Figure 3

(i) Explain what it means to say that this is a paired sample design rather than a two sample design. Explain also why this paired sample design is preferable to a two sample design. [3 marks]

(ii) Carry out a suitable hypothesis test, using a 5% significance level, to investigate whether pain-killers P and Q differ in their effectiveness. [5 marks]

(iii) State an assumption required for the test in part (ii) to be valid. [1 mark]

Answers

Chapter 1

Discussion point, page 1
Statistics often points the way to new knowledge, for example the health dangers in smoking. In Carlyle's time much of the statistics that you will meet in this book had not yet been discovered. There were no calculators or computers, making it almost impossible to analyse data sets of any size. So what could be achieved was distinctly limited.

Discussion point, page 4
- Are the data relevant to the problem? In an investigation into accidents involving cyclists, collecting data on accidents involving pedestrians would be of no use.
- Are the data unbiased? In an investigation into average earnings, taking a sample consisting only of householders would give a biased sample.
- Is there any danger that the act of collection will distort the data? When measuring the temperature of very small objects, the temperature probe might itself heat up or cool down the object.
- Is the person collecting the data suitable? In a survey into the incidence of drug use, a policeman would not be a suitable person to carry out interviews with individuals since people who use drugs would be unlikely to give truthful answers and so the sample obtained would not be representative.
- Is the sample of a suitable size? In an opinion poll for an election, a sample of size 500 is too small to give a reliable result.
- Is a suitable sampling procedure being followed? In an investigation into the weight of individual carrots from a large crop in a field, sampling 50 carrots from the edge of the field would not be appropriate as those at the edge of the field might grow larger or smaller than in the middle of the field.

Discussion point, page 10

Same	They are illustrating the same data	
Different	**Labels**	
	Frequency chart	The vertical scale is labelled 'Frequency'
	Histogram	The vertical scale is labelled 'Frequency density'
	Scales on the vertical axes	
	Frequency chart	The vertical scale is the frequency
	Histogram	The vertical scale is such that the areas of the bars represent frequency. The frequency density is Trains per minute
	Class intervals	
	Frequency chart	The class intervals are all the same, 5 minutes
	Histogram	The class intervals are not all the same; some are 5 minutes, others 10 minutes and one is 40 minutes.

Exercise 1.1

1. (i) E.g. Members who attend may be of a particular type.
      E.g. Absent members cannot be included.
   (ii) 156, 248, 73, 181
2. (i) Either 19th or 20th
   (ii) $Q_2 = 30$, $Q_1 = 19.5$, $Q_3 = 41$
   (iii) $1.5 \times \text{IQR} = 32.25$; $19.5 - 32.25 < 0$ so no low outliers; $41 + 32.25 = 73.24$ so 80 and 89 are both outliers.
   (iv) Positively skewed.
   (v) No. There is no reason to believe it would be representative. (As European countries tend to be affluent it is unlikely to be representative.)
3. (i) Opportunity
   (ii) The total number of children is 43 and there were 20 mothers. $\frac{43}{20} = 2.15$
   (iii) • The sample is not representative of all women because all the women are mothers; those who have had no children have been excluded.
       • The answer 0 looks like a mistake and needs to be investigated.
       • The word 'exactly' should not be used. At best it is an estimate.
       • Debbie has given her answer to 3 significant figures which implies an unrealistic level of accuracy.
4. (i) Quota
   (ii) • Opinion is almost equally divided for and against.
       • There is a strong difference of opinion between men and women with most men against the scheme and most women in favour.
       • Most adults have made up their minds but many young people don't know.
   (iii) Proportional stratified.
5. (i) The population size appears to be cyclic with period 16 years.
   (ii) 2007 and 2015
   (iii) (a) The population is decreasing sharply.
       (b) The population is steady at about 330 000 animals.
   (iv) A systematic sample is very unsuitable for data with a cyclic pattern. It may select items at about the same points in the cycle and so give misleading information.
6. (i) Cluster
   (ii) The total number of the birds is unknown.
   (iii) $\frac{887}{734} = 1.21$ (2 d.p.)
       There is considerable variability between the clusters and with only four of them it would be surprising if the mean was very close to the true value.
   (iv) The highest rate is D with $1.613\ldots$ fledglings per nest.
       The lowest rate is C with $0.362\ldots$ fledglings per nest.
       These figures suggest outer limits of about 43 000 and 194 000 fledglings.
7. (i) Proportional stratified. Number in group divided by 5 and rounded to nearest integer.
   (ii) Writing the names on sheets of paper and drawing them from a hat.
       Listing and numbering the three groups and then using random numbers to make the selections.
   (iii) Not all samples can be selected, for example one with 20 shops.
   (iv) £395 000
   (v) There are major differences within two of the groups; these need to be understood and addressed in the stratified sampling.
8. (i) Self-selected or opportunity
   (ii) All pregnant women, or all pregnant women in the UK.
       It may well be that only the fitter women agreed to take part.
       People taking part must be volunteers otherwise it would be unethical.
   (iii) 167
   (iv) (a) 3.0% (b) 6.0% (c) About 9.2%
       (d) The categories are not exhaustive as there are others which are not counted in (e.g. day 276).
9. (i) (a) Simple random samples
       (b) A $\bar{x} = 81.52$, $s = 10.02$,
           B $\bar{x} = 77.71$, $s = 10.16$,
           C $\bar{x} = 80.75$, $s = 10.27$
           Statistical variability
       (c) Overall $\bar{x} = 79.88$, $s = 10.20$. Should be more accurate with the larger total sample size.

(ii) $\Sigma(x-\bar{x})^2$ and $\Sigma x^2 - n\bar{x}^2$

$$= \Sigma x^2 - 2(\Sigma x)\bar{x} + n\bar{x}^2$$
$$= \Sigma x^2 - 2(n\bar{x}) \times \bar{x} + n\bar{x}^2$$
$$= \Sigma x^2 - n\bar{x}^2$$

10 (i) 0.667, 1.25
(ii) Mean is reduced by 12.4%, standard deviation by 21.4%
(iii) 0.57%
(iv) 51

Chapter 2

Exercise 2.1

1 (i)

r	1	2	3	4	5
P(X = r)	0.2	0.2	0.2	0.2	0.2

(ii)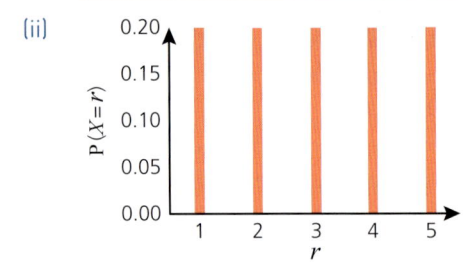

(iii) (a) 0.6 (b) 0.4 (c) 0

2 (i)

r	1	2	3	4
P(X = r)	k	2k	3k	4k

(ii) $k = 0.1$
(iii) (a) 0.4 (b) 0.6

3 (i)

r	3	4	5	6	7	8	9
P(X = r)	$\frac{1}{27}$	$\frac{1}{9}$	$\frac{2}{9}$	$\frac{7}{27}$	$\frac{2}{9}$	$\frac{1}{9}$	$\frac{1}{27}$

(ii)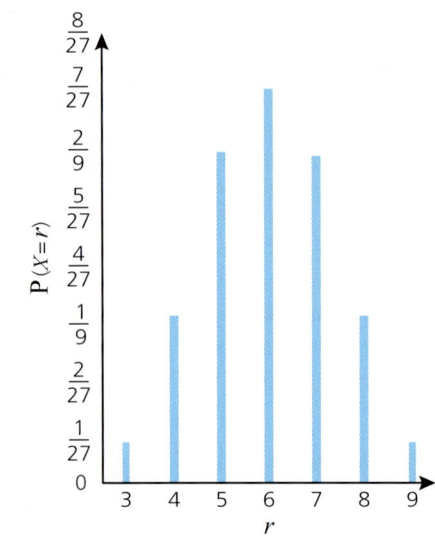

Symmetrical

(iii) (a) $\frac{10}{27}$ (b) $\frac{14}{27}$ (c) $\frac{10}{27}$

4 (i)

r	0	1	2	3	4
P(Y = r)	$\frac{1}{5}$	$\frac{8}{25}$	$\frac{6}{25}$	$\frac{4}{25}$	$\frac{2}{25}$

(ii)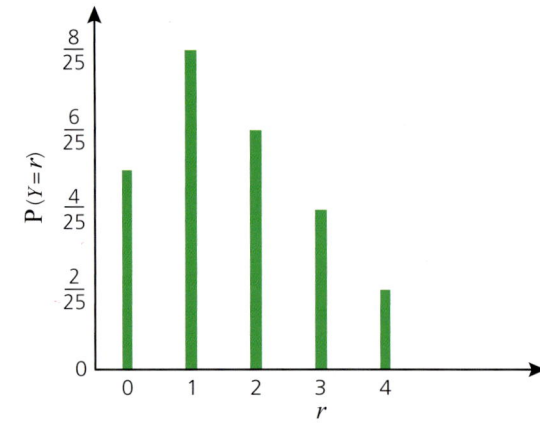

Positive skewed

(iii) (a) $\frac{13}{25}$ (b) $\frac{13}{25}$

5 (i)

r	1	2	3	4	5	6	8	9	10	12	15	16	18	20	24	25	30	36
P(X = r)	$\frac{1}{36}$	$\frac{1}{18}$	$\frac{1}{18}$	$\frac{1}{12}$	$\frac{1}{18}$	$\frac{1}{9}$	$\frac{1}{18}$	$\frac{1}{36}$	$\frac{1}{18}$	$\frac{1}{9}$	$\frac{1}{18}$	$\frac{1}{36}$	$\frac{1}{18}$	$\frac{1}{18}$	$\frac{1}{36}$	$\frac{1}{18}$	$\frac{1}{36}$

(ii) $\frac{3}{4}$

6 (i) $k = \frac{2}{7}$

r	2	3	4	5
P(X = r)	$\frac{1}{7}$	$\frac{3}{14}$	$\frac{2}{7}$	$\frac{5}{14}$

(ii) (a) $\frac{27}{98}$ (b) $\frac{71}{196}$

7 (i) $k = \frac{1}{100}$

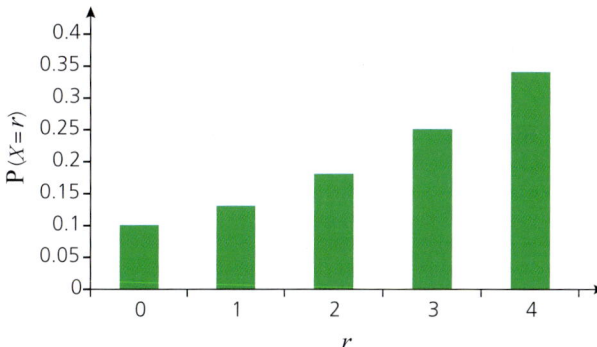

8 (i)

r	0	1	2	3	4
P(X = r)	$\frac{1}{16}$	$\frac{1}{4}$	$\frac{3}{8}$	$\frac{1}{4}$	$\frac{1}{16}$

(ii)

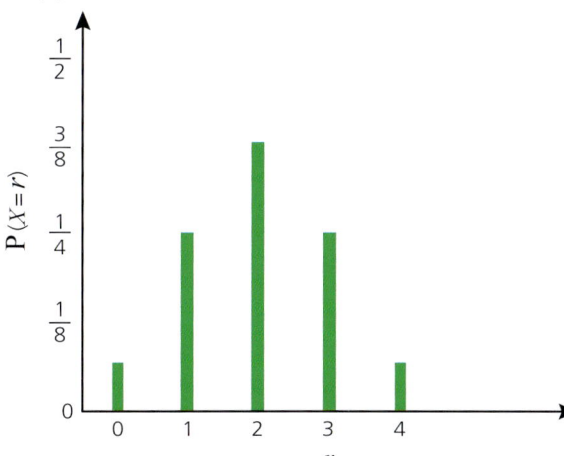

Symmetrical

(iii) $\frac{5}{16}$

(iv) Greater because you cannot have equal numbers of heads and tails, so the probability of more heads than tails will be equal to the probability of fewer heads than tails and both will be equal to $\frac{1}{2}$.

9 (i) $k = 0.014$

r	0	1	2	3	4	5
P(X = r)	$\frac{3}{10}$	$\frac{49}{250}$	$\frac{49}{250}$	$\frac{21}{125}$	$\frac{14}{125}$	$\frac{7}{250}$

(ii) $\frac{1}{25}$

10 (i) $a = 0.1$
(ii) $k = \frac{32}{31}$

(iii)

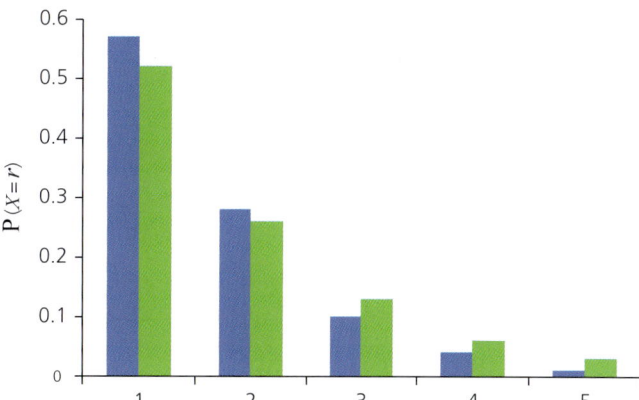

Note: actual values in blue, Model in green. The model is fairly good although it suggests that there are rather fewer cars with 1 or 2 people and rather more with 3, 4 or 5 people.

11 (i) $\frac{1}{1296}$

(ii) $\frac{16}{1296}$ or $\frac{1}{81}$

(iii)

r	1	2	3	4	5	6
P(X ⩾ r)	1	$\frac{625}{1296}$	$\frac{256}{1296}$	$\frac{81}{1296}$	$\frac{16}{1296}$	$\frac{1}{1296}$
P(X = r)	$\frac{671}{1296}$	$\frac{369}{1296}$	$\frac{175}{1296}$	$\frac{65}{1296}$	$\frac{15}{1296}$	$\frac{1}{1296}$

(iv)

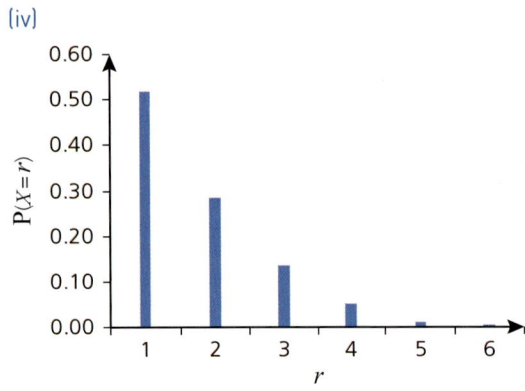

Highly positively skewed

12 (i)

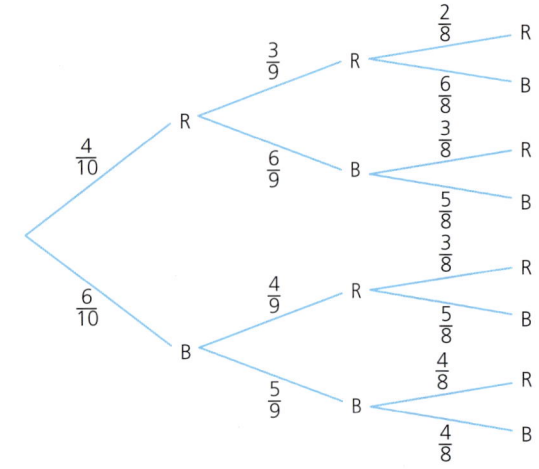

(ii)

r	0	1	2	3
P(X = r)	$\frac{1}{30}$	$\frac{3}{10}$	$\frac{1}{2}$	$\frac{1}{6}$

Discussion point, page 26

By comparing the mean scores in each competition (and perhaps also the standard deviation to get an idea of variation).

Discussion point, page 28

The mean score after the relaxation session is 1.855 as compared with 1.49 before, which suggests that the relaxation session has helped the archers to achieve higher scores. The variance after the relaxation session is 2.084 as compared with 1.97 before, which suggests that the relaxation session has made very little difference to the variation in performance.

Discussion point, page 29

The first method is preferable if the mean is not a round number, since in that case all of the values $(r - \mu)$ will be decimals resulting in a lot of work in the calculations. However, if the mean is a round number the second method may be a little quicker than the first.

Exercise 2.2

1 $E(X) = 3$;
   Because the distribution is symmetrical so take the mid-range value.

2 Mean = 3.2, Variance = 3.26

	A	B	C	D
1	r	P(X = r)	r × P(X = r)	r^2 × P(X = r)
2	1	0.20	0.20	0.20
3	2	0.30	0.60	1.20
4	3	0.10	0.30	0.90
5	4	0.05	0.20	0.80
6	5	0.20	1.00	5.00
7	6	0.15	0.90	5.40
8	SUM	1.00	3.20	13.50

3 (i) $E(X) = 3.125$
   (ii) $P(X < 3.125) = 0.5625$

4 (i) $p = 0.8$

r	4	5
P(Y = r)	0.8	0.2

   (ii)

r	50	100
P(Y = r)	0.4	0.6

5 (i) $E(Y) = 1.944$; $Var(Y) = 2.052$
   (ii) (a) $\frac{5}{9}$ (b) $\frac{1}{18}$

6 (i) $E(X) = 1.5$
   (iii) $Var(X) = 0.75$
   (v) $E(Y) = 5$, $Var(Y) = 2.5$

7 (i)

r	−5	0	2
P(X = r)	0.25	0.25	0.5

   (ii) $E(X) = -0.25$, $Var(X) = 8.188$;
        Negative expectation means a loss
   (iii) £1 less (i.e. £4)

8 (i) $P(X = 1) = 0.4 \times 0.85 + 0.6 \times 0.1$
   (ii)

r	1	2	3	4
P(X = r)	0.4	0.3	0.21	0.09

   (iii) $E(X) = 1.99$, $Var(X) = 0.97$

9 (i) $p = 0.3$, $q = 0.4$

r	3	4	5
P(X = r)	0.3	0.4	0.3

   (ii)

r	20	50	100
P(Y = r)	0.7	0.2	0.1

10 (i)

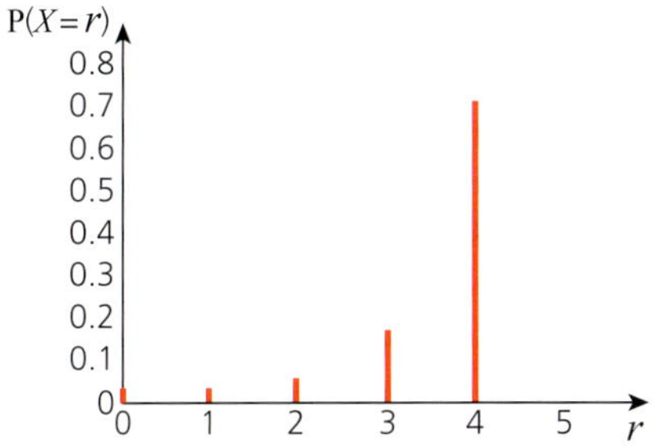

(ii) $E(X) = 3.5, Var(X) = 0.897$ (to 3 s.f.)

(iii) $P(X_1 = X_2) = 0.535$ (to 3 s.f.)

(iv) 0.932 (to 3 s.f.)

11 (i)

r	$P(X = r)$
0	0.03
1	0.10
2	0.13
3	0.16
4	0.18
5	0.17
6	0.12
7	0.09
8	0.02

(ii) $E(X) = 3.92, Var(X) = 3.83$ (to 3 s.f.)

(iii) $k = \frac{1}{84}$

(iv) $E(X) = 4, Var(X) = 3$

(v) The expectations are almost the same and the theoretical variance is only slightly less than the experimental variance, so the model is quite good.

Exercise 2.3

1 (i) (a) $E(X) = 3.1$
(b) $Var(X) = 1.29$

2 (i) (a) $E(X) = 0.7$
(b) $Var(X) = 0.61$

3 (i) $E(A + B + C) = 90, Var(A + B + C) = 23$
(ii) $E(5A + 4B) = 295, Var(5A + 4B) = 353$
(iii) $E(A + 2B + 3C) = 170$, $Var(A + 2B + 3C) = 95$
(iv) $E(4A - B - 5C) = -15$, $Var(4A - B - 5C) = 302$

5 (i) $E(2X) = 6$
(ii) $Var(3X) = 6.75$

6 (i) 10.9, 3.09
(ii) 18.4, 111.24

7 (i) 2
(ii) 1
(iii) 9

8 (i) $P(X = 2) = 0.7 \times 0.3 = 0.21$
(ii) $E(X) = 2.77, Var(X) = 2.42$
(iii) $E(X) =$ £1.36 million, $Var(X) = 1.83 \times 10^{11}$

9 (i)

r	1	2	3	4	5	6
$P(X = r)$	$\frac{11}{36}$	$\frac{1}{4}$	$\frac{7}{36}$	$\frac{5}{36}$	$\frac{1}{12}$	$\frac{1}{36}$

(ii) $E(X) =$ 73.2p, $Var(X) = 1232$
(iii) Make money since cost £1 and average winnings 73.2p.

10 (i) (a) $E(X) = 2.3$ (b) $E(Y) = 1.3$
(c) $Var(X) = 0.81$ (d) $Var(Y) = 0.61$

(ii)

r	1	2	3	4	5	6
$P(Z = r)$	0.04	0.14	0.28	0.31	0.18	0.05

(iii) (a) $E(Z) = 3.6 = 2.3 + 1.3$
(b) $Var(Z) = 1.42 = 0.81 + 0.61$

(iv)

r	−1	0	1	2	3	4
$P(W = r)$	0.1	0.26	0.31	0.22	0.09	0.02

(v) (a) $E(W) = 1 = 2.3 - 1.3$
(b) $Var(W) = 1.42 = 0.81 + 0.61$

11 (i) E (Score) = 3.5 Var (Score) = 2.917
(ii) E (Total Score) = 7.0 Var (Total Score) = 5.833
(iii) E (Difference) = 0 Var (Difference) = 5.833
(iv) E (Total Score) = 35 Var (Total Score) = 29.17

12 E(Amount) = 65 Var (Amount) = 591.7

13 (i) $E(X) = 1$ $Var(X) = 0.5$
(ii) $E(10X - 5) = 5$ $Var(10X - 5) = 50$
(iii) E (50 obs of X) = 50 Var (50 obs of X) = 25
(iv) E (50X) = E(50 obs of X) = 50
but Var (50X) = 1250 = 50 × Var (50 obs of X)

14 (i) E (Length) = 123.5 Var (Length) = 15.25
(ii) $4 \times 0.5 \times 0.2^3 + 4 \times 0.3 \times 0.2^3 + 6 \times 0.3^2 \times 0.2^2 + 0.2^4 = 0.0488$
(iii) E (Length) = 494 Var (Length) = 61

15 (i)

r	−2	0	5
$P(X = r)$	0.6	0.2	0.2

(ii) $E(X) = -0.2$, $Var(X) = 7.36$
(iii) Pay 0.33 pounds less, so pay £1.67.

16 (i) $E(X) = 1.05$, $Var(X) = 0.6475$

(ii)

r	0	1	2	3	4
$P(Y = r)$	0	0.1	0.7	0.2	0

(iii) $E(Y) = 2.1$
(iv) Because X_1 and X_2 are not independent.

Chapter 3

Discussion point, page 44

On average, you would expect $0.25 \times 50 = 12.5$ questions.

Exercise 3.1

1. (i) B(10, 0.5) (ii) 0.2051 (iii) 0.6230
2. (i) 6 (ii) 0.0634 (iii) 0.9095
3. (i) B(10, $\frac{1}{6}$) (ii) $\frac{5}{3}$
   (iii) 0.3230 (iv) 0.2248
4. (i) 3.501×10^{-5}
   (ii) 28 560 (iii) £3.77
5. (i) (a) 25 (b) 0.1 (c) 0.9
   (ii) 0.9020 (iii) $a = 6$
6. (i) 0.8178 (ii) 0.8504 (iii) 0.6689
   (iv) Because in part (ii) you can put 0 in first and 8 in second, 1 and 7, … whereas in (iii) you can only have 4 and 4
7. (i) 0.2637 (ii) 0.6328 (iii) 16.25 times
8. (i) B(20, 0.05) (ii) 0.3585
   (iii) 0.0026 (iv) 1, 0.95
9. (i) 0.4967 (ii) mean 1.6, variance 1.28
11. (i) 0.0301 (ii) 0.0445

Discussion point, page 53

(i) Births would need to occur independently of one another and at a constant average rate throughout the day.

(ii) The 'independence' condition would seem to be reasonable. There is some evidence that more births occur in the morning than at other times of the day.

(iii) This is a complex question and there are many issues to consider. They include:
- Does the expected number of births differ from one month to another?
- For how long is each bed needed?
- How many births are likely to take place at home?
- Is the expected number of births in Avonford likely to change much during the period for which the decisions are being made?
- What happens if there are too few beds, or too many beds?

You may well be able to think of other issues.

Exercise 3.2

1. (i) 0.266
   (ii) 0.826
2. (i) 0.2237
   (ii) 0.1847
   (iii) 0.4012
3. X may be modelled by a Poisson distribution when cars arrive singly and independently and at a known overall average rate; 0.4416
4. (i) 0.058
   (ii) 0.099
5. (i) 25
   (ii) 75
   (iii) 112
   (iv) 288
   Errors occur randomly and independently.
6. (i) 3
   (ii) 27.5
   (iii) 460
7. (i) Aeroplanes land at Heathrow at an almost constant rate (zero at some times of the night), so the event 'an aeroplane lands' is not a random event.
   (ii) Foxes may well occur in pairs, in which case the events would not be independent. (The fact that some regions of area 1 km² will be less popular for foxes than others may be an issue but it is at least partially dealt with by choosing the region randomly.)
   (iii) Bookings are likely not to be independent of one another, as there may be groups requiring more than one table.
   (iv) This should be well modelled by a Poisson distribution (although it would be necessary to assume that the rate of emission was not so high that the total mass was likely to be substantially reduced during a period of 1 minute – i.e. that the half-life of substance is substantially larger than 1 minute).
8. (i) The mean is much greater than the variance therefore X does not have a Poisson distribution.
   (ii) Yes, because now the values of the mean and variance are similar.
   (iii) 0.012
9. (i) 0.175
   (ii) 0.560
   (iii) $\lambda = 10$ 0.1251
   (iv) 0.5421
   (v) 0.0308
   (vi) 10

10 (i) Poisson(1.5) (ii) 0.2510
   (iii) 0.4689 (iv) 0.1847
11 $X \sim$ Poisson(1); 0.014; 0.205

Exercise 3.3
1 (i) 0.0906
  (ii) 8
  (iii) 0.0916; This is very similar to the binomial probability.
2 (i) $\lambda = 5$
  (ii)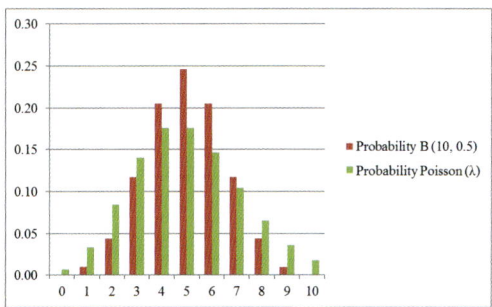
  (iii) The probabilities in the chart showing B(10, 0.5) and Poisson(5) are not very similar at all for each value of X and so the Poisson distribution (λ) is a not good approximation to the B(10, 0.5) distribution.
  (iv) $\mu = 5$
  (v) The probabilities in the chart showing B(100, 0.05) and Poisson(5) are very similar for each value of X. The probabilities in the chart showing B(10, 0.5) and Poisson(5) are not very similar at all for each value of X. This is to be expected because in the former case n is large and p is small whereas in the latter case n is small and p is not small.
3 (i) 0.7153
  (ii) 0.7149
  (iii) Because if n is large and p is small then a Poisson distribution with mean np is a good approximation to a binomial distribution.
4 (i) 0.102
  (ii) 0.3586
  (iii) 0.2545
5 (i) n is large (e.g. $n > 50$) and p is small (e.g. $p < 0.2$)
  (ii) 0.0001
  (iii) 0.00098
  (iv) 10, 9.9
  (v) 0.870
6 (i) (a) 0.180
      (b) 0.264
      (c) 0.916
  (ii) 0.296
  (iii) 0.05

7 (i) 0.135
  (ii) 0.5947
  (iii) 0.125
8 (i) 0.233
  (ii) 0.123
  (iii) 0.262
9 (i) (a) 0.007
      (b) 0.034
      (c) 0.084
  (ii) $T \sim$ Poisson (5.0)
10 (i) (a) 0.134
       (b) 0.848
   (ii) 0.086
   (iii) 0.673
11 (i) 0.531
   (ii) 0.065
   (iii) 0.159
12 (i) (a) 0.257
       (b) 0.223
       (c) 0.168
   (ii) 0.340
13 (i) (a) 0.209 (b) 0.313
   (ii) 0.454
   (iii) North-bound 2, south-bound 3
14 (i) 0.161
   (ii) 0.554
   (iii) 10
   (iv) 0.016
   (v) 0.119

Exercise 3.4
1 (i) $\frac{1}{3}$
  (ii) $\frac{5}{9}$
2 (i) $\frac{1}{3}$
  (ii) $\frac{16}{243}$
  (iii) $\frac{32}{243}$
3 (i) $\frac{1}{2}$
  (ii) $E(X) = 3.5$
  (iii) $\text{Var}(X) = 1.25$
4 (i) $E(X) = 3.5$, $\text{Var}(X) = \frac{35}{12}$
  (ii) $E(X) = 17.5$, $\text{Var}(X) = \frac{175}{12}$
5 $E(X) = 10$, $\text{Var}(X) = 4$
6 (i) $E(X) = 2$, $\text{Var}(X) = \frac{2}{3}$
  (ii) $E(Y) = 20$, $\text{Var}(Y) = \frac{200}{3}$
  (iii) $E(Z) = 20$, $\text{Var}(Z) = \frac{20}{3}$

(iv) Means are equal but variance of Y is larger because it is $10^2 \times \text{Var}(X)$ rather than $10 \times \text{Var}(X)$

Chapter 4

Discussion point, page 70

They should ideally choose a random sample of customers. However, this will be difficult, so they could instead ask, for example, 1 in every 10 customers to fill in a questionnaire. They should do this over a period so that they sample customers who are watching different types of films.

Discussion points, page 78

- the expected frequencies
  The first figure in each cell in the table
- the contributions to the X^2 statistic
  The second figure in each cell in the table
- the degrees of freedom
  Next to df below the table
- the value of the X^2 statistic
  Next to X^2 below the table
- the p-value for the test
  Next to p below the table

The observed frequencies which is the third figure in each cell in the table.

Exercise 4.1

1 (i) Walk 78, Cycle 66, Bus 113, Car 73, Age 13 172, Age 16 158
   (iv) No, because, for instance, younger students might be less likely to be allowed by their parents to cycle.

2 (i)

	Pass	Fail
Less than 10 hours	21.31	9.69
At least 10 hours	33.69	15.31

   (ii)

	Pass	Fail
Less than 10 hours	3.2421	7.1327
At least 10 hours	2.0511	4.5125

3 $X^2 = 35.87$
   $v = 12$
   Reject H_0 at 5% level or above; association

4 $X^2 = 2.886$
   $v = 1$
   Accept H_0 at 5% level or below; independent

5 $X^2 = 0.955$
   $v = 1$
   Accept H_0 at 10% level or below; no association

6 $X^2 = 7.137$
   $v = 4$
   Accept H_0 at 10% level or below; independent

7 $X^2 = 10.38$
   $v = 1$
   Reject H_0 at 1% level (c.v. = 6.635): related

8 $X^2 = 13.27$
   $v = 6$
   Reject at 2.5% level or above: not independent

9 $X^2 = 11.354$
   $v = 4$
   Reject H_0 at 5% level (c.v. = 9.488): association. The cells with the largest value of $\dfrac{(f_0 - f_e)^2}{f_e}$ are medium/induction and long/induction so medium and long service seem to be associated, respectively, with more than and fewer than expected employees with induction-only training.

10 (i) $X^2 = 5.36$
       $v = 2$
       Reject H_0 at 10% level (c.v. 4.605): association
   (ii) Two degrees of freedom, since once the urban/none and urban/one values are fixed, all the other cell values follow from the row and column totals.
   (iii) It appears that fewer rural residents than expected read more than one newspaper.

11 $X^2 = 22.48$
   $v = 9$
   Reject H_0 at 5% level (c.v. = 16.92): association
   Considering the values of $\dfrac{(f_0 - f_e)^2}{f_e}$ for each cell, shows that rural areas seem to be associated with more reasonable and excellent and less poor or good air quality than expected.

12 (i) H_0: Grades in Mathematics are independent of grades in English, H_1: Some association
       Expected frequencies 20, 27.5, 12.5 in each row, so $\chi^2 = 3.95$. Critical value for χ^2 at 10% level with 2 d.f. = 4.605, 3.95 < 4.605 so accept H_0, insufficient evidence to suggest an association between grades in Mathematics and grades in English.
   (ii) May have some cells with expected frequency < 5

13 (i) Some of the expected frequencies may be less than 5.
   (ii) H_0: no association, H_1: association
   (iii) $X^2 = 1.445$, $v = 2$, c.v. = 5.991. Accept H_0 not enough evidence to suggest association.

(iv) Best to combine small and large businesses as these have more similarities than any other group.
H_0: no association, H_1: association
$X^2 = 0.0384$, $v = 1$, c.v. = 3.841. Accept H_0 not enough evidence to suggest association.

14 (i) 4.136, 5.091
(ii) Because some of the expected frequencies are less than 5.
(iii) Combine 'Under 20' and '20–39' and combine '40–59' and '60 or over'.
(iv) H_0: no association, H_1: association. $X^2 = 13.22$, $v = 2$, c.v. = 5.991. Reject H_0 there is enough evidence to suggest association.
(v)

Contribution to test statistic			
	Pop	Classical	Jazz
Under 40	4.060	0.077	3.375
40 and over	3.086	0.058	2.565

The values of 4.06 and 3.09 show that under 40s have a strong positive association with pop, whereas 40 and over have a strong negative association with pop. The values of 3.37 and 2.56 show that under 40s have a strong negative association with jazz, whereas 40 and over have a strong positive association with jazz. The observed frequencies for classical are much as expected.

Discussion point, page 85
The data are real. The model is just your theory.

Discussion point, page 85
One cannot be certain, but it is definitely possible. Estimated $p = 0.0683$, $X^2 = 0.55$, $v = 4$, critical value $= 9.488$; accept H_0, the data fits this distribution.

Discussion point, page 95
The fit looks suspiciously good but see the text that follows.

Exercise 4.2

1.

x	0	1	2	3	≥ 4
Exp	12.13	24.00	22.55	13.38	7.95

2.

x	0	1	2	3	4	≥ 5
Exp	12.11	19.38	15.51	8.27	3.31	1.42

3. H_0: the distribution of the observations can be modelled by the Poisson (2) distribution.
H_1: the distribution of the observations cannot be modelled by the Poisson (2) distribution.
$X^2 = 14.887$, $v = 5$, significant

4. (i) $\bar{x} = 1.725$
H_0: the number of mistakes on a page can be modelled by the Poisson distribution.
$X^2 = 36.3$, $v = 4$, significant
(ii) The mean rate may not be constant, for example, she may make more mistakes when she is tired. The mistakes might not be independent if, for example, some sections are about things she cannot spell.

5. (i) Binomial, $B\left(5, \frac{1}{4}\right)$.
(ii) H_0: the number of white flowers in each tray can be modelled by a binomial distribution, $B\left(5, \frac{1}{4}\right)$.
H_1: the number of white flowers in each tray cannot be modelled by this distribution.
$X^2 = 0.343$, $v = 2$, not significant

6. (i) 1.05 (ii) 36.743, 19.290
(iii) H_0: Poisson model appropriate, H_1: Poisson model not appropriate. Merge last two rows, $\chi^2 = 4.355$. Critical value for χ^2 at 5% level with 2 d.f. = 5.991, 4.356 < 5.991 so accept H_0, Poisson model may be appropriate, manager's claim is justified

7. H_0: the students guessed the answers at random.
H_1: the students did not guess the answers at random.
$X^2 = 18.8$, $v = 4$, significant

8. (i) H_0: the size of rocks is distributed evenly on the scree slope.
H_1: the size of rocks is not distributed evenly on the scree slope.
$X^2 = 7.82$, $v = 2$, significant
(ii) The test shows that rocks of different sizes are not evenly distributed. Another, different test will be needed to determine whether the larger rocks are nearer the bottom of the slope.

9. (i) 3.303. Yes since the sample variance and mean are similar.
(iii) p-value > 0.05 so accept H_0 at 5% level: Poisson model may be appropriate or 3.303 < 9.488 so accept H_0 at 5% level: Poisson model may be appropriate

10. (i) Mean = 0.933, variance = 0.929
(ii) $p = 0.1333$
H_0: these data can be modelled by the binomial distribution.
H_1: these data cannot be modelled by the binomial distribution.

$X^2 = 0.28$, $v = 1$, not significant

(iii) Poisson, because of general spread of data in table and because mean ≈ variance.

11 (i) Several observed frequencies are too small. In order to have $f_e \geq 5$ in each class there would be only two classes. There are two constraints and so no degrees of freedom, therefore the χ^2 test cannot be used.

(ii) $\bar{x} = 0.92$
H_0: the occurrence of Morag's spelling mistakes may be modelled by the Poisson distribution.
H_1: the occurrence of Morag's spelling mistakes may not be modelled by the Poisson distribution.
$X^2 = 2.46$, $v = 1$, not significant

(iii) Spelling mistakes occur singly, randomly and independently. This could be realistic.

12 Reject H_0 because the test statistic is much larger than the critical value, 9.488, at the 5% significance level when $v = 4$.

13 (i)

No. of children	0	1	2	3	4	5+
f_e	246.6	345.2	241.7	112.8	39.5	14.2

(ii) H_0: Number of children per household can be modelled by the Poisson (1.40) distribution.
H_1: Number of children per household cannot be modelled by the Poisson (1.40) distribution.
$X^2 = 32.17$, $v = 5$, significant

14 (ii) 31.17, 1.30

(iii) H_0: binomial model appropriate, H_1: binomial model not appropriate. Merge the first two and last two cells, $\chi^2 = 19.0$. Critical value for χ^2 at 5% level with 3 d.f. = 7.815, 19.0 > 7.815 so reject H_0, binomial model is not appropriate.

15 (i) Constant $n = 7$ seeds in each row, two possible outcomes - each seed either germinates or not, constant probability that a seed germinates, independence between seeds.

(iii) $s = 9.29$, $t = 34.84$

(iv) Merge first three cells and last two cells, $\chi^2 = 10.2$. Critical value for χ^2 at 1% level with 3 d.f. = 11.345, 10.2 < 11.345 so accept H_0, binomial model may be appropriate.

16 (i) H_0: the two dice used in the casino are fair H_1: the two dice used in the casino are not fair.

(iii) Conclusion: $X^2 = 10.1$, $v = 11$, not significant so accept H_0, there is insufficient evidence to suggest that the dice are loaded.

17 (i) H_0: the distribution of the distances customers travel to the store is the same as at the manager's previous store.
H_1: the distribution of the distances customers travel to the store is not the same as at the manager's previous store.
$X^2 = 7.44$, $v = 2$, significant
There are more customers in the 5–10 miles category and fewer in the other two.

(ii) It is unlikely as all the customers were sampled at a similar time on a particular day.

18 (ii) 10.74, 30.2, 3.28

(iii) H_0: B(10, 0.2) model appropriate,
H_1: B(10, 0.2) not appropriate.

(iv) Merge last two cells so $v = (6 - 1) - 1 = 4$

(v) Critical value for χ^2 at 5% level with 4 d.f. = 9.488, 4.17 < 9.488 so accept H_0, B(10, 0.2) model may be appropriate.

Practice questions: set 1 (page 104)

1 (i) Proportional stratified

(ii) Writing the names on sheets of paper and drawing them from a hat.
Listing and numbering the three groups and then using random numbers to make the selections.

(iii) £407 000

(iv) There are major differences within two of the groups; these need to be understood and addressed in the stratified sampling.

2 (i)

W (£)	0	1	10	100
Probability	$\frac{125}{216}$	$\frac{75}{216}$	$\frac{15}{216}$	$\frac{1}{216}$

$$E(W) = \frac{(0 \times 125 + 1 \times 75 + 10 \times 15 + 100 \times 1)}{216} = 1.50 \text{ (2 d.p.)}$$

$$\text{Var}(W) = \frac{(0^2 \times 125 + 1^2 \times 75 + 10^2 \times 15 + 100^2 \times 1)}{216} - (E(w))^2$$
$$= 51.324$$

$SD(W) = 7.16$ (2 d.p.)

(ii) Stall holder's expected profit: $50(2 - E(W))$
£24.77

3 (i) (a) 0.134
(b) 0.848

(ii) 0.086

(iii) 0.673

4 (i) There are relatively few only children and middle born children when compared with first and last born children. This suggests that the majority of families have 2 children.

(ii) The 150 for first born and 90 for last born suggest that first born children may be more likely than last born children to go to university.

(iii) H_0: There is no association between birth order and subject studied at university.
H_1: There is an association between birth order and subject studied at university.

(iv) B9: $= \dfrac{E2 \times B6}{E6}$

B15: $= \dfrac{(B2 \times B9)^2}{B9}$

(v) Degrees of freedom 6.
5% critical value is 12.59.
Observed result is in the critical region.
There is sufficient evidence to suppose that birth order is associated with subject studied at university.

(vi) 4.53: far more middle born children than expected (under H_0) study humanities.
2.24: fewer middle born children than expected (under H_0) study other subjects.

5 (i) The mean number passing per day is 2.53
The mean pass rate is $\dfrac{2.53}{6} = 0.4217$

(ii) Hypotheses
H_0: Binomial is a good fit to the data
H_1: Binomial is not a good fit to the data
Calculate expected frequencies using binomial distribution, $n = 6$, $p = 0.4217$

Number who pass	0	1	2	3	4	5	6
Observed frequency	9	12	33	18	20	7	1
Expected frequency	3.73	16.37	29.83	29.00	15.86	4.63	0.56

Combine cells:

Number who pass	0,1	2	3	4	5,6
Observed frequency	21	33	18	20	8
Expected frequency	20.11	29.83	29.00	15.86	5.19

Calculate test statistic: 7.151
Degrees of freedom: 5 cells, less 1 for totals, less 1 for mean: 3
Critical value, chi-squared 3, 10% tail: = 6.251

Reject H_0. Sufficient evidence at the 10% level of significance to suggest that the binomial is not a good fit to the data.

6 (i) Faulty pixels occur independently of one another and at a uniform average rate.

(ii) (a) 0.3012 (b) 0.9662

(iii) $P(Y = 0) = \dfrac{0.3012}{0.9662} = 0.3117$

(iv) $P(Y = 1) = 0.3741$, $P(Y = 2) = 0.2244$, $P(Y = 3) = 0.0898$

(v) $E(Y) = 1.092$
$E(Y^2) = 2.0798$;
$Var(Y) = 0.8867$;
$SD(Y) = 0.9417$

Chapter 5

Discussion point, page 107

1 There is a negative association (or correlation) between life expectancy and birth rate. This is a situation where correlation does not imply causation. Neither directly causes the other but there are other factors that influence both, such as the level of economic development in a country.

2
- The life expectancy in that country
- The average age of mothers and fathers when they have their first child
- The number of children that people have

Discussion point, page 112

All the diagrams seem to show outliers. There is some evidence of positive correlation in the top right diagram and negative correlation in the bottom right diagram once outliers are removed.

Exercise 5.1

1 (i) 0.8
(ii) −1
(iii) 0

2 (i) (a)

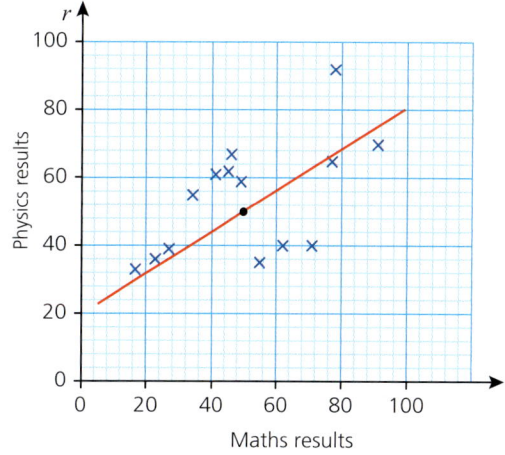

There appears to be positive linear correlation.

(b) 0.560; This confirms there is a positive linear correlation.

(ii) (a)

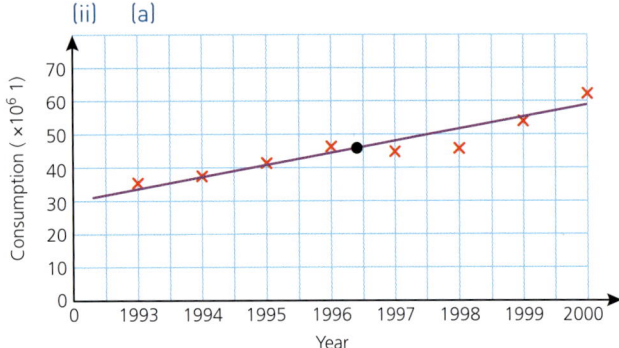

There appears to be strong positive linear correlation.

(b) 0.940; This confirms there is strong positive linear correlation.

(iii)

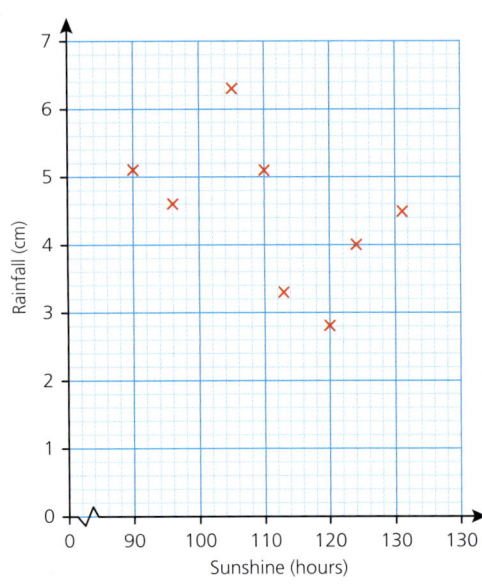

There appears to be no linear correlation (or very weak negative linear correlation).

(b) −0.461; This confirms there is weak negative linear correlation.

3 (i) 0.704
   (ii) −0.635
   (iii) −0.128
   (iv) −0.924
4 (i) 0.221
   (ii) 0.802
   (iii) 0.806

5 (i) 0.913
   (ii) $H_0: \rho = 0, H_1: \rho > 0$
   (iii) Accept H_1
6 (i) 0.380
   (ii) $H_0: \rho = 0$ (Jamila), $H_1: \rho > 0$ (coach)
   (iii) Accept H_0; there is not enough evidence to reject H_0. r needs to be > 0.4973 to reject H_0 at the 5% significance level.
7 (i) Although these data items are a little way from the rest of the data, the sample is very small and with more data it could be that the distribution is bivariate Normal.
   (ii) $H_0: \rho = 0$, $H_1: \rho > 0$
   Accept H_0; there is not enough evidence to reject H_0. r needs to be > 0.7155 to reject H_0 at the 1% significance level.
8 (i) 0.850
   (ii) $H_0: \rho = 0, H_1: \rho > 0$
   (iii) $0.015 < 0.05$ so reject H_0. There is sufficient evidence to suggest that there is positive correlation between chess grade and bridge grade.
9 A −0.79 as x increases y decreases
   B 0.08 no pattern, random scatter
   C 0.68 as s increases t increases
10 (i) 0.946
   (ii) $H_0: \rho = 0, H_1: \rho > 0$
   (iii) Accept H_1; there is very strong positive correlation.
11 (i) The effect size involving A and I is large and positive suggesting that districts with higher values of A have higher values of I. The effect size involving L and M is medium and positive suggesting that districts with higher values of L have higher values of M. The effect size involving A and L is medium and negative, suggesting that districts with higher values of A have lower values of L and vice versa. The other effect sizes are all small.
12 $H_0: \rho = 0, H_1: \rho > 0$; $r = 0.901$; Andrew
13 (i) $H_0: \rho = 0, H_1: \rho \neq 0$. 5% sig. level
   (ii) 0.491, accept H_1

(iii)

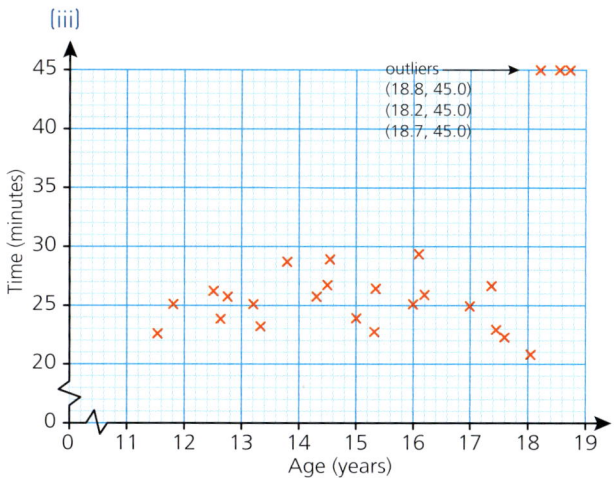

Outliers: (18.8, 45), (18.2, 45), (18.7, 45), it seems as though these girls stopped for a rest.

(iv) The scatter diagram should have been drawn first and the outliers investigated before calculating the product moment correlation coefficient. With the three outliers removed, $r = -0.1612$, accept H_0.

14 (i) The data starts in row 2 and finishes in row 61, so there are 60 rows of data.
(ii) =SUM(B2:B61)
(iii) $S_{xx} = 5006.6$ $S_{yy} = 5862.9$ and $S_{xy} = 1823.3$
(iv) $r = 0.3365$
(v) $H_0: \rho = 0, H_1: \rho > 0$;
(vi) $0.3365 > 0.2144$ so reject H_0; there is sufficient evidence to reject H_0.
(vii) The effect size is medium.
(viii) It suggests that the two scores are somewhat related. This suggests that the things they are measuring, although not identical, have some similarities. The company should therefore carry on giving both tests as the effect size is only medium.

15 $r = 0.59$. Diagram suggests moderate positive correlation which is confirmed by the fairly high positive value of r.
After eliminating the high and low values, $r = -0.145$. Discarding high and low values of x seems to produce an uncorrelated test.

16 (i) -0.816
(ii) Houses are generally cheaper the further they are from the station
(iii) -0.816

17 (i) 1800, 10 (ii) -0.749
(iii) -0.749 (iv) H

18 (i)

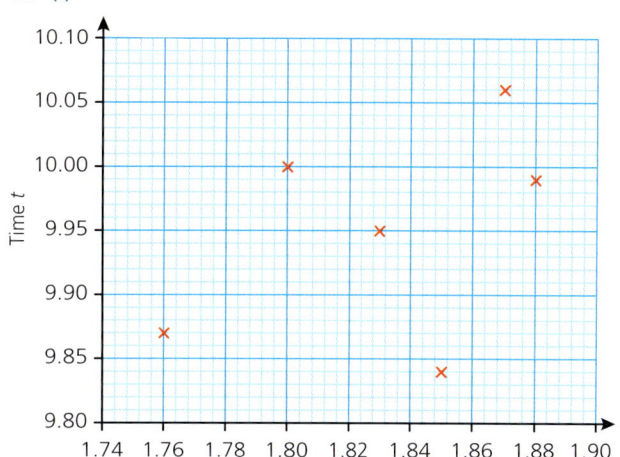

(ii) 0.4386
(iii) $H_0: \rho = 0, H_1: \rho > 0$;
$0.4386 < 0.7293$ so accept H_0; there is not sufficient evidence to reject H_0.
(iv)

Rank h	2	3	5	6	4	1
Rank t	5	3	6	4	1	2

(v) 0.3143
(vi) the correlation coefficient for the ranked data is quite a bit lower than for the unranked. This is probably because ranking loses information.

Exercise 5.2

1 0.576
2 0.20
3 0.50
4 H_0: no association between judges' scores, H_1: positive association between judges' scores.
$r_s = 0.6667 > 0.6429$ accept H_1 at 5% significance level. There is positive association between judges' scores.
5 (i) $0.766, -0.143$
(ii) $H_0: \rho = 0, H_1: \rho > 0$, accept H_1.
(iii) Product moment correlation coefficient is more suitable here because it takes into account the magnitude of the variables.
6 (i) 0.619
(ii) H_0: no correlation between marks H_1: positive correlation. Critical value at 5% (one tailed) is 0.6429. $0.619 < 0.6429$ so accept H_0, insufficient evidence to suggest positive correlation between the judges, sample supports competitor's claim.

7 (i) 0.680
  (ii) $H_0: \rho = 0, H_1: \rho > 0$, $0.680 > 0.6694$ so accept H_1.
  (iii) 0.214
  (iv) The two correlation coefficients measure different quantities. The product moment correlation coefficient measures linear correlation using the actual data values. Spearman's coefficient measures rank correlation using the data ranks. Mr Smith ought to have used Spearman's coefficient.

8 (i)

January	July
1	1
3	2
8	3
2	4
4	5
5	6
9	7
10	8
7	9
6	10

0.636
  (ii) H_0: no association between January and July temperatures, H_1: positive association between January and July temperatures. Accept H_0
  (iii) It is more appropriate to use the product moment correlation coefficient since it utilises the actual data values.

9 (i) 0.806 (or −0.806)
  (ii) H_0: no correlation, H_1: correlation between finishing position and BMI. Critical value at 5% (one tailed) is 0.5636. $0.806 > 0.5636$ so reject H_0, the lower the BMI the higher the position in the race, data support doctor's belief.
  (iii) Finishing positions are not Normally distributed.

10 (i) 0.5
  (ii) H_0: no correlation, H_1: correlation (positive or negative) between h and c. Critical values at 5% level (two tailed) are ±0.7857, $0.5 < 0.7857$ so accept H_0. Data supports councillor's claim.

11 (i) 0.1608
  (ii) H_0: no association between expenditure and score, H_1: positive association between expenditure and score. $0.5035 > 0.1608$ so accept H_0. There is insufficient evidence to suggest positive association between expenditure and score.
  (iii) If $r_s = 0.15$ for the whole population then there is an association but it is weak.

12 (i) $r_s = 0.952$; H_0: no association between prosperity and death rate, H_1: positive association between prosperity and death rate. $0.952 > 0.4637$ so reject H_0. There is insufficient evidence to suggest positive association between prosperity and death rate.
  (ii) It is not justified; a strong association exists but this does not imply that poverty causes a higher death rate.
  (iii) Death rates depend on age distributions, for example an area with many old people will have a higher death rate.

13 (i) 0.881
  (ii) 0.888
  (iii) 0.622
  There is significant correlation at the 5% level in all three cases. There are outliers in (i) contraceptives, (ii) contraceptives and (iii) nuclear power.

14 (i) $H_0: \rho = 0, H_1: \rho \neq 0$; 2-tailed test is used because the analyst does not specify a positive or a negative correlation.
  (ii)

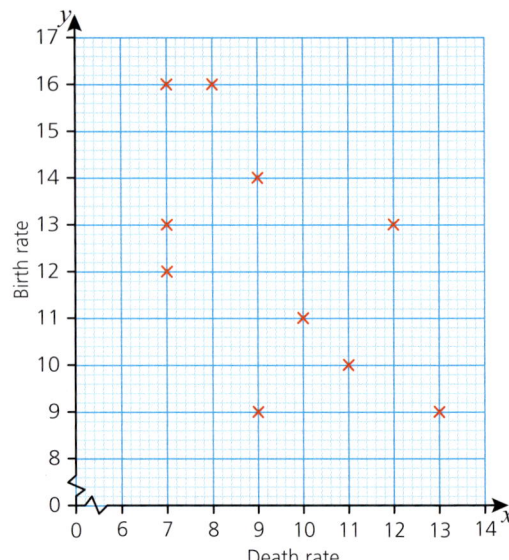

(iii) -0.574
(iv) Since $-0.574 > -0.6319$, accept H_0; the death rates and birth rates are not correlated.
(v) There is weak negative correlation between death rates and birth rates.
(vi) No, the additional evidence shows that H_0 should be rejected.

Chapter 6

Exercise 6.1

1 (i) 15, 26.6
 (ii) 250, -95
 (iii) $y = 0.38x + 32.3$
2 $1.446x + y = 114.38$, 53.7
3 (i) (a) 0.788 (b) 0.758 (c) 0.546
 (ii) 0.048
 (iii) May not be filling a car, may not be buying unleaded petrol
4 (i)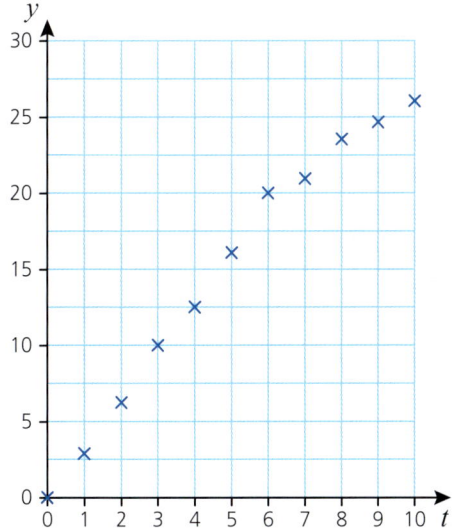
 (ii) $v = 2.68t + 1.54$
 (iii) 0.988
 (iv) Modelling data by a single straight line assumes that the correlation is linear. However, looking at the scatter diagram there is a possibility that r is proportional to a power of t thus making the correlation non-linear.
5 (i) $y = 50.2 - 3.2x$
 (ii) % charge when phone has not been used
 (iii) decrease in % charge each hour of use
 (iv) a because this suggests that the phone does not fully recharge

6 (i) rate of inflation increases and then decreases as years go by, minimum wage increases every ten years
 (ii) wage = 0.088(year) $- 172$
 (iii) The recent values for rate of inflation have been decreasing, but there is no basis for claiming 0%. The recent values have roughly halved every ten years, so 0.8% may be a better prediction. The value $5.76 comes from using the linear regression line. The product moment correlation is $r = 0.947$, so a linear model is a good fit and although the line has been extrapolated the year 2020 is not far outside the data. However, the minimum hourly wage in 2010 was $7.20 and it is unlikely that a government would be allowed to reduce the minimum wage, so the linear model may have ceased to fit.
7 (i)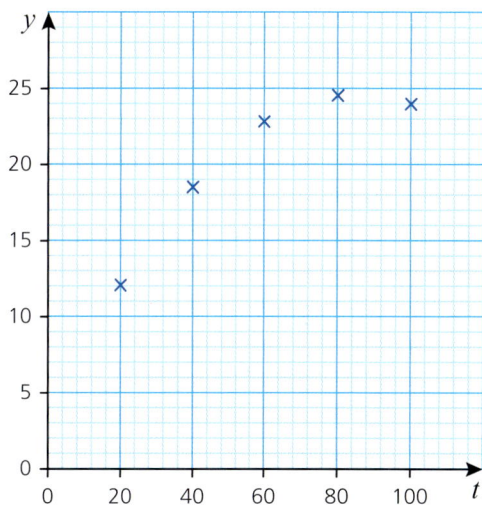
 (ii) $y = 0.1495t + 11.43$
 (iii) When $t = 30$, $y = 15.915$; likely to be accurate.
 When $t = 160$, $y = 35.35$; extrapolation very unlikely to be accurate.
 (iv) The regression line does not provide a good model; the data appear to be distinctly non-linear. Perhaps there is a curved relationship or perhaps a levelling off in y.
8 (i) A straight line seems good for x at the low end of the range, but is suspect at the high end of the range.
 (ii) $y = 77.289 - 0.374x$

(iii) When $x = 145$, $y = 23.06$; likely to be accurate as it is within the range of the data. When $x = 180$, $y = 9.97$; may not be accurate as it is beyond the range of the data.

9 (i)

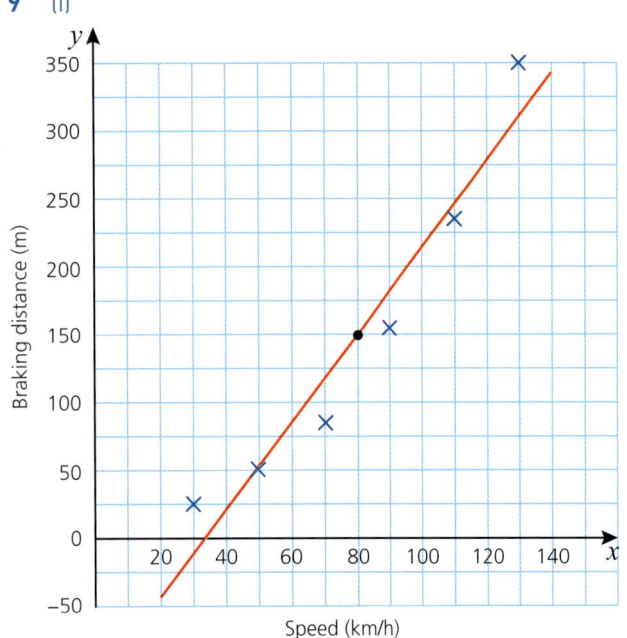

(ii) $y = 3.124x - 107.1$

(iii) When $x = 100$, $y = 241$; seems reasonable since it is within the data range and fits with the neighbouring points.
When $x = 150$, $y = 375$; probably unreliable since the value $x = 150$ lies outside the data range and there is no indication that the linear relationship can be extrapolated.

(iv) The regression line does not seem to provide a good model; a curve may fit the data better.

10 Regression line of y on x: $y = 0.78x + 24.0$ or approx. $y = x + \frac{1}{4}(100 - x)$, thus $\rho \approx \frac{1}{4}$.

Exercise 6.2

1 (i) $y = 3.728x + 4.078$ (ii) 25.80
2 (ii) $y = 3.211x + 84.249$ (ii) 1139.0
3 (i) $y = -0.425 + 0.395x$
　(ii) $f = 0.735 + 0.395m$
　(iii) 93.6
4 (i) 19.9, 153.9
　(ii) $y = 8.89 + 7.73x$

(iii) A typical car will travel 7730 miles each year
(iv) 48 thousand miles

5 (i) $H_0: \rho = 0, H_1: \rho \neq 0$
(ii) $r = 0.661$ (3 s.f.), c.v. = 0.5614, reject H_0; the population must be bivariate Normal
(iii) Regression line is $y = 0.9453x + 18.24$. Estimates are 52.3 cm and 45.7 cm.
(iv) The first is likely to be fairly accurate as interpolation and the value of r^2 is 0.437. The second is less likely to be accurate as the head circumference is well below any occurring in the data (extrapolation). RSS = 49.42

6 (i) 120.8 mm
(ii) Fairly accurate because interpolation and r^2 is fairly close to 1.
(iii) Because it is extrapolation.
(iv) Because you should use the line of regression of x on y. RSS = 197

7 (i) 11.42, 108.3
(ii) $y = -10.7 + 9.48x$
(iii) Every hour spent using the revision course increases the improvement by about 9.48 marks
(iv) 20.6
(v) x varies from 0.5 to 4.0 so extrapolating to $x = 8$ is unreliable, Lee's claim cannot be justified.

8 (i) $p = 1.92 + 0.293v$ (ii) 4.3 tonnes
9 (ii) Points lie close to a straight line
(iii) 24.2, 49.1
(iv) $f = 15.0 + 2.03d$
(v) The fare increases by £2.03 for every 100 km travelled
(vi) $t > 505$

10 (i) 191.6, −5.03
(ii) −0.908
(iii) $w = 11.6 - 0.026t$
(iv) t, for any coin its age is constant but the weight may reduce as the coin gets worn, so the weight is a function of the age
(v) (a) 11.5 grams,
　　(b) decrease of about 0.1 grams
(vi) r will decrease (become closer to −1) because the data will be a better fit to a downward sloping line.

11 (i) Speed is the controlled variable. You can tell this because the speeds are clearly chosen in advance of the test.

(ii)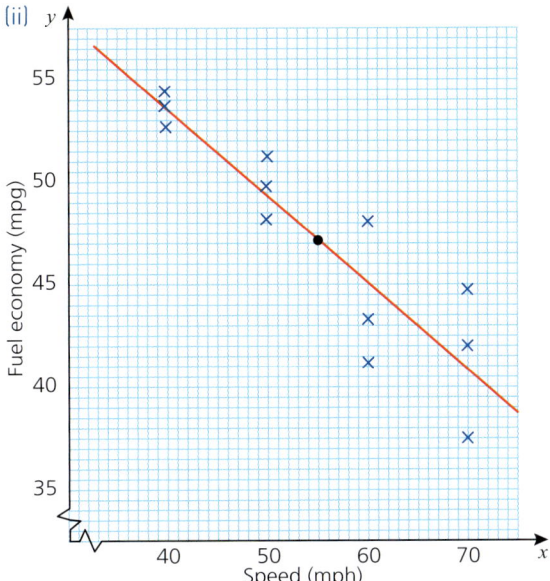

(iii) $y = 70.527 - 0.424x$

(iv) (a) 51.45 mpg
(b) 43.0 mpg

(v) The prediction for 45 mph is more reliable than that for 65 mph since the regression line is a better fit at lower speeds; as the speeds increase, the values for the fuel economy have a wider spread and so the use of the regression line to make predictions, especially at the upper end, is dubious.

(vi) 62.16

12 (i)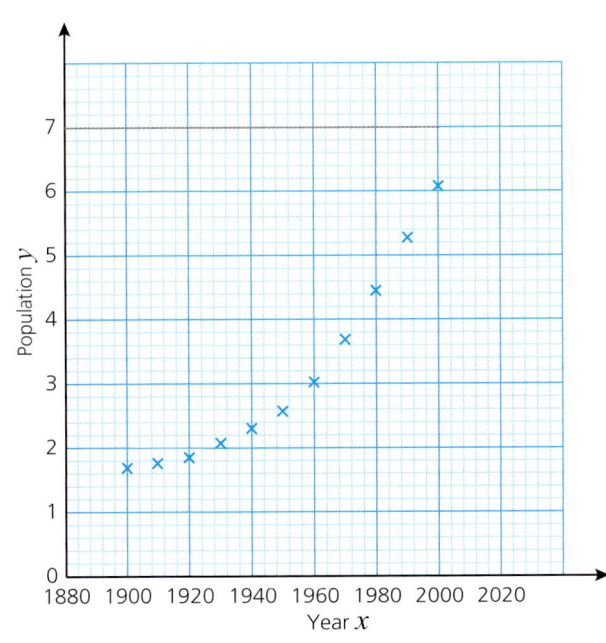

(ii) Because the scatter diagram suggests that there is a relationship but that it should be modelled by a curve.

(iii)

z	0	10	20	30
z^2	0	100	400	900
y	1.67	1.75	1.86	2.07
z	40	50	60	70
z^2	1600	2500	3600	4900
y	2.30	2.56	3.04	3.71
z	80	90	100	
z^2	6400	8100	10000	
y	4.45	5.28	6.08	

(iv)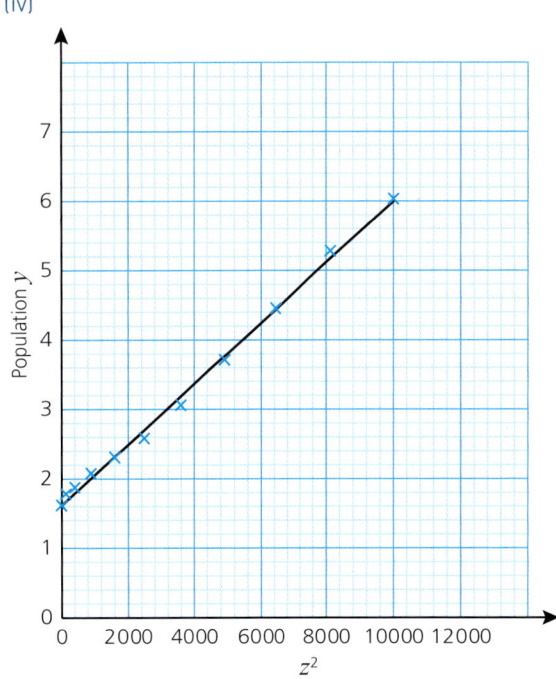

(v) $y = 0.000441z^2 + 1.62$

(vi) 5.60, 19.27

(vii) The first is very reliable as it is interpolation and the line is a very good fit. The second is very unreliable as it is extrapolation, well beyond the data range.

Chapter 7

Discussion point, page 165
It is reasonable since the wave heights are not predictable.

Discussion point, page 170
A symmetrical, B positively skewed, C negavtively skewed, D bimodal.

Exercise 7.1

1. (i) $k = \dfrac{2}{35}$

   (ii)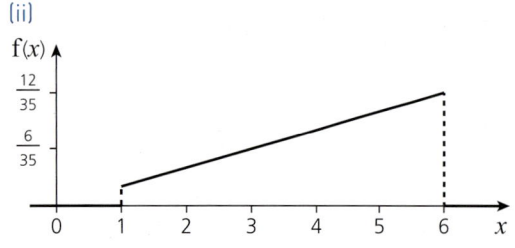

   (iii) $\dfrac{11}{35}$

   (iv) $\dfrac{1}{7}$

2. (i) $k = \dfrac{1}{12}$

   (ii)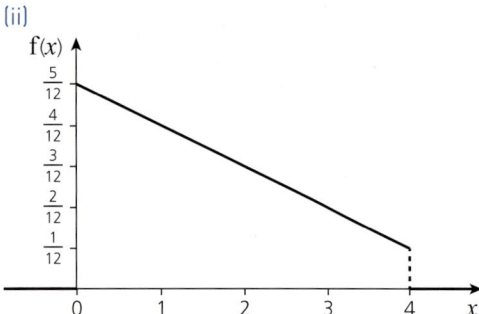

   (iii) 0.207

3. (i) $c = \dfrac{1}{8}$

   (ii)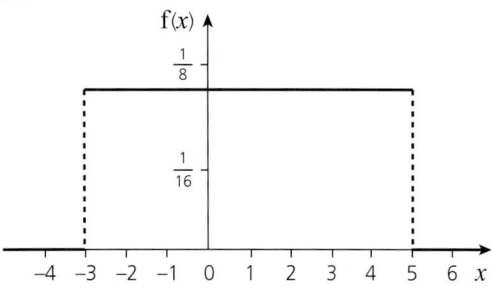

   (iii) $\dfrac{1}{4}$

   (iv) $\dfrac{3}{8}$

4. (i) $k = \dfrac{1}{4}$

   (ii)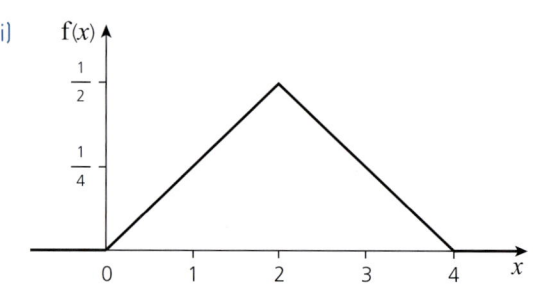

   (iii) $\dfrac{27}{32}$

5. (i) $a = \dfrac{4}{81}$

   (ii)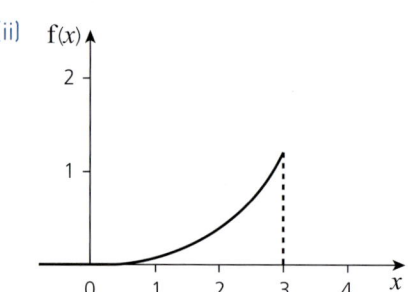

   (iii) $\dfrac{16}{81}$

6. (i) $k = 0.048$

   (ii)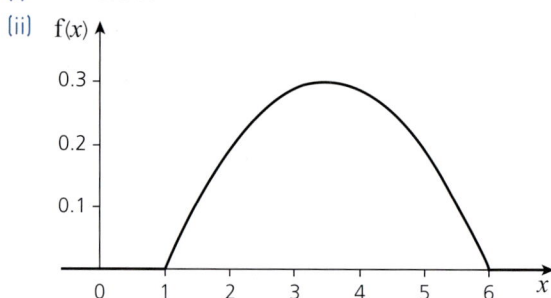

   (iii) 0.248

7. (i) $a = \dfrac{5}{12}$

   (ii)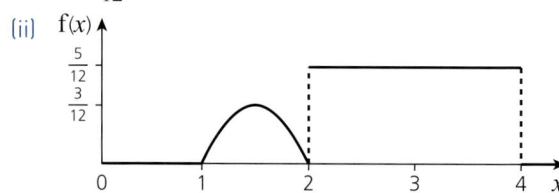

   (iii) 0.292

   (iv) $\dfrac{7}{12}$

8. (i) $k = \dfrac{2}{9}$

   (ii) 0.067

9 (i) $k = \dfrac{1}{100}$

(ii)
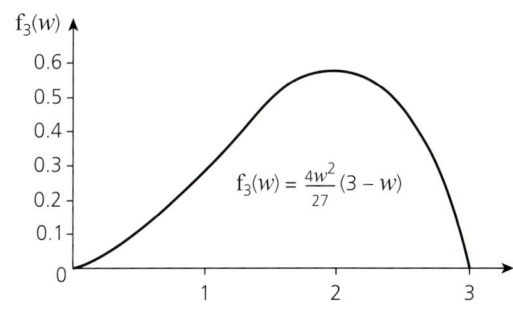

(iii) 19, 17, 28, 36

(iv) Yes

(v) Further information needed about the group 4–10 hours. It is possible that many of these stay all day so are part of a different distribution.

10 (i)
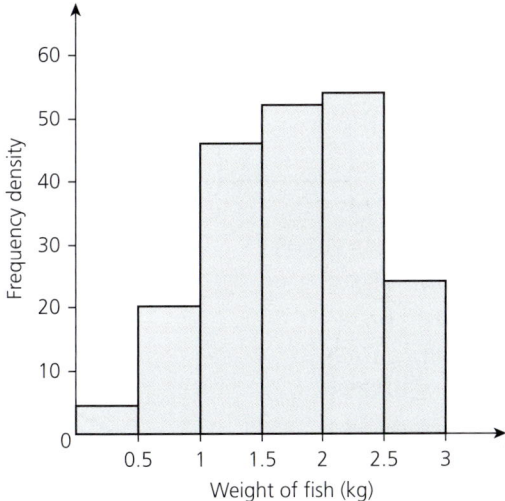

Negative skew

(ii)
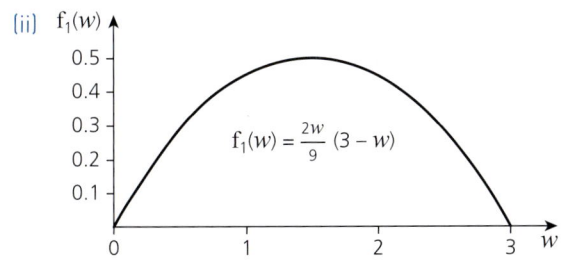

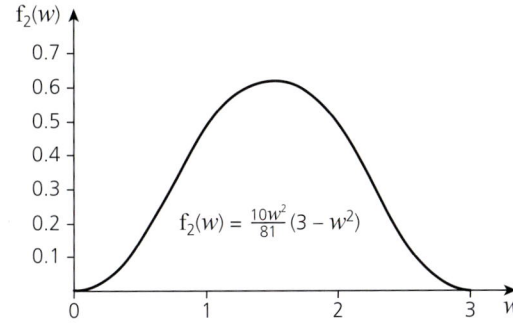

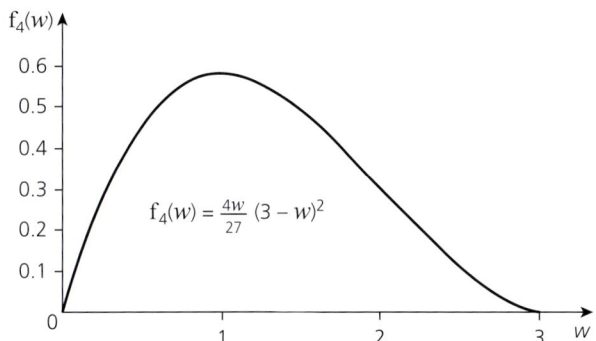

$f_3(w)$ matches the data most closely.

(iii) 1.62, 9.49, 20.14, 28.01, 27.55, 13.19

(iv) Model seems good.

11 (i) $a = 100$

(ii) 0.045

(iii) 0.36

13 (i) 0, 0.1, 0.21, 0.12, 0.05, 0.02, 0

(ii) 0.1, 0.31, 0.33, 0.17, 0.07, 0.02

(iii) $k = \dfrac{1}{1728}$

(iv) 0.132, 0.275, 0.280, 0.201, 0.095, 0.016

(v) Model quite good. Both positively skewed.

Discussion point, page 178

b, c, d, e

Exercise 7.2

1 (i) 2.67

(ii) 0.89

(iii) 2.828

(iv) 0.5, it is lower quartile.

2 (i) 2

(ii) 2

(iii) 1.76

3 (i)
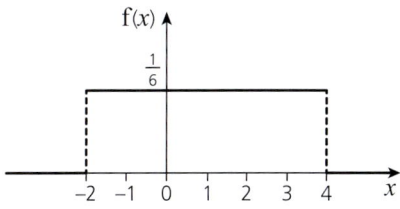

(ii) $\dfrac{2}{3}$

(iii) 1
   (iv) $\frac{1}{2}$
4. (i) 0.6
   (ii) 0.04
5. (i) 1.5
   (ii) 0.45
   (iii) 1.5
   (iv) 1.5
   (v)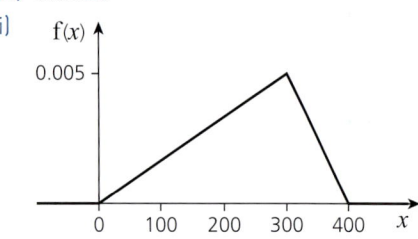

   The graph is symmetrical and peaks when $x = 1.5$ thus E(X) = mode of X = median value of X = 1.5.
6. (ii) 1.083, 0.326
   (iii) 0.5625
7. (i)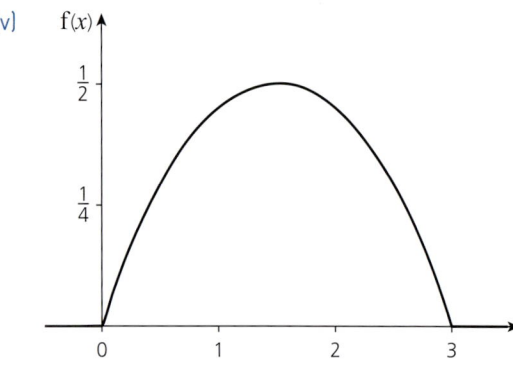

   (iii) $233\frac{1}{3}$ hours
   (iv) 7222.2
   (v) 0.083
8. (i) $k = 1.2 \times 10^8$
   (ii)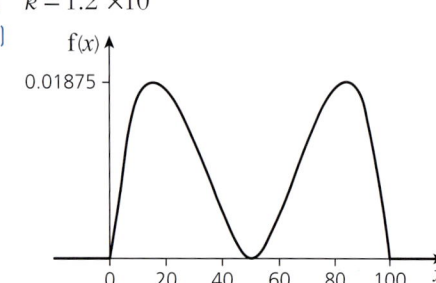

   (iii) The distribution is the sum of two smaller distributions, one of moderate candidates and the other of able ones.
   (iv) Yes, if the step size is small compared with the standard deviation.

9. (i)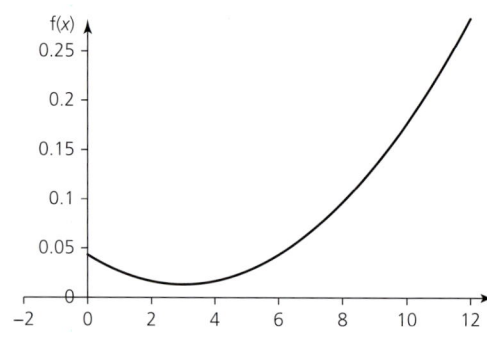

   (ii) 8.88, 2.88; 0.724
   (iii) $m^3 - 9m^2 + 39m - 450 = 0$
10. (i) 3
    (ii) 0.0556
    (iii) 5
    (iv) negative skew so E(X) < 3
11. (i) $a = 1.443$
    (ii)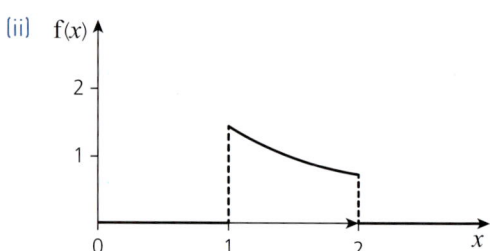

    (iii) 1.443, 0.083
    (iv) 41.5%
    (v) 1.414
12. (i)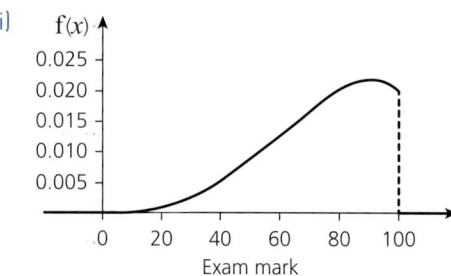

    The model suggests that these candidates were generally of high ability as a large proportion of them scored a high mark.
    (iii) 12.5%
    (iv) No; 91

Exercise 7.3

1. (i) $k = 0.2$
   (ii)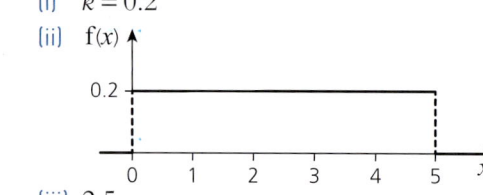

   (iii) 2.5
   (iv) 7

2 (i) 0.8
  (ii) 0.6̇
  (iii) 0.026̇
3 0.8, 0.16, £8
4 (i) 1.5
  (ii) 2.7
  (iii) 0.45
  (iv) 13.9
  (v) 0.45; both are the variance of Y.
5 (i) 0.6
  (ii) −3.4
  (iii) 0.2
  (iv) 0.64
6 (i) $3\frac{2}{3}$
  (ii) $66\frac{1}{6}$
  (iii) $14\frac{5}{6}$, $66\frac{1}{6}$
7 (i) $f(x) = \frac{1}{6}$ for $2 \leq x \leq 8$
  (ii) $a = \frac{\sqrt{3}x^2}{4}$
  (iii) 0.352
  (iv) 12.12, 57.6
8 (i) $E(X) = 3.2$
  (ii) p.d.f.

The model implies that all of the doctor's appointments last between 2 and 10 minutes, the mean time being 5.2 minutes and the variance of the distribution being 2.56 minutes².

9 $k = \frac{1}{36}$, mean = 3, probability = $\frac{5}{32}$
10 (i) $2.5b$
   (iii) $3b^2$
   (iv) $F(x) = \begin{cases} 0 & x < 1 \\ \frac{x-1}{3} & 1 \leq x \leq 4 \\ 1 & x > 4 \end{cases}$
   (v) 2.5

Discussion point, page 190

£150 is quite a large prize, but probably small in comparison to the total entry fees so it may be worth it if the model is very good. The model should work for any number of runners and should be adjustable if the nature of the entry changes.

Exercise 7.4

1 (i) 2.5
  (ii) $F(x) = \begin{cases} 0 & \text{for } x < 0 \\ \frac{x}{5} & \text{for } 0 \leq x \leq 5 \\ 1 & \text{for } x > 5 \end{cases}$
  (iii) 0.4
2 (i) $k = \frac{2}{39}$
  (ii)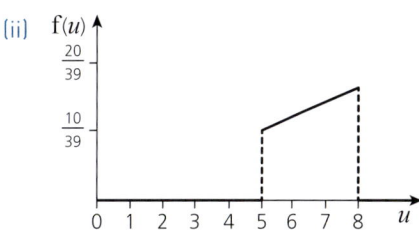
  (iii) $F(u) = \begin{cases} 0 & \text{for } u < 5 \\ \frac{u^2}{39} - \frac{25}{39} & \text{for } 5 \leq u \leq 8 \\ 1 & \text{for } u > 8 \end{cases}$
  (iv)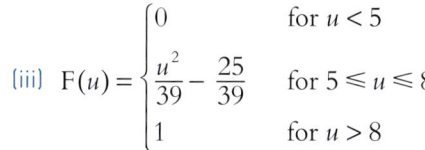
3 (i) $c = \frac{1}{21}$
  (ii) $F(x) = \begin{cases} 0 & \text{for } x < 1 \\ \frac{x^3}{63} - \frac{1}{63} & \text{for } 1 \leq x \leq 4 \\ 1 & \text{for } x > 4 \end{cases}$
  (iii) 3.19
  (iv) 4
4 (i) $F(x) = \begin{cases} 0 & \text{for } x < 0 \\ 1 - \frac{1}{(1+x)^3} & \text{for } x \geq 0 \end{cases}$
  (ii) $x = 1$
5 (i) $\frac{1}{4}$
  (ii) 0.134
  (iii) $f(x) = 2 - 2x$ for $0 \leq x \leq 1$

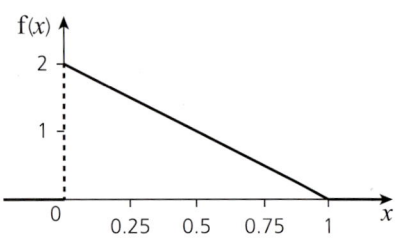

6 $E(X) = \frac{3}{4}$, $Var(X) = \frac{19}{80}$

$$F(x) = \begin{cases} 0 & \text{for } x < 0 \\ \frac{3x}{4} - \frac{x^3}{16} & \text{for } 0 \leq x \leq 2 \\ 1 & \text{for } x > 2 \end{cases}$$

7 $\frac{3}{5}$, 0.683

8 (ii) $F(t) = \begin{cases} 0 & \text{for } t < 0 \\ \frac{t^3}{432} - \frac{t^4}{6912} & \text{for } 0 \leq t \leq 12 \\ 1 & \text{for } t > 12 \end{cases}$

(iv) 0.132

9 $F(x) = \frac{k}{2} - \frac{k\cos 2x}{2}$; 0.146

10 (i) (a) 0.3935   (b) 0.2231

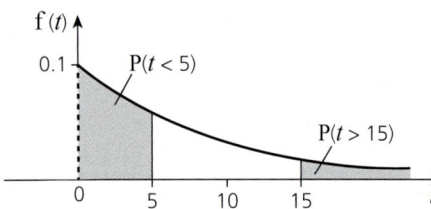

(ii) $F(t) = 1 - e^{-0.1t}$; median = 6.93

(iii) 0.183

11 (ii) $1.25\left(1 - \frac{1}{m}\right) = \frac{1}{2}$

(iii) 0.495

(iv) $f(x) = \frac{1.25}{x^2}$ for $1 \leq x \leq 5$

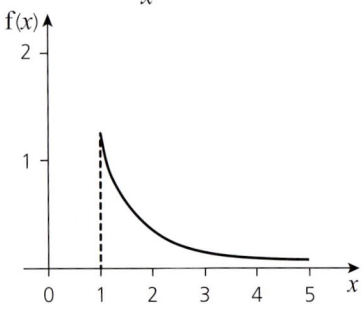

(v) $m = \frac{2b}{b+1}$, which is always less than 2

12 (i) $M = 3.568$; 5.335

(ii) $f(x) = \begin{cases} \frac{324}{x^5} & \text{for } x \geq 3 \\ 0 & \text{otherwise} \end{cases}$

(iii) $\frac{81}{256}$

13 $F(x) = \int f(t) \, dt$ and so $F'(x) = f(x)$.

Uniform distribution with mean value $\frac{a}{2}$.

$F(x) = 1 - \frac{(a-x)^2}{a^2}$ for $0 \leq x \leq a$.

Mean of sum of smaller parts = a

14 (i) (a) Validates p.d.f. form of Z.
   (b) Demonstrates that $E(Z) = 0$.
(ii) (a) $E(Y) = 1$
   (b) $Var(Y) = 2$

Exercise 7.5

1 (i)

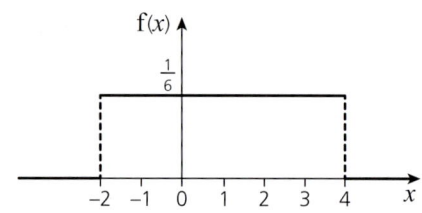

(ii) $\frac{2}{3}$ (iii) 1   (iv) $\frac{1}{2}$

2 (i) $m = 0.5$, $c = -1$
(ii) $f(x) = 0$ for $x < 2$, $f(x) = 0.5x - 1$ for $(2 \leq x \leq 4)$, $f(x) = 1$ for $x > 4$
The distribution is uniform (rectangular).
(a) $\frac{1}{2}$ (b) $\frac{1}{2}$ (c) 1

3 (i) 2.5

(ii) $F(x) = \begin{cases} 0 & \text{for } x < 0 \\ \frac{x}{5} & \text{for } 0 \leq x \leq 5 \\ 1 & \text{for } x > 5 \end{cases}$

(iii) 0.4

4 (i) $f(x) = \frac{1}{3}$ for $4 \leq x \leq 7$

(ii) 5.5   (iii) $\frac{3}{4}$ (iv) 0.233

5 3, 9, $\frac{1}{6}$

6 (i) $f(x) = \frac{1}{10}$ for $10 \leq x \leq 20$

(ii) 15, 8.33

(iii) (a) 57.7%   (b) 100%

7 (i) $k = 0.2$

(ii)
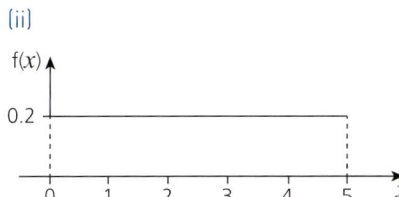

(iii) 2.5 (iv) 7

8 (i) $f(x) = \frac{1}{6}$ for $2 \leq x \leq 8$

(ii) $a = \frac{\sqrt{3}x^2}{4}$

(iii) 0.352 (iv) 12.12, 57.6

9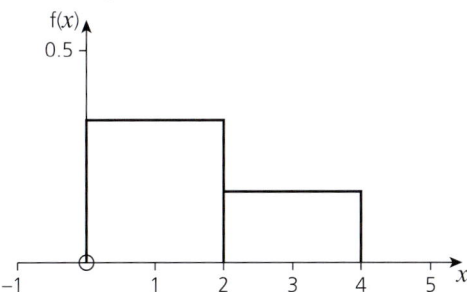

(i) $\frac{2}{3}$ (ii) $\frac{2}{3}$ (iii) $\frac{x}{3}$

(iv) $\frac{x+2}{6}$

(v) 0

For $x < 0$, $f(x) = 0$

$0 \leq x \leq 2$, $f(x) = \frac{x}{3}$

$2 < x \leq 4$, $f(x) = \frac{x}{6}$

$x > 4$, $f(x) = 0$

Mean = $\frac{5}{3}$

10 (i) $\mu = \frac{b+a}{2}, \sigma^2 = \frac{(b-a)^2}{12}$

(ii) $E(\overline{X}) = \frac{b+a}{2}, \text{Var}(\overline{X}) = \frac{(b-a)^2}{300}$

(iii) Estimate for $\mu = 49.7$, estimate for $\sigma^2 = 9.61$, which gives $a = 44.3$ (3 s.f.), $b = 55.1$ (3 s.f.)

11 (i) p.d.f. is $f(x) = 0.25$ for $1 \leq x \leq 5$, 0 otherwise. Median = 3.

(ii) c.d.f. is $F(x) = 0$ for $x \leq 1$, $\frac{x}{4} - \frac{1}{4}$ for $1 \leq x \leq 5$, and 1 for $5 \leq x$.

(iii) $P(Y \leq y) = P(X \leq \sqrt{y}) = \frac{\sqrt{y}}{4} - \frac{1}{4}$.

Differentiating, $g(y) = \frac{1}{8\sqrt{y}}$ for $1 \leq y \leq 25$.

(iv) Median = 10.333

(v)
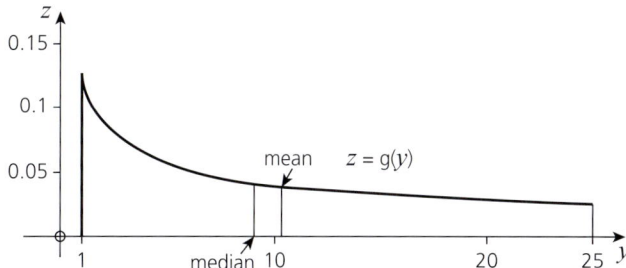

Discussion point, page 207

(a) Scale. The curve intersects the vertical axis at $f(t) = 0.2$.

(b) The vertical axis should be labelled 'frequency'

(c) The vertical scale on a frequency polygon represents the frequencies of the mid points so the label should 'Mid-point frequency'. The scale the same as in graph B. However no meaning can be assigned to the intermediate points on the various line segments.

(d) The label on the vertical axis should be 'frequency density' and the scale should be such that, in order, the heights of the bars are 4.8, 2.4, 1.4, and 0.7.

(e) The vertical axis should be labelled 'probability density' and the scale should be such that, in order, the heights of the bars are 0.096, 0.048, 0.028 and 0.014. The frequency polygon is the worst display since the intermediate points on the line segments have no meaning (and so are confusing). Choosing between B, D and E is to an extent a matter of opinion, but D has the advantage of being closest in style to the theoretical distribution shown in A.

Chapter 8

Discussion point, page 211

$P(Y \cap C)$ = Probability tests positive and has the condition = 0.20.

$P(Y \cap C')$ = Probability tests positive and does not have the condition = 0.15.

$P(Y' \cap C)$ = Probability tests negative and has the condition = 0.05.

$P(Y' \cap C')$ = Probability tests negative and does not have the condition = 0.60.

Discussion point, page 212

The four numbers in the body of the table occur at the ends of the branches of the trees. Those in the margins do not appear.

The final answers would be the same, although the tree would look different.

Discussion point, page 214

When DNA is collected, the probability of two people having the same DNA profile is very small, but not zero. If only a partial DNA sample has been collected from a crime scene, then this probability could be rather larger. Let us suppose that the probability of two different people having the same partial profile as that found at a crime scene in London is only 1 in one million. The population of London is roughly 9 million. So there will be approximately 9 people who share this profile. So without any other evidence, the probability that the sample comes from a person who has this DNA profile is $\frac{1}{9}$.

Exercise 8.1

1. (i) 0.33
   (ii) 0.385
   (iii) 0.2
2. (i) 0.24
   (ii) 0.44
   (iii) 0.136
3. (i) 0.033
   (ii) 0.545
4. (i) 0.05329
   (ii) 0.0650
   (iii) 0.9350
   (iv) It has a high false positive rate so it is not very effective unless combined with further testing.
5. (i) 0.07
   (ii) 0.0476
6. 0.6087
7. (i) 0.01465
   (ii) 0.662
8. (i) 0.025
   (ii) 0.4
9. (i) 0.1825
   (ii) 0.8

Practice questions: set 2 (page 219–22)

1. (i) (a) 0.3935 (b) 0.2231

   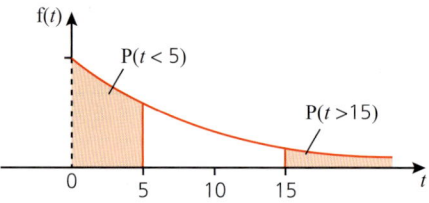

   (ii) $F(t) = 1 - e^{-0.1t}$; median = 6.93
   (iii) 0.183
2. (i) $M = 3.568$; 5.335

   (ii)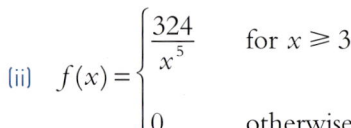

   (iii) $\frac{81}{256}$
3. (i) 64.1

   This is higher than each of the values at $t = 120$ and $t = 180$, so very unlikely to be a good estimate.

   (ii)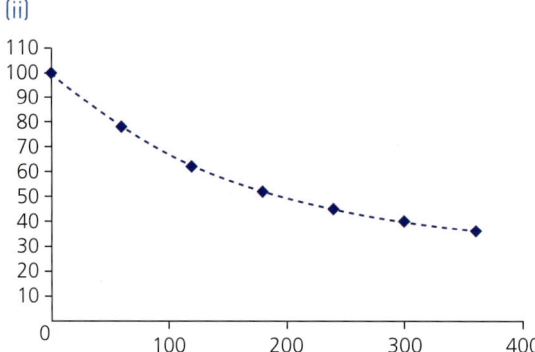

   Clearly a smooth curve, not linear. The pmcc is misleading.

   (iii) Always draw a graph of data.
   High values of the pmcc do not always indicate linearity.
4. (i) Vertical sections through the weight on age scatter diagram show similar means (except perhaps for the very oldest patients). Spread is greatest for younger patients because there are more with very high weight. Vertical sections through the height on age scatter diagram show decreasing means from the twenties onwards. The spread decreases too.
   (ii) (a) Positive skew (as the data points are more spread out above the mean).
   (b) Little or no skew, though the small number of short patients would give slight negative skew.

(iii) Using the raw data, the mean weight is about 74 kg so accept 69–79 kg. Using the raw data, the mean height is about 162 cm so accept 159–165 cm. Mean height (about 5'4") strongly suggests that these are women. Mean weight (just over 11 stone) sounds a little high for women, but it is consistent with recent figures in the United States.

(iv) The scatter diagrams for weight and age and for height and age show negative correlation. Height and weight will be positively correlated, so 0.31.

(v) There is no suggestion that these individuals form a random sample. They may differ from the population in having a condition that has brought them to the clinic for treatment.

5 (i) No evidence of bivariate Normality.
(ii) $\Sigma d^2 = 90$, $r_s = \frac{5}{11} = 0.45$.
(iii) H_0: There is no association between age of female and age of male.
H_1: There is some association between age of female and age of male.
Critical value for $n = 10$, two tailed 5% test is 0.6485.
The observed value is less than the critical value, and so not significant at the 5% level. Insufficient evidence to suppose that there is any association between age of female and age of male.

6 (i) The elliptical data cloud suggests an underlying bivariate Normal distribution.
(ii) The pmcc has magnitude 0.708. It is clearly negative, so −0.708.
(iii) $y = 1.0569$.
This is an estimate of the mean value of y when $x = 25$; that is, an estimate of the mean body density for people with BMI 25.
(iv) Calculating x when $y = 1.08$ requires the regression line of x on y. That is quite different from the regression line of y on x.

7 (i) $C|G$: the defendant is convicted given that the defendant is guilty.
That is, a defendant who is guilty is convicted.
$C \cap G$: the defendant is both convicted and guilty.
(ii) $P(C \cap G') = 0.005$, $P(C' \cap G) = 0.18$, $P(\text{wrong conclusion}) = 0.185$
(iii) $P(C \cap G) = 0.72$, $P(C \cap G') = 0.005$, $P(C) = 0.725$. $P(C|G) = 0.993$.

Chapter 9

Discussion point, page 223

(i) You would expect the probability of someone selecting either door to be $\frac{1}{2}$.
However, given a choice between left and right, most people are observed to choose right. So the probability of left is less than $\frac{1}{2}$ and the probability of right is greater than $\frac{1}{2}$.

(ii) You would expect a uniform probability distribution with 0.1 for each of the ten possible choices. However, it is observed that, when confronted with a range of possibilities, many people are reluctant to choose the extreme values and opt for those in the middle. So the probabilities for 1 and 10 are less than 0.1 and those for the middle numbers, e.g. 4, 5. 6 and 7, are greater than 0.1.

Exercise 9.1

1 (i) $\frac{1}{6}$ (ii) $\frac{25}{216}$ (iii) $\frac{11}{36}$ (iv) $\frac{125}{216}$

2 (i) From 1 to 5, respectively: $\frac{1}{3}, \frac{2}{9}, \frac{4}{27}, \frac{8}{81}, \frac{16}{243}$
(ii) $\frac{32}{243}$ (iii) 3 (iv) 6

3 (i) (a) $\frac{256}{625}$ (b) $\frac{369}{625}$ (c) $\frac{1024}{3125}$ (d) $\frac{256}{3125}$
(ii) (b), (c) and (d) add to 1. They cover all possible outcomes.

4 (i) $\frac{1}{6}$ (ii) 6 (iii) $\frac{3125}{46656}$

5 (i) $\frac{243}{1024}$ (ii) $\frac{781}{1024}$ (iii) $\frac{81}{1024}$ (iv) $\frac{27}{1024}$

6 (i) 0.2 (ii) 0.08192 (iii) 0.5904
(iv) 5 (v) 4

7 (i) 0.0791
(ii) 0.2373
(iii) 0.4375
(iv) 4

8 (i) 0.0504
(ii) 0.1681
(iii) $3\left(\frac{1}{3}\right)$
(iv) 0.1235

9 (i) $\dfrac{32}{243}$ (ii) $\dfrac{16}{243}$ (iii) $\dfrac{5}{3}$ (iv) $3\dfrac{1}{3}$

11 $\dfrac{125}{1296}$

10 (i) Geo$\left(\dfrac{1}{6}\right)$
    (ii) (a) 0.0965 (b) 0.4213 (c) 0.4822
    (iii) (a) 0.4923 (b) 0.0934
    (iv) 0.12

12 (i) Tossing a coin until it comes up heads
    (ii) Rolling a die until it comes up 6.

Exercise 9.2

1 (i)

Number of tosses	3	4	5	6	7	8
Probability	0.125	0.1875	0.1875	0.15625	0.1171875	0.08203125

   (ii) 0.85546875. The table could be extended infinitely far to the right; the sum of the infinite number of probabilities would be 1.

2 Expectation = 30; Variance = 270; Standard deviation = 16.4.

3 (i) 6 and 7 are equally the most likely
   (ii) The expectation is 12
   (iii) The probability of each of 6 and 7 is 0.06698 (to 4 s.f.)
   The probability of 12 is 0.04935 (to 4 s.f.)

4 5

5 (i) 0.09261
   (ii) 4.29

6 $n = 8$, $r = 4$, p can be either $\dfrac{4}{11}$ or $\dfrac{7}{11}$

7 $p = 0.2$, $r = 4$.

8 (i) E(R) = E(S) = E(T) = 6; geometric distribution; E(R) + E(S) + E(T) = 18.
   (ii) E(P) = 18, the negative binomial distribution
   (iii) Paul's experiment is the equivalent of those of the other three taken together.
   E(P) = E(R + S + T) = E(R) + E(S) + E(T).
   (iv) It is false. The correct result is for the variances and not the standard deviations.
   Var(P) = Var(R + S + T) = Var(R) + Var(S) + Var(T). The random variables R, S and T are independent.

9 (i) $\dfrac{1}{1296}$ (ii) 0.0296 (iii) 0.5845

10 0.0172

Discussion point, page 234

Because he would have been unlikely to have queried the value if it had been consistent with his experience.

Exercise 9.3

1 (i) $H_0: \lambda = 2.4$, $H_1: \lambda \neq 2.4$
   (ii) c.v. = 7, accept H_0; data consistent with $\lambda = 2.4$

2 (i) $B(20, p)$
   (ii) $H_0: p = 0.10$, $H_1: p > 0.10$
   (iii) c.v. = 5, reject H_0; evidence suggests that the percentage of plants that will grow into chilli nagas is greater than 10%

3 (i) Geo (p)
   (ii) $H_0: p = 0.02$, $H_1: p < 0.02$
   (iii) c.v. = 149; accept H_0; insufficient evidence to suggest that the probability of a red bead is smaller than 0.02

4 {6, 7, 8, …}

5 5

6 c.v. = 2 or $P(X \geq 1) = 1 - 0.8858 = 0.1142$; getting 1 cracked egg is unfortunate but not statistically significant at the 5% level

7 (i) c.v. = 0 or $P(X \leq 0) = 0.05$; evidence supports claim that mean is less than 3
   (ii) Aaron's experience as a fisherman, whether this result is typical, etc.

8 10

Chapter 10

Discussion point, page 240

You would expect the distribution of the values of x to be centred on the middle value of 0.525 but spread between the extremes of 0 and 1.05. You would also expect peaks at $x = 0$ and $x = 1.05$, corresponding to balls that went outside the bowling lane.

You would expect the distribution of the means for the 20 balls per evening also to have mean 0.525 but to be much less spread out with some balls going to the left of centre and others to the right.

Exercise 10.1

1. (i) N(60, 8)
   (ii) N(300, 40)
   (iii) N(3000, 400)
2. (i) Cannot because the distribution of the parent population is unknown
   (ii) N(6000, 27 000)
   (iii) N(20 000, 90 000)
3. N(540, 270). Assume that the times are independent.
4. (i) 2.4, 2.64
   (ii) N(2.4, 0.0264)
   (iii) 0.731
5. 0.1030
6. 0.9772
7. (i) 120
   (ii) 0.7977
   (iii) If the claim is true, the probability of a result as low or lower than this is 0.0000023. This suggests very strongly that the claim is wrong.
8. (i) 0.5945
   (ii) 200
9. $n = 80$
10. (i) 0.1367
    (ii) 0.2635
    (iii) 0.5176
11. (i) 0.2420
    (ii) 0.01343
12. (i) 1.6
    (ii) (a) 0.06227 (b) 0.0144 (c) 0.6515
    (iii) P(Accept H_0 | $p = 0.09$) = P(X ⩾ 2 | $p = 0.09$) = 0.601.

Chapter 11

Exercise 11.1

1. (i) P(longer than 1 hour) = p,
       P(X ⩽ 1 | $p = 0.2$) = 0.0480 = 4.8%.
   (ii) 0.0480.
2. (i) A Type I error here would mean you say that less than 20% of the beans are red when in fact the manufacturer's claim is correct.
   (i) P(X ⩽ 1 | $p = 0.2$) = 0.167

3. (i) 3
   (ii)

Number x	0	1	2	3	4
P(X-x)	0.0498	0.1494	0.2240	0.2240	0.1680
Number x	5	6	7	8	9
P(X-x)	0.1008	0.0504	0.0216	0.0081	0.0027

   (iii) $H_0: p = 0.02$, $H_1: p > 0.02$ Critical region is $X \geq 7$
   (iv) 0.034
   (v) 4.5, 0.169
   (vi)

P	0.025	0.03	0.04
λ	3.75	4.5	6
P (Type II error)	0.914	0.831	0.606
Power	0.086	0.169	0.394

P	0.05	0.06	0.08	0.10
λ	7.5	9	12	15
P (Type II error)	0.378	0.207	0.046	0.008
Power	0.622	0.793	0.954	0.992

   (vii)
   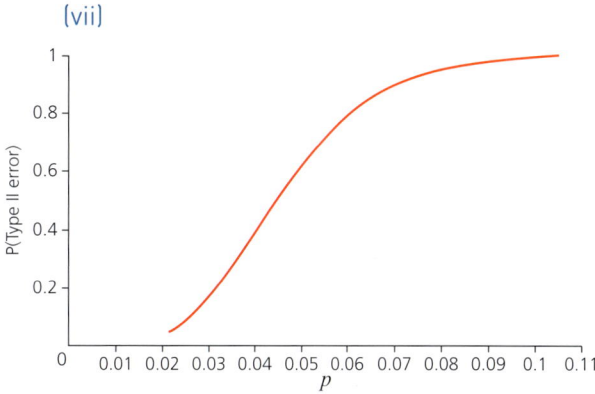

4. (i) 3 (ii) 27.5 (iii) $\lambda < 4$
   (iv) 0.005 (v) 0.542
5. (i) 0.9 (ii) 1.9
   (iii) 1.645, critical value = 26.48
   (iv) 26.71 > 26.48 so H_0 should be rejected.
   (v) (a) 0.282 (b) 0.718
7. 0.156, 0.332
8. (i) 0.175 (ii) 0.560 (iii) $\lambda = 10$ 0.1251
   (iv) 0.5421 (v) 0.0308 (vi) 10
   (vii) Critical value = 16, P(X ⩽ 16 | $\lambda = 25$) = 0.0377, so the value of 15 is significant; there is evidence to suggest the part is needed less than 5 times a day.
   (viii) P(X ⩽ 16 | $\lambda = 25$) = 0.0377 = P(Type I error). P(X ⩾ 17 | $\lambda = 20$) = 0.779 = P(Type II error).
   (ix) Power of test = 1 − P(Type II error) = 0.221.

9 (i) $B(100, \frac{1}{30})$ (ii) $\frac{19}{30}$

(iii) $H_0: p = \frac{1}{30}, H_1: p > \frac{1}{30} \frac{1}{30}$.
$P(X \geq 7) = 0.050... < 10\%$, so evidence suggests the calculator is biased towards 29.

(iv) P(Type I error) = 0.0501(3 s.f.), P(Type II error) = 0.766(3 s.f.)

10 (iii) $p < 0.170$

11 (ii) $10(1-\alpha)^{\frac{1}{3}}$

(iii) $\frac{1000(1-\alpha)}{m^3}$

Chapter 12

Exercise 12.1

1 $G(t) = \frac{1}{36}t^2 + \frac{2}{36}t^3 + \frac{3}{36}t^4 + \frac{4}{36}t^5 + \frac{5}{36}t^6 + \frac{6}{36}t^7$
$+ \frac{5}{36}t^8 + \frac{4}{36}t^9 + \frac{3}{36}t^{10} + \frac{2}{36}t^{11} + \frac{1}{36}t^{12}$

2 $G(t) = \frac{1}{6} + \frac{5}{18}t + \frac{2}{9}t^2 + \frac{1}{6}t^3 + \frac{1}{9}t^4 + \frac{1}{18}t^5$

3 $G(t) = \frac{1}{8} + \frac{3}{8}t + \frac{3}{8}t^2 + \frac{1}{8}t^3$

4 $G(t) = \frac{3}{5}t + \frac{3}{10}t^2 + \frac{1}{10}t^3$

5 $G(t) = \frac{3}{5}t + \frac{2}{5} \times \frac{3}{5}t^2 + \left(\frac{2}{5}\right)^2 \times \frac{3}{5}t^3$
$+ \left(\frac{2}{5}\right)^3 \times \frac{3}{5}t^4 + ... + \left(\frac{2}{5}\right)^{n-1} \times \frac{3}{5}t^n + ... = \frac{3t}{5-2t}$

6 $k = \frac{1}{153}; G(t) = \frac{1}{153}t + \frac{2}{153}t^2 + \frac{6}{153}t^3$
$+ \frac{24}{153}t^4 + \frac{120}{153}t^5$

7 $G(t) = \frac{1}{3} + \frac{1}{2}t + \frac{1}{6}t^3$

8 $G(t) = \frac{12}{35} + \frac{18}{35}t + \frac{1}{7}t^2$

9 $G(t) = \left(\frac{pt}{1-(1-p)t}\right)^2$

Exercise 12.2

1 $E(X) = 2\frac{1}{6}; Var(X) = 2\frac{23}{36}$

2 $E(X) = 2\frac{2}{3}; Var(X) = \frac{8}{9}$

3 $E(X) = 3\frac{1}{2}; Var(X) = 2\frac{11}{12}$

4 $E(X) = 1\frac{1}{2}; Var(X) = \frac{3}{4}$

5 $Var(X) = nk(1-k)$

6 $G(t) = ^4\left(\frac{12}{13} + \frac{1}{13}t\right)^4; E(X) = \frac{4}{13}; Var(X) = \frac{48}{169}$

8 $a = \frac{1}{3}, b = \frac{1}{6}, c = \frac{1}{2}$

9 $G(t) = \frac{1}{36}t + \frac{3}{36}t^2 + \frac{5}{36}t^3 + \frac{7}{36}t^4 + \frac{9}{36}t^5$
$+ \frac{11}{36}t^6; E(X) = 4\frac{17}{36}; Var(X) = 1.97$

10 (i) $G(t) = \frac{1}{3} + \frac{1}{2}t + \frac{1}{6}t^3$

$G'(t) = \frac{1}{2} + \frac{1}{2}t^2$ $G''(t) = t$

$E(X) = \frac{1}{2} + \frac{1}{2} = 1$ $Var(X) = 1 + 1 - 1^2 = 1$

(ii) (a) Because there are 4! different possible arrangements of four objects. Only one is complete because all four have to match.

(iii) (a) Because there are is only one way in which all can match. If $n - 1$ match, then the final one must match too, so $P(X = n - 1) = 0$.

(iv) $P(X = 5) = \frac{1}{120}$ $P(X = 4) = 0$

$P(X = 3) = \frac{1}{12}$ $P(X = 2) = \frac{1}{6}$

$G(t) = a + bt + \frac{1}{6}t^2 + \frac{1}{12}t^3 + \frac{1}{120}t^5$

$G'(t) = b + \frac{1}{3}t + \frac{1}{4}t^2 + \frac{1}{24}t^4$

$b = 1 - \left(\frac{1}{3} + \frac{1}{4} + \frac{1}{24}\right) = \frac{3}{8}$

$a + \frac{3}{8} + \frac{1}{6} + \frac{1}{12} + \frac{1}{120} = 1$

$a = \frac{11}{30}$

$G(t) = \frac{11}{30} + \frac{3}{8}t + \frac{1}{6}t^2 + \frac{1}{12}t^3 + \frac{1}{120}t^5$

(v) $P(X = 6) = \frac{1}{720}$ $P(X = 5) = 0$

$P(X = 4) = \frac{1}{48}$ $P(X = 3) = \frac{1}{18}$

$G(t) = a + bt + ct^2 + \frac{1}{18}t^3 + \frac{1}{48}t^4 + \frac{1}{720}t^6$

$G'(t) = b + 2ct + \frac{1}{6}t^2 + \frac{1}{12}t^3 + \frac{1}{120}t^5$

$G''(t) = 2c + \frac{1}{3}t + \frac{1}{4}t^2 + \frac{1}{24}t^4$

$$2c + \frac{1}{3} + \frac{1}{4} + \frac{1}{24} = 1$$

$$c = \frac{3}{16}$$

$$= b + \frac{3}{8} + \frac{1}{6} + \frac{1}{12} + \frac{1}{120} = 1 \quad b = \frac{11}{30}$$

$$a + \frac{11}{30} + \frac{3}{16} + \frac{1}{18} + \frac{1}{48} + \frac{1}{720} = 1$$

$$a = \frac{53}{144}$$

$$G(t) = \frac{53}{144} + \frac{11}{30}t + \frac{3}{16}t^2 + \frac{1}{18}t^3 + \frac{1}{48}t^4 + \frac{1}{720}t^6$$

(vi) (a) The constant term for $n-1$ is equal to the t coefficient for n.

The t coefficient for $n-1$ is equal to 2 times the t^2 coefficient for n.

The t^2 coefficient for $n-1$ is equal to 3 times the t^3 coefficient for n.

The t^3 coefficient for $n-1$ is equal to 4 times the t^4 coefficient for n.

Etc.

(b) Generalise the above to the t^k coefficient for $n-1$ is equal to $k+1$ times the t^{k+1} coefficient for n and then use the fact that the coefficient of t^{n-1} is zero and the coefficient of t^n is $\frac{1}{n!}$.

Exercise 12.3

1. (ii) $G(t) = \left(\frac{t}{6-5t}\right)^2$

    (iii) $E(X) = 12$; $Var(X) = 60$

2. (i) $G_X(t) = t(0.6 + 0.4t)$;
    $G_Y(t) = t(0.2 + 0.5t + 0.3t^2)$;
    $G(t) = G_X(t) \times G_Y(t)$
    $= t(0.6 + 0.4t) \times t(0.2 + 0.5t + 0.3t^2)$

    (ii) $E(X+Y) = E(X) + E(Y) = 3.5$;
    $Var(X+Y) = Var(X) + Var(Y) = 0.73$

3. (i) $G_X(t) = e^{3.4(t-1)}$, $G_Y(t) = e^{4.8(t-1)}$
    (iii) $G_{X+Y}(t) = e^{8.2(t-1)}$

4. $G_Z(t) = G_X(t) \times G_Y(t) = \frac{t(1-t^6)}{6(1-t)} \times \frac{t(1-t^4)}{4(1-t)}$

5. (i) Note: Let $G(t) = p_1 t + p_2 t^2 + p_3 t^3 + \ldots$ and find $G(1) + G(-1)$.

    (ii) (a) $G(t) = \frac{1}{36}(t^2 + 2t^3 + 3t^4 + 4t^5 + 5t^6 + 6t^7 + 5t^8 + 4t^9 + 3t^{10} + 2t^{11} + t^{12})$; 0.5

    (b) $G(t) = \frac{2t}{3-t}$; 0.25

6. (i) $\frac{3}{4} \times \frac{1}{5} = \frac{3}{20}$

    (ii) $\frac{3}{4} \times \frac{4}{5} \times \frac{1}{4} = \frac{3}{20}$

    (iii) $\frac{1}{4} + \frac{3}{5} \times \frac{1}{4} + \left(\frac{3}{5}\right)^2 \times \frac{1}{4} + \ldots = \frac{1}{4}$

    $\times \frac{1}{\left(1-\frac{3}{5}\right)} = \frac{1}{4} \times \frac{5}{2} = \frac{5}{8}$

    (iv) $G(t) = \frac{1}{4}t + \frac{3}{4} \times \frac{1}{5}t^2 + \frac{3}{5} \times \frac{1}{4}t^3 + \frac{3}{5} \times \frac{3}{4}$

    $\times \frac{1}{5}t^4 + \ldots$

    $= \frac{1}{4}t \times \frac{1}{\left(1-\frac{3}{5}t^2\right)} + \frac{3}{4} \times \frac{1}{5}t^2 \times \frac{1}{1-\frac{3t^2}{5}}$

    $= \frac{\frac{1}{4}t + \frac{3}{4} \times \frac{1}{5}t^2}{1-\frac{3t^2}{5}} = \frac{5t + 3t^2}{4(5-3t^2)}$

    $G'(t) = \frac{1}{4} \times \frac{(5-3t^2) \times (5+6t) - (5t+3t^2) \times -6t}{(5-3t^2)^2}$

    $= G'(t) = \frac{1}{4} \times \frac{2 \times 11 - (8) \times -6}{(5-3)^2} = \frac{1}{4} \times \frac{70}{4} = \frac{35}{8}$

    $E(R) = \frac{35}{8}$

7. (i) $P(X=r) = e^{-\lambda} \frac{\lambda^r}{r!}$ $G(t) = e^{\lambda(t-1)}$

    (ii) $k = \frac{1}{1 - e^{-\lambda}(\lambda+1)}$

    (iii) $G(t) = \frac{e^{-\lambda}\left(e^{\lambda t} - \lambda t - 1\right)}{1 - e^{-\lambda}(\lambda+1)} = \frac{e^{\lambda t} - \lambda t - 1}{e^{\lambda} - \lambda - 1}$

    (iv) $G'(t) = \frac{\lambda e^{\lambda t} - \lambda}{e^{\lambda} - \lambda - 1}$

    so $E(X) = \frac{\lambda e^{\lambda} - \lambda}{e^{\lambda} - \lambda - 1} = \frac{\lambda(e^{\lambda} - 1)}{e^{\lambda} - \lambda - 1}$

    $G''(t) = \frac{\lambda^2 e^{\lambda t}}{e^{\lambda} - \lambda - 1}$

    $Var(X) = \frac{\lambda^2 e^{\lambda}}{e^{\lambda} - \lambda - 1} + \mu - \mu^2$

    $= \mu \left(\frac{\lambda^2 e^{\lambda}}{\lambda(e^{\lambda} - 1)}\right) + \mu - \mu^2$

    $= \left(\frac{\mu}{\lambda(e^{\lambda}-1)}\right)(\lambda^2 e^{\lambda} + \lambda(e^{\lambda}-1) - \mu\lambda(e^{\lambda}-1))$

    $= \left(\frac{\mu}{\gamma}\right)(\lambda^2 e^{\lambda} + \gamma - \mu\gamma)$

8 $3\frac{6}{7}$

9 (i) $P(X = x) = pq^x$ $G(t) = p + pqt + pq^2t^2 + \ldots$
$$= \frac{p}{1-qt}$$

(ii) $G(t) = p(1-qt)^{-1}$ $G'(t) = pq(1-qt)^{-2}$
$G''(t) = 2pq^2(1-qt)^{-3}$

So $E(X) = \frac{pq}{p^2} = \frac{q}{p}$

$\text{Var}(X) = \frac{2pq^2}{p^3} + \frac{q}{p} - \left(\frac{q}{p}\right)^2 = \frac{2pq^2 + p^2q - pq^2}{p^3}$
$= \frac{pq^2 + p^2q}{p^3} = \frac{q^2 + pq}{p^2} = \frac{q(q+p)}{p^2} = \frac{q}{p^2}$

10 (i) $P(W = w) = e^{-\lambda_1}\frac{\lambda_1^w}{w!}$

$P(X = x) = e^{-\lambda_2}\frac{\lambda_2^x}{x!}$

$P(Y = y) = e^{-\lambda_3}\frac{\lambda_3^y}{y!}$

(ii) $G_W(t) = e^{\lambda_1(t-1)}$ $G_X(t) = e^{\lambda_2(t-1)}$
$G_Y(t) = e^{\lambda_3(t-1)}$
$G_Z(t) = G_{W+X+Y}(t)$
$= e^{\lambda_1(t-1)} e^{\lambda_2(t-1)} e^{\lambda_3(t-1)}$
$= e^{(\lambda_1+\lambda_2+\lambda_3)(t-1)}$

(iii) (a) $P(W + X + Y = K)$
$= e^{-(\lambda_1+\lambda_2+\lambda_3)}\frac{(\lambda_1+\lambda_2+\lambda_3)^z}{z!}$

(b) $\lambda_1 + \lambda_2 + \lambda_3$

(c) $\lambda_1 + \lambda_2 + \lambda_3$

Practice questions: set 3 (page 282–3)

1 (i) (a) $0.6^2 \times 0.4 = 0.144$
(b) $0.6^6 = 0.0467$
(c) $0.6^6 \times (1 - 0.6^6) = 0.0445$
(d) 0.311

(ii) Null hypothesis $H_0: p = 0.4$ (where p is the probability that Esperanza wins one or more prizes in a randomly chosen month.)
Alternative hypothesis $H_1: p < 0.4$.

(iii) Let X have geometric distribution with $p = 0.4$.
Large values of X favour H1.

$P(X \geq 6) = 0.078$, $P(X \geq 7) = 0.047$.
5% critical region is $\{7, 8, \ldots\}$.
So n must be in $\{1, 2, \ldots, 6\}$.

2 (i) Poisson. Parameter 3.5.
(ii) P (no occurrences) $= e^{-3.5} = 0.0302$.
(iii) $1 - (1 - e^{-3.5})^{12} = 0.308$.
(iv) Let X have Poisson distribution with parameter 3.5.
$P(X \geq 8) = 0.027$, $P(X \geq 9) = 0.0099$.
So 1% critical region (that is, the set of values of k) is $\{9, 10, \ldots\}$
(v) Let Y have Poisson distribution with parameter 4.
$P(\text{Type II error}) = P(Y \leq 8) = 0.979$.
A very high value; the test is not sensitive to this change.

3 (i) $G_X(t) = E(t^X)$.
$G_X(t) = \Sigma\, t^x P(X = x) = \Sigma\, t^x q^{x-1} p = pt(1-qt)^{-1}$.

(ii) $G'_X(t) = p(1-qt)^{-1} + pqt(1-qt)^{-2}$.
$E(X) = G'_X(1) = p(1-q)^{-1} + pq(1-q)^{-2}$,
which simplifies to $\frac{1}{p}$.
$G''_X(t) = pq(1-qt)^{-2} + pq(1-qt)^{-2} + 2pq^2t(1-qt)^{-3}$.
$G''_X(1) = pq(1-q)^{-2} + pq(1-q)^{-2} + 2pq^2(1-q)^{-3}$, which simplifies to $\frac{2q}{p^2}$.
$\text{Var}(X) = G''_X(1) + E(X) - (E(X))^2$, which simplifies to $\frac{q}{p^2}$.

(iii) The pgf for the sum of independent random variables is the product of the pgfs. Hence $p^n t^n (1-qt)^{-n}$.

(iv) $E(Y) = n E(X) = \frac{n}{p}$.
$\text{Var}(Y) = n\,\text{Var}(X) = \frac{nq}{p^2}$.

(v) The number of eggs up to and including the first one with a double yolk.
(vi) The number of eggs up to and including the third one with a double yolk.

4 (i) Positive skew (long tail to the right).
(ii) The Normal distribution supports values in the range mean $\pm 3 \times$ SD. That gives a substantial negative tail, which is impossible as numbers of absences cannot be negative.

(iii) (Using the Central Limit Theorem) the mean number of absences will be (approximately) Normal, with mean 8.6 and standard deviation $\frac{5.3}{\sqrt{50}}$ (= 0.75).

(iv) P(total absences > 450)
= P(mean absences > 9).
Hence $P\left(Z > \frac{(9 - 8.6)}{0.75}\right) =$
P(Z > 0.5337) = 0.2968.

(v) Assumed that absences in a period of 50 days are independent of one another. This is unlikely as some absences will be caused by infections such as colds and flu.

5 (i) Let X have the binomial distribution with $n = 40, p = 0.2$.
$P(X \geq 12) = 0.0875, P(X \geq 13) = 0.0432$.
Critical region is $\{13, 14, \ldots\}$.
Size of Type I error is 4.32%.

(ii) Let Y have the binomial distribution with $n = 40, p = 0.25$.
Size of Type II error is $P(Y \leq 12) = 0.821$.
So the power is $1 - 0.821 = 0.179$.

(iii)

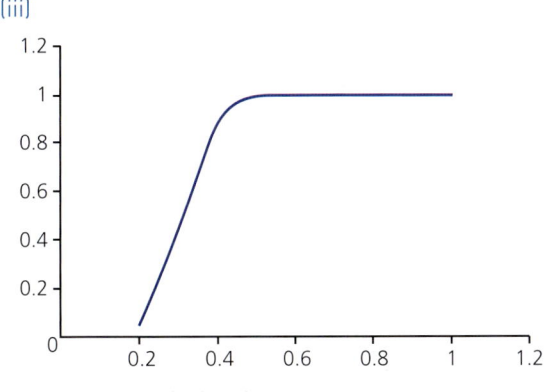

Rises rapidly, level out at 1.

Chapter 13

Exercise 13.1

1 (i) 0.8413
  (ii) 0.0548
  (iii) 0.7333
2 (i) 0.0548
  (ii) 0.6051
  (iii) 0.6816
3 (i) 4, 0.875
  (ii) 1.5, 0.167

(iii)

Main course	Dessert	Price
Fish and chips	Ice cream	£4
Fish and chips	Apple pie	£4.50
Fish and chips	Sponge pudding	£5
Bacon and eggs	Ice cream	£4.50
Bacon and eggs	Apple pie	£5
Bacon and eggs	Sponge pudding	£5.50
Pizza	Ice cream	£5
Pizza	Apple pie	£5.50
Pizza	Sponge pudding	£6
Steak and chips	Ice cream	£6.50
Steak and chips	Apple pie	£7
Steak and chips	Sponge pudding	£7.50

(iv) Mean of T = 5.5, variance = 1.042

4 (i) N(90, 25)
  (ii) N(10, 25)
  (iii) N(−10, 25)
5 0.196
6 (i) 0.0228
  (ii) 56.45 minutes
  (iii) 0.362
7 (i) 230 g, 10.2 g
  (ii) 0.1587
  (iii) 0.0787
8 5.92%
9 (i) 0.266
  (ii) No, people do not choose their spouses at random: the height of a husband and wife may not be independent.
10 0.151
11 (i) 0
  (ii) 0.0037
12 (i) 0.3533
  (ii) 60
  (iii) Using exact probability of 0.0521 gives 0.1482, using 0.05 gives 0.1426
13 (i) 0.29
  (ii) 4347
  (iii) Because the probability of scoring between 0 and 10 is about 0.99
  (iv) 0.0258
  (v) 113 or more
14 (i) (a) 0.025   (b) 0.242
  (ii) 130
  (iii) 0.161
  (iv) 0.03559

Exercise 13.2

1. (i) $N(5\mu, 25\sigma^2)$
   (ii) $N(5\mu, 5\sigma^2)$
2. (i) $N(30, 15)$
   (ii) $N(100, 500)$
   (iii) $N(110, 205)$
3. 0.1946
4. 0.0745
5. 0.1377
6. (i) $N(34, 30)$
   (ii) $N(-4, 30)$
   (iii) $N(24, 29)$
7. (i) 0.316
   (ii) 0.316
8. (i) $N(100, 26)$
   (ii) $N(295, 353)$
   (iii) $N(200, 122)$
   (iv) $N(-65, 377)$
9. (i) 0.0827
   (ii) 0.3103
   (iii) 0.5
10. (i) 0.0827
    (ii) 0.1446
    (iii) 0.5
11. (i) $N(7400, 28\,900)$
    (ii) $N(1200, 27\,700)$
    (iii) $N(600a + 1000b, 400a^2 + 900b^2)$
12. (i) 0.4546
    (ii) 93.491
13. 0.9026
    Assume weights of participants are independent since told teams were chosen at random.
14. (i) $N(2000, 1250)$
    (ii) 1942
    (iii) 0.7373
15. (i) 14%
    (ii) 0.6
    (iii) 15 m
    (iv) 0.3043
16. (i) $S \sim N(600, 105.8)$; 0.0724
    (ii) 0.839
    (iii) 0.161
    (iv) $\mu = 30.54$ g
17. (i) $N(80, 8)$
    (ii) $N(40n, 4n)$
    (iii) 0.0207
    (iv) 0.0456
    Choice of limits is ±2 standard deviations from the mean and so will include 95% of piles that contain 25 pamphlets.
18. (i) 0.3341
    (ii) 0.1469
    (iii) 394.38 mm
    (iv) 0.9595
    (v) 0.1478

Exercise 13.3

1. (i) $N(60, 8)$
   (ii) $N(300, 40)$
   (iii) $N(3000, 400)$
2. (i) Cannot because the distribution of the parent population is unknown
   (ii) $N(6000, 27\,000)$
   (iii) $N(20\,000, 90\,000)$
3. $N(540, 270)$. Assume that the times are independent.
4. 0.1030
5. 0.9772
6. (i) 120
   (ii) 0.7977
   (iii) If the claim is true, the probability of a result as low or lower than this is 0.0000023. This suggests very strongly that the claim is wrong.
7. (i) 0.5945
   (ii) 200
8. $n = 80$
9. (i) 0.1367
   (ii) 0.2635
   (iii) 0.5176
10. (i) 0.2420
    (ii) 0.01343
11. (i) 1.6
    (ii) (a) 0.06227   (b) 0.0144   (c) 0.6515

Chapter 14

Discussion point, page 309

One cannot make any firm conclusions unless it is known that the sample is random.

Discussion point, page 309

It tells you that μ is about 101.2 but it does not tell you what 'about' means, i.e. how close to 101.2 it is reasonable to expect μ to be.

Exercise 14.1

2. (i) 1021.5
   (ii) 1.94
3. (i) 5.205
   (ii) 5.117, 5.293

4 (i) 47.7
(ii) 34.7 to 60.7
(iii) 27.3 to 68.1

5 (i) 0.9456
(ii) £7790 – £8810

6 (i) (a) 0.1685
(b) 0.0207
(ii) 163.8–166.6
(iii) 385

7 (i) 6.83, 3.05
(ii) 6.58, 7.08

8 (i) 5.71 to 7.49
(ii) It is more likely that the short manuscript was written in the early form of the language.

9 (ii) $H_0: \mu = 1000$, $H_1: \mu < 1000$, $z = -1.60$, critical value at 1% level = $-2.326 < -1.60$ so accept H_0, evidence does not suggest that the mean weight is less than 1000 grams.

10 (i) 0.760
(ii) 0.984
(iii) (35.32, 35.88)

11 (ii) –3.00 to 4.58
(iii) No evidence to suggest that there is any difference since the interval contains zero.

12 (i) Players' scores cannot be Normally distributed because symmetry would require negative scores.
(ii) 11.79, 13.64
(iii) It would reduce the width of the confidence interval but the interval would be centred in the same place.
(iv) Approximately 102 500

13 0.484 to 1.016
Assume that the sample standard deviation is an acceptable approximation for σ. The aim has not been achieved as the interval contains values below 0.5.

14 25

15 (i) 1.838 mm
(ii) 1.630 to 1.910
(iii) The coach's suspicions seem to be confirmed as 4 mm is not in the confidence interval.

Exercise 14.2

1 (i) $\sqrt{\dfrac{s^2}{n}} = \sqrt{\dfrac{148.84}{7}} = 4.61$

2 (i) 11.25

3 (i) 322.9, 79.54
(ii) 278.8–366.9

4 (i) 66, 17.15
(ii) 51.7–80.3
(iii) The distribution of the yield of all the fruit farmer's trees is Normal.
(iv) Number all the trees with different consecutive integers. Copy these integers on to separate pieces of paper; put these in a hat and pick out eight at random. The numbers chosen will identify the trees to be picked for the sample.

5 (i) 18.25, 3.72
(ii) 16.32–20.18
(iii) The distribution of lengths of sentences written by the accused man is Normal and the text represented by the sample sentences is representative of the general length of sentences he writes.
(iv) The sample mean lies just outside the particular 90% confidence limits provided by this sample but could well be inside those provided by other samples. Further, it is within the 95% confidence limits. So the evidence is not sufficient to declare him not guilty.

6 (i) Monday
(ii) The 23 weekdays
(iii) 629.7–661.9
(iv) True, this distribution is not a Normal one, but it may still be accurately modelled by one. $s = 37.24$ so the step size is small compared with the standard deviation. It is very common in statistics to make a Normal approximation to a discrete distribution and the results are usually very reliable.

7 (i) 63.6–72.0
(ii) Statistical: the distribution of tyre condemnation mileages is Normal and the 12 tyres tested in the sample are representative of the distribution. Practical: the tyres are tested under genuine working conditions.

8 (i) 21
(ii) The distribution is not Normal.
(iii) No because the confidence interval obtained would be too wide to be meaningful.
(iv) 45.0–66.3; the procedure will be valid provided that the distribution of life expectancies for this group is Normal.

9   (i) 0.633–0.647
    (ii) 0.78, 0.160
    (iii) 0.616–0.944
    (iv) Large sample; no need for underlying Normality
10 (i) $1.9859 < \mu < 2.3202$
    (ii) 12
    (iii) The population of heights must be Normal and the sample must be random.
11 (i) The method must ensure that each club member has an equal chance of being chosen. Could put names in a hat and select at random or use the random number generator on a calculator.
    (ii) Mean = 4.74, s^2 = 2.74
    (iii) $3.56 < \mu < 5.92$
    Background population is Normal.
    (iv) By increasing the sample size; 13
12 (i) Because, on average, the confidence interval will not contain 55 mm in 1 in 20 batches.
    (ii) Take a larger sample and construct a confidence interval based on this sample.
    (iii) The Normal probability plot does not appear to show linearity.
    (iv) $47.31 < \mu < 55.31$.
13 (i) $-0.29 < \mu < 1.72$
    (ii) No, since the interval contains zero.
    (iii) Assumptions: random sample of areas – could well have been;
    Normally distributed differences in unemployment rates – not unreasonable.
14 (i) $-0.297 < \mu < 0.186$
    (ii) Yes, since the interval contains zero.
    (iii) Assumptions: random sample of races – apparently not since all the sprints on just one afternoon are taken;
    Normally distributed differences in timings – not unreasonable.
15 (i) $1.84 < \mu < 14.16$
    (ii) We do not know if the experimenter takes a random sample of the relevant population of rats. It seems plausible that the differences in the rats' times in the two conditions might be Normally distributed.
    (iii) So that the effect of having learned to run the maze on the second trial affects each condition in the same way.
16 1.934–2.052, 9.373–10.551; 120, halved

Chapter 15

Exercise 15.1

1   (i) 0.9
    (ii) 1.9
    (iii) 1.645
    (iv) 1.9 > 1.645 so H_0 should be rejected.
2   (i) Because the sample is large so the Central Limit Theorem applies.
    (ii) $H_0: \mu = 1.5$, $H_1: \mu < 1.5$
    (iii) 0.00142
    (iv) -1.968
    (v) -2.326
    (vi) $-1.968 > -2.326$ so H_0 should be accepted. There is insufficient evidence to suggest that the bags are underweight.
3   (i) This is the critical value for a 5% 2-tailed test.
    (ii) $0.8796 < 2.3646$ so accept H_0.
    (iii) p-value $0.4082 > 0.05$ so accept H_0.
4   (i) 167.9, 27.0
    (ii) (165.7, 170.1)
5   $H_0: \mu = 80$, $H_1: \mu > 80$ $z = 2$, critical value at 5% level (one tailed) = $1.645 < 2$ so reject H_0, evidence supports managing director's claim.
6   $H_0: \mu = 33.5$, $H_1: \mu < 33.5$, $-3.851 < -2.326$ so reject H_0. There is sufficient evidence to suggest that the mean weight has been reduced.
7   (i) $H_0: \mu = 18.6$, $H_1: \mu \neq 18.6$, Estimate of population standard deviation = 1.2438, $-1.46 < -1.645$ so accept H_0. There is insufficient evidence to suggest that the mean times have changed.
    (ii) Alternative test is t-test. Critical value for $\upsilon = 29$ not given in tables but for $\upsilon = 25$ critical value is 1.725 so using this $-1.908 < -1.725$ so reject H_0. There is sufficient evidence to suggest that the mean times have changed.
    Assumption is that the population of journey times is Normally distributed.
8   (i) $H_0: \mu = 72$ $H_1: \mu > 72$, $3.047 < 3.106$ so accept H_0. There is insufficient evidence to suggest that the batch has a higher than ideal rate.
    (ii) Statistical: the distribution of drying rates is Normal and the sample is random.

9 (i) $H_0: \mu = 750\,g$, $H_1: \mu < 750\,g$
   (ii) $t = -4.57$, significant
   (iii) The distribution of weights in the shoal is Normal, this may be reasonable. The sample is random; this is certainly not the case since all the pollack came from one shoal. The masses are independent; this may not be true when all the fish are taken from one shoal and so are likely to be of the same age.

10 (i) The distribution of total points in a hand is Normal. This assumption is not fully justified because the distribution is not symmetrical about the mean; it is positively skewed. For example, you cannot have a hand with fewer than 0 points in it but you can get a hand with more than 20 points in it.
   (ii) $H_0: \mu = 10$, $H_1: \mu < 10$
   (iii) $t = -1.16$, not significant

11 (i) If $X_1, X_2, \ldots X_n$ is a random sample from a population with mean μ and variance σ^2, then the sample mean $\bar{X} = \frac{1}{n}(X_1 + X_2 + \ldots + X_n)$ is approximately normal with mean μ and variance $\frac{\sigma^2}{n}$, for large n
   (ii) (16420, 18380)
   (iii) Distribution of sample mean is approximately normal
   (iv) The complaint is justified, 20 000 is significantly greater than 18 400 (the upper end of the confidence interval)

12 (i) 0.1290
   (ii) 0.4284
   (iii) $t = 0.7026$, not significant

13 (i) $H_0: \mu = 14.0$, $H_1: \mu > 14.0$; $t = 2.8296$, significant
   (ii) 14.197–15.763
   (iii) The sample should be random, pigs should be similar (e.g. in initial weight), pigs should be kept under controlled conditions (e.g. in respect of exercise).

14 $H_0: \mu = 9$, $H_1: \mu > 9$; $t = 3.51$, significant
   Assume the population is Normally distributed and the sample is random.

15 (i) $N(190, 5\frac{1}{3})$
   (ii) The skulls in group B have greater mean lengths and so a one-tailed test is required.
   (iii) 193.8
   (iv) Sample mean = 194.3 > 193.8 (or 1.809 > 1.645) so reject H_0. There is evidence to support the belief that the skulls belong to group B.

Exercise 15.2

1 $s = 25.27$, $t = 1.264$, $v = 63$, 1-tailed test
  Critical value at 5% level is more than 1.669.
  1.264 < 1.669 so accept H_0.
  Assumptions: independent random samples for the two groups – hard to tell, but seems unlikely; Normally distributed incomes with the same variance in each group – not reasonable, as income distribution is likely to be skewed and the variance to be different in the two groups.

2 Test statistic = 2.46, so significantly slower

3 Test statistic $t = 1.835$, $v = 9$, $p = 0.0499$, reject H_0. Data supports claim that mean at end is significantly lower than at start.

4 Test statistic $z = 1.878$, $p = 0.030$, reject H_0. Data suggests that population mean height for adults aged under 30 is greater than the population mean height for adults aged over 50.

5 (i) $s = 5.675$, $t = 2.085$, $v = 32$, 1-tailed test
     Critical value at 5% level is less than 1.694
     Assumptions: independent random samples of introverts and extroverts – you cannot tell how this selection was made; Normally distributed heights with the same variance in each group – nothing in the data to suggest this is unreasonable.
  (ii) No: causal conclusion is not justified – a confounding factor (e.g. features of childhood nurture) may produce both taller and more extrovert people.

6 (i) $s = 2.646$, $t = 1.192$, $v = 17$, 1-tailed test
     Accept H_0 even at 5% level: not heavier
  (ii) Assumptions: independent random samples on the two islands - it is hard to sample animal populations randomly – some individuals are more catchable than others; Normally distributed weights on each island – seems reasonable; same variance of weights on each island - at first glance the Daphne Major variance seems larger but the samples are small.

7 (i) $s = 5.146$, $t = 0.304$, $v = 43$, 2-tailed test
     Accept H_0 even at 10% level: same yield

(ii)
```
        9│0│8 9
    0 4 2 4│1│2 4 3 4
9 8 5 9 8 5│1│8 5 7 5 7 6 8 8
3 2 1 0 1 1│2│0 0 4 0 0 3 0 3 3
      7 7 6│2│6 9
```
│2│6 means 26 kg m^{-2}

Assumptions: independent random samples of plots for the two varieties – you cannot tell how this selection was made; Normally distributed yield with same variance in each variety – the stem-and-leaf diagram suggests this is not unreasonable.

(iii) Select plots; plant some of each variety on half of each plot. An improvement since small average differences in yield are more likely to show up when the differences attributable to the plots are eliminated.

8 (i) Test statistic $z = 1.904$, $p = 0.02872$, accept H_0. Insufficient evidence to suggest that the mean has changed.
   (ii) Test statistic $z = 2.019$, $p = 0.025$, reject H_0. Data suggests the mean has reduced.

9 Test statistic $t = 1.725$, $v = 10$, $p = 0.812$, accept H_0. Insufficient evidence to suggest that the means are different.

10 Pooled estimator of population variance $s^2 = 12.743$, test statistic $t = 1.086$, $v = 7$, critical value $= 1.895$, accept H_0. Insufficient evidence to suggest the difference in the mean times differs from 30 minutes.

Chapter 16

Activity 16.2
$E[M_1] = \frac{2}{3}m$ and $E[M_2] = \frac{114}{175}m$ so both M_1 and M_2 are biased for this distribution. They can be corrected by multiplying by $\frac{3}{2}$ and $\frac{114}{175}$ respectively.

Activity 16.3
Both M_4 and M_5 are unbiased.

Discussion point, page 373
The second method is biased because there will be more responses from people who had travelled in cars that were fuller. For example, consider just two cars, one with 1 person in and the other with 3 people in. The average number of people per car is 2. But if all 4 people were interviewed, there would be one answer of 1 and three answers of 3. The average of these four answers is 2.5.

The first method of estimation is better as it simply averages the number of people per car. The second method gives undue weight to cars with more people in.

Exercise 16.1
1 (i) $E(M_1) = E(M_2) = \mu$ so M_1 and M_2 are unbiased. $E(M_3) = \frac{4}{3}\mu$ so M_3 is biased.
  (ii) Bias $= \frac{\mu}{3}$
  (iii) $\frac{1}{3}\text{Var}(W)$, $\frac{3}{8}\text{Var}(W)$, $\frac{2}{3}\text{Var}(W)$
  (iv) M_1, it is unbiased and has the lowest variance
  (v) 1.0 m
  (vi) $\frac{1}{6}W_A + \frac{1}{6}W_B + \frac{1}{6}W_C + \frac{1}{2}W_D$.

  No, it is unbiased but its variance is $\frac{7}{12}\text{Var}(W)$ whereas $\frac{1}{4}W_A + \frac{1}{4}W_B + \frac{1}{4}W_C + \frac{1}{4}W_D$ has variance $\frac{1}{4}\text{Var}(W)$ which is less.

2 (i) M_2 as an unbiased estimator of μ
  (iv) $\text{Var}(W_k) = \frac{1}{k}\text{Var}(W)$
  (v) The value of the mean is estimated more accurately using a larger sample than it using a smaller sample.

3 (iii) T_3 has a smaller variance than the other two estimators, so is less likely to give an estimate far from the true value of μ – see the work on mean square error later in this chapter.

4 (ii) $\text{Var}[T] = c_1^2 \sigma_1^2 + (1 - c_1)^2 \sigma_2^2$
  Minimum when $c_1 = \frac{\sigma_2^2}{\sigma_1^2 + \sigma_2^2}$, $c_2 = \frac{\sigma_1^2}{\sigma_1^2 + \sigma_2^2}$,
  $\text{Var}[T] = \frac{2\sigma_1^2 \sigma_2^2}{\sigma_1^2 + \sigma_2^2}$
  (iii) T is a minimum variance unbiased estimator – it estimates correctly on average and is rarely far from the true mean – but it cannot be determined without knowing the variances of X_1 and X_2 which is unlikely in practice if μ is unknown.

5 (i) $\mu = (1 + 2 + \ldots + n) / n = \frac{1}{2}n(n+1) / n = \frac{1}{2}(n+1)$.
  (ii) $E(2M - 1) = 2 E(M) - 1 = 2 \times \frac{1}{2}(n+1) - 1 = n$.

(iii) The sample mean, m, is 5.
The unbiased estimate of n is $2 \times 5 - 1 = 9$.
(iv) This is not a possible value of n as the sample shows that n must be at least 11. Unbiasedness is not always a 'good' property for an estimator to have.

6 (i) The sample mean is the best possible estimator. It is unbiased and it has the smallest possible variance.
(ii) $a = 66.5$, $b = 8 \neq \sqrt{66.5}$.
(iii) The wording 'the mean size of fluorite squares' is not precise enough. Andrea and Bilal are estimating different things, length and area. So neither is better than the other.
(iv) The population variance. The sample variance S^2 is an unbiased estimator of σ^2, but S is not an unbiased estimator of σ.

7 (i) (a) $E(X) = p + 2q = 1 + q$.
$E\left(\dfrac{1}{X}\right) = p + \dfrac{1}{2}q = 1 - \dfrac{1}{2}q$.
(b) Setting $E\left(\dfrac{1}{X}\right) = \dfrac{1}{E(X)}$ gives $p = 0$ or 1, neither of which is valid.
(ii) (a) Umar: average journey time is 24.2 minutes. So average speed is 9.9 km/h.
Violet: average speed is 10.2 km/h.
(b) The estimates are different because $E\left(\dfrac{4}{T}\right) \neq \dfrac{4}{E(T)}$.

8 (i) $a + b = 1$
(ii) $(a^2 + (1-a)^2)\sigma^2$
(iii) $a = b = \dfrac{1}{2}$
(iv) $c = \dfrac{2}{3}, d = \dfrac{1}{3}$
$W = \dfrac{1}{3}(X1 + X2 + X3)$

9 (i) The Normal distribution is symmetrical, so the population mean and population median are equal.
(ii) The sample medians appear to be more spread out (they have a larger variance or larger standard deviation) than the sample means. This suggests that using the median to estimate the population mean will give less accurate results. (Technically, the sample median is less efficient than the sample mean.)

10 (ii) $\dfrac{n}{n-1}V$ is unbiased

12 (i) (a) L: $P(3) = \dfrac{1}{20}, P(4) = \dfrac{3}{20}, P(5) = \dfrac{6}{20}$,
$P(6) = \dfrac{10}{20}$
M: $P(2) = \dfrac{4}{20}, P(3) = \dfrac{6}{20}$,
$P(4) = \dfrac{6}{20}, P(5) = \dfrac{4}{20}$
(b) $E[L] = \dfrac{21}{4}, E[M] = \dfrac{7}{2}$
(c) $E[2M-1] = 6, E[L] \neq 6$

(ii) (a) L: $P(1) = \dfrac{1}{56}, P(2) = \dfrac{3}{56}, P(3) = \dfrac{6}{56}$,
$P(4) = \dfrac{10}{56}, P(5) = \dfrac{15}{56}, P(6) = \dfrac{21}{56}$
M: $P(1) = \dfrac{6}{56}, P(2) = \dfrac{10}{56}, P(3) = \dfrac{12}{56}$,
$P(4) = \dfrac{12}{56}, P(5) = \dfrac{10}{56}, P(6) = \dfrac{6}{56}$
(b) $E[L] = \dfrac{119}{24}, E[M] = \dfrac{7}{2}$
(c) $E[2M-1] = 6, E[L] \neq 6$

13 (i) (a) $P(\Pi = \tfrac{1}{3}) = 9p^2$; $P(\Pi = \tfrac{1}{6}) = 6p(1-3p)$;
$P(\Pi = 0) = (1-3p)^2$
(ii) $9p^2 \times \dfrac{1}{3} + 6p(1-3p) \times \dfrac{1}{6} + (1-3p)^2 \times 0 = p$

14 (i) This is the negative binomial distribution and the form of the relationship is demonstrated on page 230.

15 (iii) $Z = 3$ or 5, $E[Z] = 4$, $\text{Var}[Z] = 1$
(iv) Y is biased, but has a small variance; Z is unbiased but has a larger variance for $n = 4$ or 5.

Chapter 17

Exercise 17.1

1 $X^2_{15} = 28.89 > 27.49$; reject H_0; the evidence suggests that the variance is not 14.8.
2 $X^2_9 = 4.86 > 16.92$; accept H_0, there is not sufficient evidence to conclude that the variance is not 2.
3 $X^2_8 = 17.5 < 17.53$; accept H_0, there is not sufficient evidence to conclude that the variance is not 0.08.
4 (i) $7.29 < \sigma^2 < 20.78$
(ii) $6.73 < \sigma^2 < 23.39$
5 $0.096 < \sigma^2 < 0.673$
6 $37.5 > 36.415$, so reject H_0 and conclude that the evidence does not support the claim that σ is 1.
7 Test statistic = 28.8, c.v. = 39.364; accept H_0; data consistent with $\sigma^2 = 100$.
8 $X^2_7 = 4.18 > 1.690$ conclude variance is 9.3

9 (i) $X_9^2 = 3.99 > 3.325$ conclude variance has not reduced
   (ii) Assumes that the number of passengers is distributed Normally.
10 (i) $X_{13}^2 = 23.70 < 24.74$ conclude variance is 2
   (ii) $X_{13}^2 = 23.70 > 23.36$ conclude variance is >2
11 (i) $0.090 < \sigma^2 < 0.512$
   (ii) $0.300 < \sigma^2 < 0.716$
12 $26.84 < \sigma^2 < 136.58$
13 (i) $33.14 < \sigma^2 < 233.48$
   (ii) The standard deviation is no longer 18.4
14 (i) $S^2 = 6.67$
   (ii) (a) $X_{22}^2 = 14.68 > 10.98$ conclude variance is 10
       (b) $X_{22}^2 = 36.69 < 36.78$ conclude variance is 4
15 (i) $H_0: \sigma^2 = 6.8^2$, $H_1: \sigma^2 \neq 6.8^2$, $X_9^2 = 21.94 > 19.02$ and conclude variance has changed
   (ii) $59.96 < \sigma^2 < 305.11$
   (iii) Journey times are Normally distributed.

Exercise 17.2

1 $F_{24,19} = 1.32 < 2.45$ conclude variances are equal
2 $F_{7,9} = 1.92 < 5.61$ conclude variances are equal
3 Test statistic = 2.384, c.v. = 5.89; accept H_0; data consistent with equal variances.
4 c.v. = 2.98 which gives $s_2 = 47.7$
5 $2.0736 < 3.66$ so accept H_0 and conclude that the population variances could be equal.
6 $S^2 < 12.49$ or $S^2 > 64.38$
7 $\dfrac{S_B^2}{S_A^2} = 4.74$ but $F_{24,24}$ (critical) = 1.98 (critical) $\Rightarrow F_{59,59}$(critical) < 1.98 hence B is less consistent
8 (i) $F_{4,5} = 1.164 < 7.39$ conclude variances are equal
   (ii) Samples are independent and random and reaction times are Normally distributed.
9 (i) $F_{6,5} = 1.23 < 6.98$ conclude variances are equal
   $t_{11} = 0.626 < 2.201$ conclude no significant difference in the means at 5% level
   (ii) Samples are independent and random and egg masses are Normally distributed.
10 (i) $F_{10,7} = 2.901 < 4.76$ conclude variances are equal
   (ii) $t_{17} = 1.779 > 1.742$ conclude the new treatment has a higher mean.
   (iii) Samples are independent and random and the time taken to visit is Normally distributed. The text in (ii) assumes the two distributions have equal variances. [should run on the line]
11 (i) $0.103 < \sigma^2 < 0.480$ this assumes that the population is Normal
   (ii) (a) $F_{8,14} = 1.943 < 2.70$ conclude variances are equal
       (b) $H_0: \sigma_s^2 = \sigma_w^2$, $H_1: \sigma_s^2 > \sigma_w^2$
       (c) Samples are independent
12 (i) $F_{7,9} = 1.385 < 4.20$ conclude variances are equal
   (ii) The underlying populations are Normal.
   (iii) Yes, the unpaired t-test requires equal variances.
13 (i) $F_{7,9} = 2.13 < 4.20$ conclude variances are equal
   (ii) $235.4 < \sigma^2 < 829.2$
   (iii) The probability that σ^2 lies between l and u is either 0 or 1. If a large number of such intervals were constructed, then σ^2 would lie within about 80% of them.

Practice questions: set 4 (page 393–5)

1 (i) $\mu = 2.4 \pm 2.2010 \times 1.3 / \sqrt{12}$; lower limit 1.57, upper limit 3.23.
   (ii) The sample may be regarded as random. The data come from a Normal underlying population.
   (iii) 95% of such confidence intervals contain the true but unknown value of the parameter.
2 (i) 0.0 and 9.9 look as though they may not be genuine.
   (ii) Unbiased estimate of μ is 4.875. Unbiased estimate of σ^2 is 5.719. An estimate is unbiased if the method used to calculate it gives the correct value, on average, in the long run.
   (iii) Null hypothesis: $\mu = 6$, alternative hypothesis $\mu \neq 6$.
   Test statistic: $\dfrac{(4.875 - 6)}{\sqrt{\left(\dfrac{5.719}{8}\right)}} = -1.331$.
   Use t-distribution with 7 degrees of freedom. Critical value for 2-tailed 10% test is 1.895. Insufficient reason to reject null hypothesis; so accept $\mu = 6$.
3 (i) Boys: mean 68.375, variance 7.114. Girls: mean 65.5, variance 12.684.

(ii) Null hypothesis: the variances are equal.
Alternative hypothesis: the variances are not equal.
Test statistic: $\dfrac{12.684}{7.114} = 1.783$.
Use $F_{19,23}$ to obtain 2-tail critical value at 5% significance level: 2.374.
Insufficient evidence to reject null hypothesis. Insufficient evidence that variances differ.

(iii) 5% points on chi-squared with 23 degrees of freedom: 13.090 and 35.172.
90% confidence interval given by $13.090 < 23 \times 7.114 / \sigma^2 < 35.172$.
Solve to obtain $4.65 < \sigma^2 < 12.50$

(iv) Parts (ii) and (iii) assume that the underlying distribution of heights is normal.

4 (i) $E(W) = (a + b)\mu$.
$\operatorname{Var}(W) = a^2 \sigma_1^2 + b^2 \sigma_2^2$.

(ii) Unbiasedness gives $(a + b)\mu = \mu$. Hence $a + b = 1$.

(iii) $\operatorname{Var}(W) = a^2 \sigma_1^2 + (1 - a)^2 \sigma_2^2$.
Differentiate with respect to a and set to zero: $2a\sigma_1^2 - 2(1 - a)\sigma_2^2 = 0$.
Hence $\dfrac{a}{1-a} = \dfrac{a}{b} = \dfrac{\sigma_2^2}{\sigma_1^2}$.
Justify that this gives minimum variance (e.g. $\operatorname{Var}(W)$ is a U-shaped quadratic in a.)

(iv) $\sigma_A^2 = 4\sigma_B^2$.
Hence $a:b = 1:4$.
Estimate weight of baby as $\dfrac{(3.20 + 4 \times 3.23)}{5} = 3.224$ (kg).

5 (i) (a) = ((E4 − 1)*E3 + (F4 − 1)*F3) / (E4 + F4 − 2)
(b) = E2 − F2
(c) = E4 + F4 − 2
(d) = E7 / (E5 * (1 / E4 + 1 / F4))^0.5

(ii) Use t-distribution with 25 df
State that $t = 1.7081$ gives 5% tail (or $t = 2.0595$ gives 5% tail). The test is at 5% level.

(iii) The student is carrying out a one-tail test ('slower at midnight than at midday'). By considering the critical value or the p value, the student should reject the null hypothesis and conclude that there is evidence that reactions times are longer at midnight.

(iv) The student's samples should be random.

(v) The underlying distributions should both be Normal.
They should have the same variance.

6 (i) In a paired sample design, each patient uses tablet P and tablet Q. The relevant data are the differences in times.
In a two sample design, there would be one group of patients taking tablet P and one group taking tablet Q. The relevant data would be the individual times.
A paired sample design eliminates the variability between patients; that is, it focuses on the difference between tablets.

(ii) Null hypothesis: tablets P and Q are equally effective. Alternative hypothesis: tablets P and Q differ in effectiveness.
Test statistic: $2.51 / \sqrt{23.25 / 10} = 1.646$.
Use t-distribution on 9 df.
Critical value (5% two-tailed test): 2.262.
$1.646 < 2.262$, so insufficient evidence to suppose that tablets P and Q differ in effectiveness.

(iii) The underlying distribution of differences is Normal.

Index

A

associations 73, 109
   non-linear 123
   Spearman's rank correlation coefficient 135–7
   see also correlation
averages see measures of central tendency

B

Bayes' theorem 214–16, 218
biased estimators 364–7, 378, 432
bimodal distributions 7
binomial distribution 45–6, 68
   expectation and variance 46–7
   goodness of fit test 90–94
   link to the Poisson distribution 59–60, 277
   negative binomial distribution 230–32
   power of a test 255–6
   probability generating functions 263–4, 276, 281
   testing the values of parameters 234
   Type I and Type II errors 252
   using a spreadsheet 46–7
bivariate data 12, 144
   correlation 112–14
   linear regression 146–57
   Pearson's product moment correlation coefficient 114–22
   scatter diagrams 108–9, 110–11
box plots (box and whisker diagrams) 8

C

categorical (qualitative) data 8
causality 73, 123
censuses 3
Central Limit Theorem 242–5, 247, 301, 302–4
   and confidence intervals 309–11, 336–7
chi-squared (χ^2) distribution 73–5
chi-squared (χ^2) statistic 72–3
   notation 75–6
   properties 75–6
   unbiased estimate of population variance 380
chi-squared (χ^2) test 76–8, 103
   goodness of fit test for the binomial distribution 90–94
   goodness of fit test for the Poisson distribution 87–90, 94–6
   goodness of fit test for the uniform distribution 85–7
   left-hand tail 94–6
   using a spreadsheet 79
   using ICT 78–9
class boundaries 166
cluster sampling 5
conditional probability 218
   Bayes' theorem 214–16
   notation 210

prosecutor's fallacy 213–14
   Type I and Type II errors 250
   using contingency tables 211
   using tree diagrams 212, 215
conditions of research projects 353
confidence intervals 309, 312, 325, 339–40
   dice throwing experiment 314–16
   for the difference between two means 335–8
   known and estimated standard deviation 313
   for paired samples 313–14
   for population variance 382–3, 392
   sample size 316–17
   theory of 309–11
   using ICT 311
   using the t-distribution 322–5, 337–8
confidence levels 310
confidence limits 310
contingency tables 71–2, 76–8, 103
   for conditional probabilities 211
continuity corrections 289–90, 304
continuous random variables 9, 208
   class boundaries 166
   cumulative distribution functions 188–96
   expectation and variance 174–7
   expectation and variance of a function of X 183–5
   median and percentiles 177, 179
   mode 178–9
   probability density 164–5
   probability density functions 165–70
   uniform distribution 200–203
   see also Normal distribution
controlled variables 109
convolution theorem 273
correlation 109, 112–14
   effect size 124–5
   interpretation 123–4
   Pearson's product moment correlation coefficient 114–22
   rank correlation 134—8
   scatter diagrams 110–11
critical ratios 344–6
critical regions 342–3
critical values, chi-squared distribution 75
cumulative distribution 189
cumulative distribution functions (c.d.f.s) 189–90, 192–6, 208
   properties 191–2
   of a uniform distribution 203
cumulative frequency curves 10

D

data collection 3–7, 343, 396
data displays 9–10
   box plots (box and whisker diagrams) 8
   scatter diagrams 108–9
data distributions 7–8

data types 8–9, 18
   bivariate data 12
degrees of freedom 103, 122, 322
   and chi-squared distribution 74–5
   estimate of variance 372
dependent variables 109
discrete probability distributions 21–3, 42, 68
   binomial distribution 45–7, 59–60
   expectation and variance 26–30
   geometric distribution 224–7, 238
   negative binomial distribution 230–32, 238–9
   Poisson distribution 49–60
   testing the values of parameters 234–7
   Type I and Type II errors 252–4
   uniform distribution 64–7
discrete random variables 9, 20–21, 42–3
   conditions for 22–3
   expectation and variance of a function of X 183, 185
   linear combinations of 38
   linear functions of 33–5
   probability generating functions 261–78
   sums and differences of 35–8
   using the Normal distribution 289–90
distribution-free tests 96
distributions 7
   see also discrete probability distributions; Normal distribution
dummy variables 190, 191
   in probability generating functions 261

E

effect size 124–5
estimates 3
estimation of parameters 363–4, 378
estimators 364, 378
   accuracy of 367–8
   bias in 364–70, 432
   unbiased estimator for population mean 371
   unbiased estimator for population variance 371–2
expectation
   of a binomial distribution 46–7
   of a continuous random variable 174–7
   of a continuous uniform distribution 201–2
   of a discrete probability distribution 26–30, 42
   of a discrete uniform distribution 65–7
   of a function of a random variable 183–5
   of a geometric distribution 226–7, 238
   of linear combinations of random variables 38, 43
   of linear functions of random variables 33–4, 43
   of a negative binomial distribution 231, 239
   of a Poisson distribution 51, 53

436

from a probability generating function 266–9
of sums and differences of Normal variables 287–9, 293–7
of sums and differences of random variables 36–8, 43
see also mean
exponential distribution 206
extrapolation 123–4

F

factorials 50
false positives and false negatives 210–12, 214–15, 218
F-distribution 386–9
finite discrete random variables 21
frequency charts 9–10
frequency of a random variable 7

G

geometric distribution 224–5, 238
expectation and variance 226–7
probability generating functions 264, 277–8, 281
testing the values of parameters 235–7
total probability 226
goodness of fit tests 103
for the binomial distribution 90–94
interpretation of p-values 96
for the Poisson distribution 87–90, 94–6
for the uniform distribution 85–7
Gosset, William S. ('Student') 323
grouped data 7, 9

H

histograms 9–10
hypothesis testing
chi-squared test 72–3, 76–9
for the difference in population variances 386–9, 392
on the difference of means of two populations 353–8, 361–2
on the mean and variance of a single sample 342–8, 361
parameters of discrete distributions 234–7
for population variance 380–82
Type I and Type II errors 249–54
using a spreadsheet 347–8
using ICT 344
using Pearson's product moment correlation coefficient 117–22, 144
using Spearman's rank correlation coefficient 135–8, 144
using the Normal distribution 342–4, 354–6
using the t-distribution 346–8, 357–8
see also goodness of fit tests

I

independent variables 109
infinite discrete random variables 21, 50
information collection 3–7, 343, 396
interquartile range 8

L

least squares regression lines 147–51, 162
residual sum of squares 156–7
using a spreadsheet 151–2
left-hand tail, chi-squared (χ^2) test 94–6
linear combinations of random variables 38, 43, 295–6
linear functions of random variables 33–5, 43
linear regression 147–51, 162
residual sum of squares 156–7
using a spreadsheet 151–2

M

marginal totals, contingency tables 71, 76
mass 115
matched samples
confidence intervals 326
confidence intervals for mean difference 313–14
mean 19
of a binomial distribution 46–7
confidence intervals for 309–13
of a continuous random variable 174–7
of a continuous uniform distribution 201–2
of a discrete probability distribution 26–30, 42
of a discrete uniform distribution 65–7
of a function of a random variable 183–5
of a geometric distribution 226–7, 238
hypothesis tests on a single sample 342–8, 361
hypothesis tests on the means of two populations 353–8, 361–2
of linear combinations of random variables 38, 43
of linear functions of random variables 33–4, 43
of a negative binomial distribution 231, 239
of a Poisson distribution 51, 53
from a probability generating function 266–9
of sums and differences of Normal variables 287–9, 293–7
of sums and differences of random variables 36–8, 43
unbiased estimator for population mean 371
measures of central tendency 8, 11
see also mean; median; mode
measures of spread 8, 11

see also standard deviation; variance
median 8
of a continuous random variable 177, 179
of a cumulative distribution function 192
mode of a continuous random variable 178–9
moment generating functions 262
multivariate data 108

N

negative binomial distribution 230, 238–9
expectation and variance 231
probability generating functions 264–5, 278, 281
total probability 230–31
negative correlation 110, 114
Normal distribution 206, 241, 285–7, 305–6
Central Limit Theorem 242–5, 247, 302–4, 306
confidence intervals for the difference between two means 335–6
confidence intervals for the mean 310–12, 339
hypothesis testing for the mean 342–4, 354–6, 361
linear combinations of random variables 295–6
modelling discrete situations 289–90
notation 285–6
sums and differences of Normal variables 287–9, 293–7
tests for 327–9
Type I and Type II errors 254
Normal probability plots 327–9
notation
for bivariate data 144
chi-squared (χ^2) statistic 72, 75–6
for conditional probability 210
for discrete random variables 22
factorials 50
for the Normal distribution 285–6
p.d.f.s and c.d.f.s 191
for population parameters and estimates 3
for variance and standard deviation 12–13, 19
numerical (quantitative) data 8–9
data displays 9–10
summary measures 11

O

ogives 190
one-tailed tests 342, 345
opinion polls 5
opportunity sampling (convenience sampling) 6
outliers 8, 111

P

paired experimental design 353
paired samples, confidence intervals 313–14, 326
parent population 3
Pearson, Karl 138
Pearson's product moment correlation coefficient 114–17, 144
  degrees of freedom 122
  interpretation 117–21
  modelling assumptions 122
  using a spreadsheet 117
percentiles 177, 179
Poisson distribution 49–51, 68
  conditions for 51
  goodness of fit test 87–90, 94–6
  link to the binomial distribution 59–60, 277
  mean and variance 51, 53
  modelling with 52–5, 57–9
  power of a test 255
  probability generating functions 262, 276–7, 281
  shapes 51–2
  testing the values of parameters 235
  Type I and Type II errors 252–4
positive correlation 110, 114
power function of a test 255–6, 259
power of a test 250–54, 259
  relationship to significance level 256
probabilities, use in hypothesis testing 343–4
probability, conditional see conditional probability
probability density 164–5
probability density functions (p.d.f.s) 165–70, 208
probability distributions, discrete see discrete probability distributions
probability generating functions (PGFs) 261–2, 281
  basic properties 262
  for a binomial distribution 263–4, 276
  deriving expectation and variance 266–9
  for a geometric distribution 264, 277–8
  for a negative binomial distribution 264–5, 278
  for a Poisson distribution 262, 276–7
  of the sum of independent random variables 271–5
  for a uniform distribution 263
problem solving cycle 2, 18
  information collection 3–7
  problem specification and analysis 2–3
  processing and representation 7–13
product moment correlation 114–17, 144
  using a spreadsheet 117
  using r as a test statistic 117–21
proportional stratified sampling 5
prosecutor's fallacy 213–14
p-values 344

Q

qualitative (categorical) data 8
quantitative (numerical) data 8–9
  data displays 9–10
  summary measures 11
quartiles 8
quota sampling 6

R

random processes 3
random variables 7
  discrete and continuous 9
  see also continuous random variables; discrete random variables
rank correlation 134, 144
ranked data 8
rectangular distribution see uniform distribution
regression analysis 109
regression lines 147–51, 162
  residual sum of squares 156–7
  using a spreadsheet 151–2
relative frequency 21
research projects 353–4
residuals 148
residual sum of squares (RSS) 156–7, 162

S

sample size 4
  and Central Limit Theorem 242
  and confidence intervals 309–10, 313, 314, 316–17
  and Type I and Type II errors 254
sampling 4, 18, 396, 404
  terminology and notation 3
sampling distribution of the estimator 364–5, 378
sampling distribution of the means 242–3, 247, 300–302
  see also Central Limit Theorem
sampling error 3
sampling fraction 3
sampling frame 3
sampling techniques 4–7
scatter diagrams 108–9
  interpretation 110–11
  least squares regression lines 147–52
screening tests 210–12, 214–15
self-selected sampling 6
significance level, relationship to power of a test 256
simple random sampling 4
skewed distributions 7
Spearman, Charles 138
Spearman's rank correlation coefficient 135–6, 144
  using ICT 137
  when to use it 137–8
spreadsheets
  binomial distribution 46–7
  chi-squared test 79, 90
  dice throwing simulation 315–16
  goodness of fit test for the Poisson distribution 90
  hypothesis testing 347–8
  product moment correlation 117
  regression lines 151–2
standard deviation 11–13, 19
  estimated, and confidence intervals 313, 321–4
  estimation of 345
standard error of an estimator 378
standard error of the mean 242, 301, 303, 306
standardisation of a Normal distribution 241, 285, 305
stratified sampling 5
Student's t-test see t-distribution
sums and differences of Normal variables 287–9
sums and differences of random variables 35–8, 43
systematic sampling 5

T

t-distribution 313, 323
  confidence intervals 324, 337–8
  hypothesis testing for the mean 346–8, 357–8, 361
  tests for Normality 327–9
  use for paired samples 326
test statistics, use in hypothesis testing 344–6
tree diagrams 212, 215
two-sample experiments 314
two-tailed tests 342
Type I and Type II errors 249–50, 259
  for discrete probability distributions 250–54
  for a Normal distribution 254
  relationship to significance level 256

U

unbiased estimators 364–70, 378
  for population mean 371
  for population variance 322, 371–2, 380, 392
uniform distribution, continuous 200–201, 208
  cumulative distribution function 203
  expectation and variance 202
uniform distribution, discrete 64–5, 68
  expectation and variance 65–7
  goodness of fit test 85–7
  probability generating functions 263
unimodal distributions 7
unpaired experimental design 353–4

V

variables 7
  controlled 109
  dependent and independent 109, 147

discrete and continuous 9
see also continuous random variables; discrete random variables
variance 12–13, 19
  of a binomial distribution 46–7
  confidence intervals 382–3, 392
  of a continuous random variable 174–5
  of a continuous uniform distribution 201–2
  of a discrete probability distribution 26–30, 42
  of a discrete uniform distribution 65–7
  of a function of a random variable 183–5
  of a geometric distribution 226–7, 238
  hypothesis test for the difference in population variances 386–9, 392
  of linear combinations of random variables 38, 43
  of linear functions of random variables 34–5, 43
  of a negative binomial distribution 231, 239
  of a Poisson distribution 51, 53
  from a probability generating function 267–9
  of sums and differences of Normal variables 287–9, 293–7
  of sums and differences of random variables 36–8, 43
  unbiased estimate of 322, 371–2, 380, 392
Venn diagrams 210, 211
vertical line charts 20–21

W

weight 115

Acknowledgements

The Publishers would like to thank the following for permission to reproduce copyright material.

Practice questions have been provided by Neil Sheldon (MEI) (pp. 165–168 and pp. 319–321).

Q1 and Q2 on p.256 have been reproduced by permission of Cambridge International Examinations. Answers to these questions are on p.423. Cambridge International Examinations bears no responsibility for the example answers to questions taken from its past question papers which are contained in this publication.

Q8 on p.49, Q10 on p.56, Q5 on p.62, Q12 on p.83, Q6 on p.97, Q14 on p.100, Q15 on p.101, Q18 on p.102, Q9 on p.129, Q16 and Q17 on pp.132–33, Q6 on p.140, Q9 and Q10 on p.141, Q3 and Q4 on p.158, Q7 on p.159, Q8, Q9 and Q10 on p.160, Q10 on p.182, Q10 on p.188, Q9 and Q10 on p.318–19, Q4 and Q5 on p.349 and Q11 on p.351 have been reproduced by permission of Pearson Education Ltd. Answers to these questions are on pp.396–435. Pearson Education accepts no responsibility whatsoever for the accuracy or method of working in the answers given.

Photo credits

p.1 © Ingram Publishing Limited/General Gold Vol 1 CD 2; **p.20** © DAVIPIX – Fotolia; **p.33** © Imagestate Media (John Foxx)/Store V3042; **p.44** © Peter Titmuss/Shutterstock; **p.49** © Ingram Publishing Limited/Occupations and Trades Vol 2 CD 4; **p.57** © rigamondis – Fotolia; **p.70** © Chloe Johnson/Alamy; **p.85** © AZP Worldwide – Fotolia.com; **p.107** © Cathy Yeulet – 123RF; **p.146** © Dusan Kostic – Fotolia; **p.163** © Luciano de la Rosa Gutierrez – Fotolia.com; **p.166** © Eric Isselée – Fotolia; **p.188** © Sandor Jackal – Fotolia; **p.209** © Mark Atkins – Fotolia; **p.213** © Photographee.eu – Fotolia; **p.223** © Brian Jackson/stock.adobe.com; **p.240** © AS Photo Project / stock.acobe.com; **p.248** © Christian Horz/stock.adobe.com; **p.260** ©Vibeke Morfield/123RF; **p.284** © Ingram Publishing Limited/Animals Gold Vol 1 CD 3; **p.308** © 1997 Siede Preis Photography/Photodisc/Getty Images/ Eat, Drink, Dine 48; **p.321** © Corbis. All Rights Reserved; **p.341** © S Curtis/Shutterstock; **p.353** © Jakub Krechowicz / stock.adobe.com; **p.363** © Michael Gray/stock.adobe.com.

Every effort has been made to trace all copyright holders, but if any have been inadvertently overlooked, the Publishers will be pleased to make the necessary arrangements at the first opportunity.

Although every effort has been made to ensure that website addresses are correct at time of going to press, Hodder Education cannot be held responsible for the content of any website mentioned in this book. It is sometimes possible to find a relocated web page by typing in the address of the home page for a website in the URL window of your browser.